Pesticide Application: Principles and Practice

Pesticide Application: Principles and Practice

Edited by

P. T. Haskell

Department of Zoology,
University College, Cardiff

CLARENDON PRESS · OXFORD
1985

Oxford University Press, Walton Street, Oxford OX2 6DP

London New York Toronto
Delhi Bombay Calcutta Madras Karachi
Kuala Lumpur Singapore Hong Kong Tokyo
Nairobi Dar es Salaam Cape Town
Melbourne Auckland

and associated companies in
Beirut Berlin Ibadan Nicosia

Oxford is a trade mark of Oxford University Press

Published in the United States
by Oxford University Press, New York

British Library Cataloguing in Publication Data
Pesticide application: principles and practice.
1. Pesticides
I. Haskell, P. T.
632'.95   SB951
ISBN 0–19–854542–8

Library of Congress Cataloguing in Publication Data
Main entry under title:
Pesticide application.
Includes bibliographies and index.
1. Pesticides. 2. Pesticides—Application.
I. Haskell, P. T. (Peter Thomas)
SB951.P387 1985   632'.95   84–27390
ISBN 0–19–854542–8

Set by Latimer Trend & Company Ltd, Plymouth
Printed in Great Britain by St Edmundsbury Press,
Bury St Edmunds, Suffolk

Foreword

Like all other animals, and indeed plants, Man is in constant competition with numerous biological species. These range from the bacteria, viruses, internal and ectoparasites that 'feed' directly on him, or on his domesticated animals, to those arthropods, nematodes, bird, and mammal pests that consume his crops and food stocks; fungal, viral, or bacterial pathogens that cause disease in them and decrease their yields, or weeds that also prevent maximum crop production.

Man has attempted to combat these competitors by magical, religious, agronomic, and ecological means since medicine, horticulture, and agriculture began, but with little success until hygiene, medical and biological knowledge increased rapidly during the 19th century. Even then the benefits of increasing knowledge were very limited, leading to a greater appreciation of the losses caused and why they occurred. Limiting them was a different matter both in medicine and agriculture, where crop rotation and cultivations were the main methods, aided by a few pesticides like derris, pyrethrum, and copper mixtures.

War-time chemical research profoundly changed the situation; new drugs to combat human diseases and new pesticides to control our competitors have changed the potential for man's dominance of his biological environment more in the last 30 years than in the previous 3 million years of his development. They are the greatest beneficial advance that man has made since the adoption of agriculture. The speed of this advance was phenomenal: fewer than 20 chemical pesticides (c. 60 formulations) in 1945, 27 (c. 400 formulations) in 1955, c. 200 (c. 800 formulations) by 1975.

However, despite this success in formulation chemistry, we still lose, on a world scale, about a third of our standing crops and a fifth of our crop products to our competitors, and the imbalance between our food supplies and the population increase that drugs and pesticides have unleashed is still not resolved, particularly in the tropical developing countries where pesticides are little used yet, except on plantation crops. Pesticide use in the more advanced agriculture has shown what can be done, not only to limit damage but to modify agronomic practices and save labour. Fortunately, most pesticides have minimal or no detrimental side effects when used properly.

Nevertheless, much more research, development, and education are needed if we are to reap the full harvest from our discoveries, and this book makes a considerable contribution in the educational field. No other book covers the subject so widely or deals with both principle and practice in both public health and crop protection areas in such an integrated and comprehensive manner.

The authors are well known authorities, having between them wide experience in both temperate and tropical countries. Dr Peter T. Haskell, the Editor, is to be congratulated on both the concept and completion of this work.

Leonard Broadbent
Formerly Chairman,
Commission of European
Communities Scientific
Committee for Pesticides

Preface

Although there have been many books on pesticides and their application in recent years, they have all dealt with specialized aspects of the subject, such as application equipment, the physics or chemistry of pesticide use, or the role of pesticides in particular crops. However, anyone who has ever used pesticides knows that what is actually done in the field is always a compromise between the conflicting demands of principle and practice. The choice of compound, its formulation, its dosage, and its method of application can only be made in relation to a knowledge of the pest–crop or pest–host relationship and the ecosystem in which they exist, and ultimately, as mentioned above, the answer is always a compromise in which the optimum parameters proposed by theory are adapted to meet the inevitable demands of practice. This book has therefore been designed to outline in Chapters 1–9 the principles underlying the choice and use of pesticides and pest-control systems, and to illustrate in Chapters 10–19 how far these can be applied in a variety of practical situations. In this book, the word 'pesticide' covers insecticides, fungicides, herbicides, avicides, nematicides, rodenticides, and molluscicides, but not the behaviour-modifying chemicals or attractants and repellents used in control systems.

Such a book has itself to be a compromise over the depth and coverage of the various subjects; thus some areas of the enunciation of principles, such as, for example, environmental aspects, deserve a book to themselves and only an outline of the basic problems can be given here. The same is true, *a fortiori,* of the choice of practical illustration; clearly, for example, it is impossible to cover all field crops, and the compromise here has been to use cotton, the crop on which more pesticide is used than on any other, and contrast it with the use of pesticides in crops such as pigeonpea, essentially a small-farmer crop, where use of pesticides is greatly restricted not only by biological problems but also by practical ones, such as the fact that small farmers simply cannot afford to apply a lot of pesticide or use sophisticated system approaches. Some specialized fields of practical application, such as stored products pest control, have been omitted because usage therein does not illustrate application of general principles, but others, for example mollusc control, have been included, because they deal with use in a special medium – in this case water.

Throughout the book an attempt has been made to emphasize the paramount importance of adequate consideration of biological and ecological information in the development of practical control systems, with particular reference to the application of the Integrated Pest Management

(IPM) concept. Because it is now widely agreed that this approach offers the best avenue for the development of efficient, economic, stable, and environmentally acceptable pest-control techniques, the IPM concept is adopted as the underlying principle, but it must be realized that it is beyond the scope of the book to give details of IPM systems for particular purposes. Thus the general aim of the book has been to offer, within the above constraints, an outline of modern knowledge of pesticide application which will be of value as one source of theoretical and practical knowledge for research and field workers and students.

Cardiff   P.T.H.
September 1984

Acknowledgements

I am grateful to several colleagues in the Tropical Development and Research Institute for advice and assistance during the preparation of this book and to the Director, TDRI, for permission to reproduce Figs. 14.1 and 14.7.

My particular thanks go to Miss Margaret Murphy, who bore the brunt of the secretarial work, and to Miss Sabina Thompson and my wife Mrs. Aileen Haskell, who assisted in the typing and preparation of the index.

Contents

Contributors

Mr F. B. Barlow, c/o TDRI, Division of Chemical Control, Porton Down, nr. Salisbury, Wilts., UK.

Dr L. Brader, Shell International Research, Group Toxicology Division, Wassenaarseweg 80, PO Box 162, 2501 AN The Hague, Netherlands.

Dr J. Bridge, CIP, 395A Hatfield Rd., St. Albans, Herts. AL4 0XU, UK.

Professor J. R. Busvine, Musca, 26 Braywick Road, Maidenhead, Berks. SL6 LDA, UK.

Dr E. J. Buyckx, Plant Protection Service, FAO, Rome, Italy.

Dr M. E. Cammell, Imperial College Field Station, Silwood Park, Ascot, Berks., UK.

Dr J. F. Copplestone, Division of Vector Biology and Control, WHO, 1211 Geneva 27, Switzerland.

Dr J. C. Davies, Principal Agricultural Research Adviser, O.D.A., Eland House, Stag Place, London, UK.

Dr N. G. Gratz, Director, Division of Vector Biology and Control, WHO, 1211 Geneva 27, Switzerland.

Mr J. H. Greaves, MAFF Lab, Tolworth, Hook Rise South, Tolworth, Surbiton, Surrey KT6 7NF, UK.

Miss S. Green, TDRI, College House, Wrights Lane, London W8 5SJ, UK.

Dr P. T. Haskell, Department of Zoology, University College, Cardiff, UK.

Dr J. M. Jewsbury, Department of Parasitology, Liverpool School of Tropical Medicine, Pembroke Place, Liverpool L3 5QA, UK.

Mr D. R. Johnstone, TDRI, Division of Chemical Control, Porton Down, nr. Salisbury, Wilts., UK.

Dr P. J. Jones, Department of Forestry and Natural Resources, University of Edinburgh, Darwin Buildings, The Kings Buildings, Mayfield Road, Edinburgh, Scotland.

Mr T. Jones, TDRI, College House, Wrights Lane, London W8 5SJ, UK.

Professor R. J. V. Joyce, Maltfield, Berriew, nr. Welshpool, Powys, Wales.

Professor R. H. Kips, 151 Defrélaan, 1180 Brussels, Belgium.

Dr A. J. Lacey, Biology Department, Brunel University, Cleveland Road, Uxbridge, Middx. UB8 3PH, UK.

Dr G. A. Matthews, Imperial College Field Station, Silwood Park, Ascot, Berks., UK.

Dr G. A. Norton, Imperial College Field Station, Silwood Park, Ascot, Berks., UK.

Dr W. Reed, ICRISAT, 1-11-256 Begumpet, Hyderabad 500016 AP, India.

Dr J. M. Waller, Commonwealth Mycological Institute, Ferry Lane, Kew, Richmond, Surrey, UK.

Professor M. J. Way, Imperial College Field Station, Silwood Park, Ascot, Berks., UK.

Dr A. G. Whitehead, Rothamsted Experimental Station, Harpenden, Herts., UK.

1
Chemistry and formulation

F. BARLOW

1.1 Introduction

There are 551 separate entries in the sixth edition of *The Pesticide Manual*, edited by Worthing (1979), and almost all of these refer to compounds which have been synthesized by the chemical industry. Some will be obsolete and new ones will have been introduced by the time the next edition appears but the size of the world's stock of pesticides remains at approximately that level. About 35% of these are insecticides and acaricides, 34% herbicides, and 18% fungicides, the remainder being used against miscellaneous other pests. While the rate of introduction of new insecticides has declined over recent years, relative to herbicides and fungicides, the organophosphorus group of insecticides still contains the largest number of individual compounds with about 20% of the whole. Of course, some pesticides are more popular than others for the control of major pests for reasons such as potency, availability, cost, and safety and at any time a few members of each group tend to dominate the market. The relative importance of the different types of pesticide varies throughout the world with insecticides more prominent in tropical and subtropical regions and herbicides in temperate regions.

There is only space here for a very brief summary of the main classes of pesticides with a few examples drawn from each, and for a more extended description the other chapters in this book and recent general books on pesticides should be consulted. Examples of the latter are Green, Hartley, and West (1977), Cremlyn (1978) and Hassall (1982); a book which is arranged according to modes of action is that of Corbett (1974). The present summary is followed by a description of formulations, some interactions with pests and comments on factors which influence the lives of active ingredients in spray residues.

New users of pesticides, or readers of the literature, who may be confused by the many names and codes under which individual compounds appear should be helped by the publication of cross-referencing indexes such as the *Tropical Pest Management Pesticide Index* (1981) or that of the Australian Department of Health (1978). There are four main classes of names. Systematic chemical names are usually based on rules of the International Union of Pure and Applied Chemistry (IUPAC) or the derived system used by *Chemical Abstracts*. Common names may be international, as published by the International Standards Organisation (ISO 1981), national, for example those approved by the British Standards Institution (BSI 1969), or approved by a professional society such as the Entomological Society of America. Trade names for a compound or its formulation can be very numerous and even vary according to use and country although from the same source. Code names or numbers are given by the original manufacturers, other interested companies, and international and national agencies, and appear frequently in publications.

1.2   The main groups of pesticides

A conventional primary division according to the type of pesticidal activity, with subdivisions based on chemical groups, is used here. Other classifications, depending on such features as mode of action or the way they are applied, have been discussed (see Green *et al.* 1977), but they have disadvantages and the present arrangement is probably the least unsatisfactory. A neat separation into classes is made difficult by some individual compounds being general biocides while broad chemical groupings such as organophosphorus compounds and carbamates contain compounds with different types of activity.

The tables show the structures of a few of the long-established, most-used, or newer compounds and, where values are available, some physical properties which are important in formulation (melting-point and solubility) or persistence (vapour pressure).

1.2.1   Insecticides

The *organochlorine* group contains compounds such as DDT and dieldrin which are well-known to the general public, not least because of their association with environmental problems. Other members are also shown in Table 1.1. They were mostly discovered and developed in the period 1942–1956, and played a large part in the early successes of synthetic insecticides. A detailed description of the group has been provided by Brooks (1974).

As the name suggests, the common feature is a substantial chlorine content. They also share very low solubilities in water and partition from water into hydrocarbon solvents although the actual solubilities in either water or oils, do, in fact, cover a wide range. Such properties encourage their storage in body fats and accumulation in successive organisms of a food chain. Melting points are high enough to allow grinding and formulation as high-concentration powders except for those which are complex mixtures such as chlordane and camphechlor. Vapour pressures of some are sufficiently high to permit a strong fumigant action in closed spaces while those of others are low enough to provide a long persistence.

Although they are generally more stable than many chemicals from other groups and have a reputation for being resistant to degradation (hence their association with the adjective 'hard') their chemical properties do vary considerably. Depending on the possibilities offered by their individual structures they undergo such reactions as dechlorination, epoxidation, rearrangements, and hydrolysis, especially when biochemically and photo-chemically mediated routes are available.

They are general insecticides with a wide spectrum of activity but details of their modes of action remain obscure despite many investigations over the past 30 years. Symptoms of poisoning differ in detail but all are associated with the nervous system. DDT can bind to natural membranes, in particular those which surround the nerve axons, and is thought to interfere with the movement of ions associated with conduction.

Examples in Table 1.2 have been chosen from the important group of *organophosphorus* insecticides. Parathion was developed as early as 1947 and new compounds have continued to appear ever since. This profusion derives from their preparation from a few intermediates, which are the source of their basic activity, by combination with a wide variety of additional chemicals that are readily available in the chemical industry. The products are mostly esters of phosphoric, phosphorothioic, and phosphorodithioic acids, or the corresponding phosphonic or phosphoramidic acids, and choice of the esterifying groups modifies their chemical and physical properties and so rates of reaction, selectivity between pest and non-pest and persistence in residues or the general environment. They can be liquids or solids, have solubilities in water and other solvents which range from very low to complete miscibility and vapour pressures which cover a very wide range.

As esters they are susceptible to hydrolysis and other common reactions include various isomerizations and oxidations of sulphur when this is present. Rates vary with structure but organophosphorus compounds are in general sufficiently unstable to avoid any accumulation in the general environment.

Much more is known about the way they and the carbamates act than for other groups of insecticides. The esters themselves, or their metabolic products, combine with the enzyme acetylcholinesterase and prevent conduction at junctions in the nervous system where acetylcholine is the natural transmitter. Rates and duration of poisoning depend upon physical properties which control entry into the insect, metabolic activation and de-activation and the characteristics of the reaction between enzyme and insecticide.

Two of the recent books devoted to this group of insecticides are Fest and Schmidt (1982) and Eto (1974).

Some of the best-known *carbamates,* or esters of *N*-methyl and *NN*-dimethylcarbamic acids, are shown in Table 1.3. Activity resides with the

TABLE 1.1 *Organochlorine insecticides*

Common name (ISO)	Systematic name (IUPAC)	m.p. (°C)	Vapour pressure at 20°C (mPa)	Solubilities in				Common formulations
				(1) Organic solvents			(2) Water, (p.p.m.) at ambient temperature (20–30°C)	
				H	M	L		
Aldrin*	1,2,3,4,10,10-Hexachloro-1,4,4a,5,8,8a-hexahydro-exo-1,4-endo-5,8-dimethanonaphthalene	104	10	Ar		A	0.03	EC WDP D G
Camphechlor	'Chlorinated camphenes containing 67–9% chlorine'	Wax	—	A			—	EC WDP D G
pp'-DDT*	1,1,1-Trichloro-2,2-bis-(4-chloro-phenyl)ethane	109	0.025	K		A	0.009	EC WDP S D SC

Dieldrin*	1,2,3,4,10,10-Hexachloro-6,7-epoxy-1,4,4a,5,8,8a-hexahydro-*exo*-1,4-*endo*-5,8-dimethanonaphthalene	176	0.41	K	Ar	A	0.033	EC D	WDP G	S
Endosulfan	6,7,8,9,10,10-Hexachloro-1,5,5a,6,9,9a-hexahydro-6,9-methano-2,4,3-benzo-(e)-dioxathiepin-3-oxide	107 (α-isomer) 213 (β-isomer)	0.59	Ar	A		—	EC C	WDP G	S
Lindane	1,2,3,4,5,6-Hexachloro-cyclohexane, gamma isomer	113	1.3	K	Ar	A	6	EC D	WDP SC	S

*pp'-DDT approved by BSI but not ISO; aldrin and dieldrin refer to technical grades although used for pure compounds in some countries.

General notes on the Tables

1. Solubilities in organic solvents are only indicated as H = high (more than 200 g/l of solvent), M = medium (50 to 200 g/l) and L = low (less than 50 g/l) at ambient temperatures in solvents of the following general types which are commonly used in formulations and arranged in order of increasing solvent power. A = saturated hydrocarbons, Ar = aromatic hydrocarbons, K = ketones, and Am = amides. Even when actual figures are given in the literature they cannot be compared easily because the measurements vary in choice of example for a given solvent type, accuracy and units.

2. Low vapour pressures and solubilities in water are difficult to measure accurately and where values have been reported by different workers for the same compound they often show considerable variation. Therefore although values taken from the literature (mainly the 6th edition of the *Pesticide Manual* 1979) have been given they are only intended as an approximate guide to these properties. Where necessary, vapour pressures have been recalculated to 20°C by the equation given on page 883 of Hartley and Graham-Bryce (1980).

3. Formulations are shown as follows: EC = emulsion concentrate, WDP = water-dispersible powder, SC = suspension concentrate, WSP = water-soluble powder, WSL = water-soluble liquid, D = dust, G = granules, S = solution (in water or organic solvent), B = bait, CR = controlled-release device, SM = smoke generator, A = aerosol dispenser.

TABLE 1.2. *Organophosphorus insecticides*

Common name (ISO)	Systematic name (IUPAC)	m.p. (°C)	Vapour pressure at 20°C (mPa)	Solubilities in						Common formulations
				(1) Organic solvents					(2) Water, (p.p.m.) at ambient temperature (20–30°C)	
				H	M	K	L	Ar		
Acephate	*OS*-Dimethyl acetylphosphoramidothioate	90–1	0.12	Am		Am			650 g/l	WSP A G
Azinphos-methyl	*S*-3,4-Dihydro-4-oxobenzo-[d]-[1,2,3]-triazin-3-ylmethyl) *OO*-dimethyl phosphorodithioate	73–4	—	Ar		A			33	EC WDP D
Bromophos	*O*-(4-Bromo-2,5-dichlorophenyl) *OO*-dimethyl phosphorothioate	53–4	17	Ar		A			0.3	EC G WDP S D
Chlorpyrifos	*OO*-Diethyl *O*-3,5,6-trichloro-2-pyridyl phosphorothioate	43	1.2	A					0.7	EC S WDP G
Demeton-S-methyl	*S*-2-Ethylthioethyl *OO*-dimethyl phosphorothioate	Liquid	48	Ar		A			3300	EC
Diazinon	*OO*-Diethyl *O*-2-isopropyl-6-methyl-pyrimidin-4-yl phosphorothioate	Liquid	19	A					40	EC S WDP G

Dichlorvos	2,2-Dichlorovinyl dimethyl phosphate	Liquid	1600	Ar	A	10 000	EC / A	G / CR	S
Dimethoate	*OO*-Dimethyl *S*-methylcarbamoylmethyl phosphorodithioate	51–2	0.55	Ar	A	25 000	EC / G	WDP	S
Fenitrothion	*OO*-Dimethyl *O*-4-nitro-*m*-toluyl phosphorothioate	Liquid	0.8	Ar	A	20	EC / D	WDP	G
Jodfenphos*	*O*-(2,5-Dichloro-4-iodophenyl) *OO*-dimethyl phosphorothioate	76	0.1	Ar	A	0.1	EC / D	WDP / D	S
Malathion	*S*-1,2-Di (ethoxycarbonyl)ethyl *OO*-dimethyl phosphorodithioate	Liquid	1.3	Ar	A	145	EC	WDP	D
Parathion	*OO*-Diethyl *O*-4-nitrophenyl phosphorothioate	Liquid	0.61	Ar	A	24	EC / D	WDP / SM	G
Parathion-methyl	*OO*-Dimethyl *O*-4-nitrophenyl phosphorothioate	35–6	1.1	Ar	A	38	EC / CR	WDP	D
Phorate	*OO*-Diethyl *S*-ethylthiomethyl phosphorodithioate	Liquid	110	Ar		50	EC	G	
Phosalone	*S*-6-Chloro-2-oxobenzoxazolin-3-ylmethyl *OO*-diethyl phosphorodithioate	48	—	—	Ar	A	EC	WDP / D	
Pirimiphos-methyl	*O*-2-Ethylamino-6-pyrimidin-4-yl *OO*-dimethyl phosphorothioate	Liquid	3.6	A		5	EC / CR	D / A	S

*BSI name is iodofenphos.

TABLE 1.3. *Carbamate and pyrethroid insecticides*

Common name (ISO)	Systematic name (IUPAC)	m.p. (°C)	Vapour pressure at 20°C (mPa)	Solubilities in				Common formulations
				(1) Organic solvents			(2) Water, (p.p.m. at ambient temperature (20–30°C)	
				H	M	L		
1. Carbamates								
Aldicarb	2-Methyl-2-(methylthio)propionalde-hyde O-methylcarbamoyloxime	100	7.0	K	Ar	A	6000	G
Bendiocarb	2,3-Isopropylidenedioxyphenyl methylcarbamate	130	0.32	Am	K	Ar	40	WDP
Carbaryl	1-Naphthyl methylcarbamate	142	0.031	Am	K	Ar	120	WDP G SC
Carbofuran	2,3-Dihydro-2,2-dimethylbenzo-furan-7-yl methylcarbamate	154–5	0.54	Am	K	Ar	700	WDP G D SC

Common name	Chemical name	mp (°C)						Formulations
Methomyl	S-Methyl N-(methylcarbamoyloxy) thioacetimidate	78–9	3.4	K		58 000	Ar	WSP WSL D
Pirimicarb	2-Dimethylamino-5,6-dimethylpyrimidin-4-yl dimethylcarbamate	91	1.0	Ar		2700	A	EC WDP S
Propoxur	2-Isopropoxyphenyl methylcarbamate	89–90	0.4	K	Ar	2000	A	EC G WDP S D B

2. Pyrethroids

Deltamethrin	(S)-α-Cyano-3-phenoxybenzyl (1R,3R)-3-(2,2-dibromovinyl)-2,2-dimethylcyclopropanecarboxylate	99–100	0.00084	Ar	Ar	0.003	A	EC WDP S
Permethrin	3-Phenoxybenzyl (±)-cis,trans-3-(2,2-dichlorovinyl)-2,2-dimethyl-cyclopropanecarboxylate	34–9	0.020	A		0.04		EC WDP S
Phenothrin	3-Phenoxybenzyl (±)-cis,trans-chrysanthemate	Liquid	0.024	A		2		EC
Resmethrin	5-Benzyl-3-furylmethyl (±)-cis,trans-chrysanthemate	43–8	–	A		–		EC A WDP S

carbamate moiety and, as with the organophosphorus esters, there are many possibilities for manipulation of biological, chemical, and physical properties by changes in the esterifying groups. Vapour pressures, solubilities, and melting points are also very dependent on the presence or absence of a hydrogen on the nitrogen atom which forms part of the carbamate structure and can form hydrogen bonds. Those insecticides which are esters of *N*-methylcarbamic acid with phenols are crystalline materials with medium to high melting points and low solubilities in water and many common organic solvents, whereas the corresponding *NN*-dimethylcarbamates, or other compounds without the hydrogen, have higher volatilities and solubilities and lower melting points. In a few carbamates the hydroxyl group used in formation of the esters is provided by aliphatic oximes rather than a phenol and the products are more polar, soluble in water and alcohols and less soluble in hydrocarbons.

Hydrolysis is an important reaction of carbamates and they decompose in alkaline media more rapidly than, for example, the common organophosphorus esters used as insecticides. Again, however, *NN*-dimethylcarbamates show different properties and are much more stable.

Carbamates are general insecticides with a mode of action which is basically the same as that of organophosphorus esters although the inhibition of the enzyme is more easily reversed. Poisoned insects and animals therefore tend to recover from lower doses and these are not cumulative. A description of the group has been provided by Kuhr and Dorough (1976).

The most effective and safe natural insecticides are those present in pyrethrum flowers and there has been a continued interest in their properties and those of synthetic analogues, the *pyrethroids*. Earlier pyrethroids, such as allethrin and resmethrin, were unstable in light and air, as are the natural compounds, but newer examples are both more stable and insecticidal. These properties have greatly extended the range of use of pyrethroids in agriculture and veterinary and public health applications. Some examples are included in Table 1.3. Recent reviews are those of Elliott and Janes (1978) and volume 7 (1980) of Wegler (1970–).

Most are liquids or solids of low melting point with good solubilities in paraffinic solvents although deltamethrin is an exception. A feature of the currently available pyrethroids, again with the exception of deltamethrin, is that the technical grades usually contain a mixture of the isomers which can arise from the presence of at least two and sometimes more centres of optical and geometrical isomerism. Since the isomers have different biological activities the toxicity of a technical product will vary with the number and amounts that are present.

Interconversions of *cis* and *trans* isomers and racemizations are common reactions brought about by heat, light, and mild alkaline conditions.

Pyrethroids are esters and will be hydrolysed in the presence of stronger bases.

They are very potent contact insecticides which act on the peripheral and central nervous systems although the details are not known.

The groups of insecticides mentioned so far include most of the compounds in current use. Insects are also attacked by a variety of other chemicals and descriptions of these will be found in the general texts listed in the introduction.

1.2.2   Herbicides

There are many herbicides and only a small number of the better known are included in Table 1.4. The other books on pesticides already mentioned, the treatise edited by Audus (1976) and Ashton and Crafts (1981) should be consulted for further information.

The *phenoxyacetic acids,* 2,4-D and MCPA, were the first selective and safe herbicides to be discovered and are still used in large quantities. They simulate the action of natural hormones in the plant and produce unco-ordinated growth, but a reason for the selective effect against broad-leaved species, other than increased retention, has not been found. Phenoxypropionic acid derivatives (diclofop and trifop) have, in fact, a different spectrum of activity. Other related compounds are benzoic acids with chloramben and dicamba as well-known examples. The active molecule in all these herbicides is the free acid but this has a low solubility in water and they are usually supplied as sodium or amine salts, which are water-soluble, or as esters which are oil-soluble and formulated as emulsion concentrates. They are relatively stable compounds although variations in structure result in different rates of destruction in plants and soil.

Chlorinated-aliphatic acids such as TCA (trichloroacetic acid) or dalapon are selective for grasses but they do not interfere with the auxin system and their mode of action is as yet unknown. They hydrolyse slowly in water and are distributed as soluble powders of the sodium salts.

Insecticidal carbamates are mainly based on *N*-methylcarbamic acid. Herbicidal *carbamates,* on the other hand, include a great variety of *N*-phenyl and *N*-alkyl derivatives other than methyl, as well as the corresponding thiocarbamates. There is a comparable wide variation in their uses and properties. The *N*-phenylcarbamates are mitotic poisons and inhibit cell division while others may interfere with lipid biosynthesis. As esters, all are susceptible to hydrolysis.

Another class is formally based on the *amide* group but, as with the carbamates, the common possession of so small a unit of structure should not be taken to infer a common mode of action. Propanil, one of the most successful, is used as a contact, post-emergence herbicide and is thought to

TABLE 1.4. *Herbicides*

Common name (ISO)	Systematic name (IUPAC)	m.p. (°C)	Vapour pressure at 20°C (mPa)	Solubilities in				Common formulations
				(1) Organic solvents			(2) Water, (p.p.m.) at ambient temperature (20–30°C)	
				H	M	L		
1. Phenoxyacetic and other acids								
Chloramben	3-Amino-2,5-dichlorobenzoic acid	200–1	0.11				700	S (ammonium salt) EC (methyl ester)
2,4-D	2,4-Dichlorphenoxyacetic acid	141	0.12	K		Ar	620	S WSP (of salts) EC (esters)
Dalapon*	2,2-Dichloropropionic acid	Liquid	—				46	WSP (sodium salt)
MCPA	4-Chloro-2-methylphenoxyacetic acid	118–19	—			Ar	825	S WSP (of salts) EC (esters)
2,3,6-TBA	2,3,6-Trichlorobenzoic acid	125–6	—	Ar	A		7000	S (of salts)

2. Ureas										
Diuron	3-(3,4-Dichlorophenyl)-1,1-dimethyl-lurea	158–9	0.0042			K	42	WDP	SC	
Linuron	3-(3,4-Dichlorophenyl)-1-methoxy-1-methylurea	93–4	0.98	K	Ar	A	75	WDP		
3. Triazines										
Atrazine	2-Chloro-4-ethylamino-6-isopropylamino-1,3,5-triazine	175–7	0.040	Am		Ar	30	WDP	SC	
Simazine	2-Chloro-4,6-di (ethylamino)-1,3,5-triazine	225–7	0.00081		Am	3.5	WDP	SC		
4. Miscellaneous										
Amitrol†	3-Amino-1,2,4-triazole	157–9	—			K	280 g/l	WSP		
Chlorpropham	Isopropyl 3-chlorophenylcambamate	41	—	Ar	A		89	EC	WDP	G
Diquat	1,1'-Ethylene-2,2'-bipyridylium ion	—	—			Ar	700 g/l (dibromide)	S		
Glyphosate	N-(Phosphonomethyl)glycine	230	—				12 000	S (of salt)		
Propachlor	α-Chloro-N-isopropylacetanilide	67–76	0.32	Ar	A		700	WDP	G	
Propanil	3',4'-Dichloropropionanilide	92–3	0.054	K	Ar		225	EC	S	
Trifluralin	2,6-Dinitro-NN-dipropyl 4-trifluoro-methylaniline	49	7.2	A			1	EC	G	

*BSI name.
†BSI name is aminotriazole.

interfere with photosynthesis, whereas others, of which propachlor and alachlor are examples, are applied to the soil in a pre-emergence treatment and may inhibit synthesis of proteins. As amides, they decompose under both acidic and alkaline conditions.

Numerous *ureas* and *triazines* have been developed as herbicides. Many, including the most popular, have similar physical properties in that melting points are usually high and solubilities in water, or even common organic solvents, low so that water-dispersable powders or suspension concentrates are the usual formulations. Vapour pressures are generally low and the compounds persistent in soil so that high doses can free ground from all vegetation for long periods. Both types are stable in neutral or slightly acid or basic media but decompose at higher and lower pHs. Both also act by inhibiting photosynthesis.

Another important group of pre-emergence compounds is formed by the *2,6-dinitroanilines* such as trifluralin and nitralin. They vary considerably in physical properties such as melting point, vapour pressure, and solubility according to the substituents present. Trifluralin, for example, is relatively volatile and must be incorporated into soil immediately it is applied to reduce evaporation. As a group they are susceptible to degradation by light. They are known to inhibit mitosis and this may be the basis of their action.

Bipyridylium ion herbicides are unusual in possessing a positive charge and are strongly bound to clay minerals in soil so that they are inactivated. As salts of strong bases they are very soluble in water and supplied as concentrated aqueous or water-soluble granules. They inhibit photosynthesis by catalytic production of a toxic agent, hydrogen peroxide.

Other types of molecules with herbicidal action include nitriles, uracils, pyridazines, dinitrophenols, arsenic-containing organic compounds (which are still used on a large scale although general poisons), and petroleum oils. Glyphosate is an organophosphorus compound which, like the bipyridylium herbicides, is inactivated by soil constituents and only effective by foliar application.

1.2.3 Fungicides

These are usually divided into two groups according to the way they are used. Protective fungicides are applied to plants and other surfaces, or are incorporated in coatings on surfaces, so that they can attack spores as they arrive or as they germinate. Fungicides with systemic properties have been discovered more recently and, as the name implies, are absorbed by and circulate within living plants so that they can deal with an infection that is already established or with spores germinating at untreated sites. A comprehensive publication on fungicides and their properties has been edited by Siegel and Sisler (1977). A selection of the better-known are included in Table 1.5

TABLE 1.5. *Fungicides*

Common name (ISO)	Systematic name (IUPAC)	m.p. (°C)	Vapour pressure at 20°C (mPa)	Solubilities in (1) Organic solvents			(2) Water, (p.p.m.) at ambient temperature (20–30°C)	Common formulations
				H	M	L		
Benomyl	Methyl 1-(butylcarbamoyl) benzimidazol-2-ylcarbamate	Decomp.	—			Ar	3.8	WDP
Captan	3,4,7,7a-Tetrahydro-*N*-(trichloromethanesulphenyl) phthalimide	178	—		Am	Ar	3.3	WDP D
Chlorothalonil	Tetrachloroisophthalonitrile	250–1	167			Am	0.6	WDP SC SM
Ethirimol	5-Butyl-2-ethylamino-6-methylpyrimidin-4-*ol*	159–160	0.13			K	200	SC
Fenarimol	2,4'-Dichloro-α-(pyrimidin-5-yl) diphenylmethanol	117–19	0.0057				14	EC
Fentin hydroxide*	Triphenyltin hydroxide	118–20	—				1.2	WDP
Maneb	Polymeric manganese ethylene bisdithiocarbamate	Decomp.	—			Ar	—	WDP
Quintozene	Pentachloronitrobenzene	146	1075	Ar			—	EC WDP D
Thiophanate-methyl	1,2-Di(3-methoxycarbonyl-2-thioureido) benzene	172	—				—	WDP

*BSI name.

Those *protective fungicides* which contain *copper* have been used in agriculture and horticulture for many years. A variety of compounds and formulations have been utilized during this time but since the active component is the copper ion they are really alternative ways of supplying the ion at an appropriate rate from a durable deposit. *Mercury* and *tin* are other heavy metals with fungicidal properties. Unfortunately, mercury in both inorganic and organic forms is a general biocide and many of its compounds are now considered too hazardous for widespread use. Tin is always employed in the form of organic derivatives, and triaryls, such as fentin acetate and fentin hydroxide, are specially favoured. Many of the heavy-metal compounds have low solubilities in water and are used as water-dispersable powders. Spores concentrate copper and its action may depend upon non-specific effects on enzymes. Mercury also reacts non-specifically with thiol groups, and organotin compounds are inhibitors of oxidative phosphorylation in mammals although they have not been shown to do this in fungi.

Another important group of protective fungicides in terms of quantities used is formed by *sulphur* itself and *sulphur-containing* organic compounds. The latter include dithiocarbamates, such as ziram, and alkylene bisdithio-carbamates, such as zineb and maneb. They are reactive materials and susceptible to hydrolysis and oxidation under the influence of moisture, heat, and light. Many have low solubilities in water and are offered as water-dispersable powders and dusts. Rather surprisingly, the way sulphur itself acts is not known but the dithiocarbamates are thought to affect respiration and the alkylene derivatives to react with thiol groups.

Captan was introduced as early as 1949 as a new type of protective fungicide, and closely related compounds containing the *N*-trichloromethyl-thio group have also been developed. They are relatively stable materials and only decompose under strongly alkaline conditions. Solubilities in water and common solvents are low and they are used as water-dispersible powders or dusts. Like copper, captan is accumulated in spores and being a reactant with thiol groups it could have a number of non-specific effects. The same targets are probably attacked by chlorinated quinones like chloranil or the chlori-nated dinitrile, chlorothalonil.

Other examples with protective action are *dinitrophenols* and their esters (dinocap) and *chlorinated nitrobenzenes* such as tecnazene and quintazene.

A selective toxicity for fungal tissues is not needed by a protective fungicide because there is very little movement of the chemical from a spray deposit. They can therefore be general poisons and react with pathways of energy production, for example, which are common to both pest and host. *Systemic fungicides* must necessarily move into the plant and through its transport systems to reach the pest, and selective action becomes essential.

Most of the types of action that have been demonstrated or are suspected therefore turn out to involve inhibitors of biosynthetic reactions which are specially important to the fungus, the following are some examples. Benomyl or related compounds such as thiophanate-methyl are inhibitors of nucleic acid and protein synthesis through a common intermediate. Others inhibit synthesis of lipids (triarimol and fenarimol), chitin (kitazin P) and single-carbon transfer reactions involving folic acid (ethirimol and dimethirimol). Systemic activity has been found in a variety of chemical structures and there are corresponding variations in chemical and physical properties. Several are formulated as water-dispersable powders and suspension concentrates and a high, or even moderate, solubility in water is certainly not needed for systemic properties.

Miscellaneous pesticides include acaricides, nematicides, molluscicides, rodenticides, and chemicals used in bird control. Few of them are specific for the pests and many are also insecticides, particularly organophosphorus esters and carbamates, or general biocides such as heavy-metal compounds or halogen-containing fumigants. The general books mentioned in the Introduction should be consulted for further details.

1.3 Formulations

A favourable combination of pest, pesticide, and method of application can sometimes allow the pesticide to be used as neat, technical-grade material. Examples are fumigants vapourized directly into a confined space or liquid insecticides which can be dispersed as very fine drops and used to control flying insects.

In most situations, however, the user can only employ the pesticide conveniently and efficiently if it is supplied as a formulation which contains other, non-pesticidal, components. These have various functions which will be mentioned as the main types of formulation as described. About 30% of the formulations mentioned in Martin and Worthing (1977) are water-dispersible powders and 25% are emulsion concentrates. Granules, solutions, and dusts account for about 15, 10, and 8% respectively and the remainder is made up of minor varieties such as smokes, baits, and suspension concentrates. The particular types available for any one pesticide depend upon its properties and range of uses but there is no doubt as to which are the most popular formulations overall.

Considering the importance of formulations in pest control there are few extended accounts of the subject. The most useful are those of Flanagan (1972) and Van Valkenberg (1973). Other publications derive from symposia, for example the Society of Chemical Industry (1966) or Gould (1969). There is a chapter on formulations in Green *et al.* (1977) and Matthews (1979) and manufacturers often provide formulation manuals for their pesticides. A catalogue of the various types of formulations has been published by Gifap (1978).

1.3.1   Solutions

A few organic pesticides are very soluble in water and are available as solutions in this solvent or as water-soluble powders and granules which can be dissolved shortly before spraying. Examples are herbicides which are salts and some insecticides like methomyl and trichlorphon.

Most compounds can be considered insoluble in water from the formulation point of view, and organic solvents are needed for the preparation of solutions or emulsion concentrates. A choice among these solvents is limited to those which are available as bulk chemicals, are safe to use, do not attack containers or spray gear, have a low phytotoxicity when this property is relevant, and, in most cases, can dissolve high concentrations of the pesticide. In practice the various mixtures of aromatic hydrocarbons prepared directly or indirectly from petroleum are the ones frequently employed. There are a number of these, distinguished, among other things, by aromatic content and boiling range. Xylene fractions are too volatile for general use in sprays, and solvents with an initial boiling point of about 160°C and upwards are normally chosen. Concentrations of dissolved pesticide cover a wide range although restrictions imposed by solubility and economic transport mean that values between 200 and 500 g/l will be encountered more frequently. The less-soluble chemicals may need to be supplied at lower concentrations or a second, more powerful solvent may replace part of the main solvent.

Solutions should remain as single-phase systems until they are used and the concentration of pesticide must be kept sufficiently below the solubility level to allow for possible reductions in temperature and consequently, solubility, during storage and transport. The larger containers are opaque and separations of additional liquid or solid phases may not be noticed. Other changes include degradation of the pesticide and corrosion of the container. An example of the former is the acid-catalysed decomposition of some insecticidal esters. This can be avoided by including a small amount of epichlorhydrin to remove any acid formed during storage.

Solutions in water or organic solvents are the simplest formulations to make, add to the spray tanks, and apply because they are physically homogeneous, unlike spray liquids prepared from emulsion concentrates and water-dispersible powders. They are, however, denser than the solvent alone and should be stirred during dilution to ensure an even concentration.

Dilute solutions in organic solvents are an expensive and hazardous way of applying pesticide at the higher volume rates and after their early use with insecticides such as DDT were replaced by the water-based dispersions such as emulsions and suspensions. More recently there has been some return to solutions, usually containing high concentrations of pesticides, with the advent of equipment designed for low and ultra-low volume rates.

1.3.2 Emulsion concentrates

A few pesticides have been available as concentrated oil-in-water emulsions which can be diluted further with water for spraying but the great majority of emulsions are prepared from solutions which contain small amounts of oil-soluble emulsifying agents and can be added directly to water in the spray tank. The name long used for these solutions is emulsion concentrates although they could more appropriately be called emulsifiable concentrates (Gifap 1978) since no emulsion exists before dilution. However the older name is still used by the World Health Organization (1979) and will be retained here.

The solvents in these concentrates are normally of the type used in single solutions, although sometimes of lower boiling range, and the variety of pesticide concentrations encountered is, accordingly, very similar. Exceptionally, a liquid insecticide can be formulated with very little or no solvent and very high concentrations are possible; there is an 860 g/l emulsion concentrate of malathion for example.

A good emulsion concentrate is one which produces a suspension of fine drops, as judged by the white appearance, when poured into water. This emulsification is not truly spontaneous because some mixing is needed but this should be the minimum provided by the addition process and gentle stirring. The initial emulsion should also maintain a reasonably uniform concentration of pesticide during the time between preparation and spraying. Changes in concentration result from movement of oil drops, upwards or downwards, at rates which depend upon their size and the difference in density between the oil and the surrounding water phase. An increase in numbers of drops towards the top or bottom of a column of emulsion is known as creaming and should be kept as low as possible in a tank which has little or no agitation. If a formulation has a tendency to cream, the process can be prevented or reversed by even a moderate amount of mixing. On the other hand, if the creamed droplets coalesce and eventually produce a separate layer of oil the emulsion has broken and this situation may only be reversible with a substantial input of mechanical energy. While some creaming is normally acceptable therefore, breaking is not.

The ability of a given pesticide and solvent to provide a satisfactory emulsion with a variety of field waters and temperatures is determined by the kind and amounts of emulsifiers present. A blend of two emulsifiers with different surface-active properties is usually more effective than one acting alone. About 5 to 10% by weight is normally present in the final concentrate.

Although these formulations have been used for many years the events occurring during emulsification are not fully understood, although some progress is being made (Groves 1978).

The emulsions prepared from these concentrates are often applied at high volume rates following dilution to 5 or even 1% by volume but there have been recent trends towards the use of lower volumes of higher concentrations. It should be noted that a given formulation is not necessarily satisfactory at dilution rates for which it was not designed.

1.3.3   Water-dispersible powders

These are also known as wettable powders (Gifap 1978) but, although the processes of wetting and dispersion are both essential, the name favoured here does show that water is the medium in which they are dispersed. Apart from the active chemical the basic ingredients are one or more inorganic solids and surface-active agents. Both solid and liquid pesticides can be formulated in this way but differ in method of preparation, composition, and physical nature of the product before and after dispersion in water.

Powders of a solid pesticide are prepared by milling and the final size distribution of the particles is an important characteristic, since it will determine the behaviour of the particles in suspension, their life in spray residues and their biological activity. It is a valuable feature of this type of formulation therefore that such a property is controlled during manufacture. Inert, non-fusible solids are needed as diluents because technical-grade pesticides tend to be sticky materials, and additives such as kaolinite and attapulgite clays and synthetic silicas prevent caking during milling and storage. Most pesticides are hydrophobic, and penetration of water into the fine powder, followed by dispersion to the individual particles to give a stable suspension with gentle stirring, require the presence of a few percent of surface-active agents. A single agent may suffice but two compounds with specific wetting and dispersing properties are often used. The complete powders can contain up to 850 g/kg of a pesticide such as carbaryl which has suitable physical properties, although the range of 500 to 750 g/kg is more usual.

When the dispersion is first prepared the pesticide and diluent particles will be distributed throughout the column of spray liquid. Both, however, are heavier than water and will sink to the bottom of the container at rates which depend upon their sizes and the difference in density between the particles and the surrounding water. Sedimentation, like the creaming with emulsions, will cause a change of concentration of pesticide in a spray liquid and should be kept below a minimum value to allow for use in equipment which has no provision for mixing during spraying. Even fine particles will be unsatisfactory if they flocculate. Flocculation and the subsequent accelerated sedimentation of the aggregated particles results when too little dispersing agent is present or salts in natural water destabilize the protective layers of adsorbed surfactant molecules.

Many liquid insecticides are also offered as water-dispersible powders even though the important aspect of a controllable particle-size distribution no longer applies. The inert, inorganic solid now becomes a carrier rather than a diluent and needs to have a high oil absorptivity. Maximum concentrations of active ingredient are about 500 g/kg although a few products with very absorbent carriers have raised this to 600 g/kg. Surfactants are still needed to allow wetting and dispersion but they also encourage displacement of the liquid toxicant from the carrier. Spray liquids prepared from these powders are therefore usually emulsions of insecticide mixed with a suspension of carrier particles. Secondary emulsification of this type has, of course, not been a feature that is controlled in the development of a powder and drop sizes vary greatly with different formulations. Breaking of the emulsion is usually noticed but creaming is not easily seen in the presence of carrier and can only be inferred from suspensibility tests.

Water-dispersible powders are usually applied at high volume rates with low concentrations of the order of 1 to 2% of active ingredient but, as with emulsion concentrates, there is a recent tendency to use smaller volumes of higher concentrations.

Some of the clays which are used as diluents and carriers can have very acidic groups on their surfaces and will catalyse decomposition of pesticides unless neutralizers are included in the powders.

1.3.4   *Suspension concentrates*

These are suspensions containing up to 800 g/l of a finely ground solid in a liquid which is essentially a non-solvent for the pesticide at ambient temperatures (less than 0.01% has been quoted as a guide). They are diluted with the same or related liquids before spraying. Water is the liquid most often used and water-based systems are described here but there are a few examples where the dispersion has been made in an oil. Milling may be wet or dry and surfactants are needed to stabilize the newly formed particles during grinding and in the final suspension. Their presence is not enough in itself because during long periods of storage even very fine particles will slowly sink in the container and produce a sediment which can be difficult to redisperse. The rate of this sedimentation is reduced by thickening the liquid phase with polymers such as polysaccharide gums or by adding clays whose particles form a semi-rigid structure throughout the dispersion. These viscous mixtures must have carefully chosen thixotropic properties so that after a brief shaking or stirring they can be poured from the container and diluted with water. Freezing of the water phase would seriously damage the structure of this delicately balanced system, and anti-freeze chemicals such as glycols are often included.

Suspension concentrates share with water-dispersible powders of solid pesticides the advantage of containing particles whose sizes are determined during manufacture. The sizes are rather smaller than in many powders, being usually less than 5 µm or even 2 µm to reduce sedimentation, but this feature may result in a better biological activity in some applications, if not in others. Their liquid form also provides the additional advantages of being more compact for storage and easier to dispense than powders, without the hazards of flammability and toxicity by dermal absorption of other liquid formulations. On the other hand they may be less able to withstand adverse storage conditions, especially in tropical countries. The physical chemistry of suspension concentrates has been reviewed by Tadros (1980).

1.3.5   Dusts

In the formulations already mentioned the efficient application of a small amount of active chemical is assisted by extension with water or an organic solvent followed by atomization. These liquids are replaced in dusts by a solid, particulate diluent and the free-flowing powder is mixed with air with the aid of dusting machines. The air-borne particles are thus carried by the wind or a directed blast of air.

Concentrations of pesticide must be low since dusts can only be handled at high application rates and are usually of the order of 10 to 50 g/kg. In these circumstances even a solid pesticide would be expected to be carried by the particles of diluent and it is the size distribution of these that is important in good application. If particles are too small they do not flow well because they aggregate or, after dispersion, can cause drift problems. The density of mineral carriers is high and if particles are too large they may be difficult to lift into a stream of air or will sediment too rapidly. While there is no accepted standard the range from 50 to 100 µm is generally suitable although a smaller size can be useful for special applications, especially where better adhesion to foliage is required.

The appropriate low concentrations of pesticides may be blended with the diluent followed by milling but these dilute dusts are expensive to transport in terms of active ingredient, and dust concentrates are often made by one formulator for local dilution by another. If water-dispersible powders are available they are also, of course, concentrated dry powders and can be diluted locally to prepare dusts.

1.3.6   Granules

The main function of the formulations mentioned so far has been to aid the mechanical distribution of small amounts of pesticides directly to the pest or its immediate environment. This is done by carrying them in or on fine liquid drops or solid particles. However, in some applications, soil and water treatments for example, the pesticide may only need to be taken part of the way to the pest with the final distribution being dependent on its physical properties and operating through local transport systems. Granules are suitable means of providing such intermediate distribution.

As the name implies, this formulation is made of large particles. Their definition in terms of size varies but there is no need to be too precise. If the lower limit is taken as about 0.1 mm diameter, dusts are excluded, while if the sizes rise much above 2.5 mm they will be progressively less easy to apply evenly. The usual range is about 0.3 to 1.3 mm, although microgranules of 0.1 to 0.3 mm are favoured where adhesion to foliage is relatively important.

If granules are an effective way of distributing a chemical for control of a particular pest their other advantages can be realized. These include an improved accuracy of placement compared with dusts and sprays, which results from their larger average mass, and increased safety in handling because the concentration of active ingredient is normally 100 g/kg or less. In addition there should be reduced environmental contamination because granules do not drift and good formulations are free of dust. Finally granules are formulations in which some control over the rate of release of the chemical can be built into the design.

The development of biologically effective concentrations in the volume of soil, water, or air surrounding the granules depends upon their dose, size, content of pesticide, and release characteristics. These characteristics are determined in the first instance by the method of preparation and, secondly, by the influence of various additives. An impermeable base material of the required particle size can be simply coated with the pesticide directly or as a solution. Such granules are tough and all the pesticide is available as a surface layer and released quickly. Alternatively a porous carrier may be chosen. The pesticide will now be released over a longer period of time because it has to diffuse from the interior of the granules. Rates can be altered by varying the carrier, or by adding other materials which influence diffusion and solution. Finally a paste of water, pesticide, and powdered carrier can be extruded, cut to appropriate size, and dried. This method gives the most scope for control of release because a wide range of other fillers, binders, and miscellaneous additives can be included in the original mix. It also produces granules which tend to be more fragile than those made by other methods but the progressive break-down provides another way of releasing the chemical.

The intimate association of small amounts of pesticide with large amounts of carrier or diluent, especially when these are absorbent, may encourage decomposition and stabilizers are sometimes needed. This possibility also applies to the dilute dusts mentioned in the previous section.

Further information on granules and their properties is available in the papers given at a symposium arranged by the Society of Chemical Industry (1972).

1.3.7 Other controlled-release formulations

The notion of controlled release is very appealing. In the ideal situation a constant weight of chemical is delivered to the local environment of the pest in unit time. The most efficient use of the pesticide will therefore be obtained when, after a build up to a lethal concentration, the rate is just sufficient to offset losses caused by such processes as evaporation, decomposition, or adsorption. Conventional formulations must be applied at higher initial doses to give the same minimum lethal concentration over the same period, and much of the pesticide is wasted.

Many controlled-release devices other than granules have been developed in recent years. They range from capsules a few μm in diameter to continuous polymeric films on walls. A few examples only will be mentioned here and more detailed information can be found in the books of Cardarelli (1976) and Kydonieus (1980).

An important group uses the reservoir system where the pesticide alone, or in suspension or solution, is surrounded by an insoluble polymeric membrane. If the internal concentration of pesticide does not change, the amount diffusing through the membrane and so to the outside in a given time will also be constant. The principal sources of control over the rate of loss are choice of polymers, and their modification by cross-linking or the addition of plasticizers, membrane thickness, and, with larger types, the geometry of the device. The membrane or wall may also be a mixture of polymers or be built of more than one layer. Well-known examples are microcapsules and laminated strips.

A second group employs the simpler idea of a polymeric matrix in which the pesticide is dispersed or dissolved. Rates of loss now decrease with time but a choice of polymer and other non-active components will still give a useful control over release at any time and total life-span. Examples are rubber pellets impregnated with molluscicides and polyvinylchloride strips with volatile insecticides. If the polymer can be eroded or is biodegradable and the geometry of the device such that a change in area counterbalances the decreasing rate of loss, the rates can be more nearly constant.

In a third type the pesticide is covalently linked to a polymer and is released by reactions likely to occur during weathering or in the presence of micro-organisms. The choice of pesticide is restricted to those with suitable functional groups and phenoxyacetic acid herbicides are ones that have been used.

Specialized formulations with minor uses include pressurized aerosols and smoke generators and baits. They are described in Green *et al.* (1977). There is also a recent monograph on seed dressings edited by Jeffs (1978).

1.4 Specifications

Purchasers and users of formulations naturally wish to be sure that these conform both qualitatively and quantitatively to agreed standards as laid down in appropriate specifications. Individual manufacturers will have their own specifications for control of the quality of their products. However, there is obviously also a need for specifications which have been developed independently, preferably by international organizations so that they have a wide measure of agreement, and published so that they are available for public use. There are two such collections at the present time. That of the World Health Organization (1979) deals with the major pesticides and formulations used in public health and is mainly concerned with insecticides although a few rodenticides and molluscicides are included. Specifications for pesticides used in agriculture have been provided by the Food and Agriculture Organization (1971–) and include fungicides and herbicides as well as insecticides. Where the same compounds and formulations appear in both series the details are generally similar and all the specifications have been developed in collaboration with industry, thus ensuring that the standards set are economically feasible. Details of the methods used in carrying out the tests are given in World Health Organization (1979) but users of the Food and Agricultural Organization series are referred to CIPAC (1970). Provisional specification for some additional pesticides appear in the USAID Pesticide Manual (Von Rümker and Horay 1972). Technical grade pesticides are included even though they are not usually of concern to the user in the field or his purchasing agency because specifications for formulations stipulate that the active ingredient shall be of an agreed quality.

The formulations included in one or both of the above publications are those which are long-established and most used: solutions, emulsion concentrates, water-dispersible powders, and dusts. Those which have been developed more recently such as granules and suspension concentrates will be included in the future.

Specifications are intended to cover the following points. The identity and content of active ingredient must be as stated and the physical properties of the spray liquid prepared from a formulation should be reproducible to ensure efficient application. This latter requirement is particularly important in view of the present tendency to reduce costs and unwanted side effects by keeping to the lowest dose which gives the required biological effect. Ability to meet a specification should not change appreciably during the time between purchase and use when the formulation is stored under reasonable conditions.

The detailed clauses in specifications cannot be described here but the outlines of those which correspond to the points just mentioned can be given briefly. With all formulations there are tolerances on the content of active ingredient and maximum limits for acidic or alkaline impurities which could accelerate decomposition during storage and the corrosion of metal containers. A maximum water content for many of the non-aqueous systems is also given. Solutions and emulsion concentrates should not show separation of additional liquid or solid phases during storage at low temperatures. Those which are used as heterogeneous dispersions in water, such as emulsion concentrates and water-dispersible powders, must produce dispersions of a stated physical stability in water of defined hardness when tested both before and after accelerated storage treatments. The latter are intended to indicate the likely shelf-life under ambient conditions from shorter exposures to higher temperatures.

Finally, it should be noted that present specifications do not attempt to control the identity and chemical and physical properties of ingredients other than the pesticide. Such control would, in general, not be advisable because it would greatly increase the problems of manufacturers in dealing with fluctuations in supply of raw material and, therefore, cost. This is not to say that clauses could not be included to deal with special cases such as abrasive carriers, or impurities which are toxic to non-target animals, especially vertebrates. This generally *laissez-faire* attitude to non-pesticidal constituents may eventually change as a result of the interest currently being shown in their possible environmental effects.

1.5   Some interactions of pesticides with pests

1.5.1   Contact insecticides and flying or resting insects

Spray drops may impact directly on the insect or, more often, land on the surface of a substrate where the insect will subsequently land or walk. Absorption of insecticide by the insect from a spray deposit can occur during the time of contact or, if parts of the deposit adhere to the insect, after it has moved away. Efficiency of insecticidal action therefore depends upon the ability of the chemical to penetrate the cuticle and the ease of mechanical transfer from the deposit. The relative importance of the two processes will depend upon the particular combination of insect, insecticide, formulation, and substrate.

Good contact insecticides can pass through the epicuticular wax layer unaided and are primarily lipophilic although some solubility in aqueous phases, possibly assisted by solubilization or complexing with natural macromolecules, must be needed for subsequent movement in and beyond the cuticle. The absorption of a contact insecticide, good or bad, is improved by the presence of an organic solvent if this can disorganize the wax barrier and increase permeability (Hadaway, Barlow, and Flower 1976), allow impacting drops to cover a larger area of cuticle or help transfer from deposits (Hadaway and Barlow 1958). Oils with the low volatilities needed for persistence in deposits have been incorporated in solutions for spraying vegetation at low-volume rates in the control of cotton pests and tsetse flies, for example.

Most deposits do not contain such solvents however and the insecticide itself must have an adequate contact action in the dry residues left when water has evaporated. The extent to which it can exercise this action depends upon the physical characteristics of the deposit which, in turn are determined by the formulation used and the surface treated.

Substrates can be divided into two main classes; porous (e.g. plaster, cement, brick, air-dried soils) and those which are essentially non-porous, at least on the scale of spray particle dimensions (e.g. metal, wood, thatch, leaves). Any insecticide applied as a solution or an emulsion, or a liquid insecticide emulsified from a water-dispersible powder, is absorbed by a porous substrate and only a very small fraction of the dose is readily available to the insect as it rests or moves on the surface. The same substrate sprayed with a suspension of particles of a solid insecticide filters the particles as the aqueous phase containing dissolved surfactants is absorbed. The insect is now in direct contact with a much higher proportion of the dose. In addition, the particles which form the deposit are easily transferred and adequate kills are obtained with much shorter residence times.

The effect of a change to non-porous surfaces is in opposite directions for the two types of formulation. All the insecticide from a solution or emulsion now remains on the surface and the insect has access to a much larger dose during the time it remains on the deposit. Solid particles from suspension, while still concentrated at the surface, have lost the advantage of easy mechanical transfer to the insect because adhesive surfactant residues now remain with the particles instead of being absorbed with the water.

The particle size distribution of a solid insecticide is an important factor. Toxicity increases as the average size decreases down to an optimum value which varies with insect and substrate. There is an optimum because particles which are too fine are more difficult to detach from the substrate. Particles with a mean size of 5 to 15 μm are suitable for most substrates. They provide a good cover and are relatively easily removed and retained by insects (Hadaway and Barlow 1955). Fortunately these sizes correspond to suspensibility properties which are appropriate for efficient spraying.

A solid insecticide applied as a solution or emulsion may also crystallize on a non-porous substrate after the solvent has evaporated, and the crystal habit influences the toxicity of the residue. It varies with the insecticide, the physical nature of the substrate, and the presence of other solutes.

1.5.2   Activity of pesticides in soils

Insecticides, fungicides, herbicides, and nematicides are all applied to, and mixed with, soils to control pests. Movement of the pesticide after application is necessary to provide the uniform concentrations needed for effective control and this depends upon properties of the chemical, the soil constituents and their chemical and physical properties and soil moisture which, in turn, is determined by the weather. The various pathways for movement include diffusion in the gas and aqueous phases and bulk transport in water, as in leaching. It is not surprising that the behaviour of pesticides has been difficult to predict in such a complex system, especially under field conditions, but there are signs that the large amounts of time and effort that have been devoted to this problem under the stimulus of concern for the environment and registration requirements are now being rewarded. Properties such as solubility, adsorption, and partitioning characteristics can be included in models which explain biological activity and distribution. The work of Briggs (1981) may be cited as one example and the general reviews of Hartley (1976) and Kaufman (1977) chosen from among the many on this subject.

1.5.3.   Entry of pesticides into plants

The main barrier to foliar entry is a cuticle which is progressively hydrophobic towards its outer suface where there is layer of epicuticular wax. Oils and organic solvents move into leaves with relative ease, either through the cuticle itself or through stomatal pores. Most pesticides however are applied as aqueous sprays and these do not penetrate easily even when the surface tension is low enough to allow spreading over the surface of the leaf. An emulsion may benefit from the presence of an oil phase if this is not too volatile and can remain to facilitate entry after the water has evaporated. In the absence of a solvent as a carrier a chemical must therefore diffuse into the plant surface and across the cuticle before it can progress further by the apoplast translocation system. The situation is very similar to the movement of insecticides into insects. Lipophilic properties are an advantage for initial solution or partitioning into the cuticle but the ability to move into an aqueous phase is also needed at the inner surface of the cuticle. Penetration from an external, aqueous layer continues for a longer time if complete drying-out of the deposit is prevented and humectants are often added to sprays. Uptake is also improved with high concentrations of surfactants which may dissolve waxes or swell cuticles with a resultant increase in diffusion coefficients. The absorption of non-ionized, less-polar molecules is even favoured in roots and these may partition directly into lipid components although, again, there will have to be further partitioning into the aqueous transpiration stream. Where membranes have to be crossed against a concentration gradient active transport will be necessary and this can be influenced by the pesticides themselves.

There is a very extensive literature on the subject and among many reviews those of Price (1982) and Bukovac (1976) are noted here.

1.6   Persistence at sites of application

Persistent is an adjective which is used in two ways to describe chemicals employed for pest control. The more restricted meaning, which is used here, refers to their behaviour on the target to which they have been applied and where they are available to the pest. It is therefore related to the duration of their action against the pest although not necessarily in a simple way because factors other than the total weight of pesticide are involved. In recent years it has also been specially attached to those pesticides, such as the organochlorine compounds, which are relatively stable in the general environment after they have been redistributed from the original site of application (Edwards 1973). The general causes of loss from spray residues are evaporation, decomposition, and diffusion into porous substrates when these are treated or close at hand. Mechanical effects of wind and rain, solution in rainwater, and abrasion can be added for outdoor treatments.

A detailed survey of the various physical processes involved in pesticide use has been prepared by Hartley and Graham-Bryce (1980).

1.6.1   Evaporation

This important cause has been discussed in detail by Spencer, Farmer, and Cliath (1973). Where compounds can be compared under the same conditions the rates of evaporation are proportional to their vapour pressure and the square root of their molecular weights. In the field, the rate for an individual compound will also depend upon dose, formulation as it governs the physical nature of the deposit, the substrate, and local environmental variables such as temperature, humidity, and ventilation.

Vapour pressures of pesticides cover a wide range and although this property is only one factor in evaporative loss it is still a useful indicator of the likely persistence of a compound. Dichlorvos, a short-lived insecticide with a strong vapour action has a pressure of 1600 mPa at 20°C, while for pp`-DDT, a much more persistent compound, the corresponding figure is 0.025 or 6×10^4 times less. Even DDT can be mainly lost by evaporation after being applied as a low dose to the large surface area of a field crop.

Dilution with less-volatile solvents reduces the vapour pressure of a pesticide and prolongs its life. Improvements of 5 to 10 times are easily available with practical formulations. If the additive is non-volatile and controls the rate of diffusion from the mixture, there can be a very large reduction in rate. This also happens when a pesticide becomes incorporated in porous substrates like soils or the plaster or concrete of walls. With suspensions of solid particles an increase in particle size will reduce evaporation but at the expense of suspensibility and activity.

1.6.2 Decomposition

A pesticide may also be depleted in spray deposits because it decomposes. Rates will depend upon the structure of the chemical, the surface on which it is lying or the medium in which it is dispersed, and factors like temperature, moisture, light, and air. Non-biological degradation reactions have been described for pesticides in general by Armstrong and Konrad (1974), for herbicides by Crosby (1976), and for fungicides by Woodcock (1977). Most of this information relates to soils and there is much less for residues on or in other materials.

The main routes found are hydrolysis of esters, oxidations of sulphur and nitrogen, reductive dechlorination and other substitutions, rearrangements of multi-ring organochlorine compounds and isomerizations of phosphorus esters and *cis/trans* pairs. These reactions may occur under relatively mild conditions if they are energetically favourable or if catalytically active substances are present. There has also been a lot of interest in those which use energy absorbed from daylight and it is now evident that photodecomposition is a potentially important route for losses of pesticides from vegetation, water, and even air-borne particles.

Photodecomposition can only occur if light is absorbed by the pesticide molecules but it does not follow that the wavelength of available light must be at or near absorption maxima of the compound, as is frequently assumed. Many pesticides which have a very weak absorption in the most energetic part of the solar spectrum can still decompose, especially if naturally occurring sensitizers are available. There may equally well be natural absorbers or quenchers, particularly in deposits on vegetation, and this is one reason why decomposition measurements made in the laboratory are only of practical significance if they are confirmed in the field. Even there, variations in quality and quantity of light with season and location must be taken into account in any assessment of the overall importance of photodecomposition. The subject has been reviewed by Zabik, Leavitt, and Su (1976).

Stabilizers are generally only used in formulations to protect the active ingredient during storage, not after it has been applied. Neutralizers for acids and anti-oxidants are examples. If protectants for spray residues were considered they would have to be effective at low concentrations, stable to light, as persistent as the potential life-time of the protected pesticide, and the formulation would have to be one which allowed an intimate mixture of both chemicals in the spray deposit. In the case of a protectant against photodecomposition it would also need to absorb the damaging light. A combination of all these properties is difficult to achieve and perhaps accounts for the lack of interest in such compounds.

1.6.3 Solution

Examples are provided by pesticides dissolving in rain-water, either on vegetation or in soils, or in waxes on the surface of leaves. Protective fungicides and insecticides which are not intended to have a systemic action usually have very low solubilities in water but these low values can still range over several orders of magnitude. Some influence on persistence of deposits or leaching from soils would therefore be expected. In areas of high rainfall even the very insoluble compounds may dissolve from low doses of exposed deposits.

Solution in leaf waxes can increase persistence of an insecticide in a residue by reducing evaporation or solubility in rain but this will at the same time reduce biological activity to insects if these do not eat the leaves.

1.6.4 Mechanical processes

Dried residues of pesticide formulations can be dislodged by rain in addition to any solution of the chemical itself. Other agencies with a mechanical action include wind, attrition by wind-borne dust, and the mutual abrasion of neighbouring pieces of vegetation. Such processes become more noticeable at heavier doses because the outer layers of a deposit are less adherent than those in close contact with the surface. Their importance also varies with the type of formulation and therefore the physical characteristics of the dried spray deposits. Solid particles from dusts and water-dispersable powders are probably removed more easily than the non-particulate residues provided by solutions and emulsions although more investigations on this subject are needed.

Losses of this type can be reduced by including adhesives in the formulations (Somers 1967). These have ranged from hydrophobic polymers, particularly polyisobutenes and polyethylene polysulphides, to hydrophilic and water-soluble gums, starches, and proteins. Water solubility is not a defect because their swelling and solution from dried deposits are slow processes. Adhesives do increase persistence, as judged by laboratory measurements, but often at the expense of biological activity and their value in the field is not clearly proved even in the case of fungicides with which they have been mainly used.

1.6.5 Sorption

Pests may be controlled by chemicals as vapours in air or solutions in water. These media contain, or are bounded by, other materials which can adsorb the pesticide and so decrease the amounts which are available to the pest. Examples are provided by sorption of vapours by the porous walls of buildings and layers of silt underlying water which adsorb larvicides and molluscicides. In many of these situations there is a reduction in effectiveness and higher doses are needed to replace the losses.

Porous surfaces are also sprayed directly. A liquid spray will be absorbed immediately but even a suspension of solid particles, which is at first filtered at the surface to produce a superficial layer of dried deposit (see page 27), can be completely lost by diffusion and subsequent adsorption on the under-lying substrate. These processes normally result in a reduction of activity because an insect, for example, is no longer in close contact with a concentrated layer of insecticide (Hadaway and Barlow 1955). Sorption will also accelerate any decomposition reactions promoted by constituents of the substrate. It can have beneficial effects by prolonging the residual lives of volatile, stable compounds if these are potent enough to be biologically active in the sorbed state. The only practical way to prevent both absorption and adsorption on porous walls is to seal the pores.

Losses of pesticide from sites of application can also be brought about by biochemical means and arise, for example, from the activity of micro-organisms in soil or metabolic reactions in plants and animals which have been treated with systemic compounds. Investigations of these processes have been very extensive and summaries will be found in the specialized texts which describe the main groups of pesticides.

1.7   Future developments

The need for synthetic chemicals which have a direct toxic action against pests will continue although they are more likely to form a part of general management schemes than to be used alone.

Methods of application will be improved and the discovery of more potent and selective compounds can be expected. A high potency means lower doses and less chance of unwanted environmental effects. These are further reduced if the newer structures can combine stability at sites where chemicals meet the pests with decomposition elsewhere to products of negligible biological activity.

The control of activity, persistence, and selectivity of a chemical by a choice of formulation type and constituents has not been fully exploited until now and there are likely to be more efforts in this direction.

1.8   References

Armstrong, D. E. and Konrad, J. G. (1974). In *Pesticides in soil and water* (ed. W. D. Guenzi), p.123. Soil Science Society of America, Madison.
Ashton, F. M. and Crafts, A. S. (1981) *Mode of action of herbicides* (2nd edn). John Wiley and Sons, Chichester.

Audus, L. J. (ed.) (1976). *Herbicides* (2nd edn) Vols 1 and 2, Academic Press, London and New York.

Australian Department of Health (1978). *Pesticides: synonyms and chemical names* (4th edn). Australian Government Publishing Service, Canberra.

Briggs, G. G. (1981). *J. Agric. Food Chem.* **29**, 1050.

Brooks, G. T. (1974). *Chlorinated insecticides,* Vols 1 and 2, CRC Press, Cleveland, Ohio.

BSI (1969). British Standard 1831:1969. *Recommended common names for pesticides.* British Standards Institution, London. (Supplements are issued at irregular intervals).

Bukovac, M. J. (1976). In *Herbicides* (ed. L. J. Audus) Vol. 1, p.335. Academic Press, London and New York.

Cardarelli, N. F. (1976). *Controlled release pesticides formulations.* CRC Press, Cleveland, Ohio.

CIPAC (1970). *CIPAC Handbook,* Volume 1, Analysis of Technical and Formulated Pesticides. Collaborative International Pesticides Analytical Council Limited, Harpenden, England. An addendum, Volume 1A, was published in 1980.

Corbett, J. R. (1974). *The biochemical mode of action of pesticides.* Academic Press, London and New York.

Cremlyn, R. (1978). *Pesticides.* John Wiley and Sons, Chichester.

Crosby, D. G. (1976). In *Herbicides* (ed. L. J. Audus) Vol. 2, p.65. Academic Press, London and New York.

Edwards, C. A. (1973). *Persistent pesticides in the environment* (2nd edn). CRC Press, Cleveland, Ohio.

Elliott, M. and Janes, N. R. (1978). *Chem. Soc. Rev.* **7**, 473.

Eto, M. (1974). *Organophosphorus pesticides.* CRC Press, Cleveland, Ohio.

Fest, C. and Schmidt, K. J. (1982). *The Chemistry of organophosphorus pesticides* (2nd edn) Springer, Berlin.

Flanagan, J. (1972). In *Industrial production and formulation of pesticides in developing countries,* Vol. 1, p.75. United Nations, New York.

Food and Agriculture Organization (1971–). *FAO specifications for plant protection products.* FAO, Rome. (Booklets for individual or closely related pesticides are issued at irregular intervals.)

Gifap (1978). *Catalogue of pesticide formulation types and international coding system.* Technical Monograph No. 2 Gifap, Brussels.

Gould, R. F. (ed.) (1969). *Pesticidal formulations research. Advances in Chemistry Series* 86. American Chemical Society, Washington, D.C.

Green, M. B., Hartley, G. S., and West, T. F. (1977). *Chemicals for crop protection and pest control.* Pergamon Press, Oxford.

Groves, M. J. (1978). *Chem. Ind.* **417.**

Hadaway, A. B. and Barlow, F. (1955). *Bull. ent. Res.* **46**, 547.

—— and —— (1958). *Ann. appl. Biol.* **46**, 133.

——, ——, and Flower, L. S. (1976). *Penetration of insecticides from solutions into tsetse flies and other insects.* Miscellaneous Report No. 22. Centre for Overseas Pest Research, London.

Hartley, G. S. (1976). In *Herbicides* (ed. L. J. Audus) Vol. 2, p.1. Academic Press, London and New York.

Hartley, G. S. and Graham-Bryce, I. J. (1980). *Physical principles of pesticide behaviour,* Vols 1 and 2. Academic Press, London and New York.

Hassall, K. (1982). *The chemistry of pesticides.* Verlag Chemie, Weinheim.

ISO (1981) International Standard 1750, *Pesticides and other agrochemicals – Common names.* ISO, Geneva.

Jeffs, K. A. (ed.) (1978). CIPAC Monograph 2. *Seed Treatment.* Collaborative International Pesticides Analytical Council Limited, Harpenden.

Kaufman, D. D. (1977). In *Antifungal compounds* (eds. M. R. Siegel and H. D. Sisler) Vol. 2, p.1. Marcel Dekker, New York and Basel.

Kuhr, R. J. and Dorough, H. W. (1976). *Carbamate insecticides.* CRC Press, Cleveland, Ohio.

Kydonieus, A. F. (1980). *Controlled release technology,* Vols 1 and 2. CRC Press, Boca Raton, Florida.

Martin, H. and Worthing, C. R. (eds.) (1977). *The pesticide manual* (5th edn). British Crop Protection Council.

Matthews, G. A. (1979). *Pesticide application methods.* Longmans, London.

Price, C. E. (1982). *Linn, Soc. Symp. Sep.,* **10,** 237.

Siegel, M. R. and Sisler, H. D. (eds.) (1977). *Antifungal compounds,* Vols 1 and 2. Marcel Dekker, New York and Basel.

Society of Chemical Industry (1966). *The formulation of pesticides.* SCI Monograph No. 21. SCI, London.

—— (1972). *Pestic, Sci.* **3,** 745.

Somers, E. (1967). In *Fungicides* (ed. D. C. Torgeson) Vol. 1, p.153. Academic Press, New York.

Spencer, W. F., Farmer, W. J., and Cliath, M. M. (1973). *Resid. Rev.* **49,** 1.

Tadros, Th.F. (1980). *Adv. Colloid Interface Sci.* **12,** 141.

Tropical Pest Management Pesticides Index (1981). Centre for Overseas Pest Research, London.

Van Valkenberg, W. (ed.) (1973). *Pesticide formulations.* Marcel Dekker, New York.

Von Rümker, R. and Horay, F. (eds.) (1972). *Pesticide manual, Part III.* Agency for International Development, Washington, D.C.

Wegler, R. (ed.) (1970–). *Chemistry of plant protection and pest-controlling agents.* Springer, Berlin.

Woodcock, D. (1977). In *Antifungal compounds* (eds. M. R. Siegel and H. D. Sisler) Vol. 2, p.209. Marcel Dekker, New York and Basel.

World Health Organization (1979). *Specifications for insecticides used in Public Health,* (5th edn). WHO, Geneva.

Worthing, C. R. (ed.) (1979). *The pesticide manual* (6th edn). British Crop Protection Council.

Zabik, M. J., Leavitt, R. A., and Su, G. C. C. (1976). *Annu. Rev. Entomol.* **21,** 61.

2
Physics and meteorology

D. R. JOHNSTONE

2.1 Introduction

The pesticide applicator's particular concern is with the dispersion of formulated active material into small particles (usually droplets) and with the ensuing particulate placement, or travel and deposition. His success is measured by the immediate or cumulative impact on the pest population, evident through reduction in its damaging effects, balanced against the possible adverse side effects which can result from misplacement or from unwanted persistence of the active material in the environment. Information must be drawn from multidisciplinary sources, for, in responding to an economic pest problem, biological or ecological evaluation is required to pinpoint the target and advise on the optimum timing of single or multiple application, while toxicological study is necessary in order to recommend the amount of active ingredient required for the desired effect. The role of the physicist, supported by the chemist, is to select or devise an appropriate formulation of the active ingredient which will be effective toxicologically, as well as compatible with the requirements of dispersion and travel; to determine the most effective limits of dispersion; and, having regard to the effects of meteorological variables in field application, to recommend the appropriate limiting range of meteorological conditions and optimum mode of application. The engineers task is the provision of equipment to realize the specified requirements.

Given the nature of the target, i.e. its mobility and disposition, the problem resolves into the choice of appropriate mode of action and active ingredient, allied to suitable form, i.e. solid, liquid, or possibly vapour phase, depending in part on the use of appropriate diluent or carrier, achievement of the requisite dispersion; and this then leads on to consideration of the behaviour of small particles (droplets) in the atmosphere.

The past two decades have seen a big decline in the use of toxic dusts in favour of spraying (Hill 1975), so that attention here will be focused primarily on the application of liquids as spray droplets, rather than solids as dust particles.

2.2 Characterization of droplet distribution

Though droplet size is probably the most important individual physical factor influencing the efficiency of the spray application process, an atomizer has yet to be devised which will provide a source of uniformly sized droplets at the throughputs required for field application. Recent research efforts have been directed more towards finding means of narrowing the spread of droplet size produced by various atomizer types than in searching for possible ways of obtaining complete mono-dispersion. We need to consider what parameters express most simply the central tendency of a distribution of spray droplets, together with the dispersion of the spectrum of droplet size.

2.2.1   Droplet size measurement

Droplet size distribution is obtained by examination of samples of the sprays taken in appropriate ways (Courshee and Byass 1953; Rathburn 1970; Matthews 1975). The examination necessitates measurements, usually of only a representative fraction, from such samples, by microscope, or other appropriate technique (Cadle 1975). The raw data obtained in this way are usually in the form of the numbers of droplets assigned to each class of a series of class intervals covering the range and it is from these data that the numerical parameters which characterize the spray must be determined.

2.2.2   Distribution by number

The simplest pictorial representation of the data is the histogram, possibly smoothed to a frequency diagram, indicating the range and identifying the mode or modes (most frequently occurring diameter(s) or class interval(s)). This also indicates the form of the distribution, i.e. whether it is symmetrical (possibly Normal or Gaussian) (see Fig. 2.1(a)), asymmetrical, or skewed towards the smaller or larger droplet size classes (see Fig. 2.1(b)), or possibly bimodal (see Fig. 2.1(c)), or multimodal (having several peaks). The shape may portray a real facet of the atomization process, or may be an artifact of the sampling or collection process. An example of the former would be the bimodal distribution obtained in direct droplet formation from 'classical' rotary atomization at very low throughput, where main droplets and smaller

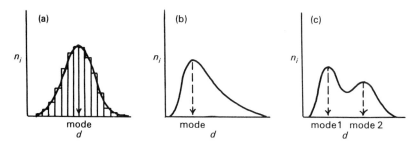

Fig. 2.1. Histogram and frequency diagrams for three forms of distribution (a), (b), and (c).

satellite droplets provide two quite distinct peaks. An example of the latter occurs where, for instance, small droplets are missed due to inadequate settlement time being allowed, or other poor collection technique so that the distribution is sampled incompletely, resulting in an artificially skewed distribution.

The alternative, cumulative plot, in which the proportion by number of droplets of less than the stated size is plotted against droplet size is usually preferred. Thus if there are n_i droplets in the ith class (h classes, labelled 1...i...h) and the total number of droplets, given by $\sum_{i=1}^{h} n_i$ (for $i=1$ to h), $=n$, the cumulative percentage (P_i) by number up to the ith class, is given by:

$$P_i \text{(by number)} = \left(\sum_{i=1}^{i} n_i/n\right) \times 100$$

The three examples of Fig. 2.1 are shown again, presented in this form, in Fig. 2.2(a–c).

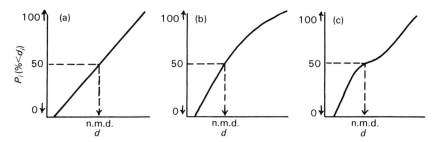

Fig. 2.2. Cumulative plot for the same three distributions (a), (b), and (c) as in Fig. 2.1.

The number median diameter (n.m.d.) is identified as that value of diameter for which 50% by number of the droplets are greater and 50% smaller in size. It is a useful index for monomodal distributions, although it can be misleading for distribution of multimodal form.

In addition to this graphical analysis, which is generally preferable to algebraic analysis initially, the data can be treated readily algebraically to determine the number average diameter (n.a.d.) (Maas 1971), where:

$$\text{n.a.d.} = \sum_{i=1}^{h} n_i d_i/n$$

d_i being the average diameter of the ith class. The dispersion can be characterized by the arithmetic standard deviation (by number), $\sigma_a(N)$ where:

$$\sigma_a(N) = [\sum_{i=1}^{h} n_i(d_i - \text{n.a.d.})^2/n]^{1/2}$$

2.2.3   Distribution by volume

Of equal interest is the distribution in terms of volume, and the raw data can be handled simply, numerically, to provide the volume distribution in similar form. The cumulative percentage (P_i') by volume up to the ith class is given by:

$$P_i' = (\sum_{i=1}^{i} n_i d_i^3 / \sum_{i=1}^{h} n_i d_i^3) \times 100$$

As with droplet numbers, the volume median diameter (v.m.d.) is identified as that value of diameter for which 50% by volume of the sampled spray is contained in droplets that are greater, and 50% by volume in droplets that are smaller than the median size.

As before, simple algebraic treatment of the data may be used to determine the volume average diameter (v.a.d.), i.e.

$$\text{v.a.d.} = [6/\pi(\sum_{i=1}^{h} n_i d_i^3/n)]^{1/3}$$

The arithmetic standard deviation of volume $\sigma_a(V)$ may also be derived as before.

The surface mean diameter (s.m.d.) given by:

$$\text{s.m.d.} = \sum_{i=1}^{h} n_i d_i^3 / \sum_{i=1}^{h} n_i d_i^2$$

(also called the Sauter diameter) is an alternative sometimes cited and useful in calculations related to evaporation losses, since it links volume with surface area.

2.2.4   Forms of the distribution function

Herdan (1953) discusses some standard forms of the distribution function, including the Normal law, the log-Normal transformation, Poisson distribution and the Rosin–Rammler curve, while Cadle (1975) refers to the Nukiyama–Tamasawa equation. For brevity only the log-Normal distribution will be examined here, as this provides a general approximation to the droplet size distributions obtained from a number of atomization processes, the major relevant exception being 'classical' direct droplet formation from rotary atomizers (which provides a readily discernible multimodal distribution).

Because the log-Normal form of distribution has such wide application, droplet assessment counts are usually performed with a geometrically spaced class interval, a $\sqrt{2}$ progression being featured in a number of eyepiece graticules used for droplet sizing (May 1965). These conveniently provide equal spacing of class intervals from a geometrical progression of droplet diameter. Data are most readily plotted as cumulative size distributions on logarithmic probability graph paper enabling n.m.d. and v.m.d. to be read off at the respective d_{50} or 50% values. For a log-Normal distribution, both number and volume plots are straight lines of identical gradient and the geometric means with respect to number and volume, ($d_g(N)$ and $d_g(V)$), defined as:

$$d_g(N) = (d_1^{n1}.d_2^{n2}...d_i^{ni}...d_h^{nh})^{1/n}$$

$$d_g(V) = (d_1^{3n1}.d_2^{3n2}...d_i^{3ni}...d_h^{3nh})^{1/3n}$$

are identical with n.m.d. and v.m.d. respectively, and likewise with the respective modes.

Partly for this reason the dispersion of a log-Normal distribution is usefully characterized by the geometric standard deviation (σ_g), where

$$\sigma_g = d_{84}/d_{50} = d_{50}/d_{16}$$

[d_{84}, d_{50}, and d_{16} are respectively the droplet sizes below which 84, 50, and 16% by number (or volume) of the spray is contained (d_{50} = median diameter).]

The ratio v.m.d./n.m.d. (r) has sometimes been preferred as a simply determined coefficient expressing the dispersion of the spray, whatever the form of the distribution. However, for a log-Normal distribution this is not unreasonable, since r is a function of σ_g (Johnstone 1978a).

Experiment suggests that for ligament and sheet (or film) atomization produced by rotary atomizers (which may be described approximately by the log-Normal distribution function), the narrowest dispersion which can be achieved may correspond to $r \sim 1.3$ ($\sigma_g = 1.35$), occurring with optimum ligamentary atomization (Johnstone and Johnstone 1976).

With hydraulic nozzles of cone or fan spray pattern, providing mainly sheet atomization, r is more usually in the range 2–3 ($\sigma_g = 1.65$–1.8), or even greater.

2.3 Spray deposition

2.3.1 The processes involved

In close proximity to the nozzle or atomizer, efflux momentum can carry spray droplets directly on to the target. For small droplets this momentum is rapidly dissipated, so that, at longer range, the intervening distance must be traversed under the influence of gravity and the forces imposed by air movement. However, when small droplets come within striking range of the target, be it a crop or weed canopy, the soil surface, or a resting or flying insect, deposition or collection may take place, either by gravitational sedimentation, or inertial impaction (including interception), or by a combination of both processes. Retention follows, provided no droplet bounce or blow-off occurs. Such collection takes place either in natural air currents, or in supplementary airstreams specially generated by the dispersal equipment, e.g. by mistblowers.

Sedimentation is the process whereby, due to gravity, the droplet is falling relative to the airstream and as a result is deposited on the target sites which possess components of horizontal aspect.

In the case of impaction, a droplet borne along by an airstream will tend to divert around an obstacle with the stream, but may strike sites normal to the flow on account of its inertia.

In the special case of interception, the trajectory of the centre of the droplet which is large in relation to the obstacle, while not itself intersecting the collecting surface, may pass close enough for the surface of the droplet to contact the collector, so that the droplet adheres.

In order to examine the nature of sedimentation and inertial impaction and the factors which determine their relative importance we may hypothesize a simple model, to which the following definitions relate. Consider an airstream moving with uniform horizontal velocity u, bearing a uniform mass concentration C_M of droplets of diameter d (the corresponding number concentration being C_N). Let the target have a horizontal area index (projected area) A_H and vertical area index A_V. Let $(Q_H/t)_M$ denote the rate of mass deposition per unit area by sedimentation [$(Q_H/t)_N$ the corresponding number rate of deposition] and $(Q_V/t)_M$ the rate of mass deposition per unit area by impaction [$(Q_V/t)_N$ the corresponding number rate]. This distinction between mass and number is very important, because coverage, so vital in the efficient application of sprays, is more readily apparent in terms of droplet number deposition than total mass deposition.

2.3.2   Deposition by sedimentation

The rate of sedimentation by mass is given by:

$$(Q_H/t)_M = C_M v_g A_H$$

where v_g, the terminal fall velocity in still air, is expressed simply by Stoke's law as:

$$v_g = (\rho - \rho_a)g d^2/18\eta$$

and ρ, ρ_a are respectively the density of the droplet and of air, η is the viscosity of air, g is the gravitational constant and d the droplet diameter. Table 2.1 indicates the variation of v_g with d for droplets of unit density in the range 1–1000 μm, falling in still air at 20°C.

TABLE 2.1. *Terminal fall velocities* (v_g) *for droplets of diameter* d *in the size range 1–1000 μm having unit density, at 20°C*

d (μm)	v_g (cm/s)
1000	385
500	200
200	70
100	25
50	7.2
20	1.2
10	0.30
5	0.078
1	0.0035

It follows that $(Q_H/t)_M \propto C_M A_H d^2$; i.e. the rate of mass deposition by sedimentation is proportional to the square of the droplet diameter. In terms of droplet number, the rate of sedimentation is given by:

$$(Q_H/t)_N = C_N v_g A_H \text{ and since}$$

$$C_M = (\pi d^3 \rho / 6) C_N$$

$$(Q_H/t)_N = (6/\pi d^3 \rho) C_M v_g A_H$$

$$\text{i.e. } (Q_H/t)_N \propto C_M A_H (1/d)$$

and the number rate of deposition by sedimentation is inversely proportional to droplet diameter.

Thus there is an early conflict, in that maximization of the rate of mass deposition appears to require the largest droplet diameter consistent with the probability of striking the proportion of area available (A_H), whereas the maximization of coverage, or number rate of deposition by sedimentation, appears to require the smallest droplet diameter consistent with the probability of striking A_H. It is not possible to resolve this dilemma without taking into account the relative importance of impaction.

2.3.3 Deposition by impaction

The rate of inertial impaction, by mass, is given by:

$$(Q_V/t)_M = C_M E u A_V$$

where E is the efficiency of collection by impaction on the given target. The impaction or collection efficiency of a target interposed in an airstream laden with small droplets, defined as the ratio of the number of droplets striking the obstacle to the number which would strike if the airstream were not deflected, varies with the airstream velocity and the relative size of the target and spray droplets in a somewhat complex way (Richardson 1960; Dorman 1966). However, it is possible to estimate collection efficiency as a function of a dimensionless particle inertia parameter, P, which links the relevant physical variables in the relatively simple expression:

$$P = (v_g/g)(u/D)[\rho/(\rho - \rho_a)]$$

where D = effective width of the target normal to the flow. The functional relationship between E and P is complicated and is best shown graphically (Fig. 2.3). The upper and lower limits of E are found with potential and viscous flow; i.e., high and low Reynolds number ($Re = \rho_a ud/\eta$), respectively. May and Clifford's (1967) experimental determination of E versus P for ribbon collectors is shown as a broken line in the figure.

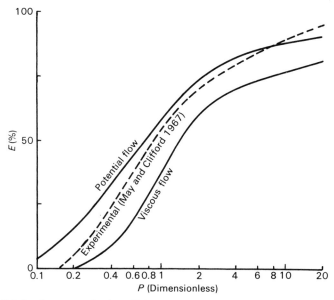

Fig. 2.3. Relation between collection efficiency (E) and particle inertia parameter (P).

Given the definition of P above and that E increases directly as some undetermined function (f) of P,

$$(Q_V/t)_M \propto C_M A_V f(ud/D)$$

i.e. the rate of mass deposition by impaction is proportional to some direct function of wind speed and droplet diameter and an inverse function of obstacle width.

In terms of droplet number, the rate of impaction is given by:

$$(Q_V/t)_N = C_N E u A_V$$

$$\text{or } (Q_V/t)_N = (6/\pi d^3 \rho) C_M E u A_V$$

$$\text{i.e. } (Q_v/t)_N \propto C_M A_V (1/d^3) f(ud/D)$$

The additional $(1/d^3)$ term suggests that the rate by number of droplet deposition by impaction will be markedly enhanced by progressive reduction in droplet size until the gain in number is more than balanced by fall-off in E, placing a lower limit to the effective diameter.

Interception is only significant when the width of the collector is of the same order as the size of the droplets, when it has the effect of increasing the rate of collection due to impaction by a factor equal to d/D.

2.3.4   Sedimentation versus impaction

Comparing the rates of deposition due to sedimentation and impaction, we have.

$$(Q_H/t)/(Q_V/t) = C v_g A_H / C E u A_V$$
$$\text{or}$$
$$Q_H/Q_V = (v_g/u)(A_H/A_V)/E$$

i.e. the importance of sedimentation as opposed to impaction as the predominant deposition process depends on the ratios of the droplet terminal velocity to the wind speed and the horizontal to vertical area indices. It is also inversely proportional to the collection efficiency. Deposition by sedimentation is thus enhanced by large droplets, very light winds, and near-horizontal wide-surfaced collectors. Deposition by impaction is of greater significance with smaller droplets, stronger winds, on targets with area predominantly normal to the airflow, but not so wide as to reduce E significantly.

Large droplets ($> 300\ \mu m$) tend to deposit primarily by sedimentation, and in very light winds this process is also predominant for considerably smaller droplets, particularly if A_H/A_V is > 1. However, it is observed that as droplet size falls below the rather arbitrary value of $150\ \mu m$ (at which size the fall velocity is about $0.5\ m/s$) natural inertial impaction becomes increasingly evident, and below a droplet size of about $50\ mm$ (fall velocity $\sim 0.07\ m/s$) inertial impaction is usually the dominant process of deposition, sedimentation only being more effective under unusually still or sheltered conditions.

The airstream from the mistblower or fan-assisted nozzle merely augments the natural wind to a greater or lesser extent, depending on distance, relative magnitude, and direction.

However, at a droplet size of 30 μm, for example, inertial collection efficiency can vary from as much as 88% (or higher) for a ribbon target of 1 mm width (or narrower) in a moderate airstream of 3 m/s, to virtually zero for a ribbon of 10 mm width in a light airstream of 0.5 m/s (Johnstone 1978b). These values, linked with large possible variations in A_H/A_V, give some indication of the transient variations which may be encountered in deposition on natural obstacles, including leaves, stems, and other plant structures, when working in the coarse aerosol droplet size range at ambient wind speeds, the wind frequently being attenuated within the structural canopy of the taller field crops.

Very small droplets of 20 μm, or less, move readily around most obstacles along with the airstream and will tend to be filtered out by thin stems, plant hairs, or protuberances of similar small dimensions, provided that the airstream velocity is sufficient for this process to occur. Alternatively, they may deposit by sedimentation in the very still air most likely to be found during intensely stable periods which occur predominantly during the night. In convective daytime conditions these droplets are very prone to drift.

In a crop whose canopy structure restricts the airflow towards the base of the plants, impaction may be the predominant process of deposition at the upper level by small droplets borne in a moderate airstream, whereas in the stiller air below the upper canopy, deposition of these droplets may be affected primarily by sedimentation.

The combined rate of collection, Q/t, by both sedimentation and impaction, is given by:

$$Q/t = Q_H/t + Q_V/t = C(v_g A_H + Eu A_V)$$

and this relationship emphasizes the complementary aspects of the two processes. Given a large horizontal area index (A_H), large droplets always maximize $(Q/t)_M$. However, in terms of efficient spraying, the significant parameter is likely to be $(Q/t)_N$ which will be maximized at some intermediate value of d, which cannot be determined theoretically at all readily. We seek a maximum of the expression

$$(Q/t)_N \propto A_H/d + A_V(1/d^3)f(ud/D),$$

which includes five independent variables.

Table 2.2 presents some comparative data for $(Q/t)_N$ derived from the initial equation above and a graphical solution for E from P. A constant value (1.0) has been assumed for C_M and the values of $(Q/t)_N$ are expressed

TABLE 2.2. *Comparative values for the rate of deposition, in terms of droplet number $(Q/t)_N$, from a continuous infinite spray cloud of fixed uniform mass concentration ($C_M = 1$), by sedimentation ($A_V = 0$, $A_H = 1$), by impaction ($A_V = 1$, $A_H = 0$) and by a combination of sedimentation and impaction ($A_V = A_H = 0.5$), for five wind speeds ($u = 0.1$–10 m/s), six droplet diameters ($d = 5$–200 μm) and three configurations (equivalent to ribbon width $D = 0.1$–10.0 cm). The values have been determined with reference to an adjusted value of 1.00 for $u = 1$, $d = 100$, $D = 0.1$ and $A_H = A_V = 0.5$; maxima have been underlined.*

		u = 0.1			u = 0.5			u = 1.0			u = 2.0			u = 10.0		
D (cm)	d (μm)	A_V=0, A_H=1	A_V=0.5, A_H=0.5	A_V=1, A_H=0	A_V=0, A_H=1	A_V=0.5, A_H=0.5	A_V=1, A_H=0	A_V=0, A_H=1	A_V=0.5, A_H=0.5	A_V=1, A_H=0	A_V=0, A_H=1	A_V=0.5, A_H=0.5	A_V=1, A_H=0	A_V=0, A_H=1	A_V=0.5, A_H=0.5	A_V=1, A_H=0
0.1	200	0.15	0.08	0.02	0.15	0.12	0.10	0.15	0.18	0.20	0.15	0.28	0.40	0.15	1.1	2
	100	0.40	0.26	0.12	0.40	0.57	0.74	0.40	1.0	1.6	0.40	1.8	3.2	0.40	8.2	16
	50	0.90	0.74	0.58	0.90	2.9	5.0	0.90	6.0	11	0.90	12	23.4	0.90	64	128
	20	2.0	1.0	0	2.0	20	38	2.0	60	118	2.0	145	288	2.0	871	1740
	10	4.8	2.4	0	4.8	2.4	0	4.8	130	256	4.8	610	1220	4.8	5600	11200
	5	18	9.0	0	18	9.0	0	18	9.0	0	18	9.0	0	18	9.0	0
1.0	200	0.15	0.08	0.01	0.15	0.11	0.08	0.15	0.16	0.17	0.15	0.27	0.39	0.15	1.1	2
	100	0.40	0.21	0.03	0.40	0.45	0.50	0.40	0.8	1.2	0.40	1.5	2.6	0.40	8.2	16
	50	0.90	0.45	0	0.90	1.2	1.5	0.90	3.3	5.8	0.90	8.4	16	0.90	55	109
	20	2.0	1.0	0	2.0	1.0	0	2.0	1.0	0	2.0	21	40	2.0	591	1180
	10	4.8	2.4	0	4.8	2.4	0	4.8	2.4	0	4.8	2.4	0	4.8	4720	9430
	5	18	9.0	0	18	9.0	0	18	9.0	0	18	9.0	0	18	9.0	0
10.0	200	0.15	0.07	0	0.15	0.10	0.04	0.15	0.13	0.12	0.15	0.22	0.29	0.15	0.2	1.7
	100	0.40	0.20	0	0.40	0.20	–	0.40	0.33	0.26	0.40	0.8	1.2	0.40	6.2	12
	50	0.90	0.45	0	0.90	0.45	0	0.90	0.45	0	0.90	0.45	0	0.90	29	58
	20	2.0	1.0	0	2.0	1.0	0	2.0	1.0	0	2.0	1.0	0	2.0	1.0	0
	10	4.8	2.4	0	4.8	2.4	0	4.8	2.4	0	4.8	2.4	0	4.8	2.4	0
	5	18	9.0	0	18	9.0	0	18	9.0	0	18	9.0	0	18	9.0	0

relative to a value of 1.0 for $d = 100\,\mu m$, $D = 0.1\,cm$, $u = 1\,m/s$ and $A_V = A_H = 0.5$. The columns for which $A_V = 0$, $A_H = 1$ represent deposition by sedimentation alone, and the columns for which $A_V = 1$, $A_H = 0$ relate to deposition entirely by impaction. For $A_V = A_H$, equal areas of the target contribute to collection by sedimentation and impaction. A similar table could be prepared for $(Q/t)_M$ by multiplying the figures in each row by the appropriate volume conversion factor (d^3). The table bears out most of the points already made, but it should be noted that the data for the rate of deposition by sedimentation of smaller droplets, in particular those $< 50\,\mu m$, may not be entirely realistic, except in especially smooth, or stable conditions. Under more normal conditions, with turbulence in the airstream generated by the frictional drag of vegetation (and also by convection on warm days), the velocities within small eddies may be similar in magnitude to the fall velocities of droplets of $50\,\mu m$ or less in diameter and may give rise to deposition by impaction on surfaces with horizontal in addition to vertical component, which will modify the simplified picture here portrayed.

2.3.5   Implications for spray cover and for spray drift

The implications with respect to drift of spray droplets can be summarized as follows:

(i) Large droplets sediment rapidly and can therefore reduce the likelihood of spray drift, but they provide relatively poor (and selective) coverage, unless the volume rate is high, so that they may well be economically ineffective.

(ii) Smaller droplets, typically those $< 200\,\mu m$ in diameter, are deposited by a combination of sedimentation and inertial processes, the relative importance of which depends not only on droplet size, but also on wind speed and the nature or form of the target, especially its horizontal or vertical collection area. Plants which have a greater effective vertical than horizontal area will filter out a higher proportion of drifting droplets by the processes of inertial impaction and interception (in this connection, the role of hedgerows as 'sinks' for drifting droplets merits consideration). It may be desirable to take advantage of this selectivity by an appropriate choice of droplet size for specific targets.

(iii) Very small droplets, i.e. say $< 20\,\mu m$ in diameter, do not readily deposit at all, unless the airstream is sufficiently fast and obstacles occur which can provide sites for impaction (and interception). Sedimentation is only possible in very sheltered sites, or under extremely stable conditions in the open. Such droplets could therefore be carried long distances before deposition is effected.

2.3.6   Spreading, wetting, and retention

Surface properties of the droplet and the target can affect the efficiency of the deposition process. The spreading power of a liquid drop on a surface, or its readiness to extend the solid/liquid interface, is displayed by the maximum angle of contact, or advancing angle (θ_A). Wetting ability, as distinct from spreading power, relates to the readiness to form persistent liquid/solid interface when excess liquid has drained, and this is made apparent by the minimum, or receding angle of contact (θ_R). A high degree of wetting is only possible as θ_R tends to zero.

Ford and Furmidge (1966) have given a detailed theoretical treatment of the dependence of droplet spread on impact energy. They have suggested that droplets of diameter $< 500\ \mu m$ are unlikely to bounce, unless $\theta_A - \theta_R$ is very small. This is uncommon, but occurs, for instance, with water droplets landing on pea leaves; however, Brunskill (1956) showed a dramatic increase in percentage retention of impinging droplets of aqueous mixtures on pea leaves as surface tension was reduced, bringing about an accompanying lowering of θ_A and increase in $\theta_A - \theta_R$.

The physics of movement of coalesced globules on inclined surfaces, and implications for high-volume run-off spraying have been examined by Furmidge (1962) and by Johnstone (1973a). The properties of surface-active additives (emulsifiers, wetters, etc.) are important in determining the maximum initial retention, in addition to droplet size. Johnstone (1973a) developed a statistical model for the build-up of deposit by droplet coalescence to predict change in maximum initial retention with target size, slope, droplet diameter, and the interfacial properties of liquid and surface (θ_A, θ_R) which indicated that minimum concentration of surfactant (to eliminate droplet bounce and maintain a finite θ_R) and a small droplet size, together favour maximum initial retention in high-volume application.

2.4 Droplet travel under wind, gravity, and electric field forces

2.4.1 Combined effects on travel of wind and gravity

The energy required for spray droplets to travel to the target may be supplied in a number of ways. If the nozzle can be passed close to the target, the efflux velocity may be sufficient to cause the droplets to impinge. In certain applications an air-carrier jet may be employed to carry droplets to elevated targets, or to increase the local impaction of small droplets. The disadvantage in both these cases is that the high droplet velocities close to the nozzle tend to produce uneven deposition as proximity to the target varies. The wake created by the passage of the sprayer can also have a marked effect on the initial droplet movement, particularly in the case of low-flying crop-spraying aircraft (Trayford and Welch 1977). Frequently, however, it is a combination of wind and gravity which determines the fate of small droplets and, when this is so, fall velocity assumes increasing importance.

If such droplets are released from above and upwind of the target, their travel may be described, approximately, with the aid of some simplifying assumptions: in particular that:

(a)   droplets issue from a line source, downwards, at terminal velocity,

(b)   air disturbance is negligible,

(c)   droplet movement takes place in stable, isothermal air, over flat ground, with zero wind shear.

With these assumptions, the movement of spray droplets released from a given height (H) becomes a simple function of wind speed (u) and fall velocity. Since the time (t) taken to fall vertically through a height (H) at velocity (v_g) is H/v_g, *the distance* ($S = ut$) travelled horizontally in that time in a wind speed (u) is

$$S = Hu/v_g$$

By adjusting the height inversely with change in wind speed it is possible to keep the product Hu constant, so that droplets of any given size should always travel the same fixed distances downwind of the source. This reasoning provided the basis for the 'Porton' method of locust spraying described by Gunn (1948) and Sawyer (1950). If the droplet size distribution is known, a graphical method may be used to predict the deposition curve (Johnstone 1972). Fig. 2.4 shows four such curves, illustrating percentage by volume deposited versus distance travelled downwind for various values of Hu (in m^2/s) for droplet spectra having v.m.d. of 35, 70, 140, and 280 μm, with $\sigma_g = 1.28$.

In interpreting the curves the limitations imposed by the simplifying assumptions must be borne constantly in mind, in particular the effect of turbulence on the spray. For quiet, stable conditions, which generally imply light winds of under 2 m/s, Hu may be about 1 or 2 m^2/s for a source height of 1 m when using portable sprayers, or about 2–3 m^2/s for the equivalent aircraft crop-spraying operation. At 280 μm v.m.d., it is predicted that 50% of the spray would deposit with less than 5 m lateral displacement from the line of the aircraft source for $Hu = 5\,m^2/s$, and with less than 2 m displacement from a portable sprayer for which $Hu = 2\,m^2/s$. Equivalent displacements for 140, 70, and 35 μm droplets from aircraft/portable sprayer sources, for Hu values of 10 and 4 respectively, are shown as 22m/8m, 75m/30m, and 270m (extrapolated) 108m, indicating the marked effect of wind displacement anticipated for droplets of less than 50 μm in diameter.

Whether or not turbulence modifies droplet behaviour in this situation will depend on the scale of turbulence present, in relation to fall velocity. The strength of turbulence may be characterized in a number of ways, one of which uses the root mean square deviation of the vertical component of wind speed. Lawson and Uk (1978), following Davies (1966) and Chamberlain (1975), adopted the friction velocity (u_*) (see Section 2.5.2) as an appropriate

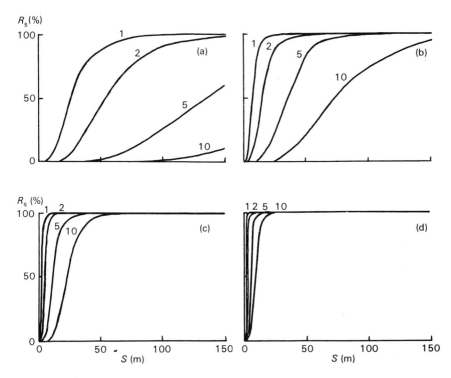

Fig. 2.4. Recovery by sedimentation (R_s), versus distance travelled downwind (S) for droplet spectra with $\sigma_g = 1.28$ and v.m.d. of (a) 35, (b) 70, (c) 140, and (d) 280 μm, at indicated Hu values of 1, 2, 5, and 10 m²/s, under conditions specified in the text.

approximation for this eddy velocity. Thus if v_g is large compared with u_*, the droplets will be little affected. By contrast, larger, stronger eddies tend to pick up those droplets having only small v_g, and the droplets then participate in the turbulent motion of the airstream and become diffused by eddy processes. The effect of instability is normally to increase the bandwidth predicted by the simplified approach, some droplets being deposited nearer to the source, while some travel much greater distances. In field trials over mature cotton for instance, Threadgill and Smith (1975) measured vertical air currents averaging about 3 cm/s (range −2 to 10 cm/s), which could markedly modify the travel of droplets having diameters of 70 µm and below. Bache and Sayer (1975) have suggested that when spraying from aircraft, flying height should be increased when operating in turbulent conditions to maintain the position of peak deposition at the same distance from the aircraft track, but give no estimate of likely reduction in short-range recovery incurred due to increased elevation.

For the more rigorous theoretical treatment of the drift of very small droplets, particularly those in the aerosol category, Pasquill (1961) and Yeo (1974) have developed Sutton's earlier (1949) statistical approach to turbulent diffusion to predict deposition over the medium range 0.1–10 km.

2.4.2   *Effect on travel and deposition of electric charge and field forces*

The effect of imposing an electric charge and field on a droplet is to introduce a third force, in addition to wind and gravity, which then modifies the fate of the droplet. In a field of strength E a droplet bearing a charge q experiences a vector force of strength qE in the direction of the field. Depending on the relative magnitude of the wind drag and gravitational force, the droplet will tend to follow more or less closely the direction of the electric field, the latter being dependent on the disposition of the source of charge (atomizer and droplet cloud) and the nature and form of the earthed target.

In general, we can say that, given both charge and field of presently realizable strengths, for droplets > about 250 μm in diameter, gravity will still tend to dominate, charge merely accelerating droplet fall. For very small droplets of about 25 μm, or less, the wind force may well predominate – unless the droplet is very close to the target and in a region of intense electric field. It is perhaps for droplets of intermediate size, of say 30 to 100 μm, that charge and field may have the most significant effect. Coffee (1979) has shown that, with a potential difference of 25 kV, acting over, say, 0.4 m, the maximum force field strength is of the order of 6.25 kV/m, and, in such a field, a 100 μm diameter droplet bearing a 3.7×10^{-12} coulomb charge should experience a vector force of 2.3×10^{-7} N, compared with a gravitational force of 5×10^{-9} N and a wind drag force of a similar order. Herein lies the promise of greater control and precision in the application of small droplets in very low volumes over short range.

2.5   Meteorological aspects

2.5.1   *Micro- and meso-scale phenomena*

Most agricultural applications are made at close range, the sprayer usually within 1–100 m of the target, and frequently to a crop which is acting as host to the pest, so that the relevant variations are those occurring over quite short distances (and time) due to interaction of wind and radiation with ground surface, foliar cover, and superficial moisture, in essence, microscale phenomena.

However, when we are concerned with the extended drift of very small droplets, deliberately, as in space- or drift-spraying against certain insect pests in flight or otherwise suitably exposed, or less intentionally, when small droplets 'escape' into the body of the mixing layer, the relevant range of travel can be considerably increased and more extensive mesoscale phenomena, e.g. orographic winds, variation in depth of the boundary layer with formation and break-up of inversion stratifications, and the movement of convergence fronts (the latter merging into the synoptic scale), may have to be taken into account to explain effects of an episodic nature (Smith and Hunt 1978) occurring at longer range [e.g. at distances of 10 to 50 km, when emitting insecticide as a coarse aerosol from aircraft in sequential night spray operations against tsetse fly (Andrews *et al.* 1983)].

In addition to the sources already cited in Section 2.4, useful descriptive micrometeorological background has been provided by Scorer (1958), Munn (1966), Geiger (1966) and Thom (1975). Essentially mathematical treatments of boundary layer airflow have been given by Richardson (1950), Prandtl (1952) and Csanady (1973).

2.5.2 *The wind profile and wind fluctuation*

Terrain, vegetation, and Man-made structures at the Earth's surface exert a frictional drag on the airflow causing turbulence and momentum transfer, with reduction in wind speed near the ground. This drag is transmitted upwards, through eddies of increasing size, within a mixing layer whose thickness (δ) can vary, but is of the order of 0.5 km.

Above the top of the mixing layer, horizontal air movement results mainly from interaction of any pressure gradient force and the Coriolis force acting at right-angles to the direction of motion. This interaction determines the direction of windflow parallel to the isobaric lines, and the wind speed at which the two forces balance, the geostrophic wind (G), can be calculated from the pressure gradient. Within the mixing layer, dimensional analysis (Tennekes 1973) has provided the following relationships:

$$G^2 = (u_*^2/k)(A^2 + [\ln(u_*/fz_0) - B)^2]$$

$$= Cu_*/f$$

and, for $z_0 \ll z \ll \delta$, i.e. near to surface,

$$u = (u_*/k) \ln [(z - d)/z_0]$$

where z is the height above the ground, $u(z)$ is the horizontal mean wind velocity at that height (note that towards the surface, the wind is deflected somewhat away from the direction of the isobars towards the direction of the pressure gradient), f is the Coriolis parameter (typically $10^{-4}\,s^{-1}$), k is Von Karman's constant (~ 0.4), and A, B, and C are constants.

The constants, u_* (friction velocity), z_0 (roughness length), and d (zero plane displacement) relate to the nature of the surface. For a given value of $u(z)$, for instance, u_* will be larger over a more aerodynamically rough surface than over a smoother one, as characterized by the value of z_0. For a plant canopy, z_0 is a measure of the capacity for momentum absorption, while d

may be considered to indicate the mean level at which momentum is absorbed by the constituent elements of the canopy. For a specific canopy, z_0 and d will tend to vary only in so far as the canopy structure is modified by the streamlining effects of strong winds (Thom 1975). u_*, z_0, and d may be determined from measurements of the wind profile near the ground, through a least-squares fit of the previous equation to the measured wind profile. z_0 is determined as the apparent height at which the mean velocity vanishes when extrapolated back towards $z = 0$, while d is typically 0.6 to 0.8 h, where h is the height of the roughness elements, e.g. the plant height (Thom 1975). u_* is derived from the profile gradient ($k/2.303u_*$) found on the appropriate logarithmetic plot (Johnstone *et al.* 1974). Typical values for z_0 (in mm) include 1–10 for short grass, 10–100 for field crops, and 200–1000 for forests.

The log-law relating wind speed with height appears adequately valid for $5z_0 < z < 0.1\delta$ over a small range of unstable, through neutral, to a surprisingly wide range of stable conditions (Webb 1970). An alternative empirical form with possibly more universal application is:

$$u(z) = u_1(z/z_1)^p$$

where u_1 is the wind speed at a particular reference level z_1 and p an exponent which depends on ground roughness and lapse rate (Munn 1966).

Within the canopy, the log-law breaks down and the wind profile is determined by the structure of the vegetation and its varying resistance to airflow (Thom 1971; Landsberg and James 1971). Thus the mechanism of deposition can change from predominantly inertial impaction in the upper portion of a crop canopy to predominantly sedimentation for those droplets which penetrate through to the lower levels.

Strong stable stratification can effectively damp out turbulent mixing, allowing flow near the ground to become virtually independent of that higher up. While, in general, neutral conditions are associated with stronger winds, lending dominance to mechanically induced turbulence, thermally induced turbulence (convective buoyancy effects) assume greater importance in light airs, or calm conditions, which coincide with strong solar radiation. Diurnal stability fluctuations account for the noticeable tendency for the surface wind to freshen to a maximum, with increasing gustiness, in the afternoon and decrease during the night, in particular when, under clear skies, surface heat loss by radiation can promote the onset of marked temperature inversions near to ground level.

The likely effects of change in u_* on the travel of small droplets have already been referred to in Section 2.4. In general, u_* is correlated with the average windspeed ($\bar{u}$); however, the deposition of small droplets shows better correlation with a combination of meteorological parameters than with any particular variable (Johnstone *et al.* 1974).

2.5.3 Temperature, humidity, and the evaporation of spray droplets

Volatility of the carrier liquid becomes an important factor in spray application as the applied volume and the spray droplet size are reduced. Evaporation of a volatile carrier can result in decrease in droplet size to the point at which the small droplets solidify. Such 'dust' particles may travel long distances or fail to adhere on impact. The evaporation of a pesticide spray mix can be described approximately by a two-fluid model in which the volatile fraction (i.e. the carrier) evaporates as if it were homogeneous, while the remaining fraction (i.e. active ingredient, emulsifiers, and other additives) remains essentially involatile. Wanner (private communication) has verified this model in laboratory evaporation measurements on suspended acqueous droplets. Amsden's data (1962) have been used here to plot the variation in lifetime of water droplets subject to evaporation at 30°C, 50% r.h. (relative humidity), also at 20°C, 80% r.h., with initial droplet size, over the range 30–200 μm, as shown by the broken lines in Fig. 2.5 (Johnstone 1972). On the same graph the variations in fall time from several heights (*H*) with droplet

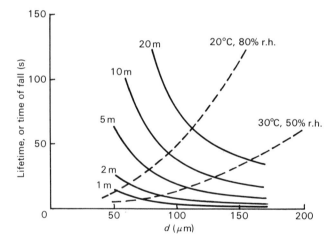

Fig. 2.5. Lifetime of water droplets (broken lines) for two ambient conditions; and fall times from indicated heights (full lines), plotted against droplet diameter (*d*).

size have been shown as a family of curves (full lines). The points of intersection of the full and broken lines define critical sizes for given heights of fall and ambient conditions, which will just reach the ground before evaporating. Under moderately severe conditions of 30°C and 50% r.h., the critical sizes for release heights of 0.5, 1, 2, 5, and 10 m are approximately 60, 70, 85, 110, and 130 μm respectively, indicating the limitations of water as a carrier for all but close range deposition under these conditions. The use of a liquid adjuvant of low volatility could prove useful in mitigating this disadvantage (Amsden 1962).

Furmidge (1968) has used the equation:

$$dm/dt = 2\pi M \mathscr{P} d(p_0 - p_\infty)/RT$$

[where m is the weight of material (molecular weight, M) evaporating in time t, $\mathscr{P}$ the diffusion coefficient, p_∞ is the ambient vapour pressure (c_∞ the concentration) in the absence of the droplet and p_0 is the pressure of the saturated vapour (c_0 the concentration) at the droplet surface] to examine the evaporation of 80 μm droplets of different carriers (see Table 2.3).

TABLE 2.3. *Evaporation of 80 μm droplets where* t *is the time in seconds to lose 10% of the initial volume*

	P_0 (mmHg)	∂ (cm²s)	Mol.wt.	(g/l)	CC_0 (g/l)	dm/dt (g/s)	t (s)
Water r.h. 50%	13	0.22	18	1.38×10^{-5}	0.69×10^{-5}	7.63×10^{-8}	0.35
Water r.h. 90%	13	0.22	18	1.38×10^{-5}	1.22×10^{-5}	1.77×10^{-8}	1.5
Xylene	5	0.077	106	6.23×10^{-6}	0	2.41×10^{-8}	1.1
Medium oil	10^{-2}	0.04	400	2.35×10^{-7}	0	4.73×10^{-10}	56.4

Jarman (1958) and Johnstone and Watts (1970) have examined the evaporation of oil-based sprays from aircraft and shown that a considerable proportion of the carrier liquid may evaporate prior to deposition. Data from the latter source is included at Table 2.4. Clearly only carriers having a component with a very low vapour pressure will be suitable for ULV drift spraying, using very small droplets. If any appreciable residual effect is desired then the same may be said for the active ingredient, but it should be emphasized that quite small amounts of impurities can reduce the predicted rate of evaporation to a large extent, especially if they are of high molecular weight and are surface active in nature. Waterless formulations may be tested for volatility in the laboratory if evaporation problems are suspected (Johnstone and Johnstone 1977).

2.5.4 Effects of stability and turbulence

Laminar or viscous flow is encountered only under very stable, or marked inversion conditions – those calm, or near-calm conditions of clear evening, night-time, or early mornings, when radiation from the ground causes the surface to cool below the temperature of the adjacent air so that buoyancy is eliminated. Strong sunlight, on the other hand, warms the soil surface and/or

TABLE 2.4. *Volume and mass recoveries* (R_v *and* R_m) *of sprays on to cotton by aircraft determined from samples sedimenting on to horizontal artificial targets above the top of the crop*

Application	Droplet size v.m.d. (μm)	H (m)	R_v (%)	R_m (%)	Evaporation $(R_m - R_v)/R_m$ (%)	t (GMT) (h)	$\bar{u}$ (m/s)	T (°C)	Cloud amount (oktas)
F	120	6.2	10	19	47	1415	0.72	27.2	0
G	110	5.2	15	42	64	0915	1.7	25.6	5
H	120	4.0	18	39	54	0830	1.4	24.4	4
I	125	3.7	22	46	52	0755	0.9	23.9	1
J	120	3.7	23	28	28	ca.0800	1.8	23.3	2–4
K	125	3.7	28	42	33	0810	0.62	23.9	3–6
L	140	2.5	34	48	29	0810	0.72	21.1	3–4
M	140	3.1	42	62	32	0810	0.9	20.6	3

Applications made from a Beaver aircraft equipped with two Micronair AU 2000 atomizers, applying a formulation of 20% DDT in heavy aromatic naphtha.

vegetation cover and with it the neighbouring air, setting free convection in motion. Hot spots and thermal plumes develop, especially in conditions of very light wind and these can be depicted by sonic echo technique (Holmes *et al.* 1976). Air movement is then perceived as a series of gusts and lulls. An effect of rising wind is to reduce the magnitude of these thermal effects, the resulting frictional mixing of the boundary layer air neutralizing any marked temperature gradients which might otherwise develop. As noted previously, convective turbulence, often associated with moderate to high super-adiabatic lapse rates in the morning or early afternoon, can result in rising air currents of sufficient velocity to overcome the fall velocity of very small droplets, especially those with diameter less than 30 μm, which may then fail to deposit or impact (Johnstone and Huntington 1977).

One function recognized as a criterion for the transition between stable and unstable conditions is the dimensionless Richardson number, *Ri,* (Richardson 1920), which expresses the ratio of buoyancy to frictional kinetic energies associated with a parcel of air in the boundary layer. A simplified definition of *Ri* may be taken as:

$$Ri = (g/T)(\Delta T . \Delta z)/(\Delta u)^2$$

where g = gravitational constant, T = temperature (K); ΔT and Δu are respectively temperature and velocity differences over the height interval Δz.

Provided temperature and wind gradients close to the ground can be measured, a characteristic Richardson number may then be determined. Values of *Ri* range from about +1 (or greater) for marked stability with temperature inversion, through zero and very small negative values (~ 0.005)

for neutral stability, to negative values of about -1 for very marked instability associated with high super-adiabatic lapse rates.

Over two decades ago Yeo *et al.* (1959) demonstrated a correlation between recoveries by sedimentation of coarse aerosol sprays emitted from an aircraft and a stability factor which was essentially equivalent to Richardson number (Christensen *et al.* 1969). More recently correlation of positive *Ri* with sedimentation and inverse correlation with inertial collection has been demonstrated for the application of insecticides as very fine sprays to cotton by portable rotary atomizers (Johnstone and Huntington 1977). The latter's results for sedimentation are summarized in Fig. 2.6, in which recoveries (R_s), estimated volumetrically, for three different formulations, have been compared under five different conditions (tests A–E). The component diagrams illustrate the magnitudes of the individual variables *Ri, T*, r.h., *u*, and the divergence in time of day from the noon datum ($|t|$). Highest recoveries have been obtained in early morning and late evening. It is apparent that factors other than *Ri* can influence drift and deposition strongly, especially if the formulation is volatile. The recoveries of the two aqueous formulations (1 and 2) are markedly affected by the two inversely correlated variables r.h., and *T* (c.f. applications D and E). The effect of wind speed is less marked in this example.

Some measurements in which total mass recoveries (R_T) of very fine spray were determined colorimetrically from individuals and groups of plants during application of insecticides on to cotton in Thailand by portable sprayers are recorded, together with stability data, in Table 2.5 (Johnstone 1977a,b; Johnstone *et al.* 1977). In these tests, mass recoveries on plants generally exceeded 70% for droplet spectra with v.m.d. ranging from 60 to 90 μm, provided that wind speed remained under about 3 m/s. The lower recovery in application R is accounted for by the combination of instability and moderate wind, a combination which also explains the significantly lower mass recoveries when the droplet size was reduced to 42 μm v.m.d.(S). Good recovery was obtained at small droplet size using a mistblower to enhance inertial impaction (T).

2.5.5   Instrumentation for meteorological monitoring during application

The instrumentation required for meteorological measurement will depend on the purpose and scope of the data collection. Simple basic equipment which can suffice for monitoring routine applications may need to be augmented with more sophisticated instruments, possibly of the recording variety, for more elaborate tests, including the field evaluation of new sprayers, formulations, and for interpreting medium-range drift.

Essential basic equipment at the site of spraying can be restricted to a magnetic compass (for determining wind direction from observation of

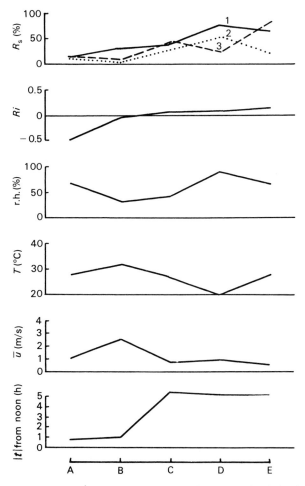

Fig. 2.6. Recoveries by sedimentation (R_s) of carbaryl sprays, measured volumetrically, for three formulations (1, 2, and 3), and the corresponding meteorological parameters in five tests A–E. (1) ULV aqueous formulation; n.m.d./v.m.d. 102/128 μm. (2) ULV aqueous formulation with 18% molasses n.m.d./v.m.d. 75/98 μm. (3) ULV formulation in isophorone; n.m.d./v.m.d. 68/83 μm.

smoke, windsock, or other indicator), a windspeed indicator and a whirling hygrometer. A cup or vane anemometer may be preferable for indication of the average windspeed, as compared with floating (ball or disc pattern) indicator, which will register the continual fluctuations, or gusts. The whirling hygrometer provides wet and dry bulb temperature readings, from which the relative humidity can be derived.

More detailed study of the effects of boundary-layer turbulence involves

TABLE 2.5. *Total mass recoveries* (R_T) *of sprays on cotton plants by portable atomizers under different meteorological conditions*

Application	Droplet size n.m.d./v.m.d. (μm)	R_T (%)	t (GMT) (h)	u (m/s)	Ri —	T (°C)
N	65/90	89	1015	0.6	−0.40	30
O	34/60	74	1100	0.3	−0.90	30
P	43/86	67	0855	3.0	−0.017	28
Q	44/80	73	1000	1.2	−0.021	31.5
R	44/80	48	1120	1.7	−0.047	31
S	26/42	19	1120	1.7	−0.047	31
T	27/59	74	1000	1.2	−0.021	31.5

All applications were made with portable rotary atomisers, except T (motorized mistblower). Application P–T were 15% solution of triazophos in 70% vegetable oil, 15% other solvents; application O was 25% triazophos in xylene; application N, dyed water only.

determination of the wind and temperature profiles, so that a series of anemometers and thermometers, preferably of the electrical recording variety, become necessary, in conjunction with a suitable portable mast system for supporting the sensors at requisite heights (Huntington and Johnstone 1973). Some system of data-logging may be required to facilitate computerized processing of a rapid sequence of observations. In certain cases, bidirectional wind vanes, or meters for measuring vertical air currents may be of use (Schwerdtfeger 1976), while in forestry, particularly in locations with difficult terrain, the use of radio-sonde may be the only effective way of obtaining the relevant information. Doppler equipment, for accurate detection of wind strength and direction from aircraft in flight (Rainey 1972), and sodar (acoustic radar), both for graphically documenting mixing conditions in the lower boundary layer (Russell and Uthe 1978) and for quantitative estimation of diffusion (Kerman 1978), have specialist applications.

2.6   Choice of droplet size and formulation for specific tasks

The empirical development of spray application technology can be traced from the use of high-volume, run-off sprays, through medium-volume, to low-volume application. During this development, droplet size has often been incidental to volume application rate and has been determined in part by the requisite delivery through hydraulic pressure nozzles. With the trend to much reduced volume application pioneered in Africa in the fields of tsetse fly and locust control and developed in agricultural applications in the guise of very low- and ultra-low-volume spraying, the importance of droplet size control has been more widely accepted and the concept of controlled droplet application is approaching realization. The philosophy of controlled droplet application maintains that the primary application criterion should be the choice of optimum droplet size determined by the nature of the target, together with other relevant application considerations. Concomitant is the use of the minimum volume rate, at the chosen droplet size, needed to effect control. Fig. 2.7 provides a graphical illustration of the relation between droplet density (number/cm^2), droplet diameter, and volume application rate. As an example, if the appropriate droplet size is 100 µm and a cover of

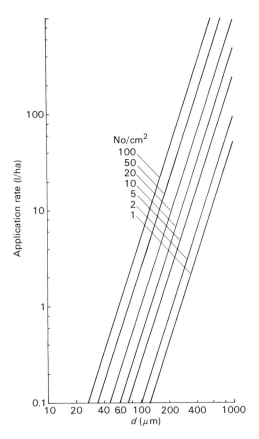

Fig. 2.7. Relation between droplet density (no./cm^2), diameter (d), and volume application rate (litres/ha).

50 droplets/cm^2 is necessary for efficient contact of the pest, a minimum application rate of 2.5 l/ha of surface is indicated. The ratio of foliar surface to ground (leaf area index) should be borne in mind when estimating the application area per plot. It is clear that controlled droplet application should also be controlled volume application and will tend towards low-, very-low-, or ultra-low-volume application as defined in Table 2.6.

TABLE 2.6. *Classification of sprays by droplet size and volume application rate*

A. *By droplet size*

Description	Range of v.m.d.* (μm)
Coarse spray	500
Medium spray	200–500
Fine spray	100–200
Very fine spray/mist	30–100
Aerosol	30

*Volume median diameter.

B. *By volume application rate*

	Application rate (l/ha)	
Description	Field crops	Bushes and trees
High volume (HV)	600	1000
Medium volume (MV)	100–600	300–1000
Low volume (LV)	20–100	50–300
Very low volume (VLV)	5–20	20–50
Ultra-low volume (ULV)†	5	20

†Essentially waterless formulation.

Note: Definitions are necessarily arbitrary, but the above seems to be a rational description of current usage with reference to ground machines and portable sprayers when applying insecticides and fungicides. (For herbicides, LV is taken as 50–200 and VLV as 5–50 l/ha respectively.) Aerial spraying must normally fall in the LV, VLV, or ULV categories. The boundaries can be considered slightly flexible, e.g. a very fine spray with v.m.d. of 40–50 μm has many of the features of, and in certain instances may be justifiably described as, a coarse aerosol.

There may still remain a few occasions when total wetting and high-volume application appears necessary, but, even in these exceptional cases, the use of controlled droplet size, and in particular smaller rather than larger droplets, should reduce the volume required to achieve the desired result, avoiding the early onset of wasteful run-off before all target surfaces have been sufficiently wetted.

The choice of formulation is closely linked with selection of droplet size and the physicochemical requirements in relation to solubility, volatility, toxicity, phytotoxicity, initial stability, spreading, wetting, retention, and ultimate degradation of the deposit. Solubility and stability are dealt with in appropriate detail in Chapter 1; however, the optimum concentration of active ingredient, in the mixture as applied, can be estimated in relation to the intrinsic toxicity of the material and its formulation, and to the development stage of the pest. For example, the larval mass of the *Pieris brassicae* caterpillar increases from about 0.25 to 250 mg between first and late Vth instar (a 1000-fold change); the toxic dose required for kill increases commensurately. An appropriate insecticide will normally have an LD_{50} of about 10 μg/g insect body weight, which means that, for instance, 5 ng would be required for 50% kill of newly hatched first-instar *Pieris brassicae* larvae, while as much as 5000 ng (5 mg) might be required to kill a mature Vth-instar caterpillar. The importance of timely spraying to combat the early stages of a pest is thus emphasized. Fig. 2.8 illustrates the relation between toxicity, droplet diameter, and concentration of active ingredient (a.i.) such that the LD_{50} will be contained in just one droplet. As an example, if we assume 10 to 100 ng is the normal range of dose for direct-contact kill, at 100 μm the concentration required to give an LD_{50} dose in just one droplet lies between 2 and 20% (the corresponding concentration range for 70 μm droplets would

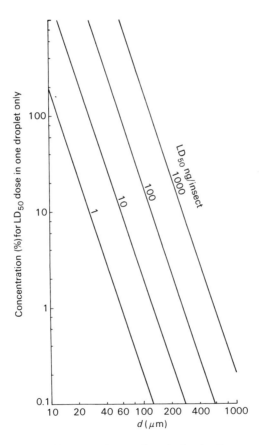

Fig. 2.8. Relation between toxicity (LD_{50} dose, ng/insect), droplet diameter (*d*), and concentration of active ingredient for one droplet to contain the LD_{50} dose.

be 5 to 50%). In the case of action resulting from pick-up of residue due to locomotion over a sprayed surface a factor describing the efficiency of the transfer process must be considered (Johnstone 1973b).

With present knowledge and experience it is a matter of debate as to whether one particular droplet size can fit all the requisite conditions pertaining to any given pest; nevertheless for a specific pest control problem an optimum droplet size range may be predicted by a combination of the theoretical considerations summarized in this chapter, possibly supported by some appraisal in laboratory wind tunnel tests (Hadaway and Barlow 1965). The final choice for field application should always be made with reference to the interaction of pest, crop, formulation, and environment.

A broad guide to the appropriate choice is indicated in Table 2.7, adapted from an outline published by EPPO (1973) for ULV aerial application.

TABLE 2.7. *Guide to choice of spray droplet size in controlled droplet application*

Approximate v.m.d. (μm)	Use	Remarks
500	Herbicide application from the air when avoidance of drift a critical consideration.	Only suitable for slow flying aircraft, in particular, the helicopter.
200–500	Larvicide application against public health pests and bait spraying against mobile pests (ULV). Crop spraying with residual sprays against most pests and diseases (MV/LV).	Suitable for spot placement. Useful with aqueous sprays in conditions of high temperature and low relative humidity, particularly in aerial application.
125–250	Crop spraying using contact and residual sprays against most pests and diseases (LV/VLV).	Good deposition, but possible loss of cover density at VLV rates.
60–120	Crop spraying with contact and residual sprays against most pests and diseases (VLV/ULV)	Good canopy penetration. Check volatility of formulation, especially in aerial application.
30–60	Contact action sprays against flying or resting adult insects, e.g., tsetse flies, mosquitos (ULV).	Poor deposition on large obstacles; prone to drift. Evening or night-time application, or within 1 h after sunrise best.
30	Contact sprays against adult mosquitos (ULV).	Little, if any, lasting effect. Best application conditions as for 30–60 μm.

Specific situations should allow closer definition. The narrowest droplet spectra currently widely available for field use are provided by rotary atomizers, which can give a ratio of v.m.d./n.m.d. of about 1.3, compared with a ratio in excess of 2 for hydraulic pressure nozzles. Until novel ways of effecting comparable or better homogeneity in atomization are developed, rotary atomizers are likely to remain the preferred equipment for controlled droplet application, although electromechanically pulsed jets and electrodynamic atomization (at least at low flow rates) appear to offer renewed promise of monodispersity (Yates and Akesson 1978; Coffee 1979, respectively).

2.7 Techniques for the assessment of depositing and drifting sprays

Reference to reviews dealing with sampling techniques for droplet size measurement is made in Section 2.2.1. The purpose of this section is merely to outline some of the principles of sampling spray droplets in the field, with pointers to basic procedures and to some techniques of use for more detailed study.

2.7.1 Basic procedures

Droplets may be collected by sedimentation, inertial impaction, or a combination of the two processes. Contaminated vegetation, either whole plants, or parts of plants, may display the results of both. Alternatively, artificial surfaces, including papers, plastic sheets, glass plates, etc., may be set out, either horizontally, on the ground, or vertically at suitable levels to sample the drift of airborne material by impaction from natural air currents. More consistency is obtained if the impaction device provides control of sampling rate, as is the case with the cascade impactor (May 1945) and the rotorod sampler (Perkins 1957). These latter are useful tools for sampling very small droplets (or particles) in very light winds (May *et al.* 1976).

2.7.2 Techniques for more detailed study

While sampling of airborne materials can be performed in several ways, ideally the technique of choice should be appropriate to the nature of the drifting material. Liquid droplets, for instance, require a different technique from the dry particles produced as a result of volatilization of carrier liquid, while the vapour of a volatile active agent calls for yet another form of sampling technique (Yule and Cole 1971). Sampling surfaces may be treated in a variety of ways to increase retention efficiency by absorption, or to facilitate identification of the particle size of the deposit (May 1949).

   Porous filters, of the paper, glass wool, or plastic variety, are useful for trapping small particles, providing that appropriate suction can be devised. Any aspirated device should ideally sample isokinetically, i.e. the speed at which the air is drawn into the device should be, as far as possible, identical to the speed at which the air is naturally approaching the intake orifice, thereby simplifying calculations of airborne concentrations.

Vapours may be taken up by drawing through scrubbers or bubblers containing appropriate solvents or active chemicals; alternatively solid sorbents, e.g. activated charcoal, may be used to adsorb the aspirated vapour (Yule and Cole 1971; Lewis 1976).

The sampling techniques must be designed in conjunction with the proposed analytical techniques, as the two processes are complementary. Analysis may depend on physical measurement, e.g. optical measurement of crater size on magnesium-coated slides, or the size of droplet stains on papers (when using a dye tracer) to estimate volumetric recoveries (Cadle 1975; Matthews 1975). A variety of physicochemical techniques are available for determining small quantities of active materials, or suitable tracers, and these include counting of fluorescent particles under UV illumination, simple colorimetry, chromatography (t.l.c., g.l.c., or h.p.l.c.), measurement of radioactivity, etc. (Lewis 1976).

In certain instances, bioassay techniques may be used to supplement physiochemical measurements, e.g. the relevant insects, housed in permeable cages, may be exposed to the drift, or may be caged subsequently on the sampled vegetation to assay residual effect. For assessment of the effects of herbicide drift, susceptible indicator plants may be pot-cultured and set out in the downwind region for exposure, and the nature of their subsequent development used as an indication of the magnitude of drift contamination (Hartley 1959).

2.8   Epilogue

In reviewing facets of the physics and meteorology of the application of pesticides as sprays it becomes clear that the performance of different spray machines can only be critically evaluated if the relevant meteorological conditions are known, and that if valid comparisons of performance are to be made, particularly for applications in the very fine spray or aerosol category, tests should ideally be carried out simultaneously, using appropriate assessment techniques. Furthermore, if consistent results are to be achieved in applying sprays of this nature, some guidelines regarding modification of operating procedures for variations in weather conditions will generally be required, accompanied by an acceptance of limitations on spraying under weather conditions which are unsuitable.

We are improving our control of droplet size, together with our understanding of drift and deposition processes and the effect of the weather. It is becoming increasingly desirable that we use this understanding, linked with choice and control of droplet size to provide more truly controlled application.

2.9 References

Amsden, R. C. (1962). *Agric. Aviat.* **4**, 88.
Andrews, M., Flower, L. S., Johnstone, D. R., and Turner, C. R. (1983). *Trop. Pest. Manag.* **29**, 239.
Bache, D. M. and Sayer, W. J. D. (1975). *Agric. Meteorol.* **15**, 257.
Brunskill, R. T. (1956). *Proc. 3rd Br. Weed Control Conf.* **2**, 593.
Cadle, R. D. (1975). *The Measurement of airborne particles.* Wiley and Son, New York.
Chamberlain, A. C. (1975). In *The movement of particles in plant communities in vegetation and the atmosphere* (ed. J. L. Monteith) Vol. 1, pp. 155. Academic Press, London.
Christensen, P., Yates, W. E., and Akesson, N. B. (1969). *Proceedings of 4th International Agricultural Aviation Congress* (Kingston, 1969). I.A.A.C. pp. 337.
Coffee, R. A. (1979). *Pests Dis.* **3**, 777.
Courshee, R. J. and Byass, J. B. (1953). *N.I.A.E. Silsoe Report No. 31.*
Csanady, C. T. (1973). *Turbulent diffusion in the environment.* Reidel, Dordrecht.
Davies, C. N. (1966). In *Aerosol science* (ed. C. N. Davies), p.393. Academic Press, London.
Dorman, R. G. (1966). In *Aerosol science* (ed. C. N. Davies), p.195. Academic Press, London.
EPPO (1973). *OEPP/EPPO Bull.* **3**,(2), 51.
Ford, R. E. and Furmidge, C. G. L. (1966). *Soc. Chem. Ind. Monogr.* No. 25, 417.
Furmidge, C. G. L. (1962). *J. Sci. Food Agric.* **13**, 127.
—— (1968). *J. Proc. Inst. Agric. Eng.* **23**(4), 173.
Geiger, R. (1966). *The Climate near the ground.* Harvard University Press, Cambridge, Mass.
Gunn, D. L. (1948). *Bulletin No. 4. Anti-Locust Research Centre.* London.
Hadaway, A. B. and Barlow, F. (1965). *Ann. Appl. Biol.* **55**, 167.
Hartley, G. S. (1959). *Report 1st International Agricultural Aviation Congress.* Cranfield. I.A.A.C. p.142.
Herdan, G. (1953). *Small particle statistics.* Elsevier, Amsterdam.
Hill, D. (1975). In *Agricultural insect pests of the tropics and their control.* p.65. Cambridge University Press.
Holmes, N. E., Lucken, A. C., and McIlveen, J. F. R. (1976). *Weather* **31**, 218.
Huntington, K. A. and Johnstone, D. R. (1973). *COPR Miscellaneous Report No. 14.* College House, London. 12 pp.
Jarman, R. T. (1958). *Br. J. Appl. Phys.* **9**(4), 153.
Johnstone, D. R. (1972). *Cotton Growing Rev.* **49**, 166.
—— (1973a). In *Pesticide formulation* (ed. Van Valkenburg), p.343. M. Dekker, New York.
—— (1973b), *Pestic. Sci.* **4**, 77.
—— (1977a). *Cotton Fibre Top.* **32**, 67.
—— (1977b). *J. Agric. Eng. Res.* **22**, 439.
—— (1978a). *BCPC Monograph No. 22. Controlled Drop Application.* p.35.
—— (1978b). *BCPC Monograph No. 22. Controlled Drop Application.* p.43.

—— and Huntington, K. A. (1977). *Pestic. Sci.* **8**, 101.

—— and Johnstone, K. A. (1976). *C.O.P.R. Miscellaneous Report No. 26.* 14 pp. College House, London.

—— —— (1977). *P.A.N.S.* **23**, 13.

——, Huntington, K. A., and King, W. J. (1974). *Agric. Meteorol.* **13**, 39.

—— and Watts, W. S. (1970). *Cotton Growing Rev.* **47**, 36.

—— Rendell, C. H., and Sutherland, J. A. (1977). *J. Aerosol Sci.* **8**, 395.

Kerman, B. R. (1978). *Atmos. Environ.* **12**, 1827.

Landsberg, J. J. and James, G. B. (1971). *J. Appl. Ecol.* **8**, 729.

Lawson, T. J. and Uk, S. (1978). *BCPC Monograph No. 22. Controlled Drop Application.* p.67.

Lewis, R. G. (1976). In *Air pollution from pesticides and agricultural processes* (ed. R. E. Lee). Chemical Rubber Publishing Co., Cleveland, Ohio.

Maas, W. (1971). *ULV application and formulation techniques.* N. V. Philips-Duphar, Amsterdam.

Matthews, G. A. (1975). *P.A.N.S.* **21**(2), 213.

May, K. R. (1945). *J. Sci. Instrum.* **22**(10), 187.

—— (1949). *J. Sci. Instrum.* **27**(5), 128.

—— (1965). *J. Sci. Instrum.* **42**, 500.

—— and Clifford, R. (1967). *Ann. Occup. Hyg.* **10**, 83.

—— Pomeroy, N. P., and Hibbs, S. (1976). *J. Aerosol Sci.* **7**, 53.

Munn, R. E. (1966). *Descriptive micrometeorology.* Academic Press, New York and London.

Pasquill, F. (1961). *Atmospheric diffusion.* Von Nostrand, London.

Perkins, W. A. (1957). *2nd Semi-Annual Report. Aerosol Laboratory, Dept. of Chemistry and Chemical Engineering,* Stanford University. p.186.

Prandtl, L. (1952). *Essentials of fluid dynamics.* Hafner Publishing Co., New York.

Rainey, R. C. (1972). *Aeronaut. J.* **76**, 501.

Rathburn, C. B. (1970). *Mosquito News* **30**(4), 501.

Richardson, E. G. (1950). *Dynamics of real fluids.* Edward Arnold and Co., London.

——(1960). In *Aerodynamic capture of particles* (ed. E. G. Richardson), p. 3. Pergamon Press, Oxford.

Richardson, L. F. (1920). *Proc. R. Soc. Ser. A* **97**, 354.

Russell, P. B. and Uthe, E. E. (1978). *Atmos. Environ.* **12**, 1061.

Sawyer, K. F. (1950). *Bull. Entomol. Res.* **41**, 439.

Scorer, R. (1958). *Natural aerodynamics.* Pergamon, London.

Schwerdtfeger, P. (1976). *Physical principles of micrometeorological measurements.* Elsevier, Amsterdam.

Smith, F. B. and Hunt, R. D. (1978). *Atmos. Environ.* **12**, 461.

Sutton, O. G. (1949). *Atmospheric turbulence.* Methuen and Co., London.

Tennekes, H. (1973). *J. Atmos. Sci.* **30**, 234.

Thom, A. S. (1971). *Q. J. R. Meteorol. Soc.* **97**, 414.

—— (1975). In *Vegetation and the atmosphere* (ed. J. L. Monteith) Vol. 1. p.57. Academic Press, London, New York, and San Francisco.

Threadgill, E. D. and Smith, D. B. (1975). *Trans. Am. Soc. Agric. Eng.* **18**, 51.

Trayford, R. S. and Welch, L. W. (1977). *J. Agric. Eng. Res.* **22**, 183.

Webb, E. K. (1970). *Q. J. R. Meteorol. Soc.* **96**, 67.

Yates, W. E. and Akesson, N. B. (1978). *Proceedings of 1st International Conference on Liquid Atomisation and Spray Systems. Tokyo.* p.181.

Yeo, D. (1974). *BCPC Monograph No. 11. Pesticide Application by ULV Methods.* p.10.

—— Akesson, N. D., and Coutts, H. M. (1959). *Nature (London)* **172,** 168.

Yule, W. N. and Cole, A. F. W. (1971). *Proceedings 4th International Agricultural Aviation Congress,* (Kingston 1969). I.A.A.C. p.346.

3
Biological considerations

M. J. WAY and M. E. CAMMELL

3.1 Introduction

Prior to the early 1940s, measures for pest control involved a limited number of chemicals together with various biological and cultural methods. The chemicals were mostly not particularly efficient nor widely applied except on certain high-value perennial crops such as apples. Although undesirable side effects, including induced resistance, were occasionally recorded, as with San José Scale to lime sulphur in 1914, there was no general awareness of the likely ecological consequences of widespread and intensive use of chemical pesticides and therefore the first modern synthetic pesticides, notably insecticides, were developed with limited regard for biological considerations other than toxicity to pests. The persistence and broad spectrum of activity of many of them, especially insecticides, often enabled a single application of one chemical to control a complex of pests. The high kill attained for many major pests also brought about notable changes in other aspects of crop production (e.g. high-yielding crop varieties, simplification of crop rotations, drilling-to-a-stand, and direct drilling), which could now be adopted irrespective of whether the new system became inherently more pest susceptible. However, our failure to recognize the damaging consequences of new technologies that pesticides have made possible soon resulted in well-known undesirable side effects such as resistance, pest resurgence, increased pest status of hitherto minor or potential pests, environmental hazards and toxicity to non-target animals and plants, and by the early 1960s we had entered what Metcalf (1980) aptly described as the 'Era of doubt'. Intense criticism of pesticides followed and there was considerable re-examination of other long-standing methods of control and much emphasis on novel new approaches. Although the latter have provided a few successes, it is evident that pest control based on available novel methods is not feasible for most major pests. Furthermore, current crop production technologies and the ever-increasing demand for higher yields often limit the role of long-standing non-chemical controls. Yet, whilst it must be recognized that chemical pesticides remain essential tools for the management of pest populations, the spectre of resistance and of other problems that undermine their use means that they must no longer be used as in the past but only be applied within the framework of sound ecological concepts.

The main problems associated with pesticide application have arisen because of the excessive use or misuse of relatively non-selective chemicals, especially persistent ones. There has been some change towards chemicals showing greater intrinsic selectivity (physiological selectivity) between target and some non-target organisms, notably to protect Man and other vertebrates, but there has been little interest in developing insecticides and fungicides that are selective for target pests and relatively inactive against arthropod natural enemies or beneficial competitor fungi, for example. In contrast, the development of herbicides has often necessitated highly selective action in relation to different plants, and in this respect has been spectacularly successful. However, few of those concerned with the development of herbicides have taken account of how herbicides may be influencing incidence and control of other kinds of 'pests'.

Rational pesticide usage therefore depends crucially upon biological criteria which include:
1. Appreciating the ecological basis for development of pesticide resistance.
2. Knowledge of damage relationships leading to rational decision-making on need for chemical controls.
3. Using pesticides selectively to protect beneficial species, ideally by use of physiologically selective chemicals but,more realistically, by using non-selective chemicals in ways which conserve selectivity (e.g. ecological selectivity).
4. Appreciating interactions between chemicals used for controlling animal pests, diseases, and weeds.
5. Fitting the chemicals into an integrated pest-management programme whereby the chemicals supplement other control methods rather than supplant them.

Fig. 3.1. indicates the biological interrelationships and other processes needed to develop sensible Integrated Pest Management (IPM) methods.

3.2   The resistance problem

3.2.1   *Mechanisms of resistance*

The development of resistance to chemicals which formerly controlled them has occurred widely in many plant and animal pathogens, invertebrates, particularly insects, and also amongst some vertebrate pests; a few weed species are also showing resistance or tolerance to some herbicides. Virtually all evidence points to resistance occurring by the process of selection of pesticide-resistant individuals that happen to be in the population: such individuals become the nucleus for subsequent populations largely comprising resistant individuals. The selective destruction of susceptible individuals in a population facilitates the 'throwing together' of different genes for resistance possessed by remaining resistant individuals and can lead to

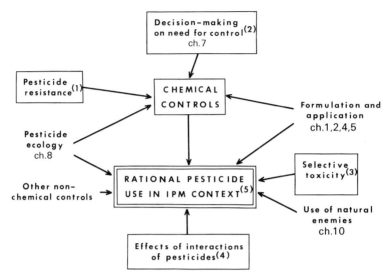

Fig. 3.1. Biological considerations of pesticide application. Aspects considered in this chapter are numbered (1–5). Other relevant chapters are designated by their numbers.

greatly enhanced resistance in the form of multiple and cross-resistance. The process is therefore a typical Darwinian one of selection of fitness to cope with a particular stress, in this case a pesticide. The resistance problem has been well reviewed by Brown and Pal (1971), Conway (1982), and Georghiou (1972). For present purposes, it is necessary to stress that the development of resistance is basically determined by the intensity of the selection pressure exerted by the pesticide and also by certain characteristics of the pest and the pest population. This means that relatively simple biological knowledge can be used to indicate conditions in which resistance is most likely to develop and hence make it possible to avoid or alleviate such conditions.

3.2.2 Factors favouring development of resistance

(a) Nature and mode of use of the pesticide. Resistance is less likely to develop to pesticides with a generalized rather than specialized mode of action. Also, persistent chemicals applied to surfaces, to soil or to other places where they remain actively exposed to pests, maintain the selection pressure favouring the selection of resistant genotypes within the pest population. This is one reason why many persistent organochlorine insecticides have failed, and it is a hazard now facing the persistent pyrethroids. Selection pressure is also maintained by using the same relatively non-persistent pesticide frequently. In these circumstances, the pressure for selecting out of resistant individuals

may be as great as for relatively persistent chemicals applied less frequently, particularly when applied routinely during periods when the pest is scarce.

(b) Qualities of the pest species. Species which are very fecund and with many generations a year can rapidly build up to large resistant populations from very small nuclei of resistant individuals remaining after pesticide treatment. Unfortunately, it is these species which often require frequent applications of pesticides to control them, so enhancing the pesticide pressure for induction of resistance; such are the characteristics of many bacteria and some fungi as well as rapidly multiplying species of insects and mites that have become resistant to controlling chemicals. This contrasts with species that breed relatively slowly and that in general have not developed resistance or have only done so after many years of intense pesticide pressure. This situation also highlights the great danger of resistance developing where conditions are favourable for multiplication throughout the year as in the equatorial and irrigated tropics. An outstanding example is that of the Diamond Back Moth, *Plutella xylostella*, which has about two generations per year in Western Europe and shows no signs of resistance, in contrast to 15–20 generations in the equatorial tropics where in Malaysia, for example, intense resistance has developed to many organochlorine, organophosphate, carbamate, and now to pyrethroid insecticides.

Another species characteristic which influences the rate of resistance development is dispersal power. Resistance can develop relatively rapidly in localized populations of poorly dispersive species where the resistant individuals are not diluted by continual immigration of non-resistant forms. Striking examples include the phytophagous mites and also coccids. The somewhat patchy distribution of resistance to acaricides of some fruit tree spider mites is no doubt associated with the initial presence of resistant individuals selected out in particular orchards contrasting with their absence from other orchards where the population has remained susceptible.

Geographical isolation as in an island, or physical isolation as in a glasshouse has a similar effect to that of species which are inherently non-dispersive – the population cannot be significantly diluted from outside nor are resistant individuals readily lost by migration.

Resistance is also likely to develop relatively quickly in situations where the pest species is confined within an ecosystem or to places in it where there is continuing pesticide pressure. For example, the eggs and larvae of the onion fly, *Delia antiqua,* may be confined to insecticide-treated onions, and spotted alfalfa aphid, *Therioaphis maculata*, to insecticide-treated alfalfa.

(c) Lack of alternative mortality factors. Where the insecticide causes a very large proportion of mortality during development of the pest compared with that of other extrinsic mortality factors such as natural enemies which do not

selectively favour resistant individuals, then resistance will unquestionably be speeded up or will develop where it would not otherwise happen. This highlights the importance of an integrated approach to pest control whereby significant mortality is caused by non-chemical factors which do not selectively favour resistant individuals. In this context, the importance of the discovery and development of natural enemies that are themselves resistant to pesticides (Croft and Brown 1975) cannot be overestimated. Conversely, where populations are often naturally controlled by intra-specific mechanisms, as with certain Diptera where populations are seemingly regulated by competition amongst larvae, this mechanism may enhance development of resistance because, following pesticide treatment, the residue of individuals is able to increase rapidly, free from intra-specific competition.

From the above criteria, and assuming that genes for resistance exist in a population, it is possible to predict species that are especially prone or not prone to development of resistance and therefore to gauge in principle where pesticides need to be used most sensitively in order to avoid over-use and also use of inappropriate chemicals that is often a first response to the appearance of resistance in a pest population. Pests with high potentiality for resistance would be non-dispersive species which form localized populations and can multiply rapidly, such as scale insects, whereas species of low potentiality are those with, for example, a long generation time, such as many wireworms, or are relatively slow developers but highly dispersive such as many locust species.

Most pest species, however, have a combination of characteristics and environmental relationships some of which favour or do not favour induced resistance. For example, the green peach aphid, *Myzus persicae,* possesses factors such as short-generation time and high fecundity and is frequently treated with insecticides at low pest densities because of need to minimize spread of transmitted viruses. Yet development of resistance is not favoured by its dispersive behaviour and wide range of hosts in many environments, often non-pesticide-treated. The last two factors are nullified in the virtually closed ecosystems of some continuously cropped greenhouses where *M. persicae* has developed resistance to a wide range of insecticides (Hussey 1965). Also outdoors in parts of southern Europe, the aphid has developed resistance to organophosphate insecticides because populations are confined largely to the primary host, the peach tree, where the progeny of the sexual population, intensively treated in spring and early summer, are powerfully selected for resistance. Although the aphid may leave peach in early summer and spend the rest of the year on many untreated plants, parthenogenesis prevents intermixing and dilution by susceptible biotypes of *M. persicae.* So the resistant population is kept distinct and, in the autumn, sexual reproduction makes it possible for genetic combinations to increase resistance. The retention of resistance by parthenogenetically reproducing strains is exemplified in southern California, where chemical control of continuously parthenogenetic *M. persicae* on sugar beet has become difficult because of annual long-distance immigration of resistant individuals from areas where the aphid survives as a resistant strain throughout the year on other intensively treated crops. Similarly in Britain, parthenogenesis has perpetuated a small degree of resistance of *M. persicae* in some areas.

Therefore there are some guidelines that indicate how to avoid or delay onset of resistance or decrease resistance in accordance with ecological conditions, the characteristics and mode of use of pesticides and the biology of the pest. However, enigmas remain, for example why *M. persicae* develops resistance so readily whereas some other intensely treated aphid species such as the cabbage aphid *Brevicoryne brassicae* have not. Similarly, the codling moth, *Cydia pomonella*, has remained susceptible throughout the world despite intensive treatment by certain organophosphorus insecticides for more than twenty years.

3.3 Decision making on the need for chemical control

3.3.1 Economic thresholds

Our first biological consideration should be whether a chemical treatment is necessary. Ideally, a pesticide should only be applied if there would otherwise be an economic crop loss. One important concept is the economic injury level which is the lowest pest level that will cause sufficient damage to justify the cost of control. Economic injury levels can be based provisionally on empirical evidence and then refined as the nature of particular pest–crop relationships become better understood (see Chapter 8).

The general relationship between crop yield and pest density is conveniently considered as three phases, namely the compensatory phase in which the plant is able to compensate for injury; the non-economic injury phase where the plant no longer compensates for injury but damage is insufficient to justify the cost of control; and thirdly the economic injury phase when the reduction in yield at least equals the cost of control measures. In a particular pest density–crop yield relationship, the importance of each phase may vary considerably depending upon the compensatory ability of the plant in relation to the feeding behaviour of the pest and the time of attack. In general, plant compensation is greater for pests which attack parts of the plant other than the crop product (Southwood and Norton 1973). Thus, heavy-tillering rice varieties in Japan can tolerate moderate levels of the rice stem borer *Chilo suppressalis* during early growth stages without loss in yield because lost tillers are rapidly replaced at this time. Moreover, in some crops such as field beans and turnips, low levels of attack by certain pests early in the season may perhaps enhance yield. In contrast, tolerance to a pest may be very low where the pest is a vector of disease, for example *M. persicae* transmitting sugar beet yellows or where the crop quality requirements are high and the mere presence of pest or damage is unacceptable ('cosmetic effects').

The pest–damage relationship may be considerably modified by environmental factors which may differentially affect compensatory plant growth and yield, and the rate of pest population increase. Therefore, the treatment level may need to be modified for differing climatic areas. Also, the relationship may vary in relation to different plant varieties and cropping practices.

Account must also be taken of the interacting effects of different pest species attacking the crop and also of effects of changed pesticide usage caused by use of economic injury data for one pest which affects the status of other pest species, previously controlled incidentally by routine chemical treatments.

In practice, defining meaningful economic thresholds has proved difficult because of the many biological and economic variables involved and, no doubt, thresholds will prove indefinable or impractical for certain pest situations. However, there is now an awareness that many useful working thresholds can be developed given the appropriate biological information, and there are many examples where the use of such thresholds can considerably reduce the amount of pesticide necessary for effective control (Cammell and Way 1977; Huffaker 1980). Further progress will also depend upon demonstrating to growers that the mere presence of a pest in a crop does not necessarily justify treatment. In fact, preservation of pest populations below damaging levels may be desirable in order to retain important natural enemies.

3.3.2   Monitoring and forecasting

Forecasts of need for chemical treatment vary from short-term forecasts, often based upon monitoring of pest populations in the crop shortly before damage occurs, to longer-term forecasts based on counts of the pest either in the field before planting, as with relatively non-dispersive soil pests, or with relatively dispersive pests at appropriate places before arrival in the crop.

Initially, the choice of forecast will depend upon availability of relevant biological information, including economic threshold data and suitable monitoring techniques. Given that such information is available, then the merits of different kinds of forecasts can be assessed using the following criteria: (a) the value placed on the time when the forecast is available, (b) the accuracy of the different forecasts in predicting need for application of pesticides and (c) the cost and convenience of operating different monitoring procedures (Way and Cammell 1977).

Work on forecasting the need for chemical control of the black bean aphid, *Aphis fabae,* attacking spring-sown field beans in southern England provides an example for assessing the relative merits of different kinds of forecasts and also highlights some of the problems and possibilities in practical forecasting (Way and Cammell 1973; Way *et al.* 1977). Three monitoring procedures with appropriate thresholds are available for *A. fabae;* short-term crop monitoring; monitoring of numbers caught by 'Rothamsted' suction traps either during dispersal to the crop in spring or during the previous autumn; and counts in winter and spring of eggs and active stages on the source host, the spindle tree, *Euonymus europaeus.*

The decision on whether to apply a pesticide based upon crop sampling may have important limitations since it may require a 'fire-brigade' response if control is needed. There is also the problem of defining when to examine the crop. In contrast, the longer-term source monitoring of *A. fabae* enables forecasts to be issued in late January, some 4–5 months before treatment may be required. Such forecasts have obvious value to pesticide suppliers, advisors, and farmers in planning long-term pesticide needs. Furthermore, sampling of the peak population on *E. europaeus* about mid-May provides information on when the main migration will occur and hence when to apply chemical treatment. The value of correct timing is considered in more detail later in this chapter.

Ultimately, the value of any forecast will depend upon the accuracy with which levels of damage can be predicted at the individual field level. Obviously, crop monitoring offers the most accurate assessment, for source and dispersal monitoring will only provide area forecasts, each of which has inaccuracies depending upon variations in pest incidence in different fields and in the various monitoring procedures used.

Forecasts based on source monitoring of *A. fabae* fall into one of three categories, each requiring appropriate action. Since 1970, results have shown that in some years a single regional forecast for southern England would have been satisfactory, either in terms of recommending treatment as in 1973, or of recommending no treatment as in 1975.

However, in other years, such as 1972, the need for treatment varies according to area, thus demonstrating the need for area-based forecasts (Way *et al.* 1977). The accuracy of area forecasts in relation to individual fields within an area was assessed for the period 1970–75 by comparing the advice based on these forecasts with plant infestations occurring subsequently in individual fields. Overall, the results showed that following a forecast of 'unlikely damage', 1% of the crop would have justified treatment, albeit only marginally, and that following a forecast of 'possible' or 'probable' damage, 18% of crop would have subsequently been treated unnecessarily. Thus inaccuracies are largely associated with recommending unnecessary treatments, which are more acceptable to the grower than risk associated with failing to recommend necessary treatment (Cammell and Way 1977). Inaccuracies associated with an area-based forecast must be considered against the background of the errors involved in establishing economic thresholds and in defining subsequent changes in the thresholds. Furthermore, crop monitoring of beans cannot at present be undertaken by advisory services, and growers are usually unwilling or unable to make accurate assessments of pest numbers, particularly if they also lack information on when it should be done. Long-term forecasts based on overwintering counts also involve simple observations during winter and spring and can be provided cheaply; for example the average cost from 1970 to 75 was only £0.14/ha. Undoubtedly, this is more convenient and more profitable than individual field inspection during a critically short and busy period in the growing season.

In conclusion, short-term forecasts based on crop sampling or on numbers of adults trapped whilst migrating into crops will no doubt continue to be usefully employed as monitoring procedures. However, long-term aims should move towards more fundamental studies on the population dynamics of pests at the regional level. In such ways, we will be able to define long-term population changes both in space and time, identify cycles of pest incidence and sources of infestation and assess the relative importance of short- and long-distance migrations. Such studies will provide the basis for regional, long-term forecasts which may then be suitably modified for local farm conditions, particularly when we have a greater understanding of the factors which influence variations of infestations in different fields of the same crop plant.

3.4   Selective toxicity

Assuming that a chemical treatment is needed, rational decisions are required on the choice, timing, and method of application of the chemical in ways which confer maximum selective toxicity between the target pest and non-target organisms. Selective toxicity is crucially important in relation to problems of human health, environmental contamination, and also as an element of many integrated control procedures. It is discussed here using insecticides as examples but otherwise in a broad sense by briefly reviewing the subject in terms of minimizing harm to man, domestic animals and crop plants and wildlife in general, particularly beneficial species such as pollinating insects and natural enemies. In general, the concepts apply to all kinds of pesticides, and for herbicides in particular many elegant means have been devised for obtaining plant-to-plant selectivity.

Selectivity can be obtained in two main ways. First, by using chemicals which are intrinsically (physiologically) selective; second, by using non-selective chemicals in ways which confer selectivity.

3.4.1   Use of intrinsically selective chemicals
In practice, almost all chemicals have some effect on non-target organisms

and whether this is acceptable depends upon the nature and degree of the side effects and the circumstances under which the chemical is to be used. Metcalf (1975) has rated some 35 different insecticides according to mammalian toxicity, toxicity to a game bird, toxicity to a fish and to a bee, as well as environmental persistence, and such data provide general guidelines on suitability.

Chemicals such as parathion and aldicarb are not only toxic to a very wide range of insect species but also to many other invertebrates and vertebrates and can also be phytotoxic. Organochlorines such as dieldrin and DDT are also relatively unselective, their persistence causing build-up in food chains and retention in non-target organisms. Some chemicals are unselective because of harmful effects on plants, for example DDT is very harmful to Curbitaceae; HCH and phorate are harmful to some plants when used as seed dressings. There is also the striking example in glasshouses where the transformation of a control procedure from one based on chemical sprays to one based mainly on biological control led to as much as 10–20% increased yield of cucumbers because previous chemical control regimes decreased yield by direct harm to plants though there was no obvious damage (Hussey 1973).

Some chemicals are relatively harmless to vertebrates and some other animals, whilst remaining relatively unselective amongst the Insecta and other arthropods. Examples are diazinon, malathion, natural pyrethrins, and some pyrethroids, although the last named are very toxic to fish. In contrast, other insecticides may be highly toxic to mammals but selective within the Insecta, for example schradan which is notably toxic only to aphids and some Diptera.

Comparatively few insecticides are narrowly selective to only a few insect groups. They include menazon which is very toxic to aphids, some Hymenoptera and Diptera, and also pirimicarb which is selectively very toxic to aphids and Diptera. Also toxins of *Bacillus thuringiensis* are very active against certain lepidopterous larvae and some Diptera but seemingly harmless to other organisms. Many viruses may be even more selective. Much better progress has been made in the development of intrinsically selective fungicides, herbicides, and acaricides. Unfortunately, the very high degree of selectivity shown by certain fungicides and acaricides seems to be associated with marked ability of the target organisms to develop resistance to them. There is now much emphasis on more broad-spectrum fungicides against which specific resistance-breaking mechanisms are less likely, but which otherwise are more likely to induce resistance by maintaining selection pressure against a wide range of pathogenic fungi.

Highly specific resistance is usually associated with a highly specialized resistance mechanism which is so often readily overcome by particular biotypes of the pathogen. In one respect, therefore, there might be a greater risk of highly specific pesticides failing, yet because of their specificity, they are otherwise the ideal supplement to biological control agents which in turn must help to prevent or delay induced resistance. In this context, it should be pointed out that the rapid development of spider mite resistance to some physiologically specific acaricides is not just associated with specialized resistance mechanisms but also with the coincident use of non-specific pesticides against other pests that destroy natural enemies which otherwise might have helped keep such acaricides effective.

In conclusion, there is probably little incentive to develop highly intrinsically selective insecticides except perhaps for a few major pests of widespread importance. In most circumstances, therefore, it is more realistic to seek means of using intrinsically non-selective chemicals in ways which confer selectivity.

3.4.2   Selective use of intrinsically non-selective chemicals

It is helpful to divide the selective uses of non-selective chemicals into two overlapping forms, physical and ecological selectivity.

(a) Physical selectivity. Such selectivity involves physically separating the pesticide and non-target organisms, whilst maintaining toxicity to the target pest. It is represented by the appropriate use of protective clothing for pesticide operators, and can also be attained, or at least approached, by using appropriate formulations and methods of application.

Granular formulations of certain non-selective chemicals can confer useful selectivity and are particularly useful for applying the most hazardous pesticides such as aldicarb, which are too dangerous to apply in spray form; also, some chemicals such as phorate, which are non-selective when applied as sprays, do little harm to beneficial insects such as bees and some natural enemies when applied as granules. The advantage of granules is clearly demonstrated in control of lesser cornstalk borer, *Elasmopalpus lignosellus,* an important pest of groundnuts in Texas. The pest feeds below the soil surface, and previously used foliar sprays applied to kill the pest also killed many beneficial arthropods and led to outbreaks of foliage-feeding secondary pests. However, granules placed in the soil gave highly effective control with considerably reduced applications and no adverse side effects (Newsom *et al.* 1976). Encapsulation of an insecticide can also protect beneficial natural enemies and other species from contact action whilst continuing to kill chewing insect pests.

Physical selectivity achieved by applying pesticides, particularly systemics, as seed dressings or soil treatments may do some harm to soil organisms, and seed dressings can harm non-target animals that eat the seed but such insecticides may do relatively little harm to organisms on aerial parts of the plant other than to the pest species which feed on the plant or live in the plant tissues. Also, there are techniques such as 'side band' treatment with soil-applied insecticides which otherwise would be phytotoxic to seedling plants if applied in the seed drills or as a seed dressing.

With herbicides, there are numerous ingenious techniques such as pre-emergence and between-row applications which provide comparable physical selectivity (Roberts 1982).

(b) Ecological selectivity. Such selectivity involves using the knowledge of the behaviour and ecology of the pest species and of pesticide action in such a way as to minimize harm to non-target organisms whilst retaining or enhancing effectiveness against the target pest.

Against certain pests, it is possible to apply the insecticide as a bait containing a suitable attractant on which only the pest will feed. This has been developed particularly for control of foraging insects, notably some ants (Cherrett and Lewis 1974; Haines and Haines 1979). Poison baits against fruit flies may also do little or no harm to vital natural enemies that provide biological control of other pests. Furthermore, the incorporation of pheromones in baits provides increased opportunities for attaining greater selectivity.

Knowledge of the behaviour of a pest may assist in correct placement of a chemical. For example, the efficiency of contact action of dieldrin seed dressings may vary considerably between different dipterous stemborer pests because of differences in the behaviour of newly hatched larvae and the position of the eggs and seeds (Way 1959). Examples of improved placement increasing control efficiency include the application of small quantities of insecticidal granules to the whorl of maize plants against some maize stemborers and also the classical example of control of some species of tsetse fly where insecticidal applications can be confined to parts of trees where it is known that the adult tsetse rests. Treatment of only part of a crop may also give effective control if the pest or stage of a pest to be controlled is dispersive and therefore is liable to contact treated surfaces even when some plants are left untreated. For example, some pests of cowpeas are effectively controlled by applying insecticide to every third or fourth row of a crop, thereby permitting beneficial species to survive on the untreated rows (Perrin 1977).

Correct timing is probably the most widely applicable method for achieving ecological selectivity. Timing in relation to persistence of harmful residues is important for avoiding harm to man and domestic animals and, in many countries, there are regulations defining intervals between pesticide application and harvesting or consumption of the crop depending on degree of persistence and toxicity of the chemical. In terms of effective pest control, timing can considerably reduce the number of applications and the dosage levels required and is particularly important for non-persistent chemicals. Correct timing involves (1) defining susceptible stages in the life cycle of the pest and (2) manipulating chemical applications to minimize adverse effects on non-target organisms and, wherever possible, enhancing the effect of natural enemies.

Many important pest species complete part of their life cycle within the non-crop environment, and numbers dispersing to crops may occasionally be considerably reduced by timely application of a pesticide to pest populations in the non-crop habitats. For example, overwintering populations of *M. persicae* may be usefully controlled on peach trees before they migrate to sugar beet and potatoes. Also *Dysdercus* spp., a serious pest of cotton can, in places, be controlled in the 'off-season' by applying the insecticide to the ground where it breeds beneath certain distinctive wild hosts. Of course, the environmental consequences of pesticide application in such areas must be very carefully assessed and also development of resistance may be favoured if the pesticide is concentrated on the sole host plant at the time.

Several important pest species migrate to crops during well-defined periods and colonization is then often followed by a period of establishment before damage occurs. Thus, a correctly-timed treatment during this period may considerably reduce the need for subsequent treatments. For example, *A. fabae* migrates to bean crops during a short period in early summer, and a single correctly timed chemical treatment at the end of migration gives effective control for the whole season (Way *et al* 1977). Also light and pheromone trapping of adult moths of the summer fruit tortricid, *Adoxophyes orana,* can be used to determine the exact time of the June and August migrations to orchards (de Jong 1980). Such information together with the known developmental period of eggs at different temperatures, enables the chemical spray to be applied against newly hatched larvae before damage occurs.

Within crops, windborne insects tend to accumulate beyond windbreaks and are often commoner on the edges of crops. Timely localized application to these areas may provide effective control. Foci of high infestation in crops may also be associated with proximity to hibernating or overwintering sites. For example, adults of the bean leaf beetle, *Cerotoma trifurcata,* first appear in relatively small areas of fields planted near hibernation sites and these areas can be treated separately before the pest disperses more widely (Newsom *et al.* 1976).

Treatments may also be timed to reduce numbers later in the season where such populations determine survival and subsequent level of attack on future crops. For example, late-season pesticides applied to cotton in parts of the southern USA reduce numbers of subsequent overwintering boll weevil, *Anthonomus grandis,* so considerably delaying or avoiding the need for insecticides the following year (Newsom *et al.* 1976).

Certain developmental stages of a pest may be particularly susceptible, so timing to coincide with them may not only increase pesticide effectiveness but also enable less chemical to be applied. Differences in susceptibility of different stages may be related to size; for example; early-instar larvae require less chemical to kill them than later-instar larvae, such that high susceptibility of the former can contrast with virtual immunity of the latter. Such differences may also be related to behaviour. This is particularly evident where newly hatched larvae are exposed to contact insecticides for only a short time before boring into the plant. In such cases, control must be critically timed towards adults or newly hatched larvae.

Host–parasite–predator relationships can also be usefully exploited by correct timing. Adult parasites are especially susceptible to chemicals, so the pesticide can be applied when most parasites are within the host, where they are protected from direct chemical exposure. For example, when a non-persistent insecticide was used against the green bug, *Shizaphis graminum,* when the parasite, *Lysiphlebus testaceipes,* was within the host, many adult parasites subsequently emerged. The application was also timed late enough to avoid killing the host when the parasite was too young to complete development if its host was killed but early enough to avoid the end of the parasites' developmental period (Lingappa *et al.* 1972). Also, young larvae of the alfalfa weevil, *Hypera postica,* are controlled with a single spray in March when the parasite, *Bathyplectes curculionis,* is hibernating within its protective cocoon from which it emerges later to control low populations of weevil surviving the chemical treatment (Metcalf 1975).

At certain times, the natural enemy and pest populations may occur in different habitats and pesticide application can be timed accordingly. For example, most adults of the predatory mite, *Amblyseius fallicis,* overwinter in debris and groundcover near the base of fruit trees, where they also multiply in spring before migrating on to fruit trees in mid-June. Afterwards, *A. fallicis* returns to feed on groundcover mites and to overwinter. Thus, chemicals toxic to predatory mites can be applied to the trees when the predator is in the groundcover, provided that compounds of negligible toxicity are used during the period of predator–prey interaction (Croft and Brown 1975). Separation of biological and chemical treatments in time has also been successfully demonstrated in an integrated control programme for various pests in glasshouses (Hussey 1973).

Applications timed before or after flowering of a crop avoid killing pollinating insects. A classic case is the use of pre- or post-blossom sprays to control certain fruit tree pests. In Western Europe, control of some pests of flowering brassica crops is dangerous because it is difficult to time pesticide application to avoid killing pollinating insects, notably honey bees.

Timing is therefore crucially important, both in terms of coincidence of treatment with the most susceptible stage of the pest and also minimizing harm to beneficial species. It also creates opportunities, insufficiently developed so far, for decreasing dosage levels below those currently recommended, with obvious significance for reduced cost, decreased environmental damage, and enhanced selectivity.

In conclusion, there are many untapped opportunities for obtaining either physical or ecological selectivity which makes non-selective chemicals relatively harmless to non-target organisms. This is a realistic approach, acceptable to industry and hence more likely to be widely applicable than the use of intrinsically selective chemicals.

3.5   Effects of interactions of pesticides for pest, disease, and weed control

Pesticides are usually separated according to their intended use, for example, insecticides, fungicides, and herbicides, though a few are used to control organisms in more than one group. Many intended solely for pests in one group can incidentally affect pests or other kinds of organisms in the other groups. Such effects may influence a non-target organism directly or by affecting its competitors, antagonists, natural enemies, or its food. There are also wider implications of weed destruction, for example, associated with the use of herbicides. Pest-control strategy within a single crop will often involve the use of several groups of pesticides and therefore we must be aware of the possible consequences of, for example, the use of a fungicide on the insect pest complex or that of a herbicide on both pest and pathogen populations.

3.5.1   *Effects of fungicides on insects, nematodes, and mites*

(a) Insects. There are several examples where insect pests are affected by a fungicide. For example, benomyl and carbendazim reduce numbers of the aphids, *Sitobion avenae, Metopolophium dirhodum, A. fabae* and *Acyrthosiphon pisum* (Bailiss *et al.* 1978; Vickerman 1977). Also the hop-damson aphid, *Phorodon humuli,* may be usefully suppressed by pyrazophos when it is used to control powdery mildew, *Sphaerotheca humuli.* In contrast, there are examples of increased pest incidence following fungicide application. A well-known example is the use of sulphur dust against scab, *Venturia inaequalis,* and mildew, *Podosphaera leucotricha,* in apple orchards which has caused outbreaks of several insect pests because of adverse effects on natural enemies (Brown 1978). Tridemorph increases the reproductive rate of cereal aphids, and several foliar fungicides have been shown to reduce infection of *M. persicae* by fungal pathogens, thus increasing aphid numbers (Nanne and Radcliffe 1971; Sagenmüller 1977). Also, the bean seed fly, *Hylemya platura,* oviposits more when seeds are treated with thiram and captan compared with untreated seeds, probably because these fungicides reduce fungal growth and thus permit greater oviposition-stimulating bacterial growth (Harman *et al.* 1978).

(b) Nematodes. Several fungicides have nematicidal properties. For example, benomyl and thiabendazole can inhibit invasion of roots of tomato, tobacco,

and egg plant by larvae of *Heterodera tabacum* (Rodriguez-Kabana and Curl 1981). Furthermore, benomyl enhances the activity of some standard nematicides, such as oxamyl and fensulfothion when used against *Tylenchorhynchus dubius*. In contrast, benomyl can increase the severity of attack of *Meloidogyne incognita* on cotton; also the soil fungicide, pentachloronitrobenzene (PCNB), can either increase or decrease nematode numbers depending on the species involved and the concentration of the chemical. Thus PCNB has been used to control *Longidorus elongatus* and *Xiphinema diversicaudatum*, but may also increase numbers of other nematode species, such as *Pratylenchus* spp. It has been suggested that these differences are related to the differing mobility of nematode species, and that increases in the numbers of some pest species are due to large decreases in the number of predatory nematodes and possibly of nematode-trapping fungi (Rodriguez-Kabana and Curl 1981).

(c) Mites. Several fungicides have acaricidal properties. For example, benomyl, carbendazim, and thiophanate-methyl decrease numbers of blackcurrant gall mite, *Cecidophyopsis ribis* on blackcurrants. However, whether a particular fungicide enhances or represses a mite pest population may depend upon its relative toxicity to the pest and natural enemies. Thus, lime-sulphur and dinocap favour development of infestations of the European red spider mite, *Panonychus ulmi,* by considerably reducing the predatory phytoseiid, *Typhlodromus pyri,* and the mirid, *Blepharidopterus angulatus.* In the integrated control of phytophagous mites in orchards, the choice of pesticides, including fungicides, depends crucially upon their relative toxicities to the mite pests and their natural enemies, particularly phytoseiid mites. Hislop and Prokopy (1981) tested twelve fungicides for toxicity against two strains of the predatory phytoseiid, *Amblyseius fallacis,* and showed karathane and glyodin were moderately toxic and most likely to encourage build-up of the spider mite pest, *Tetranychus urticae.* Also, benomyl, although of relatively low toxicity, had a strong anti-reproductive effect on *A. fallacis,* so favouring spider mites.

3.5.2 *Effects of insecticides/nematicides on fungal pathogens*

Several insecticides reduce pathogen infection. For example, aldrin reduces infection by barley foot rot, *Helminthosporium sativum,* clubroot, *Plasmodiophora brassicae,* and take-all fungus, *Gaeumannomyces graminis* (Brown 1978). Phorate reduces damage caused by *Rhizoctonia solani* but when cotton seed is treated with phorate, the seedlings are predisposed to attack by *Phythium* spp. (Erwin *et al.* 1961). The fumigant nematicide, dibromochloropropane (DBCP) inhibits growth of *R. solani* and *Pythium ultimum* but can increase incidence of disease caused by *Sclerotium rolfsii* (Rodriguez-Kabana and Curl 1981).

Mycorrhizal fungi may play an important role in the growth and nutrient uptake of many plants, and therefore chemicals that affect them may alter normal host plant development and hence susceptibility to attack by pests and pathogens. The systemic insecticide/nematicide carbofuran can cause a temporary decrease in mycorrhizae in groundnut roots whereas DBCP increases endomycorrhizal infection of cotton roots. However, the significance of this in terms of pathogen attack is unknown.

3.5.3   *Effects of herbicides on plant pathogens and pests*

Herbicides can increase or suppress both the incidence and severity of pest and pathogen attack. Such effects result either from direct action of the chemical on the pest/pathogen, its natural enemies, competitors, and host plant, or indirectly through destruction of weeds (Norris 1982; Way and Cammell 1981).

(a) Direct effects on pathogens. Altman and Campbell (1977) have listed 23 different herbicides known to increase disease incidence. In general, there are four major direct effects which may increase plant disease: (1) reduced structural defences of the host; (2) stimulated exudation from host plants; (3) stimulated pathogen growth, and (4) inhibition of micro-organisms antagonistic to the pathogen. Pebulate and pyrazon used for sugar beet weed control may increase disease incidence of *R. solani,* and mecoprop also increases *G. graminis* on spring wheat. Other examples of increased disease incidence include powdery mildew, *Erysiphe graminis,* following application of triazine and urea-type herbicides, various root rots of wheat and maize seedlings with picloram and southern corn leaf blight, *Helminthosporium maidis,* following application of 2,4-D.

Altman and Campbell (1977) also listed 17 different herbicides which decreased disease due to: (1) increased host structural defences, (2) increased host biochemical defences, and (3) decreased growth of potential pathogens. Examples include trifluaralin reducing foot root by *Aphanomyces euteiches* on peas and *P. brassicae* in cabbage, and urea and triazine herbicides reducing eyespot, *Cercosporella herpotrichoides,* and *E. graminis* of wheat.

Herbicides can induce changes in the root zone of plants which could affect pathogens by affecting mycorrhizal fungi, so influencing the plant's susceptibility to disease. For example, pichloram and 2,4,5-T are both released from roots following application (Rodriguez-Kabana and Curl 1981). However, there is little information on whether such herbicides affect disease development or yield.

Herbicides may intensify virus problems, for example simazine used on sugar cane infected with sugar-cane mosaic. Conversely, 2,4,5-T can delay symptoms of watermelon mosaic in cucumber for more than 3 weeks (Heathcote 1970).

In general, there is little evidence that herbicides used in agricultural crops in England have an important effect on disease incidence in practice (Moore and Thurston, 1970) except perhaps where the herbicide is incorrectly used or the crop is poorly grown. However, the evidence of effects on disease incidence of herbicide application illustrate the need for more understanding of herbicide–plant–pathogen interactions and the dangers of underestimating their significance.

(b) Direct effects on nematodes. 2,4-D can increase susceptibility of oats to *Ditylenchus dipsaci* (Webster 1967). However, other herbicides such as dalapon may inhibit nematode reproduction. Furthermore, application of EPTC to tomatoes reduced populations of the reniform nematode attacking them, but low dosages of the herbicide increased gall density of *Meloidogyne arenaria* on groundnuts (Rodriguez-Kabana and Curl 1981).

(c) Direct effects on insects and mites. Some herbicides are toxic to beneficial natural enemies, and the toxicity of 2,4-D to Coccinellidae was responsible for outbreaks of aphids on cereals in New Brunswick (Adams and Drew 1965). Herbicides may also increase pest populations through their influence on the growth and physiology of the host plant; for example, the pea aphid, *Acyrthosiphon pisum*, on beans (Maxwell and Harwood 1960), the stemborer, *Chilo suppressalis* on rice (Ishii and Hirano 1963), and the corn leaf aphid, *Rhopalosiphum maidis*, and European cornborer, *Ostrinia nubilalis*, on maize (Oka and Pimentel 1976).

In contrast, 2,4-D is toxic to some pests, such as wheat-stem sawfly and sugar-cane borer (van Emden 1970). Furthermore, dinoseb-amine can have a greater effect on plant bugs, *Lygus* and *Neurocolpus* spp., than on their predators (Miller and Miller 1979).

(d) Effects of weed removal on pathogen and pest populations. Herbicides have been outstandingly effective in decreasing weed populations well below levels obtained by earlier weed control practices. Such drastic removal of weeds may have wide-ranging and important effects on pest and disease incidence because weeds may act as alternate and alternative hosts for pests and pathogens and as reservoirs for natural enemies. Weed removal may also alter the crop environment and thereby affect the colonization and development of pests and pathogens. Furthermore, the use of herbicides has made possible reduced and non-tillage cropping systems which have altered pest and disease status.

There are many examples where weeds act as alternative hosts for pest and disease organisms, and their removal may considerably reduce pest and disease incidence (van Emden 1965, 1970; Franklin 1970; Heathcote 1970; Moore and Thurston 1970).

Weeds may either divert the pest species or the vector of a disease away from less-attractive crop hosts. For example, Pitre and Boyd (1970) found that more maize plants are infected with corn stunt disease in weed-free plots because the leafhopper vector, *Graminella nigrofrons,* only fed on maize in the absence of preferred weed hosts. Similarly, some nematode and insect pests may concentrate on crop plants when alternative weed hosts are destroyed (Heathcote 1970; Dunning 1971).

The removal of weeds may also increase pest incidence because the weed is either a food plant for a natural enemy or a food plant for a host on which the natural enemy develops. Zandstra and Motooka (1978) give many examples of natural enemies on weed species, but as most of these occur outside cultivated areas, the impact of herbicides is probably minimal. However, removal of weeds within cultivated areas may affect natural enemy populations; for example, in apple orchards parasitism of codling moth, *C. pomonella,* and other pests may be notably reduced; also some predators are less common after weed removal (Speight and Lawton 1976).

Weeds may decrease colonization of some dispersive pests such as aphids because they can act as camouflage and lessen the contrast between the crop and the soil – thereby reducing optomotor stimuli to, and colonization by, some pest immigrants (Smith 1976). Herbicides can, therefore, increase colonization by enhancing optomotor stimuli, but, in contrast, they also create opportunities for decreasing such stimuli by sowing crops more uniformly because wide-row spacing required for mechanical weeding becomes unnecessary. Certain weeds may also have a repellent effect on the pest (Altieri *et al.* 1977).

The use of herbicides has led to modified cropping practices such as reduced or non-tillage cultivation. Minimum cultivation favours such pests as slugs but, in contrast, may be less favourable than conventional cultivations for some insect pests such as wheat bulb fly, *Delia coarctata* (Bardner *et al.* 1971; Edwards 1975). In herbicide-treated orchards, reduced leaf burial by earthworm populations may lead to increased infections by diseases such as apple storage rot, *Phytophthora syringae* (Harris 1981).

Herbicides may also influence pest and disease incidence by altering the weed species composition. An example is the use of herbicides to control weeds in Georgia cotton fields. As primary weeds declined, there was an increase in nut sedges, *Cyperus esculentus* and *Cyperus rotundus,* and as these are good hosts for *Meloidogyne incognita,* there was an increase in root-knot disease (Bird and Hogger 1973).

Herbicides may synergize other pesticides; for example, naphthylacetic acid increases toxicity of malathion to *Dysdercus* sp. (van Emden 1970). However, there are several examples of decreased selectivity when mixed with other pesticides, and the herbicide becomes more toxic to the crop; for example, propanil is harmful to rice when applied with carbofuran (Norris 1982).

In conclusion, the interaction of pesticides on non-target pests and pathogens must continue to be explored, explained, and evaluated. In such ways, we will hopefully be able to minimize unfavourable effects and develop systems for integrating favourable effects within overall control strategies. This subject has been given too little attention, particularly the effects of herbicide usage on insects and diseases.

3.6 Rational use of chemicals for control of pest complexes

Individual crops are often attacked by a complex of pests. For example, over 100 different insect species are known to damage rice. The overall complexity is increased by the species composition and the importance of particular species within a pest complex varying considerably between different regions. Furthermore, as changes occur in, for example, crop production technologies and in pest control strategies, then changes inevitably occur in the pest complex. Therefore flexible control measures are needed within a framework of an overall crop production strategy. Such a strategy underlies the Integrated Control Approach and Integrated Pest Management (IPM) programmes in which emphasis is placed upon preservation of natural mortality factors and a consideration of all available artificial pest control actions, including chemicals, to provide an effective, economical, and ecologically based programme. The implementation of such programmes often requires considerable research and development and also perhaps a change in decision-making attitudes from the rigidity of predetermined treatments to the dynamic approach of IPM, involving frequent monitoring and evaluation. However, the failure of conventional chemical methods provides the impetus for greater emphasis on IPM. Several programmes are currently being developed for some crops and the following examples illustrate practical applications.

3.6.1 Pests on cotton.

The overuse and misuse of non-selective chemicals to control pests on cotton has probably created more pesticide-associated problems than on any other crop not only in terms of cotton pest control but also by inducing resistance in other pests and causing general environmental contamination (see Chapter 18). Such problems have led to development of IPM programmes in different parts of the world, each of which involves different strategies according to local needs.

In California, for example, pest resurgences, secondary pest outbreaks and resistance, together with evidence that certain insecticides reduced yields have necessitated a change in control strategy (Flint and van den Bosch 1981). The key pest is *Lygus hesperus,* and there are also secondary outbreak pests, namely the cotton bollworm, *Heliothis zea,* the beet armyworm, *Spodoptera exigua,* and the cabbage looper, *Trichoplusia ni.*

Initially, economic thresholds for *L. hesperus* showed that the pest inflicted serious damage only during the budding (squaring) season from early June to mid-July. Therefore, a threshold based on a ratio of *Lygus* population-to-square was established for this period and thereafter control measures are not needed. Correctly timed insecticide treatments are also made on safflower to control *L. hesperus* before it migrates from the ripening safflower to cotton. This strategy protects vital biological control of Lepidoptera which is further enhanced by possibilities of more selective chemicals such as a polyhedrosis virus against *H. zea*. An economic injury threshold has been established for *H. zea*, though not yet for the other lepidopteran pests. A field-monitoring system operating from mid-May to mid-September provides information on plant development and numbers of pests and natural enemies (Flint and van den Bosch 1981). This work demonstrates how naturally occurring biological controls and planned insecticide treatments can be successfully integrated.

Chemicals have also been used selectively in other ways to control cotton pests. For example, in Texas strips of early maturing cotton are planted as trap plants for *A. grandis* and then treated with a suitable insecticide. Late-season chemical control of *A. grandis* also kills adults that will overwinter and create the next season's attack (so-called 'diapause control').

3.6.2 Pests on rice.

The introduction of high-yielding rice cultivars has provided opportunities for considerably increasing yields. However, it has also increased pest problems because such cultivars may be intrinsically more susceptible while the required cultural conditions also tend to favour pest attack. Furthermore, the greater cost of other inputs leads to greater use of chemical control in order to protect the investment. There is an urgent need to develop and implement IPM programmes especially where widely grown high-yielding varieties have created serious problems of pest resistance, resurgence and environmental contamination following intensive pesticide use as in Japan. Now, in much of the tropics, relatively low-yielding traditional varieties and production methods, which possess in-built controls, are also being replaced by high-yielding varieties and their associated production technology.

Major insect pests of rice include planthoppers, leafhoppers, stemborers, gall midge, seedling maggots, caseworm, leaf folder, armyworm, and rice seed bugs. Economic injury thresholds have been developed for several major pests such as the brown planthopper, *Nilaparvata lugens,* rice bugs, *Leptocorisa* spp., and stemborers, and various sampling techniques are available; for example in Japan a scheme operates for advising growers when to apply chemical controls for stemborers, planthoppers, and leafhoppers.

Pesticide selectivity can be achieved in many ways. There are, for example, considerable differences in the relative toxicities of insecticides to predacious spiders, *Lycosa* spp. and the green leafhopper, *Nephotettix cincticeps* (Kiritani 1972). Timing, method, and placement of insecticides can also enhance selectivity. Correct timing is often crucially important; for example, the correct time for treating rice stemborers occurs just after peak hatching of larvae and before they disperse and bore into the rice stem. Also, treatment of the adults of rice hispa, *Dicladispa armigera,* is the most effective because eggs and larvae are encased within the leaf and are inaccessible to most foliar sprays. Selectivity of pesticides on rice may also be achieved by spot treatments, treatment of trap rice crops and of alternative crop or weed hosts. The application of chemicals to the soil or irrigation water rather than the rice plants and the increasing use of granular formulations are also important means of achieving selectivity.

Overall, the wide range of techniques available for conferring pesticide selectivity should considerably improve the conservation of natural enemies, which still provide important biological controls in many tropical areas where chemicals are still rarely used but will be increasingly needed as the new technologies are introduced. Furthermore, the current emphasis on developing pest-resistant varieties of rice and of the importance of certain cultural methods of control, such as harvesting techniques and planting time, should enable widely based IPM programmes to be developed (Anon 1979). Chemicals have a crucial role to play in these programmes; misuse, however, on vast areas of rice could be catastrophic, not only in terms of rice production but also in relation, for example, to inducing resistance in several insect vectors of human diseases which breed in rice fields (Anon 1979).

3.6.3 Pests in apple and walnut orchards

The intensive use of broad-spectrum insecticides and fungicides in orchards has led to many serious secondary pest outbreaks, notably aphids, scales, and phytophagous mites. The often high monetary value and the ever increasingly stringent quality requirements of the crop pose a considerable challenge to developing IPM programmes.

(a) Apples. In Nova Scotia, oystershell scale became a serious pest of apple orchards in the 1930s because its natural enemies were killed by sulphur fungicides; however, a simple change to copper-based or ferbam fungicides and then to more selective fungicides has enabled effective biological control of the pest to be maintained since the 1940s. Phytophagous spider mites have become serious pests in many parts of the world because the use of broad-spectrum pesticides to control pests such as *C. pomonella* seriously depleted natural enemies of the mites, notably predacious Phytoseiidae. The problem was further increased by development of spider mite resistance to many acaricides. Current approaches to control of phytophagous mites include the development of selective insecticides such as diflubenzuron against Lepidop-

tera and pirimicarb against aphids. Fungicides and acaricides are also assessed in relation to their possible effects upon the pest–predator relationship. Furthermore, biological monitoring of pest mite and natural enemy populations are undertaken and decisions on need for control are based upon economic thresholds. The discovery of predatory phytoseiid mites that are highly resistant to some organophosphorus insecticides has also transformed pest management strategy in parts of the USA (Croft and Brown 1975).

To control *C. pomonella*, treatments must be applied before the first-instar larvae enter the fruit. The use of pheromone traps and the development of predictive models have improved accuracy of determination of optimum treatment times. Consequently, both rates and numbers of applications have decreased and as a consequence, this has also reduced the need for acaricides against phytophagous mites.

In conclusion, there has been much progress towards developing IPM programmes for pests of apples, particularly in relation to phytophagous mites. The need for continuously monitoring and assessing the impact of different control strategies on the whole pest complex, however, has been illustrated by evidence that the use of more selective chemicals may increase the pest status of certain insects, such as the apple blossom weevil, *Anthonomus pomorum* (Gruys 1975). This was previously controlled incidentally by broad-spectrum insecticides. Therefore, appropriate controls for such pests need to be developed and integrated into the overall pest management programme.

(b) Walnuts. Major pests of walnuts in California are the codling moth, *L. pomonella*, the navel orangeworm, *Paramyelois transitella*, the walnut husk fly, *Rhagoletis completa*, and the walnut aphid, *Chromaphis juglandicola*. Serious secondary outbreaks of various scales and phytophagous mites may also occur.

In the late 1960s, an Iranian strain of the parasite, *Trioxys pallidus*, was successfully introduced for control of *C. juglandicola*. However, this necessitated a change in control tactics because treatment using azinophos methyl to control the first generation of *L. pomonella* also suppressed the parasite at a time when the aphid is very damaging. Chemical treatment is now directed against the second brood rather than the otherwise unimportant first brood, accurate timing being achieved with pheromone traps. The later treatment does not upset aphid control because aphids are then being suppressed by high temperatures. An alternative strategy against the second generation of *C. pomonella*, when only moderate numbers are present, involves the use of phosalone, which also suppresses the walnut aphid and controls the dusky-veined aphid, *Callaphis juglandis*, soft scales and susceptible mites. *C. juglandis* has gained in importance since the parasite for walnut aphid was introduced, probably because previously it was controlled by chemicals applied against walnut aphid. Both azinphos methyl and phosalone used against second-generation *C. pomonella* greatly suppress the *T. pallidus* population, the former by residual toxicity and the latter by eliminating aphid hosts. However, the parasite is highly dispersive, and orchards of walnut varieties which do not require codling moth control act as reservoirs from which parasites annually recolonize orchards infested with *C. juglandicola* (Riedl *et al.* 1979).

Mites are controlled with selective acaricides such as propargite and cyhexatin, which favour predator survival. *P. transitella* is a scavenger associated with poorly managed orchards, and improvements in orchard sanitation and the use of a growth regulator, ethephon, which enables earlier harvesting, can prevent serious damage by this pest.

The IPM programme for walnut pests provides an elegant example of successful integration of an introduced parasite with chemical controls. The outcome has been that chemical control of the aphid is rarely needed and opportunities to control other walnut pests with biological agents are considerably improved. Furthermore, the use of more selective insecticides, such as diflubenzuron for control of *C. pomonella*, may further strengthen the biological component.

3.6.4 Crops in glasshouses

Under glass, with short-term crops such as tomatoes and cucumbers, biological components must be artificially introduced, and this has been achieved with considerable success for the control of several important pests, notably whitefly, *Trialeurodes vaporariorum*, using the parasite *Encarsia formosa*, and the glasshouse red spider mite, *Tetranychus urticae*, using the predacious *Phytoseiulus persimilis*. This important role of introduced biological control agents in glasshouses necessitates very careful manipulation of chemical controls. Intrinsically selective chemicals such as pirimicarb may be used to control aphids but selectivity is also sought by separating the chemical and biological components in time and space. Examples include dioxathion applied to the upper leaves of chrysanthemums to control chrysanthemum leaf miner without affecting important biological controls of red spider mite and green peach aphid on lower leaves and also the use of deltamethrin on sticky polythene sheeting ('Thripstick') on greenhouse floors to kill pupating thrips (Scopes and Biggerstaff 1973; Scopes, personal communication). Selective controls have, however, led to problems such as clover mite, *Bryobia rubrioculus*, on cucumbers, now controlled by applying dicofol to walls and soil surfaces without need for control on the crop (Hussey 1973). Also, occasional attacks of the tomato moth, *Lacanobia oleracea*, not a pest where unselective chemicals were used, are now selectively controlled by *B. thuringiensis* (Burges and Jarrett 1979).

3.7   General conclusions

Ecological requirements for improved control, particularly of pest complexes, almost invariably involve more sophistication than controls based solely on chemicals. Therefore they are often less acceptable to the farmer and farm advisor except when serious pest-induced problems have necessitated a changed approach as with successful integrated control programmes on some glasshouse, cotton, and orchard crops. Notable developments in research, as in the NSF/EPA IPM programme in the USA (Huffaker 1980), continue to highlight both the obvious hazards arising from continued over-reliance on chemicals, such as induced resistance and major pest resurgencies, and also the more subtle problems. Weeds and weed control by herbicides can, for example, strikingly affect incidence, population dynamics, and control of insects and also diseases. The possible long-term consequences of these ecological effects need to be examined much more intensively than at present.

3.8   References

Adams, J. B. and Drew, M. E. (1965). *Can. J. Zool.* **43**, 789.

Altieri, M. A., van Schoonhoven, A. and Doll, J. (1977). *PANS* **23**,(2), 195.

Altman, J. and Campbell, C. L. (1977). *Annu. Rev. Phytopathol.* **15**, 361.

Anon (1979). *Guidelines for integrated control of rice insect pests.* F.A.O. Plant Production and Protection Paper, 14. Rome. 115 pp.

Bailiss, K. W., Partis, G. A., Hodgson, C. J., and Stone, E. V. (1978). *Ann. Appl. Biol.* **89**, 443.

Bardner, R., Calam, D. H., Greenway, A. R., Griffiths, D. C., Jones, M. G., Lofty, J. R., Scott, G. C., and Wilding, N. (1971). *Annu. Rep. Rothamsted Exp. Stn. 1971* part 2, 165.

Bird, G. W. and Hogger, C. (1973). *Plant Dis. Reptr.* **57**, 402.

Brown, A. W. A. (1978). *Ecology of pesticides.* Wiley, New York. 525 pp.

—— Pal, R. (1971). *Insecticide resistance in arthropods* (2nd edn), World Health Organization Monograph Series, No. 38, Geneva, 491 pp.

Burges, H. D. and Jarrett, P. (1979). *Proc. Br. Crop. Prot. Conf. – Pests Dis.* **2**, 433.

Cammell, M. E. and Way, M. J. (1977). *Ann. Appl. Biol.* **85**, 333.

Cherrett, J. M. and Lewis, T. (1974). In *Biology in pest and disease control* (eds D. Price Jones and M. E. Solomon), p. 130–146. Blackwell, Oxford.

Conway, G. R. (1982). *Pesticide resistance and world food production.* Imperial College Centre for Environmental Technology, London. 143 pp.

Croft, B. A. and Brown, A. W. A. (1975). *Annu. Rev. Entomol.* **20**, 285.

Dunning, R. A. (1971). *Proc. 6th Br. Insectic. Fungic. Conf.* **1**, 1.

Edwards, C. A. (1975). *Outl. Agric.* **8**, 243.

van Emden, H. F. (1965). *Sci. Hort.* **17**, 121.

—— (1970). *Proc. 10th Br. Weed Cont. Conf.* **3**, 953.

Erwin, D. C., Reynolds, H. T., and Garber, M. J. (1961). *J. Econ. Entomol.* **54**, 855.

Flint, M. L. and van den Bosch, R. (1981). *Introduction to integrated pest management.* Plenum. New York. 240pp.

Franklin, M. T. (1970). *Proc. 10th Br. Weed Cont. Conf.* **3**, 927.

Georghiou, G. P. (1972). *Annu. Rev. Ecol. Syst.* **3**, 133.

Gruys, P. (1975). *Proc. 8th Br. Insectic. Fungic. Conf.* **3**, 823.

Haines, I. H. and Haines, J. B. (1979). *Bull. Entomol. Res.* **69**, 77.

Harman, G. E., Eckenrode, C. J., and Webb, D. R. (1978). *Ann. Appl. Biol.* **90**, 1.

Harris, D. C. (1981). In *Pests, pathogens and vegetation* (ed. J. M. Thresh), p.429. Pitman, London.

Heathcote, G. D. (1970). *Proc. 10th Br. Weed Cont. Conf.* **3**, 934.

Hislop, R. G. and Prokopy, R. J. (1981). *Prot. Ecol.* **3**, 157.

Huffaker, C. B. (1980). *New technology of pest control.* Wiley, New York. 500 pp.

Hussey, N. W. (1965). *Proc. 3rd Br. Insectic. Fungic. Conf.* **1**, 28.

—— (1973). *Proc. 7th Br. Insectic. Fungic. Conf.* **3**, 851.

Ishii, S. and Hirano, C. (1963). *Entomol. Exp. Appl.* **6**, 257.

de Jong, D. J. (1980). *EPPO* **10**(2), 213.

Kiritani, K. (1972). *Rev. Plant Prot. Res.* **5**, 76.

Lingappa, S. S., Starks, K. J., and Eikenbary, R. D. (1972). *Environ. Ent.* **1**, 520.

Maxwell, R. C. and Harwood, R. F. (1960). *Ann. Entomol. Soc. Am.* **53**, 199.

Metcalf, R. L. (1975). In *Introduction to insect pest management* (eds R. L. Metcalf and W. H. Luckmann), p. 235. Wiley, New York.

—— (1980). *Annu. Rev. Entomol.* **25**, 219.

Miller, W. O. and Miller, C. E. (1979). *Down to Earth* **35**, 14.

Moore, F. J. and Thurston, J. M. (1970). *Proc. 10th Br. Weed Cont. Conf.* **3**, 920.

Nanne, H. W. and Radcliffe, E. B. (1971). *J. Econ. Entomol.* **64**, 1569.

Newsom, L. D., Smith, R. F., and Whitcomb, W. H. (1976). In *Theory and practice of biological control* (eds C. B. Huffaker and P. S. Messenger), p. 565. Academic Press, New York.

Norris, R. F. (1982). In *Biometeorology in integrated pest management* (eds J. L. Hatfield and I. J. Thomason). Academic Press, New York.

Oka, I. N. and Pimentel, D. (1976). *Science* **193**, 239.

Perrin, R. M. (1977). *Agro-Ecosystems* **3**, 93.

Pitre, H. N. and Boyd, F. J. (1970). *J. Econ. Entomol.* **63**, 195.

Riedl, H., Barnes, M. M., and Davis, C. S. (1979). In *Pest management programmes for deciduous tree fruits and nuts* (eds D. J. Boethel and R. D. Eikenbary), p.15. Plenum, New York.

Roberts, H. A. (ed.) (1982). *Weed control handbook: principles.* British Crop Protection Council. 7th edn. Blackwell, Oxford.

Rodriguez-Kabana, R. and Curl, E. A. (1981). *Annu. Rev. Phytopathol.* **18**, 311.

Sagenmüller, A. (1977). *Z. ang. Entomol* **82**, 293.

Scopes, N. E. A. and Biggerstaff, S. M. (1973). *Proc. 7th Br. Insectic. Fungic. Conf.* **1**, 227.

Smith. J. G. (1976). *Ann. Appl. Biol.* **83**, 1.

Southwood, T. R. E. and Norton, G. A. (1973). In *Insects: studies in population management* (eds P. W. Geier, L. R. Clark, D. J. Anderson, and H. A. Nix), p.168. Ecological Society of Australia Memoirs, Canberra.

Speight, M. R. and Lawton, J. H. (1976). *Oecologia* **23**, 211.

Vickerman, G. P. (1977). *Proc. 1977. Br. Crop Prot. Conf. – Pests Dis.* **1**, 121.

Way, M. J. (1959). *Annu. Rep. Rothamsted Exp. Stn. 1958,* 214.

—— and Cammell, M. E. (1973). *Proc. 7th Br. Insectic. Fungic. Conf.* **3**, 933.

—— and Cammell, M. E. (1977). *Proc. 1977 Br. Crop Prot. Conf. – Pests Dis.* **3**, 835.

—— and Cammell, M. E. (1981). In *Pests, pathogens and vegetation* (ed. J. M. Thresh), p. 443. Pitman, London.

—— Cammell, M. E., Alford, D. V., Gould, H. J., Graham, C. W., Lane, A., Light,

W. I. St. G., Rayner, J. M., Heathcote, G. D., Fletcher, K. E., and Seal, K. (1977). *Plant Pathol.* **26,** 1.

Webster, J. M. (1967). *Plant Pathol.* **16,** 23.

Zandstra, B. H. and Motooka, P. S. (1978). *PANS* **24,** 333.

4
Application from the ground

G. A. MATTHEWS

4.1 Introduction

The aim of pesticide application is to distribute a small amount of active ingredient to the appropriate biological target with the minimum contamination of non-target organisms. Diversity of the target – insect, plant, soil, walls of dwellings, etc. – necessitates a variety of application techniques which can be summarized in five groups:

(1) Release or propulsion through the air to the target either
   (a) in the solid state as dusts or granules, or
   (b) in the liquid state as sprays
(2) Application directly to or injection into the plant
(3) Injection into the soil
(4) Release into irrigation water
(5) Release into the air with diffusion to the target (fumigation)

Drift and inhalation hazards with fine particles less than 30 μm diameter has resulted in a decline in the use of dusts except when treating small seedlings at transplanting and for seed treatment (Jeffs and Tuppen 1978), for which specialized equipment is available for seed merchants (Elsworth and Harris 1973; Middleton 1973). Seed treatment is ideal for protecting young plants with the minimum quantities of toxicant, but phytotoxicity can be a problem so the use of granules accurately placed alongside seeds at sowing has increased. Equipment is also available for spot treatment of individual plants and granules are often broadcast, sometimes by hand, but this requires a higher dosage than with other application techniques (Kiritani 1974). Specialized equipment is needed to meter granules (Amsden 1970; Bruge 1975) (see Chapter 17) so the majority of pesticides are applied as sprays. The volume of spray liquid applied will vary depending on the size of target and on whether discrete droplets or a complete film of spray is to be distributed on the target. Some orders of magnitude for various types of application are given in Table 4.1.

Less than 0.1% of the applied dose may reach insect pests in a field crop treated with a foliar spray whereas up to 30% of an applied herbicide penetrated experimental plants sprayed in a glasshouse (Graham-Bryce 1976). This low efficiency of sprays has been largely due to the wide range of droplet sizes emitted by traditional spraying equipment. As discussed in Chapter 2 movement of droplets in the environment is so influenced by

TABLE 4.1. *Orders of magnitude of volume of spray required for various types of application*

Description of spray	Volume of spray (litres/ha)	
	Field crops	Trees and bushes
High volume	> 600	> 1000
Medium volume	200–600	500–1000
Low volume	50–200	200–500
Very low volume	5–50	50–200
Ultra-low volume (ULV)	< 5	< 50

droplet size that the trend is towards using a narrower range of droplet size with the mean size appropriate for a particular biological target. Optimum droplet sizes are indicated in Table 4.2.

The recent development of controlled droplet application, CDA, involves not only the appropriate droplet size but also selection of the number of droplets needed per unit area of target. Control of a sessile insect or application of a contact herbicide will require a greater droplet density, for example one droplet/mm^2, than is required with a mobile insect or the use of a translocated herbicide.

The minimum volume for a ULV spray can be determined from the droplet size and density, so that if even-sized droplets of 60 μm diameter were applied to provide one droplet/mm^2, then 1.13 litres would be required for each hectare treated (Fig. 4.1). The target area of foliage will vary relative to ground area according to plant growth and the nature of the pest attack. Some pests may be confined to one part of the foliage; thus control of *Heliothis* may be confined to the upper part of cotton plants, where more eggs are laid. Even when minimal volumes of spray are applied, less than 10% active ingredient is needed in the spray droplet provided that the more effective insecticides are used, namely those with an LD_{50} less than 10 ng/insect (Fig. 4.2), and the toxic dose is not carried in a single droplet but in the average number reaching the target (Graham Bryce 1976).

Increasing the concentration or volume does not improve the efficiency of

TABLE 4.2. *Optimum droplet sizes*

Target	Droplet sizes (μm)
Flying insects	10–50
Insects on foliage	30–50
Foliage	40–100
Soil (and avoidance of drift)	250–500

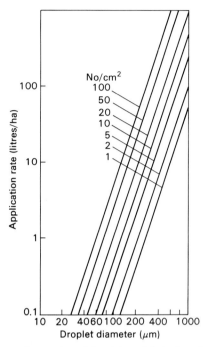

Fig. 4.1. Relation between the number of droplets and their diameter and the volume of spray applied.

sprays unless the spray is more uniformly distributed to the target (Fig. 4.3) (Courshee 1967). Recent studies with monosized droplets applied to leaf discs with *Tetranychus urticae* (Koch.) eggs showed that 1% dicofol was most effective if applied as 20-μm droplets, the LD_{50} being $5 \, ng/cm^2$ (Munthali 1981; Scopes 1981).

4.2   Production of spray droplets

The different types of nozzle are based on the energy used to break up the spray liquid into droplets and disperse them over a short distance.

4.2.1   Hydraulic-energy nozzles

A wide range of nozzles are manufactured to provide different application rates, spray angles, and patterns. In each, liquid is forced under pressure through a small opening or orifice and spreads out into a thin sheet which disintegrates irregularly into different-sized droplets (Fraser 1958) (Fig. 4.4). The volume of the largest droplets may be more than one million times that of the smallest. A typical distribution of a hydraulic nozzle (Fig. 4.5) shows a high proportion of small droplets prone to evaporation and drift, whereas

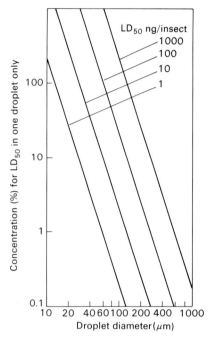

Fig. 4.2. Concentration of spray to apply an LD_{50} dosage in one droplet required for different droplet sizes and a range of dosages.

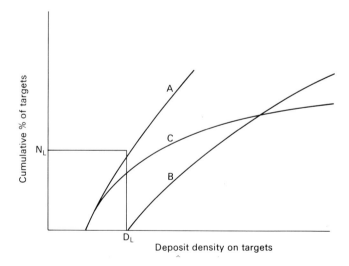

Fig. 4.3. Hypothetical deposit distribution curves on foliage. A typical distribution of doses on targets is shown by curve A. If the deposit on each target were doubled by doubling the application rate, curve B would be obtained, but in practice the heavy deposits are increased while many leaves continue to receive an inadequate deposit—curve C. The minimum deposit needed may be that indicated at point D (from Courshee 1967).

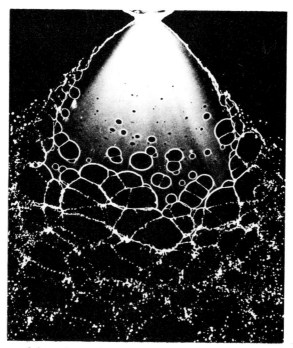

Fig. 4.4. Break up of sheet of spray from a hydraulic nozzle.

most of the volume is in a few large droplets. The average size of the droplets is halved by a fourfold increase in pressure (Ripper 1955), but this increases the proportion of aerosol droplets liable to drift. Similarly flow rate is proportional to the square root of the pressure, so a nozzle with a different orifice should be used to change the flow rate, and the pressure regulated only for a final adjustment. Careful filtration of the spray liquid is needed to reduce the risk of blocked nozzles. Ideally each nozzle should have a filter with a mesh smaller than the orifice.

Fan and cone spray patterns are achieved using elliptical and circular orifices respectively (Fig. 4.6 and 4.7) but with the latter, one, two, or four tangential slots are positioned behind the orifice to rotate the liquid so that the sheet of liquid forms a hollow cone. The slots are often in a separate swirl plate and if this has an additional central hole, a solid cone pattern can be achieved. Combining different swirl plates or cores and orifices provides a range of outputs and spray angle. The spray angle can also be reduced by increasing the distance between the swirl plate and the orifices, a feature of variable cone nozzles. Cone nozzles are ideal for spraying foliage as droplets approach leaves from more angles than from the flat fan nozzles. Cone

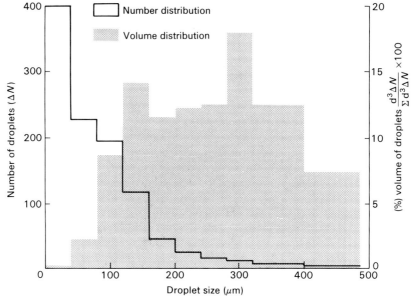

Fig. 4.5. Example of droplet distribution from a fan nozzle.

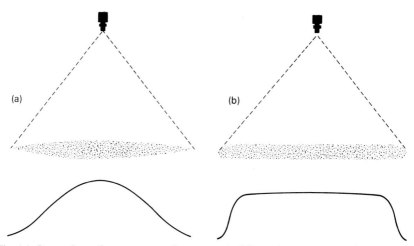

Fig. 4.6. Comparison of spray patterns from a standard fan and even spray nozzle.

nozzles are often used in orchard spraying, but for most field crops it is easier to achieve a more even deposit across a swath by mounting a series of fan nozzles on a boom so that adjacent patterns overlap. Special even-spray fan nozzles (Fig. 4.6b) are available for band applications. A fan-shaped pattern is also achieved with an impact nozzle in which liquid impinges on a flat surface close to the circular orifice. Impact nozzles are normally used at low pressures to apply herbicides in large droplets.

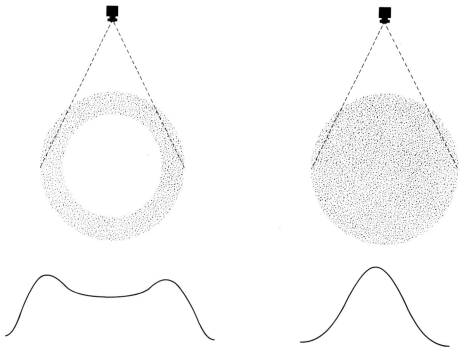

Fig. 4.7. Spray patterns from a hollow and full cone nozzle.

Recent developments of both fan and cone nozzles have attempted to reduce the proportion of droplets less than 100 μm diameter to reduce drift when applying herbicides (Bouse *et al.* 1976). On cone nozzles, a second swirl chamber is positioned immediately after the orifice (Brandenburg 1974; Ware *et al.* 1975).

When air is drawn into and mixed with the spray liquid before it reaches the orifice, foam is produced if a suitable foaming agent is added to the spray. Foam nozzles are now principally used for swath marking.

4.2.2 Gaseous-energy nozzles

An airstream is used to impact on and shatter a liquid containing pesticide into droplets. Droplet size depends on the air/liquid ratio, so larger droplets are produced by increasing liquid flow or reducing air velocity. Uniform air velocity at the liquid interface is required to reduce the variation in droplet size. Aerosol droplets can be produced very efficiently if liquid is fed into a vortex with large volumes of air at low pressure (less than 30 kPa) (Fig. 4.8).

4.2.3 Centrifugal-energy nozzles

Liquid fed near the centre of a rotating surface is spread centrifugally to the edge from which it is thrown as single droplets (Fig. 4.9). As the flow rate increases, ligaments are produced (Fig. 4.10) and then if overfed the

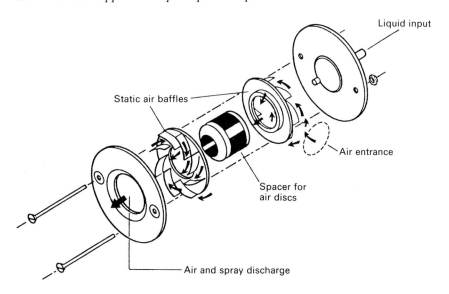

Fig. 4.8. Vortical nozzle used on aerosol generator.

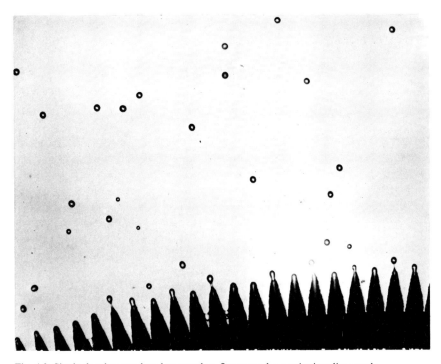

Fig. 4.9. Single droplets produced at very low flow rates by a spinning disc nozzle.

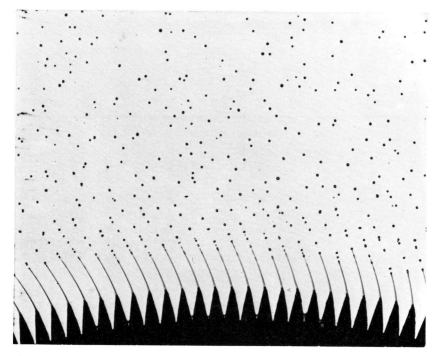

Fig. 4.10. Spray formed ligaments by a spinning disc.

ligaments join to form a sheet of liquid which breaks up in a similar manner to that obtained with a hydraulic nozzle. At a particular rotational speed, there is an optimum flow rate for single droplet and ligament formation to avoid the transitional phases when a wider droplet spectrum is produced. When ligaments are produced there is a bi-modal pattern due to the presence of very small satellite droplets, and for a given rotational speed, the attenuation of the ligament results in a smaller main droplet than when droplets are formed directly at the edge of the disc. Droplet size is inversely proportional to the angular velocity of the disc (Walton and Prewett 1949; Frost 1981), so droplet size can be selected by adjustment of rotational speed and flow rate. The performance of spinning disc nozzles has been improved by the addition of teeth, referred to as zero-issuing surfaces, to reduce the surface to which the liquid can cling by surface tension. More recently grooves leading to teeth have improved lane separation of the ligaments. The spray pattern from a disc is similar to a hollow cone nozzle, but by shrouding part of the spray, a more even distribution can be achieved (Taylor *et al.* 1976).

4.2.4   Thermal-energy nozzles

A stream of hot gas shatters the liquid into droplets but as the temperature is usually over 500°C, the droplets vaporize immediately. As the vapour meets

Table 4.3. *Details of spinning discs*

Disc diameter (mm)	Number of teeth	Disc speed (rev/min)	Flow rate (ml/min)	Droplet size (μm)
55	360	15 000	15	40
		12 500	30	50
80	360	7 000	30	70
		2 000	60	250
120	180	2 000	1 000	250
		4 000	500	125
		6 000	50	75

cool air aerosol droplets are re-formed by condensation. The visibility is reduced by the cloud of droplets many of which are less than 15 µm diameter. Some larger droplets are present, especially if the flow rate of liquid is too high to allow complete vaporization.

4.2.5   Electrostatic nozzles

Electrical energy can be used to produce electrostatically charged droplets and manipulate their trajectories (Thong and Weinberg 1971). This system has now been developed as a low energy electrodynamic (ED) nozzle (Coffee 1979), which can apply ULV formulations at less than 2.5 litres/ha. The charged aerosol/mist droplets are rapidly collected on the upper and lower surfaces of leaves with less drift downwind. Equipment using this nozzle is now being developed. The hand-held unit incorporates the nozzle in the pesticide container, and is referred to as a 'bozzle' (Coffee 1981). Electrostatic charging of droplets produced by other nozzles continues to be studied. Arnold and Pye (1980) and Carleton and Bouse (1980) have used different charging systems on spinning disc nozzles. Stent *et al.* (1981) use a pulsed Microjet nozzle (Frost and Yates 1981) while Law (1980) used a twin-fluid nozzle and Marchant and Green (1982) have used a fan type hydraulic nozzle.

4.3   Spraying machines

Sprayers using hydraulic nozzles have been the most widely used due to the interchangeability of the nozzle tips, but other types are of increasing importance with the application of ultra-low volumes. With all types, the trend is away from metal to plastic construction, especially for the spray tanks.

4.3.1   Manually operated hydraulic sprayers

The lever-operated knapsack sprayer normally has a 15-litre tank, a piston or diaphragm pump connected to a hand-operated lever, a pressure chamber, and a lance with an on/off tap or trigger valve and one or more nozzles. The pump has to be continually operated and the pressure chamber must have sufficient capacity to even out pressure fluctuations with each pump stroke. Precise direction of the nozzles is sometimes difficult and the operator usually walks into the area treated, so various adaptations have been developed to improve the distribution of spray behind the operator by means of a vertical boom attached to the tank (Tunstall *et al.* 1961, 1965), and to improve the speed of spraying with a horizontal boom (Cadou 1959; Johnstone *et al.* 1975). Compression sprayers are also widely used (Fig. 4.11) as full attention can be given to directing the lance while spraying. A cylindrical container, two-thirds full of spray, is pressurized using a small piston pump usually

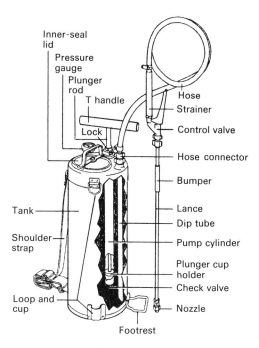

Fig. 4.11. Compression sprayer as used for treating houses to control mosquitoes in anti-malarial campaigns.

screwed into the top as part of the lid of the tank. The pressure decreases rapidly while spraying (Fig. 4.12) so the operator must stop occasionally to repressurize the tank or ideally fit a pressure-regulating valve to the tank outlet or lance. The more durable sprayers of this type have a large lid separate from the pump unit.

Although these sprayers are ideal for treating small areas, many farmers in

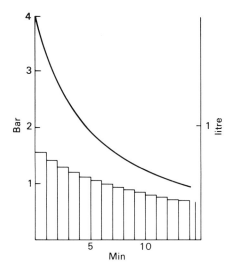

Fig. 4.12. Decrease in pressure and output as compression sprayer tank is emptied.

the tropics have been reluctant to use them because of the problems of collecting water and manually pumping in a hot climate.

4.3.2   *Power-operated hydraulic sprayer*

The most commonly used sprayer has a 200–750-litre tank mounted on the standard three-point linkage of a tractor. A pump is driven from the tractor power take-off and the nozzles are mounted on a horizontal boom with a span of up to 12 metres. Large sprayers normally have a trailed tank, or saddle tanks are mounted on the tractor. Many farmers use larger tanks and wider booms to increase the area treated when weather conditions are favourable (Adams 1978) but the proportion of time actually spraying is improved more by reducing both the time to refill the spray tank and the volume application rate. Ideally the nozzles incorporate an anti-drip device as used on aircraft.

Choice of pump depends on the total volume of liquid emitted by the nozzles and recirculated to provide agitation, the pressure required at the nozzle, and the type of formulation used. The output of piston pumps is proportional to the speed of pumping and is virtually independent of pressure. Piston pumps (Fig. 4.13) are ideal if high pressure up to 4000 kPa are required, but they are relatively expensive. Diaphragm pumps are more suitable for pumping wettable powder formulations particularly at pressures up to 1000 kPa. One or more flexible diaphragms are arranged radially around a rotating cam instead of close-fitting pistons, but care must be taken to avoid using chemicals which affect the diaphragms or valves. A pressure

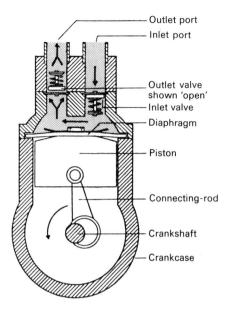

Fig. 4.13. Diaphragm type pump (section to show valves).

chamber is needed with both these pumps, but not with roller vane or centrifugal pumps.

A roller vane pump (Fig. 4.14) has a rotor with five to eight equally spaced slots in which a roller moves in and out radially to provide a seal against the wall of an eccentric case by centrifugal force. Atmospheric pressure in the spray tank forces liquid into the space between the rollers as they pass the inlet port. As the space contracts at the outlet port, the liquid is forced out. These pumps are relatively inexpensive to operate with outputs of 20–140 litres/min with pressures up to 2000 kPa. Pump life is reduced if run dry or at high pressures for prolonged periods. These pumps have largely replaced

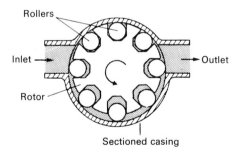

Fig. 4.14. Cross section of roller-vane pump.

gear pumps which are less suitable for applying wettable powder formulations. Centrifugal pumps (Fig. 4.15) have an impeller with curved vanes rotating at high speed inside a disc-shaped casing. Liquid is drawn in at the centre and thrown centrifugally into a channel around the edge. The volume of liquid pumped decreases rapidly when the pressure exceeds 300 kPa so centrifugal pumps are most suitable for large volumes at low pressures. A filter or line strainer should be fitted to the input side to protect the pump.

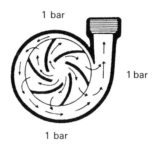

Fig. 4.15. Centrifugal pump.

Uniform application depends on both constant tractor speed and constant pressure. The latter can be achieved by fitting a pressure-regulating valve, the surplus spray liquid being returned to the bottom of the tank to provide agitation. Forward speed can vary especially with the larger trailer sprayer so systems of metering liquid to the nozzles in relation to the speed of a trailed wheel have been devised (Amsden 1970). Improved tractor cab design has stimulated development of electronic control systems (Allan 1980) with solenoid valves, which with enclosed filling systems (Akesson *et al.* 1977) reduce the risk of contamination of the tractor-driver sprayer operator.

Spray distribution may be affected by boom movements including vertical bounce and horizontal whip or both, so a stiff cantilever boom should be mounted to dampen the effect of passage of the tractor over uneven ground (Nation 1978, 1980). Boom height should allow sufficient overlap of the spray pattern of adjacent nozzles, but excessive height should be avoided because of the increased risk of spray drift. Some farmers prefer wider-angle nozzles, e.g. 110°, to reduce boom-height effect but the proportion of aerosol droplets prone to drift is increased. Correct swath matching across a field is essential to avoid under- or over-dosing. On row crops such as cotton or potatoes a field can be easily marked out to ensure that the tractor moves in the same direction between particular rows at each application to reduce the number of wheelings and mechanical damage to the crop. On closely spaced cereal crops pathways or 'tramlines' can be left by blocking off the appropriate seed coulters and adjusting their position relative to the tractor wheels. Alternatively a herbicide such as paraquat can be applied in bands to burn off the gaps for the wheels, but great care is needed to avoid spreading the herbicide elsewhere.

Other swath matching systems include the use of a foam blob, which is sometimes viewed with a periscope so that the driver can align his direction with the line of foam at the side of the boom. Automatic battery-powered wheeled markers have also been tried where the headlands are cleared of all obstructions.

Sprayers with specialized booms and nozzles are used for spraying trees and small bushes. The trend with apple orchards is towards intensive plantings of small trees which may be sprayed by vertical booms on each side of the tree and protected from the wind by a mobile tunnel shield. Spray which is not deposited on the tree is collected on the shield and recirculated so contamination of the soil under the trees is minimal (Cooke *et al.* 1977).

Boom sprayers in some countries have been fitted to an animal-drawn unit and may be used when soil conditions are too wet for passage of a tractor. Instead of a fixed boom, nozzles or lances can be fixed to a flexible hose or 'portable line' carried by operators through a field with a motorized pump unit mounted on a tractor moving along the edge of the field or stationary at one end.

4.3.3 Hand-carried, battery-operated spinning-disc sprayers

A small dc motor drives a rotating disc which is gravity-fed through a flow restrictor from a 0.5–5.0-litre bottle. The spray head is fixed to a long handle in which a number of 'D' size 1.5 V batteries are carried. Larger batteries may be carried separately in a shoulder harness. In the field, the spray head is held downwind of the operator who walks across the field to treat a series of swaths. As the operator moves progressively upwind, he walks through untreated foliage (Fig. 4.16a). The disc is held 0.5–1.0 m above the crop with the bottle vertical to provide a constant feed. This allows spread of the spray droplets before they are carried by natural air movements into the crop canopy. The disc should always be rotating when the bottle is inverted so that liquid is spun off the disc without contaminating the drive shaft and motor.

Droplet dispersal in a crop is dependent on natural air movement so a minimum wind speed of 0.75 m/s is needed. Narrow swaths (0.9–2 m) are better to overlap the deposits, compensate for variations in wind speed and direction while spraying and produce correct dosage (Fig. 4.16b). Alternatively sprays over wide swaths can be repeated so that sequential sprays under different wind conditions not only give better coverage of crops but also compensate for weathering of deposits.

The disc should not be overloaded to avoid deterioration of the droplet spectrum, and optimize the life of the battery. Flow rates are normally less

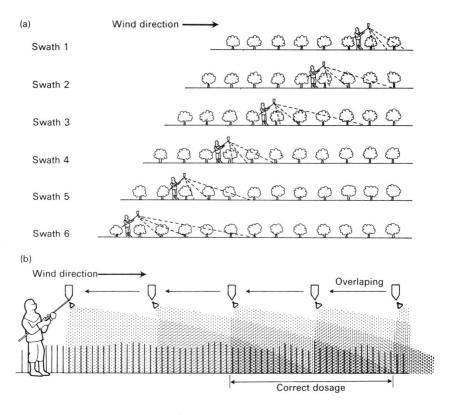

Fig. 4.16. (a) Overlapping of swaths as the operator moves upwind across a field. Note variation in swath width with changes in wind velocity. (b) Diagram showing area of correct dosage.

than 1 ml/s, the restrictor being selected according to the viscosity of the spray liquid. Disc speed will decrease as power from the batteries is used but can be regulated electrically or, for slow speed discs (2000 rev./min), by a mechanical governor. The latter is used principally for herbicide application to ensure large droplets (250 µm) are produced without drift. At high disc speeds (> 6000 rev./min) with small droplets (< 100 µm) less-volatile formulations should be used.

The technique has been widely adopted in the tropics where water is not readily available. Insecticides have been applied mainly to cotton at 2.5 litres/ha (Matthews 1971) but up to 15 litres/ha have been applied with wettable powder formulations (Mowlam *et al.* 1975). Insecticide application with this equipment on other crops has been more limited, but cowpea (Raheja 1976), wheat (Pickin 1978), rice, groundnuts, maize, and sorghum crops have been treated. Fungicides have been applied to groundnuts (Mercer 1976) and tomatoes (Quinn *et al.* 1975). Farmers who were reluctant to spend time collecting and transporting water to fields for herbicide application are adopting the controlled-droplet technique to reduce the drudgery involved in weed control. Walking at 1 m/s covering a 1 m swath, one hectare can be sprayed at 10–20 litres/ha in less than 3 hours.

4.3.4 Tractor-mounted CDA sprayers

Sprayers with a horizontal boom and vertical mast have been developed. Spray heads on the horizontal boom each have shrouded spinning discs (Bals 1976) mounted on a vertical shaft (Taylor *et al.* 1976) to apply herbicides at 15–45 litres/ha using 250 μm droplets. Results comparable with high-volume sprays have been obtained with translocated herbicides, but the droplet density achieved with contact herbicides may be the cause of variable weed control achieved in some trials. Further development indicates that the multi-disc unit can be replaced with a large spinning cup (Bals 1978; Heijne 1978) which can be operated at higher speeds, up to 6000 rev./min, to apply small droplets (70–150 μm) more suitable for insecticide and fungicide application.

The mast sprayer has a series of spinning discs on a horizontal shaft driven by an electric motor designed to operate at 9000 rev./min. The mast can be raised or lowered so that the height at which 50–100 μm droplets are released above the crop can be adjusted according to wind speed (Bals 1977; Haigh 1978). Comparatively little information is available on the performance of this sprayer (Lake *et al.* 1978) but as with the hand-carried sprayers, droplet dispersal is dependent on the wind, so good coverage is achieved by overlapping swaths sprayed with a cross-wind. Farmers have achieved acceptable results with lower dosages of certain of the less toxic insecticides and fungicides than used at high volume, possibly because the pesticides can be applied more rapidly at the onset of an infestation, and using less-volatile formulations, a higher proportion of the emitted spray is collected by the crop canopy.

4.3.5 Air-carrier sprayers

The distance droplets can be projected from a hydraulic nozzle is limited so an airstream is sometimes used to carry droplets over longer distances to the target. Air-carrier sprayers are particularly useful for projecting droplets up into trees for which droplets less than 100 μm diameter are required to minimize fall out contaminating the ground. The airstream from a powered fan may be used to produce the droplets but hydraulic- or centrifugal-energy nozzles are often mounted in the airstream. The latter allow the use of larger volumes of air at lower velocity to displace a higher proportion of the air within a crop canopy with air containing droplets, especially if the forward speed of the tractor is kept as slow as practical (Randall 1971). If droplets must be projected vertically 6 m or more a high-velocity airstream (> 60 m/ s), is needed but the airstream velocity decreases rapidly on leaving the nozzle so coverage is often difficult to achieve in the centre and top of large canopies of tall trees. Ideally the velocity should be sufficient to impact small droplets on the target with turbulence to improve distribution of the droplets, especially in dense canopies. Ultra-low volumes (< 50 litres/ha) applying lower dosages of pesticide than used at high volume have given very promising results on tree crops (Morgan 1979).

Knapsack mistblowers have a 35 cc or 60–70 cc two-stroke engine with recoil starter and centrifugal fan fitted by anti-vibration mountings to an L-shaped frame with a padded backrest. The spray tank, usually 10-litre capacity, may be slightly pressurized (20 kPa) or a pump on the engine shaft may be used to force liquid to the nozzle. Great care must be taken to use the correct petrol/oil mixture to reduce starting problems and maintain smooth running of the engine at maximum speed. The engine speed should be checked regularly with a tachometer, and the fuel tank and carburettor emptied during storage. Calibration of the flow rate should be checked especially if heat from the engine affects the viscosity of the formulation.

Hand-carried CDA air-carrier sprayers have a spinning disc mounted directly in front of a fan. Electric ac and dc motors are used on these sprayers or a two-stroke engine to drive the fan.

Tractor-mounted air-carrier sprayers usually have an axial or centrifugal fan driven from the tractor power take-off, and an array of nozzles in the air outlet. The air outlet may be modified in relation to the shape of the target and oscillation of the airstream gives improved penetration of citrus foliage (Johnstone 1970).

4.3.6   Exhaust nozzle sprayer

Exhaust gases from a vehicle engine are used to shatter the spray liquid into 70–90 µm v.m.d. droplets which are projected upwards and then drift downwind. A safety valve prevents excessive back pressure (50 kPa) which might damage the engine. Pressure of the exhaust gases is also used to force the spray liquid to the nozzle. This sprayer has been mostly used for control of locust hoppers (Sayer 1959; Watts *et al.* 1976).

4.3.7   Aerosol generators

Air from a power-driven compressor is fed to one or two sets of fixed vanes to produce a vortex into which the spray liquid is fed. Units range in size from small hand-carried units with electric motors to large vehicle mounted machines with 10–15 kW four-stroke engines. Spray liquid is delivered to the nozzle from a pressurized tank or by a displacement pump with electronic controls in the cab to ensure the application rate is related to vehicle speed. The larger sprayers are mostly used for mosquito control in urban areas where effectiveness is dependent on meteorological conditions (see Chapter 12). The smaller units are for general space spraying in buildings, warehouses, and glasshouses and may be converted to use propane gas fuel. Units have also been used experimentally to apply microbial insecticides to tree crops (Falcon and Sorensen 1976). In many circumstances these sprayers are now used instead of the thermal foggers described below because ultra-low volumes of pesticide can be applied at less cost.

4.3.8   Fogging equipment

The hot gases for a thermal-energy nozzle are usually obtained from the exhaust of a combustion chamber. Some machines based on the pulse jet require a spark from a battery-powered vibrator or a mechanically operated magneto connected to the plug initially to start the engine. Petrol is forced to the combustion chamber by pressurizing the fuel tank with a small air pump which is also used to pressurize a larger tank usually containing an oil-based solution of pesticide. When the machine has warmed up a valve is opened to allow the pesticide to flow through a restrictor to the end of the exhaust pipe where the temperature can exceed 500°C. Larger foggers have a petrol engine to operate an air-blower and two pumps; one pumps fuel to the combustion chamber while the other pumps the pesticide solution to the nozzle. The hot gases (500–600°C) from the combustion chamber pass through a flame trap to the nozzle. Convection of the hot gas from the nozzle is useful in lifting droplets up through a building or crop canopy but a light wind can disperse a fog very rapidly.

4.4   Calibration of sprayers

The correct dosage of pesticide depends on accurate calibration of equipment preferably with the formulation being applied but when the pesticide is diluted with water, sprayers can be calibrated with plain water. The level of liquid in the tank is marked, and the volume required to refill the tank to that mark is measured when the sprayer is returned to exactly the same position after spraying over a known distance. The distance travelled is usually selected by dividing the swath treated (m) into 1000, when the volume measured (1) multiplied by 10 is in litres per hectare. The sprayer must travel at a constant speed with a constant pressure over the tenth of a hectare. The tractor speed can be checked by measuring the distance travelled in 36 s which divided by 10 gives the speed in km/h. The output of each nozzle should be checked separately especially as uneven distribution could be caused by erosion of the nozzle tips by inert particles suspended in the spray. Nozzles should be changed if output is not within 5% of that required, as excessive adjustment of pressure to regulate output also affects the spray pattern and droplet size.

Knowing the output of the nozzles (litres/min) swath width (m) and forward speed (m/min) the application rate can be calculated by:

$$\frac{\text{output}}{\text{swath} \times \text{speed}} = \text{litres/m}^2,$$

which multiplied by $10\,000 = \text{litres/ha}$

4.5   Special applications

4.5.1   Injection techniques

Fungicides have been injected into trees, especially to control Dutch Elm disease. A compression sprayer is often used to force liquid through jets fixed into holes bored into the tree trunk. The technique is slow and laborious and is appropriate only for high-value trees. Protection of cut surfaces to prevent invasion of pathogens can be achieved by pruning with specially adapted secateurs with which a measured dose is applied to the blades with each cut. These secateurs have been used to apply *Trichoderma viride* to control silver leaf disease on fruit trees (Jones *et al.* 1974). Subsurface application of volatile herbicides is usually achieved by spraying on the surface followed immediately by two cultivations. Where nematicides have to be injected below the soil surface, the nozzle is fixed behind a cultivator tine or chisel plough. The operator has a flow meter to each nozzle to check that blockages have not occurred and a metering jet in each line is above ground for easy cleaning.

4.5.2   Soil fumigation

Diffusion of a fumigant into the soil requires application of the chemical under a plastic sheet sealed at the edges with soil and held in place usually for 48 hours. The centre of the sheet is usually supported over a tray into which the fumigant such as methyl bromide is discharged from a special canister. Normally the can is punctured by a special applicator connected to the tray by a length of plastic tubing. If the applicator is not available the can is placed under the sheet with a nail in a piece of wood, positioned so that when hit by a hammer, the nail punctures the can in the tray without damaging the sheet. Each can is usually sufficient to treat $10\,\text{m}^2$. Soil fumigation is used for nematode and weed control (see Chapters 19 and 17).

4.5.3   'Wipers and recirculating sprayers'

Individual weeds, especially wild oat, can be selectively treated by handling with a specially designed glove. About 1 ml of herbicide is pumped to a foam pad by a pressure bulb depressed each time a weed is gripped by the glove. The area of the weed covered by the foam pad is treated. A dye is usually added to show which plants have been treated (Holroyd 1972). Larger areas of weeds which are taller than the crop can be treated by contact with a rope wick applicator carried above the crop (Dale 1978). The wick is supplied with a translocated herbicide such as glyphosate by capilliary movement from a reservoir. Small hand-held units are also available. Alternatively jets of spray are directed horizontally across the top of a crop into a large spray-trap so that 70–90% of the emitted spray is not intercepted by the weeds and is recirculated (Lutman, 1980).

4.5.4   *Dispensers into water (see Chapters 16 and 19)*

Special dispensers can be used to add a pesticide to flowing water in an irrigation canal, or to a river, especially to control aquatic weeds and mollusc vectors of schistosomiasis. Ideally flow rate is adjustable so that dosage is proportional to water flow, but simple dispensers use a gravity feed with constant head device. Pesticides are sometimes added to irrigation water to treat crops but coverage with sprinkler equipment is not always satisfactory. In some countries with fixed irrigation equipment, the ground around citrus trees is sprayed using an impact nozzle on each side of the tree. Herbicides and systemic insecticides have been applied in this way, but great care must be taken to avoid contamination of the source of water by using non-return valves in the system.

4.6   Future trends

Increases in the cost of pesticides and greater concern about environmental pollution are necessitating improved efficiency of pesticide application. Already there is a trend away from use of large volumes of liquid to ultra-low-volume application with greater control of droplet size. This enables pesticides to be applied more rapidly when needed in relation to pest attack and when weather conditions are favourable. Improvements in nozzle design to provide a narrow droplet spectrum have been made, but research in progress indicates that electrostatic charging of droplets results in less downwind movement of the spray cloud and greater deposition on the crop (Morton 1982; Arnold and Pye 1980; Law 1980). In particular droplets less than 100 µm diameter will be more effectively used with less risk of downwind drift. Advances in nozzle design need to be combined with faster transport (Cussans and Ayres 1978) over the target and more accurate metering of smaller volumes. Light-weight specialized high-clearance sprayers will no doubt replace the tractor sprayer as used today. Such equipment could possibly be remote-controlled to reduce the need for the operator being close to the spray.

Closed systems of mixing pesticides will be necessary to reduce hazards to the operator, but the need for mixing will be reduced or eliminated by more specialized formulations, especially for electrostatic spraying. Much reduced volumes will enable the pesticide to be distributed in containers which are easily attached directly to the spray system and, after use, returned for refilling by the chemical company.

4.7 References

Adams, R. J. (1978). *Proc. 1978 Br. Crop. Prot. Counc. Conf. – Weeds* **2,** 625.
Akesson, N. B., Yates, H. E. and Boos, S. W. (1977). In *Pesticide management and insect resistance* (eds D. L. Watson and A. W. Brown) p.607, Academic Press, London.
Allan, J. R. McB. (1980). *BCPC Monogr.* **24,** 201.
Amsden, R. C. (1970). *Br. Crop. Prot. Counc. Monogr.* **2,** 124.
Arnold, A. J. and Pye, B. J. (1980). *BCPC Monogr.* **24,** 109.
Bals, E. J. (1976). Controlled droplet application of pesticides (CDA). Paper presented at symposium *Droplets in Air* organized by SCI.
—— (1977). *Proc. 1977 Br. Crop Prot. Counc. Conf. – Pests Dis.* 523.
—— (1978). *Proc. 1978 Br. Crop. Prot. Counc. Conf. – Weeds* **2,** 659.
Bouse, L. F., Carlton, J. B., and Merkle, M. G. (1976). *Weed Sci.* **24,** 361.
Brandenburg, B. C. (1974). *Am. Soc. Agric. Engng.* Paper No. 74-1595.
Bruge, G. (1975). *Phytoma* **27** (265), 9.
Cadou, I. (1959). *Cot. Fib. Trop.* **14,** 47.
Carleton, J. and Bouse, L. F. (1980). *Trans ASAE* **23,** 1369.
Coffee, (1979). Personal communication.
—— (1981). *Outl. Agric.* **10,** 350.
Cooke, B. K., Herrington, P. J., Jones, K. G., and Morgan, N. G. (1977). *Proc. 1977 Br. Crop. Prot. Counc. Conf.* **1,** 323.
Courshee, R. J. (1967). In *Fungicides: an advanced treatise* (ed. D. C. Torgeson), Vol. 1, p. 239. Academic Press, New York.
Cussans, G. W. and Ayres, P. (1978). *Proc. 1978 Br. Crop. Prot. Counc. Conf. – Weeds* **2,** 633.
Dale, J. E. (1978). *Proc. South. Weed Sci. Soc.* **31,** 322.
Elsworth, J. E. and Harris, D. A. (1973). *Proc. 7th Br. Insectic. Fungic. Conf.* 349.
Falcon, L. A. and Sorensen, A. A. (1976). *PANS* **22,** 322.
Fraser, R. P. (1958). In *Advanced pest control research* (ed. R. L. Metcalfe), Vol. II, p.1. Interscience, New York.
Frost, A. R. (1981). *J. Agric. Engng. Res.* **26,** 63.
Frost, A. R. and Yates, W. E. (1981). *J. Agric. Engng. Res.* **26,** 357.
Graham-Bryce, I. J. (1976). *Proc. 8th Br. Insectic. Fungic. Conf.* **3,** 901.
Haigh, J. (1978). The Lockinge ULVAMAST. Unpublished paper at BCPC CDA Symposium Reading April 1978.
Heijne, C. G. (1978). *Proc. 1978 Br. Crop Prot. Counc. Conf. – Weeds* **2,** 631.
Holroyd, J. (1972). *Proc. N. C. Weed Cont. Conf.* **27,** 74.
Jeffs, K. A. and Tuppen, R. J. (1978). In *Seed treatment* (ed. K. A. Jeffs), CIPAC Monograph No. 2.
Johnstone, D. R. (1970). *PANS* **16,** 146.
—— (1973). *Pestic. Sci.* **4,** 77.
—— Huntington, K. A., and King, W. J. (1975). *J. Agric. Engng Res.* **20,** 379.

Jones, K. G., Morgan, N. G., and Cook, A. T. G. (1974). *Long Ashton Annu. Rep. 1974*, 107.
Kiritani, K. (1974). The effect of insecticides on natural enemies. Particular emphasis on the use of selective and low rates of insecticides. Paper submitted to the International Rice Research Conference, IRRI, Philippines, April 1974.
Lake, J. R., Frost, A. R., and Lockwood, A. (1978). *Proc. 1978 Br. Crop Prot. Counc. Conf. – Weeds* **2**, 681.
Law, S. E. (1980). *BCPC Monogr.* **24**, 85.
Lutman, P. J. W. (1980). *Br. Crop Prot. Counc. Monogr.* **24**, 291.
Marchant, J. A. and Green, R. (1982). *J. Agric. Engng. Res.* **27**, 309.
Matthews, G. A. (1971). *Cotton handbook of Malawi.* Amendment 2/71. Agricultural Research Council of Malawi.
Mercer, P. C. (1976). *PANS* **22**, 57.
Middleton, M. R. (1973). *Proc. 7th Br. Insectic. Fungic. Conf.* 357.
Morgan, N. G. (1981). *Outlook Agric.* **10**, 342.
Morton, N. (1982). *Crop Prot.* **1**, 27.
Mowlam, M. D., Nyirenda, G. K. C., and Tunstall, J. P. (1975). *Cott. Gr. Rev.* **52**, 360.
Munthali, D. C. (1981). *Biological efficiency of small pesticide droplets.* PhD Thesis, University of London.
Nation, (1978). *Proc. 1978 Br. Crop Prot. Counc. Conf. – Weeds* **2**, 649.
Nation, H. J. (1980). *BCPC Monog.* **24**, 145.
Pickin, S. R. (1978). *Br. Crop Prot. Counc. Monogr.* No. 22, 237.
Quinn, J. G., Johnstone, D. R., and Huntington, K. A. (1975). *PANS* **21**, 388.
Raheja, A. K. (1976). *PANS* **22**, 327.
Randall, J. M. (1971). *J. Agric. Engng. Res.* **16**, 1.
Ripper, W. E. (1955). *Ann. Appl. Biol.* **2**, 288.
Sayer, H. J. (1959). *Bull. Entomol. Res.* **50**, 371.
Scopes, N. E. A. (1981). *Proc. 1981 Brit. Crop Prot. Conf. – Pests Dis.* **3**, 875.
Stent, C. J., Taylor, W. A., and Shaw, G. B. (1981). *Trop. Pest Manag.* **27**, 262.
Taylor, W. A., Merritt, C. R., and Drinkwater, J. A. (1976). *Weed Res.* **16**, 203.
Thong, K. C. and Weinberg, F. J. (1971). *Proc. R. Soc. Lond. Ser. A.* **324**, 201.
Tunstall, J. P., Matthews, G. A., and Rhodes, A. A. K. (1961). *Cott. Gr. Rev.* **38**, 22.
Tunstall, J. P., Matthews, G. A., and Rhodes, A. A. K. (1965). *Cott. Gr. Rev.* **42**, 131.
Walton, W. H. and Prewett, W. C. (1949). *Proc. Phys. Soc.* **B62**, 341.
Ware, G. W., Cahill, W. P., and Estesen, B. J. (1975). *J. Econ. Entomol.* **68**, 329.
Watts, W. S., Thornhill, E. W., Davies, A. L., and Matthews, G. A. (1976). *COPR Misc. Rep.* 28.

5
Application from the air

R. J. V. JOYCE

5.1 History and development of aerial application

The advantages of a pesticide application system which permits entry into an area without damage to the vegetation and irrespective of the condition of the ground, seized the imagination of many engaged in crop production and protection as soon as airborne vehicles looked like becoming a practical proposition. The efforts of the German forester, Alfred Zimmerman, who was awarded a patent for applying pesticides to forests by aircraft, and of John Chaytor in New Zealand, who used a hot air balloon for seeding, are described by Maan (1965) and Akesson and Yates (1974). Viable commercial systems of aerial application did not, however, develop until after World War I, the first documented operation being carried out by Dr Houser and others of Ohio Agricultural Experimental Station in 1921 using a Curtis JN6 (Jenny) to apply dusts to catalpa trees (1922).

The first aerial spraying appears to have been in the USSR in 1922 against a serious locust invasion (Azar'Yan 1966), and the success of these trials led to the development of aerial applications against mosquitoes, forest insects, and cotton pests.

A leading country in the development of the use of aircraft in agriculture was New Zealand where the aerial application of fertilizers was pioneered, first by Government Agencies and later by commercial operators (Gibson 1958).

Since World War II the use of aircraft in agriculture has expanded dramatically because of two over-riding factors – namely, the availability of surplus military aircraft and pilots skilled in low-level flying, and secondly, the need, dictated by post-war policies, for the increased food production which could be achieved by the new synthetic pesticides that became available in increasing quantities and types. Indeed the successes of the new pesticides set off a self-generating momentum, in which human mortality from diseases carried by insect vectors was reduced by insecticides and the consequent population increases dictated the need for more food which could be supplied by improved crop-production systems in which crop protection by pesticides was an essential input, and in which aerial application played an important role.

Regular, reliable statistics on the use of aircraft in agriculture are not available. The last comprehensive figures, published by FAO (Akesson and

Yates 1974), listed nearly 19 000 aircraft in 62 countries treating annually nearly 200 million hectares of crop. Of these aircraft, USSR with about 42% and USA with about 32% accounted for the biggest fleets, only 26% working in the remaining 60 countries.

5.2 The applicability of aerial application

Airborne speed is the aircraft's unique characteristic which their use in pesticide application seeks to exploit. Aircraft as vehicles for transporting and distributing materials for crop production or protection are most competitive when the areas to be treated are large, and when operations on the ground are either not possible or will cause damage. The latter criterion is a decisive one, for example, in the case of forests and many crops approaching maturation, but the former is a variable which must be quantified. In the final analysis, this quantification must be based on the economics of crop production – that is financial return for unit of investment. Compared with alternative vehicles, the capital cost and operating costs of aircraft are high. Aircraft can be made competitive only if these costs are as widely spread as possible over the revenue-earning term, namely the area treated. Thus, work output per hour must be maximized in all aerial application operations. The variables of which such productivity is a function have been expressed in several equations (e.g. Amsden 1962; Akesson and Yates 1974; Van Bemmel 1953).

The variables to be considered in calculating productivity and costs may be classified as in Table 5.1. Those variables pertaining to the aircraft are fixed once the operator has selected his aircraft, so that, then, productivity is related to the mission and the application method selected. These variables may be linked as follows (see IAAC 1973):

$$t \text{ (s ha}^{-1}) = 10^4 \times \left(\frac{T_r Q}{Q_f} + \frac{1}{Vb} + \frac{T_w}{bL} + \frac{2aQ}{VQ_f} + \frac{C}{VF} \right)$$

where
T_r = loading and taxing time (s)
Q = application rate (litres or kg m^{-2})
Q_f = aircraft load (litres or kg)
V = flying speed (m s^{-1})
b = swath or lane separation (m)
T_w = turning time at end of swath (s)
L = length of field or spray run (m)
F = average field size (m^2)
C = average distance between fields (m)
a = average distance from airstrip to fields (m)

TABLE 5.1. *Classification of variables affecting productivity and cost of an aerial application mission*

Variables pertaining to the aircraft
  Maximum take off weight: special Ag. rating
  Normal landing weight
  Empty weight
  Maximum payload
  Useful tank capacity, normal tanks
  Fuel consumption at selected power
  Maximum endurance, normal tanks
  Maximum hopper/spray tank capacity
  Landing distance, take-off distance
  Ferry cruise speed
  Swath speed
  Wing span
  Operating costs per hour

Variables pertaining to mission
  Ferry distances to target fields from unloading points
  Length of run
  Area of individual fields

Variables pertaining to application method
  Application rate
  Swath or lane separation
  Loading time

Study of this equation shows that the most important factors influencing productivity are field size, application rate, and lane separation, so that aircraft are most economically operated when these factors are optimized. Aerial operators must be always seeking those operations and techniques in which application rates can be minimized and swaths and run-lengths maximized to the greatest extent compatible with biological objectives.

At the same time, the unique characteristics of aircraft carry penalties derived from the hazards of airborne speed and the discharge of materials at high speed. Aerial operations are accordingly strictly regulated in most countries by constraints imposed, on the one hand, by civil aviation authorities concerned with safe flying, and, on the other hand, by agricultural and environmental authorities concerned with the safe use of the materials discharged. Regulations concerning the working conditions and work load of pilots, mandatory regulations concerning the operation of aircraft, aircraft design, and manufacturer's instructions concerning airframe and engine maintenance and operational life of components, combine to transform an operation potentially dangerous to pilots into one in which risks are low, provided that regulations are strictly followed. Moreover, the risk to pilots of chemical contamination from the loads carried and discharged has proved in practice to be small. Data on accidents amongst light aircraft in the USA (Table 5.2) provide an example of the typical risk to which pilots engaged in agricultural work are exposed. A mean annual accident rate of 22 per 100 000 hours flown, of which less than 10% were fatal (and in which probably over 80% resulted in no physical injury to the pilot), represents less than 1 fatality per 2 000 000 ha treated. It is on such data that insurance companies base their premiums and which provide, in this respect, edge for aircraft over competitive systems, for example, the use of tractors, in which the accident rate is much higher.

On the contrary the most serious hazard facing the aircraft operator and that which is of greater public concern is derived from the discharge into the air of materials which may be highly toxic. The operator is usually most concerned with ensuring that the biggest possible fraction of the material discharged arrives in the target area. The fraction which does not is commonly called 'drift', an ill-defined term which embraces not only the fraction remaining airborne but also that which is collected by surfaces outside the target area. It is this latter fraction that the operator is under great constraint to minimize, particularly if the crop adjacent to the target field is one which may be damaged by the pesticide applied. The regulations contained in the Aerial Application Permissions of most countries reflect this concern over the 'drift hazard'.

The problem of 'spray drift', and conversely, the means of maximizing chemical recovery within the target area, dominate the aerial application of pesticides, and, to a large extent, determine what type of crops, what type of agriculture and what type of conditions are suitable for aerial application. It is this problem which has concentrated attention on such fundamental questions as those concerning biological objectives, routes of entry and action of toxicants, the physics of liquid breakup and dispersal, the role of the crop boundary layer and aircraft-induced turbulence, and which has generated the analytical approach to the chemical control of pests and diseases, which is now influencing pesticide application from ground-based sources. The great advantages of good, and the serious hazards of bad, aerial application, provide aircraft operators with an economic incentive to surmounting problems which are common to all methods of chemical control but which are less pressing to those engaged in application from ground-based sources – that is, how to achieve the efficient transfer of toxic doses from a container (which is airborne even if mounted on a ground-based vehicle), with the minimum loss en route, to a population of biological targets, selected because their contamination will achieve a biological goal.

The approach to the problem of pesticide transfer has traditionally been empirical, and this approach is the one commonly adopted in aerial application. It must therefore be described. The hazards, however, of the misuse of pesticides, particularly acute in aerial application, as well as the

TABLE 5.2. *Accident rates among light aircraft in the USA*

| Type of flying | Total accidents per 1 000 000 aircraft hours flown | | | | |
	1971	1972	1973	1974	Total
Instructional	15.18 (1.17)	13.69 (1.13)	10.62 (0.89)	11.71 (1.11)	51.20 (4.30)
Pleasure	30.83 (4.78)	26.45 (4.71)	26.69 (4.90)	23.82 (4.40)	107.79 (18.79)
Business	8.46 (1.51)	8.23 (1.84)	5.11 (1.21)	4.95 (1.02)	26.75 (5.58)
Corporate/executive	2.64 (0.28)	2.85 (0.41)	2.76 (0.71)	2.15 (0.41)	10.40 (1.81)
Aerial application	25.85 (2.62)	21.20 (2.14)	20.54 (2.38)	23.11 (1.77)	90.70 (8.91)
Air taxi	6.65 (1.44)	5.75 (1.64)	5.32 (1.37)	5.25 (1.10)	22.97 (5.55)
Total	89.61 (11.80)	78.17 (11.87)	71.04 (11.46)	70.99 (9.81)	309.81 (44.94)

Fatal accidents are shown in parentheses.

economics of aircraft operations, have dictated the pursuit of a more analytical approach to the problem of transfer from an airborne source. This analytical approach, first pursued in relation to the special problems of locust, tsetse, and mosquito control, has generated application techniques which are far more efficient than those conventionally practised from both ground and airborne sources, and from which are emerging more target-specific methods of pesticide application.

5.3 The empirical approach to aerial application

Nearly all the pesticides consumed in the World (some hundreds of thousand tons annually) are bought by farmers who wish to use them as a medicine to cure their crop of a sickness or to prevent a sickness establishing itself. They therefore want the pesticide not only to be applied exclusively to the field at risk, but also to be distributed evenly over the field, or, if necessary, over the crop, so that no pockets of sickness remain untreated to reinfect the crop. Pesticide application methods have developed to meet the simple requirement of providing maximum recovery within the target area and a minimum variability of deposit density. The problem is considered as one for agricultural engineers working where treatments designed to provide an even distribution can be accurately controlled and results are reproducible.

The system which satisfies these requirements with regard to liquids applied from aircraft is that of the boom and nozzle, the most widely practised method of applying pesticides to arable crops (Akesson and Yates 1974) where most are at present used, thus employing the same technology as ground-based machines. Its efficiency is judged by the same criteria, namely recovery within the swath treated and the coefficient of variation of deposit across the swath. Methods for measuring recovery in this way are set out clearly and comprehensively by US Department of Agriculture (e.g. Barry *et al.* 1977).

These criteria for efficient application are also used to assess the efficiency of distribution of pesticides applied as solids by means of spreaders and dusters. The use of granules avoids 'spray drift' and provides ways of achieving high recovery and even distribution. Most granular materials are made by absorbing liquid pesticides into the particles of the granules, the sizes of which are selected by sieving (Table 6 of Akesson and Yates 1974). Even the smallest, with a diameter of 250–500 μm, have a terminal fall velocity of 1–2 m s^{-1} so that loss from a swath is very low. On the contrary, in dusts (Table 5 of Akesson and Yates 1974) over 90% of the particles may have a diameter of less than 30 μm and consequently there is a high risk of aerial transport from the treated area. Thus granules lend themselves to accurate placement by aircraft whilst dusts do not.

In practice, dusts have proved inefficient, costly, and environmentally harmful. Granular formulations, however attractive, are suitable for only those pesticides which can be absorbed by the roots and translocated through the plant's system to the zones where pests are feeding. They are, moreover, usually effective only against sucking pests. Consequently, most pesticides are applied in liquid form and application must be based on a method of breaking up the liquid so that small volumes may be spread evenly over the large surfaces occupied by a crop or represented by crop surfaces (often 3–4 ha ha^{-1}). This process of liquid breakup is known as atomization, and the application of liquids in this way is spraying (see Chapter 2).

In the development of pesticide spraying from ground-based sources there were no powerful economic constraints on the volume of liquid to be used to give a good and even cover. It was assumed that the greater the volume, the better the cover, and spraying to 'run-off' was always desirable, even though this might involve application rates in excess of 1000 litres ha^{-1}. This extravagance was impractical for aerial spraying, so the question had to be asked – what is the minimum cover acceptable in commercial practice? It was found empirically to vary from between about 20 to 100 droplets cm^{-2} and also that large droplets fell beneath the aircraft track whilst small ones were caught up in the wing-tip vortices and lost to the swath. The requirements of high recovery and even cover at an acceptable density could be determined by placing an array of collecting surfaces, such as glass plates or cards, or even rolls of toilet paper, perpendicular to the track flown by an aircraft, to record the distribution of droplets across the swath. The distortion in the pattern of droplet deposit which was a consequence of their displacement by the propeller air was corrected by careful placement of nozzles along the boom (IAAC 1973; Parkin and Wyatt 1982). Avoidance of droplet entrainment in the wing-tip vortices was achieved by shortening the boom to about 80% of the wing-span.

This empirical approach to the problem of accurate placement of aerially applied spray continues to dominate the aerial application of pesticides, and has generated the specialized agricultural aircraft and system which now characterize the aerial application field (Fig. 5.1).

The greatest problem is 'spray drift'. In commercial practice 'drift' losses may vary from a few per cent to over half the chemical applied, and this can represent a hazard not only to neighbouring crops but also to the general public not involved in the operation. More subtle damage may occur to the general agricultural ecosystem by the effect of these wasted chemicals on non-target organisms, such as insect parasitoids and birds (Brown 1978).

Solutions to the drift problem are sought in the elimination of 'drift-prone' droplets from the spectrum (that is, droplets of less than about 150 μm diameter) by manipulation of pesticide formulations, and the design of new atomizers. The following methods have been accepted in commercial practice:

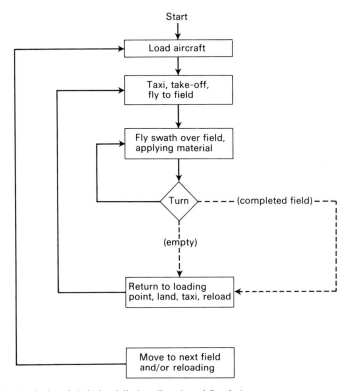

Fig. 5.1. An Agricultural Aviation Mission (Razak and Snyder).

(1) Invert emulsions, with oil as the continuous and water the discontinuous phase. These have been widely used with phenoxy herbicides for the control of bush species, for example, along power lines, ditches, and roadsides, and provide sprays which have a volume median diameter of 2000–5000 μm. In order to achieve the required droplet density, application rates of up to 200 litres ha^{-1} have to be employed.

(2) Viscosity additives act similarly to invert emulsions and produce a droplet spectrum with a large v.m.d. and correspondingly require high application rates.

(3) Special nozzles, such as the microfoil, which produce uniform droplets of 800–1000 μm diameter and solid jet nozzles with a v.m.d. of 600–900 μm. Again, minimum volume rates of application are about 100 litres ha^{-1}.

(4) Other spray additives, such as foam producers, which permit large droplets to spread over plant surfaces and thus reduce the number needed to provide the appropriate cover.

5.4   An analytical approach to aerial application

5.4.1   Definition of the target surface

Since pesticides are used to protect crops from insects and pathogens and to destroy weeds, it is generally assumed that the target for the pesticide is the entire crop surface (or ground surface in the case of pre-emergence herbicides), over which the chemical must be evenly spread, the appliance being expressed as kg or litres per ha. Economy and efficiency, however, demand a more rigorous definition of the target. Knowledge of the ecology and behaviour of the pest species will focus attention on a target stage, i.e. a particular stage in the life history when a reduction in numbers will regulate the density of the pest in the crop. The target surface, which is the surface to be contaminated with pesticide, will be chosen because from it the pesticide may be transferred with minimum loss to its site of action, e.g. the nervous system of an insect or the meristematic tissue of a weed. A systemic herbicide which is translocated in the apoplastic tissue of the crop species, and is effective only when it reaches the growing points, will be most efficiently employed when the target surface is the stem. If it falls on the canopy it will remain in the leaves (Sargent 1976). Systemic aphicides are similarly constrained. An insecticide acting by residual contact or stomach action will be effective only when applied to those surfaces used by the pest species. The choice of target surface may be affected by the easy redistribution of some pesticides by secondary processes, for example, copper fungicide by dew and rain (Courshee 1967). The contribution of redistribution in the vapour phase of insecticides, fungicides, and herbicides may also be important (Brooks 1976; Wain and Smith 1976). Pesticide granules rely on secondary redistribution from soil to roots, and appliance must take account of the very great loss en route from chemical, physical, and biological causes.

Precise definition of the target surface leads to a precise definition of target doses, which need to be described in terms of quantity (e.g. p.p.m.) and/or distribution (e.g. droplets or $\mu g\ cm^{-2}$). For example, the first-instar larva of *Heliothis armigera* is killed when it feeds on cotton foliage contaminated with 2 p.p.m. of monochrotophos, the half-life of which is 24 hours in the Sudan Gezira. Thus a target dose of 64 p.p.m. on the growing points where most of the eggs are laid will kill first-instar larvae over a period of 6 days, but four applications of the same total quantity of active ingredient each 4 days will provide protection for 16 days (Joyce 1974).

Precise definitions of the target stage, the target surface, and the target dose make possible greater economy and efficiency in the use of pesticides. Once these definitions are made, and only then, can consideration be given to the physical problem of transfer, that is, spraying efficiency.

5.4.2   *Droplet capture*

Having defined the target at which the pesticide must be aimed, it is necessary to determine the droplet size which makes the biggest contribution to the target dose under the conditions which exist in the field. This is a function of the droplet collection efficiency of the target, the number of droplets available, the time during which they are in the vicinity of the target, and the wind velocities in the vicinity of the target.

A droplet moving relative to the air which surrounds it experiences a force resisting its motion – the aerodynamic drag. This drag derives from the inertial forces arising from the air moving round the particles and the viscous nature of the air.

If a droplet falls freely under gravity through still air, it will accelerate until its drag equals the gravitational force. Its rate of fall then is constant and is known as the terminal (or sedimentation) velocity (V_s). The relationship of V_s to the forces acting on the droplet is known as Stokes Law which predicts that:

$$V_s = \frac{\rho_d\, g\, d^2}{18\mu}$$

where ρ_d is density of the droplet, g is acceleration due to gravity, d is diameter (or typical length) of the droplet and μ is viscosity of the air.

This equation predicts accurately the terminal velocity of droplets at low Reynolds numbers, but experimental data (Davies 1966) in Fig. 5.2 show that deviations occur when the droplet diameter exceeds about 50 µm. V_s increases with d, so that a droplet of 10 µm diameter has a terminal velocity of about 3 cm s^{-1} and a 260 µm droplet a terminal velocity of 1 m s^{-1}. A consequence of this is that a droplet, suddenly exposed to a fluid moving relative to it, experiences a drag which is big relative to its own inertia if the droplet is small, and vice versa. Thus a 10 µm diameter droplet will take only 0.003 of the distance of a 260 µm diameter droplet to reach its new velocity although both started with the same velocities. This is expressed by the equation:

$$S = \frac{V_s\, V_0}{g}$$

where S is stop distance and V_0 is initial velocity. The shorter the stop distance the greater is the ability of a particle to maintain the velocity of the air in which it travels, so that it will be less inclined to impact on an obstacle placed in the path of the air (Fig. 5.3). Thus the impaction of droplets on an

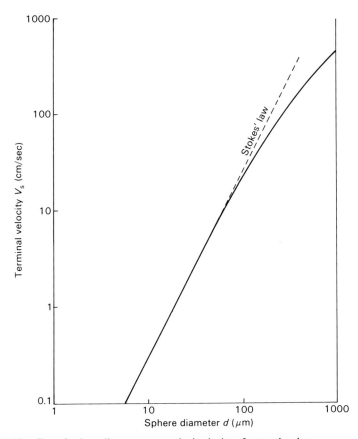

Fig. 5.2. The effect of sphere diameter on terminal velocity of water droplets.

obstacle placed in their path is a function of the aerodynamic shape of the obstacle, the diameter of the particle and the velocity of the droplet relative to that of the obstacle. This may be expressed as Sell's relationship (Latta *et al.* 1947).

$$E = \frac{d^2 V}{S}$$

where E is collection efficiency, d is drop diameter, V is relative velocity, and S is size of obstruction, e.g. diameter. It is shown graphically in Fig 5.4 (May and Clifford 1967) and 5.5 (Akesson and Yates 1974).

5.4.3 *The number and sizes of droplets*

The probability of a droplet being 'caught' by a target is a function of not only the catch efficiency of that target but also the life-time 't' of a droplet in the space occupied by the target. 't' is a function of V_s, the terminal velocity

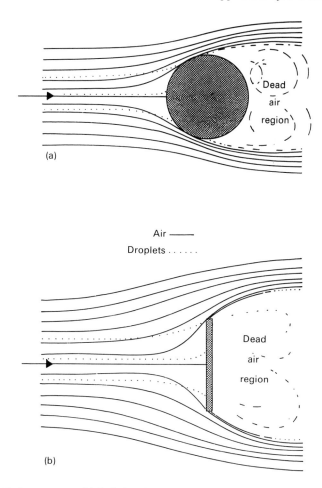

Fig. 5.3. Air flow patterns. (a) Cylinder, (b) plate (Spillman 1978).

of the droplet, and diffusive effects of turbulence. For zero turbulence conditions 't' is inversely proportional to V_s. The effects of turbulence are complex, on the one hand reducing 't' by diffusion and, on the other, repeatedly returning the particle to the airspace occupied by a target, if the scale of turbulence is small in relation to the volume occupied by the population of targets.

Fig. 5.6, taken from Spillman (1976), relates the probability of droplet catch E/Vs, to the impaction parameter of various parts of the body of a flying insect. The impaction efficiencies of other surfaces could equally well have been considered; for instance, pine needles might fall in the range of 5000–6000 V_1/d see Fig. 5.6 and V_s be substituted for V_1 (velocity of insect relative to the air). Fig. 5.7 shows the variation of E/V_s expressed in terms of a 50 µm diameter droplet for various-sized, non-volatile droplets, as a

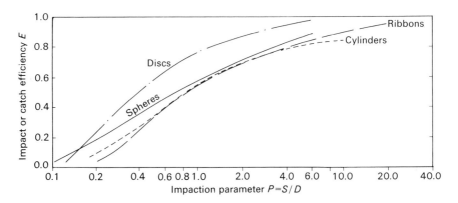

Fig. 5.4. Variation of catch efficiency with impaction parameter for discs, spheres, cylinders, and ribbons (May and Clifford 1974).

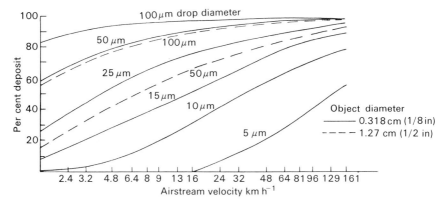

Fig. 5.5. Theoretical deposit versus airstream (liquid drop or particle velocity) for several drop sizes and two object sizes (Akesson and Yates 1974).

function of V_1/d. It will be seen that the relative catch probability increases greatly with increase in the impaction parameter for small droplets, but decreases to zero with low impaction parameter which would be characteristic of small droplets on big target surfaces.

Such calculations assume, however, that each size range of droplet is equally available and this is never so in aerial spraying operations. What is emitted from a nozzle or an atomizer is a droplet spectrum. It is usually assumed that the frequency distribution of droplet sizes given by the atomizer in general use is log normal and described by the volume median diameter (v.m.d.) and the variability, the geometric standard deviation (σg),

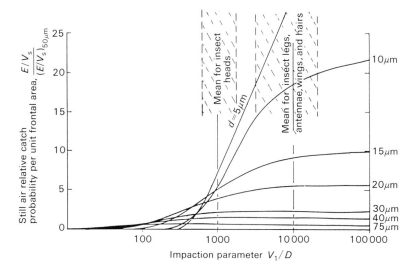

Fig. 5.6. The probability of droplet catch E/V_s to impaction parameter by various parts of an insect body (Spillman 1976).

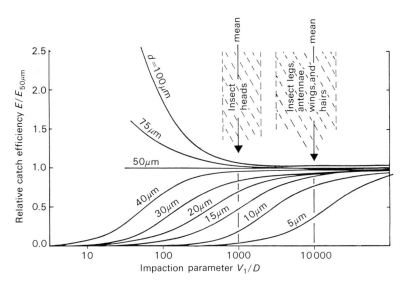

Fig. 5.7. Relative catch efficiency of various parts of an insect body relative to that of a 50 μm droplet (Spillman 1976).

about this median. This can be misleading and the term v.m.d. is often employed to describe the size of the droplet in a spectrum, whilst, in fact, droplets in the range of the v.m.d. rarely represent as much as 10% of the total number in the spectrum. When the distribution is log normal, the other parameters such as the number median diameter (n.m.d.), the volume average diameter (v.a.d.) and the mode (d.m.), or the most frequent droplet size can be derived from the v.m.d. and its geometric standard deviation (see Table 5.3).

The relations are true only if the frequency distribution is in fact log normal. There is now accumulating evidence that this is not always so, and assumption of log-normality, though providing a useful approximation, can be dangerous when considering problems of accurate transmission.

TABLE 5.3. *Log normal Z table. Relationship of various log-normal parameters (with acknowledgements to S. Parkin).*

$\dfrac{x}{y}$	n.m.d.	n.a.d.	v.a.d.	s.m.d.	v.m.d
n.m.d.	0	$\frac{1}{2}$	$1\frac{1}{2}$	$2\frac{1}{2}$	3
n.a.d.	$-\frac{1}{2}$	0	1	2	$2\frac{1}{2}$
v.a.d.	$-1\frac{1}{2}$	-1	0	1	$1\frac{1}{2}$
s.m.d.	$-2\frac{1}{2}$	-2	-1	0	$\frac{1}{2}$
v.m.d.	-3	$-2\frac{1}{2}$	$-1\frac{1}{2}$	$-\frac{1}{2}$	0

Where: $x/y = e^{z\ln^2\sigma_g}$
or: $\log x/y = z \times 2.3026 \log^2 \sigma_g$
or: $\ln x/y = z\ln^2\sigma_g$
where σ_g = geometric standard deviation.

Consider the curves of droplet frequency distribution determined by Sayer (1969) (Fig. 5.8). Each one departs from the linearity which a log normal distribution should express on the log probability scale. Departures from linearity could, until recently, be attributed to sampling errors, the large and small droplets being under-represented, the former because of their infrequency, and the latter because their occurrence in a sample reflects the collecting efficiency of the surface employed and the meteorological conditions at the time of release

By freezing droplets emitted into a wind tunnel using liquid nitrogen, Dean and Preston (1973) were able to collect true samples of droplets produced by Micronair spinning cages in air velocities from 20 to 50 m s^{-1}. Parkin (1978) used this cryogenic isokinetic probe to calibrate an Optical Array Cloud Droplet Probe (Knollenberg 1970).

The use of these devices established that previous sampling methods had

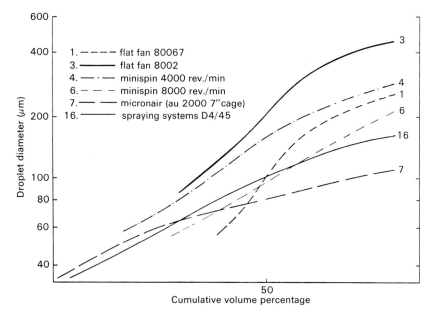

Fig. 5.8. Droplet spectra produced by several atomizers (Sayer 1969).

greatly underestimated the number of small droplets produced by atomizers and indicated that the mode of a droplet spectrum is a characteristic of the atomizer and is independent of the v.m.d.: i.e. when the distribution is log normal, σg varies with the v.m.d. Increasing the v.m.d. certainly results in a bigger proportion of the volume of a spray being contained in large droplets, but in no way alters the fact that the majority of droplets may continue to be of 'drift-prone' size (Fig. 5.9).

The droplet spectrum is clearly a complex concept and in selecting it for use in aerial spraying we must be clear about our goal, namely the transfer of

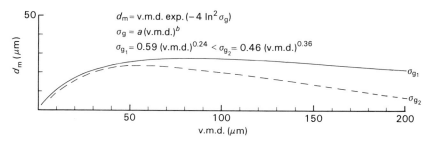

Fig. 5.9. Micronair AU 3000. Relationship of mode of droplet spectrum to change in volume median diameter (v.m.d.) (Uk 1978).

toxic doses from an airborne tank to a population of targets. We must therefore know what droplet sizes make the biggest contribution and provide the required density of cover. This is a function of the targets particle collection efficiency, under field conditions, where they are shielded by other surfaces and surrounded by air, the structure of which is determined by the crop boundary layer of the atmosphere. The preferred droplet sizes must therefore be measured experimentally in the field.

5.4.4 *Preferred droplet sizes*

Owing to the difficulty in devising techniques for visualizing the sizes of droplets collected by natural targets, most workers have been content to try to calculate the efficiency of spray application by using artificial surfaces, such as cards, spheres, tubes, ribbons, and papers, placed on the ground or on the foliage. The significance of the data thus collected cannot be evaluated until their relationship to the deposit on the natural target has been established.

Hadaway and Barlow (1965) investigated the droplet-collection efficiency of dead tsetse flies and found that oil droplets smaller than 60 μm diameter deposited more readily on flies put to rest on obstacles simulating the shape of branches and leaves than on the obstacles themselves, and, under some conditions, deposition occurred on all flies irrespective of position although the obstacle remained free of deposit.

Himel and his co-workers (Himel and Moore 1969; Himel 1969a) introduced the fluorescent particle (FP) method whereby a spray of known concentration of evenly suspended fluorescent zinc cadmium sulphide leaves a permanent record of impacted droplets on plant and insect surfaces. The sizes of the droplets can be estimated by counting the FPs and applying statistical evaluation (Uk, 1977) (Fig. 5.10).

Using this FP technique and a soluble fluorescent mixture which permanently 'dyes' the leaf surface with droplet spots, Uk (1977) found that cotton leaves collected more droplets smaller than 80 μm, and fewer droplets bigger than this, than horizontal cards placed at the same level (Fig. 5.11). Cotton leaves were thus more efficient in collecting small droplets than artificial surfaces. This difference is enhanced by pubescence.

Barry *et al.* (1977) released Zectran TM (Mexacarbate), formulated as a 'dry liquid' with a mass median diameter of 37 μm, over a forest to kill larvae of the Western Spruce budworm *Choristoneura occidentalis*. Whilst 40% of the particles collected by horizontal glass plates exceeded 40 μm diameter, spruce needles and budworm larvae collected no particles of this size, in fact 95% of the particles collected by needles were smaller than 15 μm and by larvae smaller than 20 μm. Other experimental data on the droplet collection efficiency of natural surfaces have been reported by LaMer *et al.* (1947),

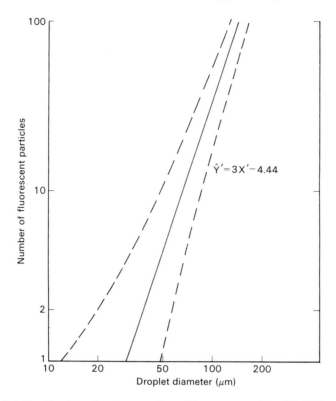

Fig. 5.10. Relationship of droplet size to number of fluorescent particles (Uk 1977).

Yeomans *et al.* (1949), MacCuaig (1962), Mount (1970), and Lofgren *et al.* (1973).

Distribution of droplets is a factor of over-riding importance in some cases. An elegant example of this fact is reported by Fisher *et al.* (1973) who investigated the mortality of the European Red Mite *Panonychus ulmi* in relation to the dispersal of Dicolol (Kelthane[R] 18.5% WP). They found that the insecticide was used more efficiently when the dosage was between 0.6 and 1.4 µg cm^{-2} and density of droplets between 70–250 cm^{-2} (Fig. 5.12).

On the other hand when a chemical is transmitted from the site of contact to the site of action by a secondary process, selection of droplet size may be less critical. The formamidine, chlordimeform, for example, is absorbed by cotton leaves and released slowly through the stomata (Dietrich 1979), where sufficient concentrations accumulate in the stagnant 1–3 mm of the boundary layer to be toxic to the larvae of the cotton leaf worm *Spodoptera littoralis* which emerges from egg clusters deposited on lower surfaces of cotton leaves.

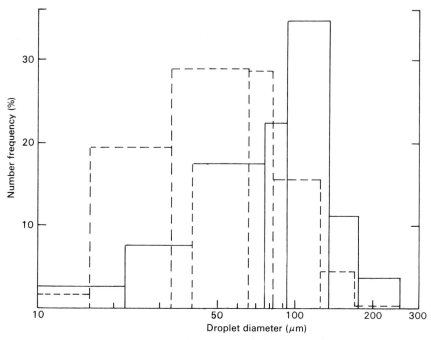

Fig. 5.11. Histogram of Saturn Yellow droplets deposited on cotton leaves and on artificial surfaces placed in the same location (Uk 1977). —, Horizontal cards; ----, leaves.

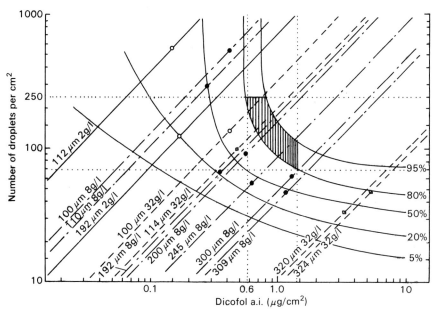

Fig. 5.12. Droplet density and the control of the European Red Mite *Panonychus ulmi* (Fisher *et al.* 1973).

Thus chlordimeform can be dispersed in big droplets which are collected by cotton leaves by sedimentation.

5.4.5 Droplet transmission

The influences on droplet dispersal from aircraft are: (a) the aircraft-induced turbulence; (b) the atmospheric boundary layer above the crop; (c) the atmosphere within the crop.

(a) Aircraft-induced turbulence. A parcel of air encountered by an aircraft is set into violent circulation which produces drag on the aircraft and provides lift to the wings. This lift is due to differences in pressure between the top and bottom surfaces of the wings and is zero at the wing-tip. Conversely, the velocity components, which are inwards on the upper surface and outwards on the lower, are greatest at the wing-tips. At the trailing edge of the wing the top and bottom flows merge to form a sheet of vortices, and those at the wing tip are dominant, so that, downstream of the wing the sheet of vortices rolls up into one strong vortex trailing from each wing at about 20% inboard from the tip.

The trailing vortex system moves downwards at a speed, for an aircraft such as a Cessna Ag-Wagon, of about $0.5\,\mathrm{m\,s^{-1}}$. If generated near the ground, the vortices move outwards, and can even rise and finally decay because of friction with the ground or the crop.

The significance of these vortices is that droplets released into them will be entrained if their terminal velocities are less than the vortex velocities. Thus, small droplets released near the centre line, where the vortices are weak are less affected than those released nearer to the wing-tip where vortices are stronger. Trayford and Welch (1977) calculated these effects together with those produced by the rotation of the propeller (Fig. 5.13). Virtually all droplets smaller than 100 μm diameter are shown to become entrained in the vortex, wherever released, though the calculations are for water-based droplets and include a large effect of evaporation (Fig. 5.14). Lawson and Uk (1978) have shown that the effect of vortices on non-volatile droplets less than 70 μm released within 2 m of the ground is as if the droplets were released from a greater height.

(b) The atmospheric boundary layer above the crop. Two processes are involved in the dispersal of droplets released from aircraft, namely, sedimentation due to gravity and diffusion due to turbulence – the former providing downward movement only and the latter both upward and downward. As the spray cloud moves downwind, the action of turbulence is to dilute the concentration of droplets, and the action of sedimentation to reduce the height of maximum concentration. Near the crop, droplets are removed as a result of capture.

The theory of dispersal by gravity was presented by Sawyer (1950) in

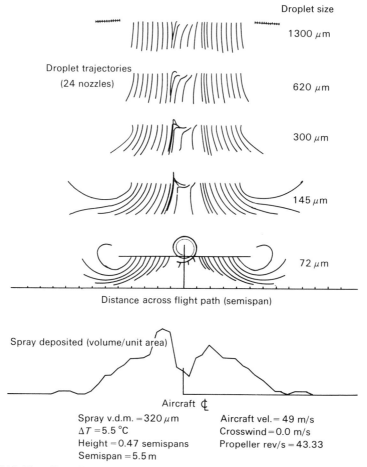

Fig. 5.13. The effect of aircraft-induced turbulence on the distribution of water droplets of various sizes (Trayford and Welch 1977).

relation to aerial curtain spraying against flying swarms of locusts. Sawyer showed that the destination of a droplet dispersing by sedimentation could be expressed as:

$$P = \frac{HU}{V_s}$$

where P is position on ground downwind of the point of release, H is height of release, V_s is terminal velocity of the droplet, and U is speed of cross-wind component. Thus to maintain P constant the height of release had to be lower when the cross-wind was strong than when it was weak. When sprays

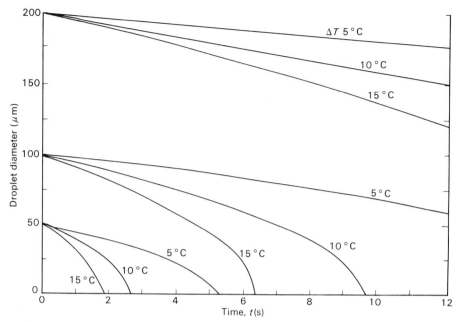

Fig. 5.14. Effect of diameter and wet-bulb depression on evaporation of water droplets falling freely in air (Trayford and Welch 1977).

containing a wide range of droplet sizes are released in still air, the aircraft-induced turbulence tends to disperse the droplets so that the deposit on the ground is approximately Gaussian or normal. If successive swaths of such a distribution are overlapped by $3\frac{1}{2}$ standard deviations, the variation in dosage is of the order of 40% (Amsden 1972). When there is a strong cross-wind the distribution becomes skew, with larger droplets falling near to the aircraft track, and smaller droplets shifted downwind. The same overlapping of $3\frac{1}{2}$ S.D. produced a wider swath for the same percentage variation in dose, and the position of the peak was predicted by the HU/V_s equation (where V_s is the terminal velocity of the droplet having the v.m.d. of the droplet spectrum) except for low wind speeds (e.g. $1–2\,\mathrm{m\,s^{-1}}$ at height of release of $3\,\mathrm{m}$) when the horizontal component of the wing-tip vortices could be responsible for lateral spread.

The concept of HU is widely employed in aerial crop spraying, as it is a valid and useful principle in relation to the dispersal of large droplets, that is those whose terminal velocity is considerably in excess of the velocity of air in the turbulent eddies. A droplet spectrum, however, almost always contains a very large number of small droplets, and we must enquire whether sedimentation or turbulence dominates dispersal. In practice it is found that, under turbulent conditions, small droplets reach the ground very much nearer to the point of release than is predicted from the *HU* relationship. This phenomenon was investigated by Bache and Sayer (1975) who found that the position of the peak of deposit (P) could be expressed by the following empirical equation:

$$P = \frac{H}{\sqrt{2bi}}$$

where *H* is height of release, *b* is a constant ($0.77 \times \sqrt{2} \sim 1.0$), and *i* is a measure of turbulence.

$$i = \frac{\sqrt{\bar{W}^2}}{\bar{u}} \approx 1.25 \frac{U^*}{u}$$

where U_* is frictional velocity, $\bar{u}$ is mean wind velocity, and $\sqrt{\bar{W}^2}$ is standard deviation of the vertical wind component. The turbulence term, *i*, varies from 0.06 over grass to 0.15 over forests. As *i* increases (due to surface roughness or to other factors which change the variability of the vertical component of the wind), *H* must also increase if P is to remain constant. Though not very sensitive to wind speed, *i* tends to increase with increased instability of the air. Accordingly release height for small droplets tends to have to be increased during the heat of the day. Since U^* ranges from $0.06 \, \bar{u}$ to $0.15 \, \bar{u}$ with a mean of about $0.1 \, \bar{u}$ over many arable crops, the destination of droplets with terminal velocity less than one tenth of the wind speed will be determined more by turbulent diffusion than by sedimentation. For instance, this applies to drops less than 60 μm released into a $1 \, \text{m s}^{-1}$ wind, or drops less than 250 μm released into a $10 \, \text{m s}^{-1}$ wind. Accurate transmission of droplets from an airborne source to a target near the ground requires an understanding of the dispersal process determining the destination of the droplets selected as having the preferred size range. Some experimental data illustrating this are given by Spillman (1982).

(c) The atmosphere within the crop canopy. Crops with an open canopy (typically young crops) have an atmosphere within the canopy closely linked with the atmosphere above the crop. On the contrary the atmosphere in crops with a dense canopy is only weakly connected with the atmosphere above the crop and is dominated in daylight hours by the fact that the highest temperatures are in the upper foliage which receives most sunshine. Below this level the air is extremely stable during the day, but often turbulent at night when there is radiant cooling at the top of the crop.

Droplet capture in young, open crops, therefore, is often by impaction as well as sedimentation, whilst in dense crops it is almost entirely by sedimentation for droplets of 30 μm and larger (Bache 1975; and Uk 1975; Bache and Unsworth 1977).

(d) Evaporation. Accurate transfer of pesticides from an air-borne source is dependent on control of droplet size, both from the dictates of the droplet-collection efficiency of the target and from those of choice of method of dispersal. If water is employed as a carrier, the effect of evaporation on droplet size must be considered. Joyce and Beaumont (1969), for example, found that the drops collected by needles of Lodgepole pine had an average diameter of about 30 μm. Since these were derived from a formulation containing 98.5% water, and had travelled some tens of metres they must have had a diameter of about 120 μm at formation, and droplets in this class constituted less than 10% of the total droplets produced, 90% being of smaller diameter. That is to say 90% of the original droplets were too small to be useful and contributed to 'drift' or environmental contamination.

Generally water is an unsuitable carrier for aerial sprays except when at least 90% of the volume in a spray is in droplets in excess of 150 μm diameter – that is a v.m.d. of at least 450 μm diameter.

5.5 Application equipment

For the most part, application equipment has been designed to meet the empirical requirement of high recovery and even distribution of particles across a swath, for the application of pesticides both as liquids and solids.

5.5.1 Liquid application

Liquid application systems consist of a tank, a pump, which is usually wind-driven, and a spray boom which is fitted with nozzles, and fed by a three-way valve, by the operation of which the pilot can choose how much of the output of the pump is directed to the boom and how much is returned to the tank. The rate of delivery is determined by the type and number of nozzles and the pressure maintained by the pump. The pilot calibrates his output with a reading on his pressure gauge so that he can deliver over his measured swath the quantity of chemical required by his estimated ground speed. Typically the swath is about 15 m and ground speed 2.5 km min^{-1}, giving a work rate of 3.75 ha min^{-1}. An emission rate set up at 100 litres min^{-1} provides an application rate of 26.7 litres ha^{-1}, and if the formulation contains less than 5% active ingredient, the errors introduced by the application rate (through changes in ground or air speed) varying ± 12.5% are likely to be acceptable, in terms of active ingredient applied.

Though wasteful of chemical and effort, the empirical system is a very forgiving one, and acceptable biological results can be achieved despite gross errors in application. For this reason no data are provided by manufacturers, nor required, for example, by British Standards, on the droplet spectra produced by spray nozzles. All that is needed is a statement of throughput (of water) in relation to pressure, although this may vary considerably along the length of the boom.

In contrast with the empirical approach, target-specific application methods relying on accurate choice of preferred droplet sizes and estimates of the total number of droplets needed to transmit the required doses at the required density, rarely demand an application rate of more than 5 litres ha^{-1}, and sometimes less than 0.5 litres ha^{-1} of a formulation which may contain between 10 and 100% active ingredient. Moreover, when small droplets, dispersed by turbulence, are the preferred size, swaths of more than 100 m may be employed, thus achieving work rates of 25 ha min^{-1} or more. Under these conditions accurate delivery of the chemical to the atomizer becomes essential and it is usual for aircraft engaged in this type of spraying to be fitted with flowmeter so that the pilot can have positive control over his emission rate, and vary it, if necessary, in flight.

Control of droplet size is less easily obtained and is usually sought in rotary devices. The sole equipment for this purpose which has operationally proven reliability is the Micronair, now available as the AU 3000, in which the final process of atomization takes place on a rotating gauze cage which is wind-driven. Similar rotating cages are employed by the Minispin and the Acumist atomizers though these have not been widely employed. Rotating cages suffer from the disadvantage that some of the liquid fails to obtain the peripheral speed of the cage, and this causes widening of the droplet spectrum.

Other rotary devices include the Turbaero aerial atomizer, an electrically driven stack of toothed discs essentially similar to that used in the ULVA hand-held sprayer. The Beecomist is an electrically driven perforated metal or sintered metal cylinder which is claimed to produce a very narrow droplet spectrum. The throughput of both discs and cylinders, however, is small. The fine holes of the Beecomist necessitate a very efficient filtration system if blockages are to be avoided. The stack of spinning discs can fail to provide the expected efficiency in atomization because of the difficulty of feeding each disc evenly.

Various types of twin fluid nozzles have been developed, in which air, travelling at high velocity, creates big shearing forces over the liquid travelling at a relatively low velocity. Twin fluid nozzles probably represent the best way of producing very fine sprays (v.m.d. of less than 30 µm), but have been little used in aerial spraying. A successful prototype was built by Parker *et al.* (1971), and a very promising operational one christened the Bifoil, recently introduced by Parkin and Newman (1977), may be described as a linear venturi. These are externally mixing nozzles.

Several types of internal mixing nozzles are available, but the fine orifices which they involve make them very susceptible to blocking. The pentagram pneumatic atomizer has been used successfully on Turbine helicopters, the air being supplied from the engine compressor (Parkin 1981).

A further method of atomization has been employed to produce large droplets of a narrow range of sizes, a notable example being the experimental vibrating jet nozzle of Wilce *et al.* (1974), which has chemically etched holes of 125 μm diameter and produces uniform droplets of 250 μm.

The Microfoil atomizer consists of a series of hypodermic needles, about 300 μm internal diameter, which eject into the airstream. The latter breaks the fluid ligament into droplets of uniform size of 800–1000 μm diameter. The needles, however, require a good filtration system to avoid blockage, and the large droplets produced, which are broken up by air travelling at more than 25 m s^{-1}, make the Microfoil suitable for helicopters. The physics of the atomization of liquids and the effect of viscosity, surface tension, and density on droplet size are described in detail by Parkin (1981).

5.5.2 *Application of solids*

The most widely used equipment for distributing pesticides in granular form are collectively known as ram-air spreaders, which depend on the release of the particles into a ducted airstream, accelerated by venturi action, to impart to them a velocity perpendicular to the aircraft track. The object is to obtain a trapezoidal or triangular distribution pattern with as wide a spread as possible. The overlap provides an even ground deposit which is tolerant to imperfect spacing of successive runs provided that the lane separations do not exceed about 3 standard deviations (Fig. 5.15). The excessive drag of ram-air spreaders, however, not only reduces the work output of the aircraft (Smith 1969) but has sometimes created a safety hazard (Stephenson 1975). Moreover, the optimum sizes of particles for efficient distribution by such apparatus have terminal velocities of 15–20 m s^{-1}, equivalent for most granular materials to particles of 5–6 mm diameter. Economic use of pesticides applied by aircraft requires a smaller size (Lee 1975). These defects led Trayford *et al.* (1972) to design the Tetrahedron Spreader, and Lee (1975) a wing-ducting system. The former gave a speed penalty on a Beaver aircraft of only 2 km and an increase in dispersal of 95% for application of 40 kg ha^{-1}, though only 13% for rates of 200 kg ha^{-1} (the latter being rates for fertilizer rather than pesticides), when particles of 2.5 mm diameter were employed. The Transland Swathmaster is the only practical device on the market which employs the ducted-wing principle.

Lateral velocity may also be provided by rotary devices, and rotating disc spreaders, driven by a power-take-off, are employed by the Cimelak 237, Bumble Bee. The same principle is employed by helicopters where the hopper and the rotating discs are underslung, the disc being driven either electrically from power generated by the helicopter or by its own engine. This system, however, does not provide a very stable platform and is not widely used for the application of pesticides by helicopters.

In the UK the Agricultural (Poisonous Substances) Regulations 1965 require that the more toxic pesticides in granular form shall have no more than 4%, and no more than 1% by weight, passing 250 μm and 150 μm sieves respectively. The US and ECE regulations are similar. These are far smaller particles than those which can be thrown laterally to provide an economic swath by ram-air spreaders, and even ducted-air spreaders are most efficient when particles are not smaller than 1000 μm diameter.

Thus, although granules are widely favoured as one of the safest means of aerial application of toxic pesticides, their use is limited because few are suitable for application in this form, they are expensive to manufacture, and the spreaders available do not permit economic work rates to be achieved. Moreover solid formulations are highly abrasive. Consequently, world wide, less than 8% of the pesticides used are applied in solid form. A detailed account of the aerial application of solids has been given by Amsden (1978).

5.6 Application methods

The goal of pesticide application is to provide a high degree of chance (>0.95) that each target surface collects at least a defined dose and the minimum chance that it collects much in excess of this. Conventionally efficiency in this respect is measured by the coefficient of variation of the deposit on the ground across the swath. The ground deposit from a single passage of the aircraft can be considered Gaussian, although skewed, when released in a strong cross-wind. The variability of deposit is a function of the amount of overlapping of individual swaths and may change to less than 10% if the lane separations are less than 3 standard deviations, but increase rapidly towards 90% if they exceed 4 standard deviations (Fig. 15). Such lane separations, connoted the effective swath, are essentially a function of the wing configuration of the aircraft and the distribution of nozzles along the wing, and are derived from deposits of droplets falling by gravity, care having been taken to avoid the production of small droplets and their entrainment in wing-tip vortices. The system is known as placement spraying and employs medium to coarse sprays with a v.m.d. of usually about 400 μm, but not less than 250 μm. Such sprays are unlikely to have more than 5% of their volume in droplets of less than 150 μm diameter, these being the droplets most likely to be lost to the swath by evaporation or turbulence, so that with placement spraying more than 95% of that applied can be recovered in the swath. Placement spraying is therefore the system most frequently employed and often demanded by farmers and government authorities, despite the fact that this method, constrained to lane separations seldom more than 20 m and often as little as 15 m, is expensive and fails to exploit adequately the aircraft's air-borne speed. It is, however, probably the most efficient method for residual herbicides, the target of which is the soil, and those insecticides which are redistributed by a secondary process, such as provided by the vapour phase (Spillman 1982).

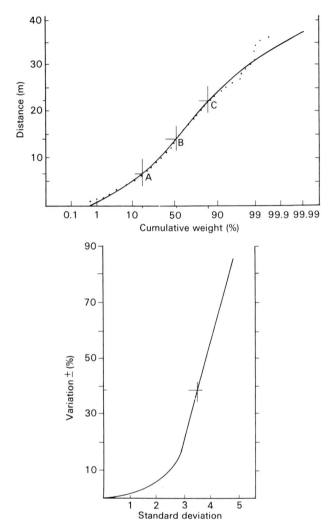

Fig. 5.15. The effect of lane separation (expressed as standard deviations of mean deposit) on the coefficient of variation of deposit (Amsden 1972).

Placement spraying is the method of choice when the empirical approach to pesticide application is pursued. The inadequacies of this approach have already been discussed. It would seem that often the spray recovered in a swath is a measure of that fraction which has failed to achieve its biological goal (Himel 1969b; Joyce *et al.* 1969). Efficiency in pesticide application is best measured as the fraction collected by a population of closely defined targets (Joyce *et al.* 1977).

As has been pointed out, the probability of a target collecting a spray particle is a function of its particle-collection efficiency, and the biological objective is not achieved by selecting a droplet size merely because it falls below the aircraft track. In practice, the spray target is frequently an insect or a specific plant organ, and the biggest fraction of the dose collected by such surfaces has been found experimentally to be particles which are dispersed by turbulence rather than gravity. The distance from the line source of particles emitted from a spray aircraft, over which it can be expected that there is a high probability of the target dose being exceeded, is thus no longer a function of the aircraft but of the air into which the droplets are released. Swath thus defined is a biological, rather than a physical, concept (Joyce *et al.* 1969), and may vary from several hundreds of metres, as in drift spraying for Desert Locust control (e.g. Sayer 1959; Joyce 1962) and in the control of the army worm *Spodoptera exempta* (Brown *et al.* 1970), to zero, as in the case of whitefly on mature cotton (Joyce *et al.* 1969). When effective swaths are achieved by a single pass, the procedure is called drift spraying. When the effective swath is achieved by the overlapping of one or more cross-wind dispersal patterns, so that the target dose is made up from one or more successive increments, the term incremental spraying has been coined (Joyce *et al.* 1970). The importance of employing the smallest-size droplet to build up the largest dose is stressed by Spillman (1980).

Conventional aerial spraying is usually conducted using not less than 20 and not more than 100 litres ha^{-1}, these volume rates being necessary to provide the density of the big droplets which have been found to relate to the desired biological effect. Sometimes, however, it has been possible to show that this effect is simply a correlation between the big droplets monitored and the associated small droplets, only the latter having been collected by the biological target (Himel 1969b; Joyce and Beaumont 1979). These may constitute only a small fraction of the volume emitted. The analytical approach to aerial application of pesticides is to determine what these droplets are and to produce the right number of them. When this is done it has been found that very much smaller quantities of spray liquid are required, normally less than 5 litres ha^{-1}. Thus the term ULV (ultra-low-volume rates of application) has emerged. ULV was developed primarily for locust control, but was employed on a large scale between 1952 and 1955 for cotton spraying in the Sudan Gezira (Joyce 1955). Regarding its introduction to grasshopper control, Skoog (1965) emphasized the low-volume rates of application as an economic goal rather than as a consequence of an analytical approach (Joyce 1974). This emphasis led to the development of ULV placement spraying (Courshee 1974), but it has been found in practice that the aircraft-induced turbulence is rarely adequate to propel small droplets efficiently to their target, so that ULV placement spraying is suitable only for those rare occasions when the route of action of the pesticide permits a low density of large (gravity-controlled) droplets.

The economic advantage of ULV spraying lies in the improved timing of applications achieved by the increased work-rate. Thus an aircraft loaded with 500 litres of spray liquid can treat 25 ha at 20 litres ha^{-1} and 500 ha at 1 litre ha^{-1}. ULV rates thus save, in this case, 20 ferry trips and 20 take-offs and landings. In large-scale commercial practice aircraft operating at ULV rates achieve 4–10 times the work output of those operating at conventional rates (e.g. Joyce and Beaumont 1979).

It must be emphasized, however, that such high rates of work employ wide lane separations, and are suitable for only large and extensive areas. They are particularly useful when pest biology dictates the application of pesticides as nearly synchronized as possible over very large areas.

5.7 Guidance systems

A vehicle moving with a velocity V_1, relative to the ground in a medium, say, the air, with a cross-wind vector of V_2 relative to the ground, will be deviated from its track at an angle, the drift angle θ, where

$$\tan \theta = \frac{V_2}{V_1}$$

To maintain his track the pilot must correct his direction of travel (his heading) by θ degrees. In visual flying he does this by directing himself towards a fixed marker, which, in aerial crop spraying, may be a man holding a flag, or some other aid, such as coloured or fluorescent flags, streamers, balloons, or kytoons. This is the system for track guidance most commonly employed in the aerial application of pesticides and thoroughly described in *Expert Flagging* (Haley 1973).

When the length of run exceeds 1–2 km, markers are difficult to position accurately on the ground so that they are visible to the pilot at the beginning of his spray run, and an alternative means of guidance becomes necessary. Simple devices, such as the automatic flag-man, by which paper streamers are ejected from the spray aircraft, sometimes suffice, but large-scale operations, such as those employing ULV rates of application, aimed at treating hundreds or thousands of hectares per day, have called for the development of special electronic techniques.

5.7.1 *Inertial navigation systems*

Though widely employed by airlines, INS are normally too costly, heavy, and complex to be employed in spray aircraft. The Litton LTN-51 INS was chosen by the Province of Quebec for their operation against Spruce Budworm from 1973 onwards. The operation of this system, mounted on a Douglas DC 7-B aircraft, is described by Boivin (1975) and Randall (1975). In these operations, 12 inertially equipped aircraft executed over 600 missions during which the average error was about 3.5% of the 900 m lane separation flown. The technical aspects of inertial navigation, and the sources of error, lie outside the scope of this chapter.

5.7.2   Doppler navigation

Air-borne Doppler operates by transmitting three (or four) beams from a special antenna array, two directed forward and one rearward. These beams are reflected by the ground back to the aircraft, and the frequency shift provides a measure of the velocity across the track flown by the aircraft (drift) and along the track (ground speed). These data, with inputs from the aircraft's standard instruments, provide information by which the aircraft's position relative to the ground can be calculated and displayed. With the Decca Doppler 72, the aircraft's position is traced on a roller map. The Doppler system has been extensively tested and used for track guidance by the tsetse fly control teams of the Government of Zambia. In single runs, Doppler has been found to provide 1% or less of cross-track error to distance flown, and this includes a major contribution from the final reference system – the Gyro compass. Errors arising in turns from both the Gyro and the Doppler can be reduced by careful but rather prolonged turns. The Doppler system has also been employed by Rainey (1972, 1974 etc.) for exploring wind-fields for flying insects.

5.7.3   Hyperbolic systems

Short-wave hyperbolic systems, such as the Decca Main Chain, operating at 100 Hz with ranges of 200–400 miles, Hi-Fix (2 MHz, range of 100 miles), and Agri-Fix (2 MHz, range 40–100 miles) have been successfully employed in the aerial application of pesticides.

It is usual in crop spraying for a pilot to know where he must begin and end his spray run, and require only guidance along the track he has to follow. The hyperbolic system provides this with only two stations, a Master and a Slave radiating phased-locked signals. In the Agri-Fix system both stations are portable (< 20 kg). The aircraft receiver is a phase-measuring device driving a light bar display which indicates where the aircraft is in relation to the pattern lanes. The system can be used to maintain the pilot on a predetermined track, each light along the bar representing 1/50 of a lane. At a frequency of 1764.4 kHz the lane width on the baseline is 85 m so that each illumination represents a deviation of 17 m from the chosen track. Hyperbolic systems suffer from the fact that only in a region in the centre of the lane pattern are the hyperbolic lane widths sufficiently straight to permit accurate guidance. With a base line of 30 km this permits an operation over some 2000 ha before the stations have to be moved. Moreover, the equipment at present available provides guidance only along and not across the hyperbolic lanes, so that the direction of flight is constrained by the position of the ground stations (CIBA-Pilatus 1979).

5.7.4   Micro-wave systems

A new track guidance system recently introduced has been called the Flying Flagman, which is based on the Trisponder (Mitchell 1979). The Trisponder is an electronic positioning system which provides accurate line of sight information on the distance between a master and a remote station. A digital distance measuring unit (DDMU) working in conjunction with the Trisponder measures the total time for a signal to make the round trip between master and remote station, and from it computes, averages, and displays the two ranges. Each measurement is completed in about one millisecond and the indicator average (10 or 100) is updated once per second.

In operation two slave stations are set up in positions where they are in line of sight of each other, and the DDMU and Trisponder are carried in the aircraft. The DDMU measures and computes the range to each ground station and provides this information to a left–right computer. The latter processes all input data and provides outputs in the form of a display, which might be the light-bar used in Agri-Fix, so as to guide the pilot along a track which remains a constant ratio of distances between the two stations. The system promises to provide greater flexibility, simplicity, and economy, compared with other systems (Smerdon 1981).

5.8   The aerial application system

If aircraft are to make their full contribution to the application of pesticides their use must not be confined to the empirical systems which have evolved to meet less-demanding conditions. The transmission of toxic doses from an airborne source to a population of specifically defined biological targets which have spatial and temporal dimensions, with minimum loss en route, involves complex problems which this chapter has discussed. The aim must be to develop target-specific methods, the efficiency of which is measured in terms of the fraction of the chemical released which achieves its biological goal. Such target-specific methods involve the integration into a single system of inputs from a variety of disciplines in which all components are interdependent. The key to such systems is the accurate definition of the biological target in relation to the route of entry of the chosen pesticide. Such definitions call for far more fundamental knowledge of the bionomics, behaviour, and physiology of the target organism than is normally available, as well as of the toxicology, mode of action, and route of entry of the pesticide. Once these data are provided by the biologist and the chemist, the physical and engineering problems of transmission are capable of solution – but only then. These latter aspects cannot be considered in isolation.

5.9   References

Akesson, N. B. and Yates, W. E. (1974). The use of aircraft in agriculture. FAO, Rome. Agricultural Development Paper No. 94.

Amsden, R. C. (1962). Aircraft applications. Chesterford Park Research Station Report. Fisons, UK.

—— (1972). *Agric. Av.* **14**, 103.

—— (1978). Aerial application of solid material. *Cranfield short course on the aerial application of pesticides.* Cranfield Institute of Technology.

Azar'Yan, M. B. (1966). The application of aviation in agriculture and forestry. Washington D.C. US Dept. Commerce. Translation FTD-MT 24, p. 101.

Bache, D. H. (1975). *Agric. Meteorol.* **15**, 379.

—— Sayer, W. J. D. (1975). *Agric. Meteorol.* **15**, 257.

—— Uk, S. (1975). *Agric. Meteorol.* **15**, 371.

—— Unsworth, M. H. (1977). *Q. J.R. Met. Soc.* **103**, 121.

Barry, J. W., Tysouski, J. R., Orr, F. G., Ekblad, R. G., Maisales, R. L., and Ciesia, W. M. (1977). *J. Econ. Entomol.* **70**(3), 387.

Boivin, G. (1975). *5th Int. Agric. Av. Congr.,* 350.

Brooks, G. T. (1976). In *Insecticide biochemistry and physiology* (ed. C. F. Wilkinson), p.3, Plenum Press, New York.

Brown, A. W. A. (1978). *Ecology of pesticides.* Wiley-Interscience, New York.

Brown, E. S., Stower, W. J., Yeates, M. N., and Rainey, R. C. (1970). *E. A. Agric. For. J.* **35**(4).

CIBA-Pilatus (1979). Private communication.

Courshee, R. J. (1967). In *Fungicides: an advanced treatise* (ed. D. C. Torgeson), Vol. 1, p.240. Academic Press, London.

—— (1974). *Br. Crop. Prot. Counc. Monogr.,* No. 11.

Davies, C. N. (1966). In *Aerosol science* (ed. C. N. Davies), p.393, Academic Press, London.

Dean, R. B. and Preston, R. G. (1973). A cryogenic sampling probe for droplet size analysis in a high speed air flow. Unpublished Report. AARU (CIBA-GEIGY). Cranfield, England.

Dittrick, V. (1979). Private communication.

Fisher, R. W., Menzies, D. R., Herne, D. C., and Chiba, M. (1973). *J. Econ. Entomol.* **67**(1), 124.

Gibson, E. A. (1958). *J. R. Aer. Soc.* **62**, 423.

Hadaway, A. B. and Barlow, F. (1965). *Ann. Appl. Biol.* **55**, 267.

Haley, J. (1973). Expert flagging: a training manual for aerial application ground crews. University of North Dakota Press, ???

Himel, C. M. (1969a). *J. Econ. Entomol.* **62**(4), 912.

—— (1969b). *Proc. 4th Int. Agric. Av. Congr.* 275.

—— Moore, A. D. (1969). *J. Econ. Entomol.* **62**(4), 916.

House, J. S. (1922). *Ohio Agric. Exp. Stn. Bull.* **7**, 126.

IAAC (1973). *Handbook for agricultural pilots.* International Agricultural Aviation Centre, The Hague.

Joyce, R. J. V. (1955). In *Annual report of the research division, Ministry of Agriculture. Sudan, 1952–1953,* p. 90.

—— (1962). *Report on the desert locust survey 1955–1961 EASCO, Nairobi, Kenya.*
—— (1974). *Br. Crop. Prot. Counc. Monogr.* **11**, 29.
—— (1976). In *Insect flight* (ed. R. C. Rainey), p.135, Blackwell Scientific Publications, Oxford.
—— Beaumont, J. (1979). In *Control of pine beauty moth by fenitrothion in Scotland* (eds. A. V. Holden and D. Bevan), Forestry Commission, UK.
—— Marmol, L. C., and Lucken. J. (1969). *Proc. 4th Int. Agric. Av. Congr.* 128.
—— Uk, S., and Parkin, C. S. (1977). *Efficiency in pesticide application on insecticide resistance* (eds D. L. Watson and A. W. Brown), Academic Press, London.
—— Marmol, L. C., Lucken, J., Bal, E., and Quantick, R. (1970). *PANS* **16(2)**, 309.
Knollenberg, R. G. (1970). *J. Appl. Meteorol.* 86.
LaMer, V. K., Hodges, K., Wilson, I., Fales, J. A., and Latta, R. (1947). *J. Coll. Sci.* **2**, 539.
Latta, R., Randall *et al.* (1947). *J. Wash. Acad. Sci.* **37**(1) 397.
Lawson, T. J. and Uk, S. (1978). *Br. Crop. Prot. Counc. Monogr.* 22.
Lee, K. C. (1975). *Proc. 5th Int. Agric. Av. Congr.* Stoneleigh, 328.
Lofgren, C. S., Anthony, D. W., and Mount, G. A. (1973). *J. Econ. Entomol.* **66(5)**, 1085.
Maan, W. J. (1965). The use of aircraft in the mechanisation of agricultural production. FAO, Rome. Informal Working Paper No. 26.
MacCuaig, R. D. (1962). *Bull. Entomol. Res.* **53**, 111.
May, K. R. and Clifford, R. (1967). *Ann. Occup. Hyg.* **10**, 83.
Mitchell, H. W. (1979). *Agric. Av.* **20(2)**, 77.
Mount, G. A. (1970). *Mosquito News* **30(1)**, 70.
Parker, J. D., Collins, B. G. P., and Kahumbura, J. M. (1971). *Agric. Av.* **13(1)**, 24.
Parkin, C. S. (1978). The Particle Measuring Systems (Knollenberg) Droplet Measuring System. Unpublished Report AARU (CIBA-GEIGY) Cranfield, UK.
—— (1981). Physics of droplet production. In *Cranfield short course on the aerial application of pesticides.* Cranfield, UK.
—— Newman, B. W. (1977). *Agric. Av.* **18(1)**, 15.
—— Wyatt, J. C. (1982). *Crop. Prot.* **1(3)**, 309.
Randall, A. P. (1975). *5th Int. Agric. Av. Congr.* 336.
Rainey, R. C. (1972). *Aero. J.* **76**, 740.
—— (1974). *Br. Crop Prot Counc. Monogr.,* **11**, 20.
Razak, K. and Snyder, M. H. (undated). A computer operational analysis of an Ag-Plane. Soc. Automobile Engineers Inc. No. 770480.
Sargent, J. A. (1976). In *Herbicides* (ed. L. J. Audus), Vol. 2, p. 303. Academic Press, London.
Sawyer, K. F. (1950). *Bull. Entomol. Res.* **41(2)**, 439.
Sayer, H. J. (1959). *Bull. Entomol. Res.* **50**, 371.
—— (1969). *Agric. Av.* **11(3)**, 78.
Skoog, F. E. (1965). *J. Econ. Entomol.* **58**, 559.
Smerdon, R. G. (1981). Electronic guidance systems. In *Cranfield short course on the aerial application of pesticides.* Cranfield, UK.
Smith, M. R. (1969). *Proc. 4th Int. Agric. Av. Congr.* 83.
Spillman, J. J. (1976). *Agric. Av.* **17**, 28.
—— (1977). Aerodynamics of droplet capture. In *Cranfield short course on the aerial application of pesticides.* Short Course Notes: Cranfield Institute of Technology.
—— (1980). *Aer. J. Feb. 1980* 60.
—— (1982). *Crop. Prot.* **1(4)**, 473.
Stephenson, J. (1975). *Proc. 5th Int. Agric. Av. Congr.* 321.
Trayford, R. S. and Holt, G. E. (1972). *CSRIO Div. Mech. Eng. Rep.* No. 113.

—— and Welch, L. W. (1977). *J. Agric. Eng. Res.* **22**, 183.

Uk, S. (1977). *Pestic, Sci.* **8**, 501.

—— (1978). Analysis of drop-size spectra in agricultural sprays. *Cranfield short course on the aerial application of pesticides*. Cranfield Institute of Technology.

Van Bemmel, P. M. (1953). *Crop spraying by air*. Delft Shell Research Laboratory.

Wain, R. L. and Smith, M. S. (1976). In *Herbicides* (ed. L. J. Audus), Vol. 2, p. 279. Academic Press, London.

Wilce, S. E., Akesson, N. B., Yates, W. E., Christensen, P., Cowden, R. E., Hudson, D. C., and Weigt, G. T. (1974). *Agric. Av.* **16(1),** 7.

Yeomans, A. H., Rogers, E. E., and Ball, W. H. (1949). *J. Econ. Entomol.* **42(4),** 591.

6
Field experimentation

W. REED, J. C. DAVIES, and S. GREEN

6.1 Introduction

A short chapter such as this cannot provide a comprehensive survey of Field
Experimentation, with adequate coverage of sampling methods, experimental design, and statistics. Therefore, a basic knowledge of these subjects or a
willingness to learn from the recommended texts is assumed, and the chapter
has been used to emphasize the particular problems that are encountered in
field experimentation with pesticides, rather than to embark upon a basic
account of agricultural experimentation as a whole.

A literature survey revealed many useful comprehensive accounts of field
experimentation but most are primarily concerned with agronomic trials e.g.
Cochran and Cox (1957), and only a few mentioned the particular problems
of experimentation with pesticides. Much of this account is derived from
personal experience of research in the control of the insect pests on cotton,
sorghum, and groundnuts in Africa and India over the past two decades.

6.2 The evolution of pesticide experimention

The origins of field experimentation on insect control are shrouded in
history. Probably simple comparisons of treatments were made by farmers
comparing their crops with those of their neighbours and attributing
differences to their various practices, including rudimentary pest control.
Dyke (1974) included a chapter on the history of field experimentation in his
very useful book and he traces relatively sophisticated experimentation in
agriculture back to Lawes and Gilbert in 1852, soon after they had laid the
foundations for Rothamsted Experimental Station.

The systematic testing of various inputs in agriculture really became a
science in the 1930s when R. A. Fisher, F. Yates, and others gave their
attention to field experimentation and developed replicated testing in block
designs. Further developments can be attributed to very many workers but
Cochran and Cox (1957), Steel and Torrie (1960), and Bailey (1981) have
provided invaluable texts for trial design and analysis. Useful basic works for
statistical sampling are Yates (1981) and Cochran (1977).

In the decade from 1955 to 1965, there was a flood of new 'miracle'
pesticides that had to be tested by economic entomologists particularly in the
tropics, where insects pests generally cause the greatest losses, mainly on the

cash crops such as cotton and where the monetary returns were sufficient to pay for such expensive inputs. In that era, much of the field work of cotton entomologists was concerned with pesticide experimentation. Trials of new insecticides, dosages, concentrations and application frequencies were demanded by the agricultural administrators, and the emphasis was on short term research that would pay dividends at the farmer level. There is a diverse and substantial literature on insecticide trials, particularly on cotton in the 1960s and 1970s. As a result of these trials, recommendations were made to farmers, sometimes without verification of efficacy outside the research stations. Most field experimention with pesticides in the future must be within the context of Integrated Pest Management. Pesticides are but one element within an overall pest management strategy. This requires a critical re-evaluation of the methods of pesticide experimentation, for we are now far more concerned about the interactions between pesticide use and very many other factors including agronomic practices, crop cultivars, and natural control elements. The days of the randomized block design trials with small plots, comparing ten or more differing pesticides applied according to the calendar for the control of the pest complex on any crop are now hopefully part of our less enlightened past. Pesticide experimentation in the future will be much more difficult and complex and will require inputs by many scientists in addition to the entomologists, pathologists, and herbicide specialists.

6.3   Information required before field experimentation

Before embarking upon field experimentation we must first determine:
   (a)   What do we need to know?
   (b)   What is already known?
   (c)   What experimentation is required to satisfy our immediate require-
          ments and what is possible within the limitations of time, funds, and
facilities available?

It is essential that the tentative answers to these questions should be committed to paper and then adequately discussed with colleagues. Most pesticide experimenters will be members of teams or organizations that include workers in several disciplines. Ideally, each member of the team should be aware of the problems faced by his colleagues and the possibilities of interactions, and interdisciplinary discussions at seminars and planning sessions are of great value. Continued personal contact in the course of the experimentation is very important. The answer to 'What do we need to know?' must be precise and comprehensive. The field target of an experiment may be the single life stage of an individual pest or a whole pest complex. The environment, for which the solution to the problem is sought, will often be the average farmer's field, but we have long since realized that average farmers are hard to find and that the results from most field experimentation will be site and season-specific, at least to some degree. Such complexity should be recognized but not allowed to overwhelm the planning. The required experimentation will normally be part of a greater series, originating in the synthetic chemists' laboratory and reaching a temporary conclusion in the farmers' fields. We will find no permanent solutions in the open biological systems across which we have chosen to work. Normally, the field experimentation will progress from small-plot observations, to larger-scale testing on research station fields, then to a multilocation experiment on the fields of co-operating farmers, as a prelude to a recommendation for the farmers of an area. At each stage of testing, efficiency and relevance will be of prime concern.

There will always be data relevant to the problem somewhere in the literature. Seldom is the experimenter pioneering in a virgin field and adequate time should be spent in pre-experiment reading. There will generally be a bank of basic data concerning any particular pesticide available from the manufacturer. In addition, there will usually be some independent reports on the pesticide and the target pests, if only on other crops or in other other environments. Such data will usually enable the scientist to build experimentation on a firm foundation. But we must always continue to question whether the previous work was sufficiently conclusive and carried out in circumstances relevant to the current problem. Confirmatory experimentation will generally be cheaper than the original exploration and some duplication need not be wasteful. Currently, use of abstracting and literature retrieval services greatly assists in planning of trials. The time for reading is before the experimentation, not afterwards merely to assemble a list of references to accompany the published report. Many useful points to consider are given in Statistical Checklists 1 and 2 (Jeffers 1978) a,b, which cover respectively design of experiments and sampling.

Success in field experimentation only comes through adequate planning and preparation, intensive personal supervision of the field work, and comprehensive analysis of data. Failures are experienced sometimes and we will generally be able to point to factors that were beyond our control. However, by learning from one's own and others' mistakes it is soon discovered that the difference between sucess and failure in field experimentation is often within one's own control and capability, rather than lack of available facilities or bad luck.

Unterstenhöfer (1976), in a comprehensive review of field trial techniques in crop protection, warned that it is necessary to acquire an exact knowledge of the biology, ecology, and epidemiology of the target pest before embarking on any pesticide trial, for it is these factors, in conjunction with the mechanism of action and other properties of the pesticide, which should determine the planning and implementation of the experimentation. To this formidable list of prerequisites we must add the need for a basic knowledge of the crop and environment, within which the pest is to be tackled. For example, the management of a polyphagous pest such as *Heliothis armigera* will require very different strategies on the several crops that it attacks, partly because of the differing crop environments but also because of the differing economic returns from those crops. Thus on cotton, this pest may be profitably managed with the aid of expensive selective pesticides but on pigeonpeas the only practical management may be to grow cultivars that can compensate for early damage, after the buildup of natural control elements has reduced the pest population. Much of our pesticide experimentation will continue to be concerned with the identity of the best pesticide to use, quantities needed, what is the best means of delivery to the target and when? These simple phrases conceal complexity, as many factors have to be considered when deciding on the best pesticide. The prime consideration is of course its effectiveness against the pests and increasingly whether it is safe in use, not only during application, but also in the longer term, particularly if the crop is to be used as food or feed. Environmental factors are becoming increasingly important and the ideal pesticide will kill the pests but leave the crop, the beneficial insects, and other animals unharmed. A further major consideration is the cost/benefit ratio for a particular pesticide and crop, which largely determines the dosage. All too often dosages are quoted in terms of kg of active ingredients or product per hectare, but where the pesticide is being used to give coverage of a crop that varies in size during different growth stages, such a fixed dosage may well be impractical. For example, cotton may have to be protected at the seedling stage and near maturity but the crop surface to be protected will vary from 0.1 ha to more than 4.0 ha per hectare of field, so involving a fortyfold difference. Additionally, different crop cultivars of plant species can vary enormously in growth habit, stature, and rapidity of growth through different stages. Many modern cultivars are short-statured with high harvest indexes, whereas traditional cultivars are often tall and have low indexes. Pesticides are generally applied as granules, powders, or liquids, on or in the soil or crop, and the quantity of pesticide plus carrier that has to be applied may range from 1 to 1000 litres or kg per hectare. To all of this complexity, we have to add the interactions with the many other factors that affect the crop pests and environment. (See Chapter 3)

Obviously, it is impractical to experiment with each and every factor independently. It is essential that the sequence of field experimentation is planned so as to maximize acquisition of relevant information.

6.4   Research station anomalies

Many experimenters will conduct much of their field testing of pesticides on research station farms as a prelude to making recommendations for use by farmers. It is important to consider whether such testing will give results that are valid.

In general, research stations will be ecologically very different from the real world outside the station fence. On research farms in the tropics, the breeders of annuals will attempt to grow more than one crop in each year to obtain a more rapid generation advance, agronomists will have sowing date trials, entomologists will have pesticide-free plots, pathologists will have sick plots, and weed specialists their weed nurseries. In the semi-arid tropics, in particular, research stations generally stand out as green oases in the middle of the dry season partly because they have irrigation facilities, but also because they are not overstocked by cattle, sheep, and goats that will have consumed almost all the green vegetation in farmers' fields during this time. Thus, the research stations will present an abnormal opportunity for pests to feed actively and 'carry over' through the dry season. This factor, combined with a continuous availability of flowering and fruiting crops throughout the rest of the year, will generally ensure that the pest situation on research stations is atypical and more severe than on representative farmers' fields, in spite of more intensive pesticide use which in turn brings other problems and abnormalities.

There is a clear need to list and quantify the differences that exists between any given experimental crop on the research farm and that in typical farmers' fields, with regard to the crop itself, the pests, and their natural enemies. Such differences may be sufficiently great to make results from some pesticide experimentation on the research farm of no relevance whatsoever to the real world outside. Simple comparative experiments of the killing power of several pesticides, at differing dosages, concentrations, and with differing application methods, will give valid data if the target pests are present and not too mobile. Data from pest management studies involving interactions of cultivars, cultural practices, pesticides, pests, and natural enemies and any trials involving yield benefits from pesticide use, may well give totally misleading results. Such experimentation may only be relevant if carried out in farmers' fields.

On most research stations there will have been no conscious attempt to keep an area of the farm free from pesticides. Any pesticide-free crop on an area previously treated may be atypical for pesticide residues in the soil and changes of the soil fauna and flora will have long term and far reaching effects. DeBach (1970) when considering insect pest natural enemies, stated that a 5-year freedom from pesticide use may be needed in tree crops to reattain a material balance after pesticide use.

6.5 Interactions of pesticides with other factors

It is of the upmost importance not to consider pesticide use in isolation from other agricultural practices. We often see experiments reported from research stations where pesticide use has increased yields twofold. In the same report, we may find that the plant breeder has produced cultivars greatly increasing yield, as has the agronomist with better plant populations and judicious fertilizer use. These gains may be interdependent and the yields quoted for research stations in the tropics are often four or five times greater that those achieved by peasant farmers. Before making revolutionary claims or utilizing the results, however, it is generally wise to study the details of the research station trials and the relevance of the reported gains. Pesticide-use trials are often comparisons of pesticide-treated and pesticide-free crops with close spacing, at high soil fertility from a new improved cultivar with protective irrigation. Many workers, including Adkisson (1958), have found that close spacing and increasing fertility increased the pest problems. New cultivars are often bred on the research stations under umbrellas of protection and yield well when fully protected and grown on highly fertile soils. In such cases, the reported increases from pesticides are unrealistic for planning exercises, for nobody will grow a close-spaced, high-fertility, pest-susceptible crop without protection from pests. Similarly, the increased yields reported from trials of fertilizer and new cultivars will have been obtained under protected conditions but this is seldom considered to be worth mentioning in the report.

When considering inputs, it is relevant to compare the protected, high-fertility, close-spaced, new cultivar and the unprotected local cultivar grown at the traditional spacing and at the natural fertility level. The difference in gross return can then be estimated, costing all inputs including pesticide, fertilizer, seed, labour costs, and perhaps irrigation. Such comparisons may well reveal an unattractive investment from pesticides, particularly where the farmer has little capital, where credit, if available, involves very high interest rates, and where the risks of flood, drought, starvation, and a market collapse for that commodity and certain socioeconomic factors have to be considered. At some stage, the pesticide scientist must relate his findings to those of other disciplines and together with economists make appropriate observations and experiments on farmers' fields. Such experimentation is the acid test of the reseach team's individual efforts and claims. (see also Chapter 7).

6.6   Experimenters and statisticians

If a statistician is available then he should be consulted at the planning stage. He is often only called in after the experiment has been completed and when the experimenter has admitted defeat in analysing the data. Then the statistician may or may not be able to help, but the information obtained will often be less than it could have been.

Finney (1960) considered that: 'The statistician can produce good designs only if he understands something of the particular field of research, and the experimenter will receive better help if he knows the general principles of design and statistical analysis. Indeed the two roles can be combined when an experimenter with a little mathematical knowledge is prepared to learn enough of the theory of design to be able to design his own experiments.'

Most experimenters will have learnt some statistics and will be familiar with the design and analysis of simple trials which are often the most appropriate to use for pesticide work. However, it is still worthwhile discussing plans with a statistician since this will entail defining objectives precisely and listing out the special constraints involved in pesticides and pest experimentation before field work starts.

6.7 Interplot effects

A major and generally underestimated problem to be dealt with in field trials of pesticide usage is the 'interplot effect' as described by Joyce (1956).This was defined as, 'The interaction of one plot with the insect population on an adjacent plot. This effect is considered to operate in three possible ways: (a) spray drift,(b) insect (pest) movement and (c) insect (parasite and predator) movement.' Pearce (1976), while considering the design of experiments on pests and diseases of fruit trees, noted that 'trials of this sort need fairly large plots because of the difficulty of confining the treatments within a limited area. Also, a fairly high standard of guarding is called for because pests and diseases can spread as well as treatments'.

In trials on cotton in the Sudan, Joyce and Roberts (1959) studied the extent of the interplot effect with regard to various components of the pest complex. They found that insect populations and yields of 4.2 ha plots were measurably affected by the treatments in plots 150 metres and even 450 metres distant! They considered that spray drift was unlikely to have been involved and that the effects were due to insect movement.

This report appeared to attract little attention and most pesticide workers continued with small-plot trials. Work in Uganda (Davies, unpublished) measured marked effects on insect numbers and damage by both *Taylorily-gus* and *Heliothis armigera* for 21 feet either side of sprayed/unsprayed boundaries and small effects thereafter for up to 75 feet. In Tanzania and Uganda, Reed (1972, 1976) queried the use and validity of untreated control plots in small-plot pesticide trials. A 14-year series of pesticide trials at Namulonge in Uganda, using randomized block designs, including unsprayed controls with plot sizes of up to 0.1 ha, gave yield differences ranging from a loss of 11% to a gain of 70%, but with an average increase of 11%, which was insufficient to pay for the pesticides used.When large blocks of pesticide-treated and -untreated cotton were compared over 4 years on the same farm yields were increased in a range from 43 to 228 %, with an average of 79%. The major pest in the complex was *Taylorilygus vosseleri*, a very mobile insect. In the unsprayed plots of the small plot trials, populations of this pest were greatly reduced by dispersion into neighbouring treated plots where they fell victim to the pesticides; conversely, dispersion from unsprayed to sprayed plots after the pesticide ceased to act would cause yield underestimate. Thus, the damage caused by mobile pests such as *T. vosseleri* in unsprayed controls within small-plot pesticide trials would be much less than that in larger blocks of unsprayed crop. In this way, small-plot trials will tend to underestimate the yield increases that would result from pesticide use against mobile pests. There is also a possibility that untreated controls, when surrounded by treated plots, may be shielded from pest invasion, thus leading to an over-estimate of the untreated crop yield and a subsequent underestimate of the pesticide benefit. With more static pests, such trials may overestimate the yield benefits. Here, the parasites and predators will be more mobile than their hosts and these natural control elements will be reduced by dispersal and death in the treated plots, so leaving the pests to thrive unchecked in the untreated plots. Reed (1972) considered that trials on cotton in Tanzania provided evidence of such a situation where *Heliothis armigera* was the major pest.

6.8   Experimental design and block allocation

For most pesticide trials the simplest designs such as completely randomized or randomized blocks are best if there is room for all treatments together. Missing plots may well occur and if the original design is unbalanced vital comparisons may be lost even though, using a computer, statistical analysis is possible.

The randomized-block design is the most frequently used lay-out for field comparisons of pesticides, although Pearce (1978) has suggested that completely randomized designs might sometimes be preferable particularly if the variability can be reduced by utilizing data from neighbouring plots to make adjustments.

However, blocks provide administrative advantages, such as making it easy to check that all treatments are replicated correctly and in providing suitable subgroups of plots for harvesting or treating on the same day; so, if they can be chosen reasonably, they do offer advantages.

Blocking in most agricultural trials is normally done on the basis of obvious soil differences, thus knowing the variation in crop yield over the area in previous years is very useful.

6.9   Pesticide drift

With some pesticide application methods, such as the use of granules in soils, there will be little danger of drift, but even so the movement of rain or irrigation water across the soil surface, or seepage, may carry the pesticide from one plot to another. There is clearly a much greater problem with drift when pesticides are applied from the air, particularly in small-droplet sprays or dusts. The use of ultra-low-volume, or controlled-droplet applications, either from aircraft, or ground sprayers, where pesticide drift by air movement is used to provide adequate coverage in the form of overlapping swaths, will give particular problems in comparative tests in the field (see Chapters 4 and 5).

The use of barriers between plots may help to reduce the drift to neighbouring plots. Portable screens are often used for this purpose as these are most convenient and cause fewest problems. Living barriers of taller crops between plots have also been used, but these may interfere with the ecology of the test crop and its pests. Living barriers may deprive the nearby crop of light, nutrients, and water, and affect the distribution of wind-carried pests (Lewis 1965a,b).

If controlled-droplet or ultra-low-volume sprays are to be used, and there are prevailing winds, then oblong plots with the greater axis lying along the wind direction will probably be most convenient. Where wind directions are variable then oblong plots may still be of advantage, but the treatments should only be applied when the wind is in a convenient direction.

For most pesticide application, drift can be minimized by treatment at a time when air movement is slight. Such precautions, and the provision of discards (which are treated similarly to the enclosed plot), will generally reduce the drift problem to a level that can be ignored. However, anybody who has seen the effect of 2,4-D ester drift on sensitive crops such as cotton will be well aware of the dangers of underestimating the drift problem.

6.10   Control plots

The validity of 'unsprayed control' plots has been questioned in a previous section, but there are other reasons for caution when including 'controls' in pesticide trials and interpreting the results from these. However, there is a case for maintaining 'standard' unsprayed crop areas well away from sprayed areas on an annual basis to build up information on seasonal fluctuations in pest numbers over several years.

In field trials of pesticides, untreated and ineffectively treated plots will act as foci for the reinfestation of the treated plots after the residual effects of the treatments have disappeared. In the absence of such foci of infection, the treatments could be expected to provide longer-lasting pest reduction in the treated plots. Van der Plank (1963) coined the term 'representational error' to describe this effect, where plots differ from the fields which they are intended to represent. Jenkyn (1977) noted that this is not a major problem where soil-borne fungi are the target pests, but he produced evidence to show that such errors can be substantial with air-borne pathogens such as *Erysiphe graminis*. Similar problems may be expected in herbicide trials where weed seeds are wind-dispersed.

Such objections to control plots can be partly overcome by replacing the untreated check plot by a known standard control (Jenkyn, 1977). In many cases, the benefits of pesticide use on a crop will have long since been established. Current experimentation will be concerned with the search for improved pesticide usage, either by comparing new chemicals, or testing differing dosages and timings and new methods of application. In such comparative trials, the pesticide application currently recommended for use on that crop should be used in the check plots making untreated controls unnecessary. This should greatly reduce the interplot effects and representational errors, but not remove them entirely. In herbicide trials, hand-or machine-weeded plots will serve as checks, since weedy control plots are of little purpose, unless of course the experimenter needs to use them as a standard.

6.10.1   Guard rows and infestor rows

Interplot effects can be reduced by the provision of guard rows around each plot. Although it is essential that these rows be given treatments similar to those of the enclosed plot, they should not be included in sampling for pests nor for yield data. It may be convenient to bend away, uproot, or cut these rows just before harvest, thus providing convenient pathways, and reducing errors by clearly delineating the areas to be harvested. However, in trials such as spacing comparisons, where the ecoclimate may affect pathogen and insect damage, care must be taken to leave the guard rows as long as possible before the plot is harvested. Extra guard rows along the edge of trials may be of value in filtering out dust from a nearby road or in providing a buffer from the depredation by cattle, goats, or people.

When pesticides are being evaluated against sporadic pests, it may be necessary to boost the natural populations during the trial. This may be done by breeding the pest and then releasing or spreading it evenly through the trial field. For bacteria and fungi which affect leaves it may be sufficient to collect a sack or two of infected leaves at the end of each season, keep them in store and then scatter them through the plots in the next season. Alternatively, it may be necessary to prepare cultures of the organisms which are then sprayed evenly across the plots. In addition, infestor or infector rows can be grown, either replacing the guard rows or as a proportion of the rows within the plots. These infestor rows may be of the same crop cultivar but sown earlier and artificially inoculated so that they will act as foci of infection, or they may be of a particularly susceptible cultivar that will attract and build up pest populations that will then disperse across the test plots (Starks 1970; Williams and Singh 1981).

Beware of overestimating yields from small plots where guard or infestor rows have been uprooted or killed well before harvest. Many crop plants have a remarkable capacity to grow quickly and fill space, both above ground and below. A two-row plot from which the adjacent rows have been removed or have died may well yield almost twice as much as a similar-sized plot subject to competition from plants in those adjacent rows.

6.10.2 *Plot size and shape*

Earlier, we cited an example (Joyce and Roberts 1959) where 4.2 ha plots separated by 150 m gave measurable interplot interference. Few experimenters will be afforded the luxury of such areas for their pesticide trials so there is no way in which most will be able to eliminate interplot effects by resorting to large plots and massive discards. Choice of plot size will be entirely dependent upon the objective of the experiment, the pesticide application method, the nature of the pests to be controlled, and the area available.

If the objective of a trial is to compare the efficiencies of differing pesticides in killing relatively static pests, then very small plots even down to single plants may be sufficient. In the instance of pests such as aphids, coccids, and bacteria or fungi causing leaf spots, even individual branches on large plants may serve as plots, provided the pesticide is non-systemic.

Alternatively, if the objective is to estimate the economics of pesticide use and its overall effect on the pest and natural enemies, then it may not be possible to design a conventional replicated trial that will give valid estimates in a single area of practical size. In such cases, the objective must be to compare pesticide-treated and -untreated areas in plots large enough to eliminate interplot effects. This will often involve the use of whole fields or even groups of fields as plots. Each plot should ideally form an ecological unit, where the pests and natural enemies in one plot are shielded from the effects of pesticide use or non-use in the comparison plots. Where pests are migratory, as the locusts, countries or years may well have to be considered as plots.

Wishart and Sanders (1955) recommended the use of long narrow plots, for these have advantages in sowing and accessibility and they also cut across spot differences in soil variability and so reduce variance between plots. Such a recommendation was, however, primarily intended for agronomic and breeding trials. For most pesticide trials, the squarer the plot the better, for this cuts down on the perimeter length and reduces the area that has to be sacrificed to guarding. The obvious exception to this has already been mentioned, where sprays are distributed by wind so requiring long plots with the axes along the prevailing wind direction.

6.11 Replication

The most important question to answer, when we are choosing the number of replicates, is what size differences between treatments are important in practical terms? If we then have a measure of variability from previous trials we can use tables from Cochran and Cox (1957) to tell us how many replicates we need to show up such differences. Clearly for large differences or low variability we need less replicates than if we are interested in small differences or have high variability.

In theory, with unlimited replication we could show significant differences between all treatments since they are unlikely to be identical but this would be a waste of effort. In practice we usually have a limited budget of land, cash and recording ability and have high variability so, without a measure of the latter, we should have as many replicates as possible without cutting plot size.

Steel and Torrie (1960) cautioned that replication will not reduce error due to faulty technique and they also advised that if you cannot run an experiment with enough replication to obtain the required precision, then it should be postponed, or the number of treatments reduced to allow more replication. However, a possible alternative is to set up other replicates in subsequent years if some treatment effect looks interesting but is not statistically significant. Since we wish to generalize from experimental results it will be useful to have information from different years.

Finally let us not be obsessed with statistical significance. Reed (1976) cited an example where the search for statistical significance hindered progress and the comparison of unreplicated large blocks gave practical information not given by a series of well-replicated small-plot trials with plots that were too small. A possible compromise is to have large blocks replicated twice so that at least we can see that there is consistency in the results.

6.12   Randomization

Most trial designs and sampling techniques involve the use of random sampling from all available possibilities. The concept of random observation is fundamental to the theories of probability and statistical estimation. Random allocation of treatments to plots and random collection of samples are valuable safeguards against conscious and subconscious bias. It is recommended that randomization should be achieved with the aid of tables of random numbers such as those of Fisher and Yates (1963), computer programs, or drawing of random numbers blindly.

Most experimenters having randomized a trial will study the layout hoping that it appears to be random and convenient. Many will be unable to resist the temptation to 'rerandomize' layouts where there appears to be an unfortunate juxtaposition of plots within and across blocks. Statisticians consulted generally agree that such rerandomization is a lesser evil than that of proceeding with a trial where the unfortunate juxtaposition of plots may present problems in interpretation. If the relative positions of plots are of crucial importance, however, then layouts involving restricted randomization such as the row and column designs will largely overcome this problem.

6.13 Ensuring validity of treatment effects

There are several factors that can promote spurious treatment effects. The actual application of pesticide treatments in trials is crucial to the validity of the comparisons, and gross errors can be made at this stage. If the experimenter is not physically applying all of the treatments himself, then he must be present and very alert in supervision. In many developing and tropical countries, the application of pesticides to field trials and indeed to farmers' crops is by hand sprayers, wielded by low-paid and often illiterate labourers. In such cases, all stages of the operations, including measuring of the pesticides and diluent, mixing, spraying, and cleaning must be double-checked, both to protect the health and safety of the workers and to ensure that the intended treatments really are applied on the correct plots.

It is essential that treatments be applied block after block, rather than treatment after treatment. If treatment A is applied first, over all the replicates, followed by B, C, and D, then changes in weather and the increasing skill, or flagging enthusiasm, of the applicators, may themselves become important factors in the coverage of the plants or areas and so produce differences between treatment effects that are divorced from the intended treatments. Ideally, when applying treatments in a trial, all plots in each block should be applied simultaneously by one man using one sprayer, but this is obviously impossible. A good second best is to have available a number of sprayers and spraymen equal to the number of treatments to be applied. All the spraymen should have been previously trained so that their speed of walking is appropriate to the intended application rate and all sprayers should have been checked to ensure that their outputs and spray characteristics are as near identical as possible. If 10 litres of spray mix is intended to be applied to each plot, then the rate of walking should be such that this amount is exhausted just as the end of the plot is reached. Spraymen and sprayers should be randomly or systematically interchanged between blocks, otherwise the treatment effects may become confounded with differences between spraymen or sprayers. A good sprayman who has taken the trouble to ensure good coverage of the target, if assigned to one treatment across all the blocks, could be responsible for a significant increase in the pest reduction in that treatment, particularly if another treatment has been applied by an inefficient man who misdirected applications.

Such precautions may be impractical in some pesticide trials. Economics will usually dictate that only one spraycraft will be available for aerial spraying trials, so the simultaneous spraying of all treatments in each block will be impossible. It is better to spray each plot in a block before moving on, but to minimize the number of flights it may be necessary to spray all the plots requiring the same treatment and then fill up with the next pesticide. If the weather conditions change, there is an obvious danger of the intrinsic treatment effects being obscured by the effects of the differences in coverage obtained. As in all trial work, the ideals are often impracticable and then we must be fully aware of the limitations when interpreting the results.

Some experimenters insist that the 'control' plots should also be treated with the carrier – water or dust – because the movement of the sprayer through the crop, or the carrier, may in itself affect the pests or crop. We consider such precautions to be fallacious as the treatment itself must be considered to be a combination of the pesticide, its carrier, and the passage of the applicator through or over the crop. If with a high-volume spray, the water content of the treatment has a beneficial effect on the crop, then that is as much a treatment effect as the pesticide effect on the pests. Such component effects may be elucidated if inert treatments and nil treatments are included in the trial design. Note, however, our previous comments that 'control' plots can be advantageously replaced by standard treatments in many instances.

On occasion, there will be a question of whether or not a treatment should be repeated if rain falls during or soon after application, for the pesticide may have been washed off the plants and so be ineffective. Most pesticides are formulated to be relatively 'rainproof' and there will generally be no need to repeat treatments, particularly if the spray has had time enough to spread and dry. However, it is worth emphasizing that notes should be kept of weather conditions to aid in interpretation of data. Normally, of course, treatments would be delayed if rain is likely. It is important to stress that all plots including the control should receive identical treatment and if weather, e.g. hail, affects plots unequally it is sometimes necessary to discard a replicate or even a trial.

6.14   Data collection

In most trials the aim will be to compare pest and damage reduction by different pesticides. There are three ways of doing this. The first, and the most important, is to measure the crop yield and value since pesticides are applied to prevent loss. However, sometimes it may be preferable to measure the damage itself by, for example, counting deadhearts in sorghum or flared squares in cotton. Finally we might have to measure the pest population just before and at intervals after treatment applications to quantify the latter's initial effectiveness and the subsequent rate of reinfestation. This will also allow comparisons between pesticides when the pest population is too low to cause much damage.

We must first decide how to make our measurements. For yield it will be easiest to harvest and weigh the whole crop within the guard rows. However, probably for damage estimates and almost certainly for pest population estimates, we shall have to take samples so we need to decide on sampling units. This will depend on the crop, the pest, and the variable we decide to measure but will be chosen to keep the variability as low as possible. In situations with continuous groundcover, such as a pasture or a crop sown broadcast, it is probably best to use a fixed-area sampling unit preferably defined by a frame. For row crops either fixed lengths of row, fixed numbers of adjacent plants or individual plants may be used as the sampling unit. Thus for counting shootfly eggs on sorghum seedlings the sampling unit could be a 5-metre length of row or 50 adjacent seedlings. In crops where plants are unevenly spaced, samples from set small numbers of plants will tend to be variable because of uneven growth. In cotton, for example, plant size, leaf area, boll numbers, and pest-carrying capacity may vary by a factor of four and then fixed-length-of-row samples will give lower sampling variance. However, for large plants a fixed number of plants may be better because whole plants being included or just excluded will increase variability. When the sampling unit is a fixed number of plants the plant density should also be estimated since the treatment may change it.

Sometimes sampling units may only be small parts of plants if, for example, we are counting leaf spots on apple trees or thrips on groundnuts. Here the problem is more difficult because we must ensure that the samples are representative of the whole. It is most convenient to count the leaf spots on the leaves at eye level, but will these tell us anything of the infestation at the top of the tree? We need to do some preliminary sampling to ensure that we are taking samples that are either representative of the whole, or a fixed proportion of the whole, so that they give valid comparisons of treatment effects.

Sometimes our sampling unit may be completely independent of the crop if we have to estimate the pest population by using traps, baits or soil samples. Southwood (1978) gives a good summary of possible sample units and the variables to measure.

We can either make complete counts of pests or damage on the sampling units to give absolute population estimates or we can give relative population estimates, such as the number of insects on a fixed number of twigs on a tree, both enabling us to compare treatments. For relative estimates a simple scoring system, usually using digits between 0 and 9, may suffice. For analysis by computer 0 should be reserved for no information and 1 used for no pests or damage. If pests are to be scored four or five classes, probably defined on a logarithmic scale such as 1,5,25..., will be about the most we can distinguish. Note that scores will probably require special analysis techniques as they are not continous and usually not normally distributed. They are also prone to subconscious bias so it is best for the scorer not to know the treatment or to have two independent scorers.

Having chosen our sampling units and variables we now have to decide how many samples to take and how to make them representative. Two useful books on sampling are by Yates (1981) and Cochran (1977).

Theoretically the sample should be chosen so that all sampling units are equally likely to occur. In practice this may not be possible, particularly if a series of samples has to be taken, since many sampling methods are destructive and may affect future samples. An obvious example is when tomato plants are dug up to count nematode cysts on the roots when comparing nematicides. Less obvious destructive sampling may occur if handling plants while counting pests affects plant growth and future pest populations. To offset the effects of destructive sampling we may have to allocate portions of a plot for such samples and then avoid them for subsequent samples and yield determinations. However, samples small in relation to the total population may have little effect; thus insect pests collected from less than 5% of the crop will probably not affect the overall population much.

Sampling units can be selected in many ways but the most common is to choose a completely random sample. To ensure that this really is representative of the plot, random numbers should be chosen in advance so that the chosen units are precisely defined. Any method allowing subjective choice such as taking the row next to a 'randomly' thrown hat or rejecting abnormal plants should be avoided if possible. The data in Table 6.1 (unpublished) from a spacing trial of chickpea at ICRISAT provides an extreme illustration of bias caused by the latter. The small plants which were numerous in the closest spacing were eliminated from the samples, the recorder's reasoning being that such plants did not contribute to the yield and so could be ignored. He was convinced that this sampling was still at random! Occasionally, such as when the sample is small, abnormal plants may be avoided, since this will reduce the variability, but only when the abnormality is known to be independent of the treatment.

Another way of choosing a sample is to divide the plot up into different areas or strata and sample at random from each one. This may be because there are several types of habitat in the plot and making separate population estimates from each and combining them appropriately will reduce variability. Also, because pests tend to be clumped, the population may vary across

TABLE 6.1.

Plant density	$33/m^2$	$8.3/m^2$	$2.9/m^2$
Yields from sample plants (kg/ha)	735	663	640
Actual yields from plots (kg/ha)	396	626	645

the plot and again variability will be reduced, as well as a better coverage obtained, if the plot is subdivided before random samples are taken.

The drawback to taking predefined completely random or stratified random samples is that it will be time-consuming to find the sample locations. A practical alternative is to take a systematic or semi-systematic sample. For example, we might decide to take four samples on each of three rows. We could either choose the three rows at random or if we had, say, 30 rows we could choose a random number between 1 and 10 to get our first row and then take the rows 10 and 20 on to space them over the plot. The positions along the rows can be chosen in advance in the same way. Such systematic samples should be avoided if there is some plot effect which occurs cyclically, making the sample unrepresentative. Since in most trials with pesticides between-plot errors will be used to test treatment differences, completely random samples within plots are not theoretically necessary. The advantages of systematic samples are that it is easy both to locate the sampling units without subjective bias and to get good coverage of the plot without excessive sampling. Also, if the results are recorded appropriately the pest distribution within the plot will be obtained.

We must now decide how many samples to take. Obviously the more we take the better will be the pest population estimate on each plot. However, if, as is likely, there is a large plot-to-plot variability, the plot estimates need not be very precise. The number of samples has to be balanced against the size of the sampling unit. If the pest is aggregated then, for a sample of fixed area, the sampling variability will drop as the unit size decreases and the number of units sampled gets larger. It is then usually useful to run some pilot trials to help decide what is a suitable balance between precision and convenience.

Whatever variable we measure and whatever sampling unit and sample size we take the basic aim is to obtain objectively a representative sample and not what we feel, subjectively, is a representative sample.

If we are sampling to see whether a pest population has reached a particular density before spraying it may be inefficient to take a pre-determined number of samples. In most cases, after relatively few samples, the counts will be so far below, or above, the threshold level that further counts will be superfluous; only in marginal situations will it be necessary to complete a full survey (Ingram and Green 1972). In sequential sampling, appropriate here, one of three decisions is made after each sample—to spray, not to spray or to carry on sampling. These decisions are based on fixed rules derived from previous samplings of the same pest on similar crops. A sequential sampling package has been developed for cotton arthropods in Texas (Sterling and Pieters 1974) which might well be more generally applicable.

6.15   The analysis and interpretation of data

Data collected from a pesticide trial will have to be analysed and are usually presented in the form of estimated means for the various factors. It is generally useful to quote not only such means but also a statistic that will give an indication of the confidence that can be placed both upon the means themselves and upon the differences between the means that are to be compared. This normally necessitates using analysis of variance, and details of how to calculate this are given in statistical text books such as Bailey (1981) and Steel and Torrie (1960).

When analysing pest population data it may be necessary to transform counts by taking logarithms, or square roots of them, since they tend not to have symmetrical distributions. However, since transformed means are less easy to interpret, this should be avoided unless the distributions are very skew.

When interpreting the results definite recommendations should usually only be given if treatment differences are shown to be statistically significant by the analysis. However, it may be well worth doing more trials when apparent differences occur which are not statistically significant, since it is often difficult to set up enough replicates to show up differences big enough to be of practical interest.

6.16   Economic thresholds and compensatory growth

Much has been written on the determination of economic threshold levels for pest damage which is so important to the development of integrated control strategies (see Chapter 7). The problems in determining such levels in the field are formidable, since although laboratory feeding and breeding tests and computer modelling may be of help, damage and pest numbers in the field are subject to a whole host of often interconnected factors.

Threshold determination will only be simple where the pest is univoltine, can be counted at a particular stage before it causes damage, and is on a crop which is uniform, of known value, and non-compensating. Then a simple randomized block trial using artificially introduced pest numbers or selective destruction of the natural population can be used to find the relationship between pest numbers and crop loss. The value of the crop lost at different pest populations can then be compared to the cost of effective treatment to establish a threshold level.

Unfortunately, few pests or crops are as accommodating. Mobility of pests is a problem, and, though caging is possible, experience has shown that both pests and plants are unlikely in these situations to be representative of those in the open field. Overlapping pest generations, age of the insect pest population, particularly if this has an important effect on consumption damage, and presence of other pests add to the complexity. It may well therefore be a necessity to calculate thresholds at more than one stage for reasons associated with the pest or plant. The latter requirement is related to the ability of many plants to compensate for losses; examples can be cited for a range of crops from sorghum (Davies and Jowett 1970), to cotton (Brown 1965), and pigeonpea (Reed *et al.* 1981). In many instances, weather plays an important role in determining if adequate compensation for early loss occurs. Economics, through fluctuating prices, often greatly alter pest thresholds seasonally. Crude thresholds can often be determined from trials not specifically set up for this purpose. Such crude estimates can be refined by subsequent experimentation/experience.

Much 'controlled' research station experimentation is ultimately aimed at farmers' fields and profit for the producers, who themselves vary enormously in ability to farm and in sophistication of production methods. The dangers of misleading estimates based on small-plot comparisons have already been stressed. In the final analysis, the only convincing evidence must come from target farmers' fields and from more than one season, to take account of weather and pest fluctuations.

The benefit or detriment of any pesticide may extend beyond the crop to which it is applied and this must not be lost sight of when assessing the economics. For example, pesticide use on cotton in the Punjab can permit the crop to mature more quickly and so facilitate the sowing of another crop, such as wheat, on the same land in that year, and such benefits may equal or exceed the simple yield benefit produced by pesticide use on the cotton. Again, persistent pesticide use may kill much of the soil fauna, compounding soil compaction problems in future years, and destruction of natural enemies by pesticide on one crop may be detrimental to natural control on other crops.

The answer to all such complexity may be to forsake the search for statistically significant differences at the farmer comparison level. We first of all have to demonstrate that a new treatment is more profitable both in the short and long term, and then convince the farmer; most will be able to assess whether or not a treatment is worthwhile when they see its effect in their own or their neighbours' field.

6.17 Reduction of errors

Variance is inherent in all biological systems and particularly so in field trials of pesticides. Large coefficients of variation, in pest counts and plot yield, are in some cases unavoidable and trial work can be particularly unrewarding where treatment differences of 10% are sought in field trials where coefficients of variation exceed 30%. However, we have found that careful supervision of each and every trial stage and operation, from sowing to data analysis, can greatly reduce the errors. In developing countries, it is often tempting to delegate mundane tasks such as laying out trials, measuring and mixing pesticides, sampling pest populations, harvesting and subsequent weighing to subordinates, but in practice such mundane tasks can go badly wrong in many unforeseen ways. Hastily written figures, mixed up plot yield bags, harvesting the wrong rows, wrong transposition of records from field notes to computer sheets, and simple arithmetical errors can all contribute to the large coefficients of variation which are mistakenly believed to be results of unavoidable error. The only formula for successful pesticide field experimentation is constant scrutiny and double-checking at every stage. Plot areas should be carefully measured and rechecked, both at sowing and at harvest, log books should be standardized and all operations carefully recorded as they are undertaken. If field counts have to be transposed from the field data sheets to the log book and from there to the summary sheets or to the computer, then they should be double-checked, if possible by more than one person. Sacks used for harvesting plot yields should have labels inside the bag and outside. Balances or weighing scales should be checked before and after weighing. The raw data of pest counts and yields should be summarized by the scientist and means for treatments computed by hand and head before they are entrusted to the computer, and preferably immediately after the record has been taken so that anomalies can be checked by a return visit to the field or to the weighing store. Occasionally human error or natural calamity, such as the application of the wrong treatment on a plot, loss of a bag at harvest or a record sheet or even a lightning strike will render results from a plot or block useless. Statistical techniques exist for allowing for such justifiable 'missing plots'—but they must be used only where elimination is based clearly on reasons unconnected with the treatments.

6.18   Conclusions and recommendations

When trials have been carried out efficiently, analysed correctly and significant differences between treatments are found, questions must still be posed, e.g. will a 95% probability that pesticide treatment A is better than treatment B be sufficient justification to change a recommendation immediately, or should the trial or trials be repeated for another season and over more sites? Are we sure that the differences that we have measured are really caused by the treatment applied? Was the trial carried out in a manner that is likely to indicate what will happen in the real world in the farmers' fields? In most cases, of course, the recommendations for pesticide usage will be based upon a series of trials, probably starting with trials of toxicity in the research station and progressing to extensive trials in farmers' fields. Wishart and Sanders (1955) recommended the practice of repeating trials over areas and years, keeping plot size, replications, and treatments the same, but with differing randomizations in each trial. The data may then be combined to good effect. Sometimes, however, pressure to produce a recommendation for the farmers' in as short a term as possible, particularly to cope with a novel but severe pest situation, will not afford the luxury of a well-planned series of trials over several seasons. In such cases, we have to be very confident that our experimentation is as well organized as the circumstances permit.

At the other extreme, we should not assume that treatments which do not give significant differences in our trial or trials are therefore equal. Such reasoning may blind us to quite important differences or treatment benefits. The lack of significant differences between treatments in any trial may result from an inability, usually forced upon us by economics, to replicate sufficiently and to keep the environmental and sampling errors to a minimum.

6.19   References

Adkisson, P. L. (1958). *J. Econ. Entomol.* **51**, 757.

Bailey, N. T. J. (1981). *Statistical methods in biology* (2nd edn) Hodder & Stoughton, London.

Brown, K.J. (1965)*Empire Cott. Gr. Rev.* **42**, 279.

Cochran, W.G. (1977). *Sampling techniques* (3rd edn). Wiley, New York.

—— Cox, G. M. (1957). *Experimental designs* (2nd edn). Wiley, New York.

Davies, J. C. and Jowett, D. (1970). *E. Afr. Agric. For. J.* **35**, 414.

De Bach, P. (ed.) (1970). *Biological control of insect pests and weeds* (3rd edn). Chapman and Hall, London.

Dyke, G. V. (1974). *Comparative experiments with field crops.* Butterworths, London.

Finney, D. J. (1960). *An introduction to the theory of experimental design.* University of Chicago Press, Chicago.

Fisher, R. A. and Yates, F. (1963). *Statistical tables for biological, agricultural and medical research* (6th edn). Oliver & Boyd, Edinburgh.

Ingram, W. R. and Green, S. M. (1972). *Cott. Gr. Rev.* **49**, 265.

Jeffers, J. N. R. (1978a). *Statistical checklist 1: design of experiments.* Institute of Terrestrial Ecology, Cambridge.

—— (1978b). *Statistical checklist 2: sampling.* Institute of Terrestrial Ecology, Cambridge.

Jenkyn, J. F. (1977). *Pest. Sci.* **8**, 428.

Joyce, R. J. V. (1956). *Nature (London)* **177**, 282.

—— Roberts, P. (1959). *Ann. App Biol.* **47**, 287.

Lewis, T. (1965a). *Ann. App. Biol.* **55**, 503.

—— (1965b). *Ann. App. Biol.* **55**, 513.

Pearce, S. C. (1976). *Commonw. Bur. Hort. Plant Crops. Tech. Commun.*, no. 23.

—— (1978). *Trop. Agric. (Trinidad)* **55**, 97.

Reed, W. (1972). *Cott. Gr. Rev.* **49**, 67.

—— (1976). In *Agricultural research for development (the Namulonge contribution)* (ed. M. H. Arnold), p.123. Cambridge University Press, Cambridge and London.

——Lateef, S. S. and Sithanantham, S. (1981). *Proc. Int. Workshop Pigeonpeas,* **1**, 99.

Southwood, T. R. E. (1978). *Ecological methods with particular reference to the study of insect populations* (2nd edn). Chapman and Hall, London.

Starks, K. J. (1970). *J. Econ. Entomol.* **63,** 1715.

Steel, R. G. D. and Torrie, J. H. (1960). *Principles and procedures in statistics.* McGraw-Hill, New York.

Sterling, W. L. and Pieters, E. P. (1974). *Tex. Agric. Exp. Stn., Dept. Entomol. Tech. Rep.,* 74.

Unterstenhöfen, G. (1976). *Pfschutz-Nachr. Bayer* **29,** 83.

Van der Plank, J. E. (1963). *Plant diseases, epidemics and control.* Academic Press, London.

Williams, R. J. and Singh, S. D. (1981). *Ann. App. Biol.* **97,** 263.

Wishart, J. and Sanders, H. G. (1955). *Tech. Commun. Commonw. Bur. Plant Breed. Genet,* no 18.

Yates, F. (1981). Sampling methods for censuses and surveys (4th edn.). Griffin, London.

7
Economics of pest control

G. A. NORTON

7.1 Introduction

The central role of ecology in the theory and practice of pest management is now generally accepted. What is not so well recognized is the equally important role of economics – the science concerned with the allocation of scarce resources for the satisfaction of human wants. Since pest problems are determined as much by social perceptions, constraints, and values as by the state of the natural world in which we live, the whole concept of a pest, and its control, is meaningless without resort to socioeconomic considerations.

This chapter deals with the two major contributions an economic approach to pest control can make. First, we consider the way in which decisions on pest control are made. An understanding of this process is essential if research and extension effort is to be effectively directed towards feasible techniques of control. Second, we consider the evaluation of control strategies and, in combination with other disciplines, show how economic principles can be employed to assess best strategies of control.

7.2 Pest control decision making

As with other natural hazards (Kates 1970; Slovic *et al.* 1974), the decision problem associated with pest attack can be resolved to two questions:

(a) What form of adoption to pest attack should I take? For instance, should I rely solely on cultural methods of control, or should I use resistant varieties or pesticides as the main control strategy?

(b) When and how should I adopt these measures? If I am to apply pesticide, for example, when is the best time to apply it? How often should I apply it? In what form should I apply it – in granular form or as an ultra-low-volume spray? and so on. To answer these questions, decision makers – whether individual farmers or regional control agencies – will take account of three factors:

(1) the nature of pest attack and damage,
(2) the range of control measures available, and
(3) their objectives.

7.2.1 Pest attack and damage

Two important features of a pest affect the nature of the decision problem:

the life cycle of the pest and the damage it causes. All pests, whether animals, diseases, or weeds, can be separated into two categories on the basis of their life cycle (cf. Bunting 1972). Endogenous pests, such as nematodes, graminaceous weeds, and soil-borne diseases, remain for most or all of their life cycle within the decision-maker's realm. These pests all have the potential for buildup within the crop or farm area although, by the same token, control measures taken against them are likely to affect subsequent levels of attack. Exogenous pests, in contrast, spend only part of their life-cycle within the decision-maker's domain and control of these pests – typified by migratory locusts – is taken on a short-term, 'one-off' basis. It is clear that with this definition, exogenous pests can be converted to an endogenous form by transferring decision making to a regional or national level.

A second classification of pests can be made according to the form of their damage relationship (Southwood and Norton 1973). For certain pests (Fig. 7.1a), there is a threshold level of pest attack below which damage does not occur. This situation arises where the crop is tolerant to pest attack or is able to compensate for some of the injury caused. In contrast, the situation shown in Fig. 7.1b occurs where the crop is highly susceptible, is unable to compensate, or is attacked by a disease vector or by a pest that causes serious loss in the marketable quality of the product. In such cases, a high level of control is to be expected, with the emphasis on chemical rather than biological control agents.

7.2.2   Control methods available

The complexity of a pest control decision problem is determined by the range of methods available and the decisions involved in applying them. The

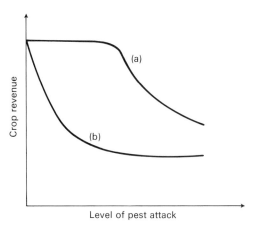

Fig. 7.1. Two categories of the pest damage relationship—(a) with threshold, (b) without threshold.

hypothetical decision tree for a crop farmer, shown in Fig. 7.2, sets out the broad range of measures technically available to him. There are two important points to be made with respect to such a decision tree.

First, the range of methods technically possible has to be distinguished from that which is feasible for a particular farmer. Particular control methods may be 'unavailable' for a variety of reasons; for example, the lack of an adequate water supply, no appropriate application machinery, a shortage of capital, lack of 'know-how,' seasonal labour constraints, or incompatibility with the cropping system. For many subsistence farmers,

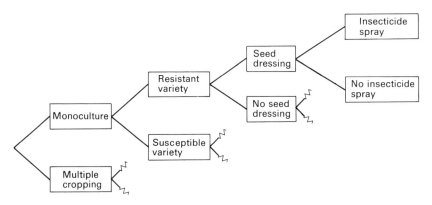

Fig. 7.2. Hypothetical decision tree for a crop farmer.

these constraints are such that cultural control methods, including the spatial and temporal diversification of planting associated with multiple cropping, are the only methods available (Norman 1974; Norton 1975; Norton and Conway 1977). Apart from on-farm constraints, restrictions may be imposed from outside, through inadequate marketing systems or as the result of legislation on pesticide use. Consequently, the range of feasible control strategies can be increased not only by developing new and improved techniques but also by reducing or removing on- and off-farm constraints.

Second, as well as identifying the range of methods, decision trees can indicate the order in which decisions are made. The sequence of decision making in Fig. 7.2 moves from left to right. Before sowing, a relatively wide range of measures is possible. As the crop matures, the range of choice narrows, the farmer's ability to adapt to pest attack becoming progressively limited. The dilemma that many farmers face is that while information on pest attack increases with time, available options decline.

To illustrate this point further, consider a particular pest problem. The aphid, *Myzus persicae,* is a vector of virus yellows, a disease of sugar beet in the UK. Faced with this pest problem, sugar beet growers can choose one of five options (Mumford 1978), as shown in Fig. 7.3. Since in-furrow treatment, with a systemic granular insecticide, is applied at planting time, the decision to apply this control measure has to be taken before the level of pest attack can possibly be monitored. By contrast, foliar sprays, which only work when in direct contact with the pest, offer opportunities for spraying only when needed, as determined by monitoring pest attack.

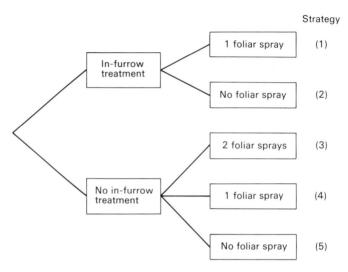

Fig. 7.3. Decision tree for sugar beet pest control (after Mumford, 1978).

Having identified the range of pest attack and determined the decision options, a pay-off matrix can be constructed (Table 7.1), providing a useful framework for assessing the outcome of different strategy–pest attack combinations. For the particular example shown in Table 7.1, outcomes have already been assessed and expressed in monetary terms. Whether this criterion is fully appropriate will depend on the third decision-making factor – the farmer's objectives.

7.2.3   Objectives

Since a variety of socioeconomic goals determine overall farming practice, pest-control measures are likely to be assessed on the same basis. For true subsistence farmers, the priority is clearly to safeguard the production of sufficient food to meet family requirements. In such cases, subsistence farmers are locked into their traditional farming systems as much by their inability to withstand losses as by the various constraints mentioned earlier. These farmers are said to be risk-averse, making pest-control decisions

TABLE 7.1. *A pay-off matrix for sugar beet pest control. These outcomes are for the Ely region in the UK and are expressed in terms of net revenue ($£s\ ha^{-1}$); that is, the revenue obtained from the crop (including remaining pest damage) minus the cost of the control strategy. (Source: Mumford, 1978.)*

Potential pest attack (expressed as classes of potential % loss)	Strategy (see Fig. 7.3)				
	1	2	3	4	5
0%	685	693	704	712	720
15%	631	623	639	633	612
45%	523	482	510	475	396

according to a maximin rule. That is, they attempt to find that strategy that maximizes food production under the worst pest conditions.

Although risk-aversion is still likely to be important for commercial farmers, since the main concern is crop revenue rather than yield *per se,* the cost of risk-aversion is likely to be more apparent. To illustrate the dilemma, consider a sugar beet grower who is choosing from two of the options shown earlier – namely strategy 1 or strategy 4 (Table 7.2). Strategy 1 provides good protection against pest attack but is expensive: strategy 4 is cheaper but gives less protection. The problem is to decide whether £27 ha^{-1} saved with strategy 4 when low attack occurs offsets the possibility of a considerably lower net revenue (£475 ha^{-1} compared with £523 ha^{-1}) if a high level of attack occurs. Although the probability of this occurring is important, as we will see shortly, the decision ultimately rests on the decision maker's subjective attitude to risk.

TABLE 7.2. *A two strategy pay-off matrix for sugar beet pest control (net revenue in $£s\ ha^{-1}$).*

Potential pest attack (expressed as classes of % potential loss)	Strategy	
	1	4
0%	685	712
15%	631	633
45%	523	475

7.2.4   Decision making in practice

Although this rather theoretical view of the pest-control decision making process provides a valuable viewpoint, can it explain what happens in practice? The short answer appears to be yes, provided we recognize that real-world decision making is based, not on objective assessments of pay-off but on farmers' subjective perceptions and estimates (Fig. 7.4), which may not be the same thing at all.

From a survey of sugar beet growers in two regions of the UK, Mumford (1981) found that growers' estimates of typical losses due to virus yellows were close to 'objective' field assessments but their perception of the worst losses that had occurred in their area showed far less agreement. Thus, while Mumford found these growers were always rational in their decision making,

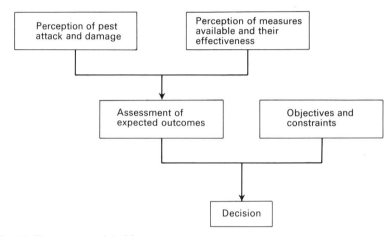

Fig. 7.4. The pest-control decision process.

according to their own perceptions, objectives, and constraints, poor perception of losses, combined with an overestimate of pesticide efficiency, means that growers may well be making poor choices. In such cases, where pay-off is incorrectly assessed, better pest control may be achieved by improving farmers' perceptions.

On the other hand, where farmers' objectives are the over-riding determinant of pest-control decisions, as Tait (1977) appears to have found for vegetable growers in the UK, and Turpin and Maxwell (1976) for Indiana corn farmers, a change in pest control inputs may require 'alternative institutional arrangements'. Since a perceived increase in risk is likely to be a major impediment to the adoption of novel control strategies, an associated crop-loss insurance scheme, that provides a reliable safety net, could meet this constraint. However, as yet, insufficient economic analysis appears to have been carried out to appraise this option fully (Carlson 1979). Perhaps a more promising alternative lies with pest management consultants. From studies of farmers who engage private consultants, Norgaard (1976) finds evidence not only that substantial reductions in pest management costs are possible, often with increased yields, but those who adopt consultants also tend to be the more risk-averse.

7.3 Evaluation of control strategies

In the previous section, we concentrated on obtaining an understanding of the way in which pest-control decisions are made. Here we focus on the evaluation of control strategies and, in particular, on the decisions involved in pesticide application. A generalized representation of the problem is shown in Fig. 7.5, the initial decision being whether to apply pesticides as a prophylactic, or in response to information on pest attack. After discussing the form of economic analysis appropriate to each of these strategies, brief consideration is given to a comparative model, and to the problem of pesticide resistance.

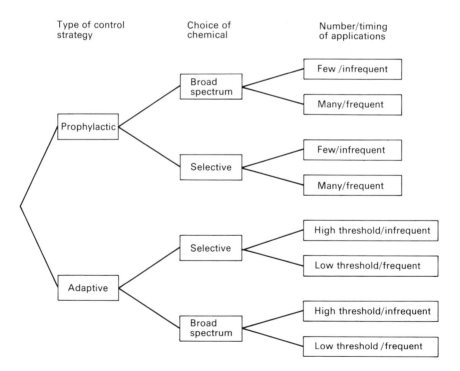

Fig. 7.5. A generalized decision tree for pesticide application.

7.3.1   Prophylactic control

Since prophylactic measures, by definition, are taken without information on the current level of attack, it is to decision theory that we turn for an appropriate form of analysis. This technique has now been applied to a number of pest control problems (Carlson 1970; Gilmour and Fawcett 1973; Norton 1976; Webster 1977; Mumford 1978). To illustrate its use, consider the decision problem outlined in Table 7.2.

Since we do not know what level of pest attack is going to occur, the best we can do is to consider the probability of each level of attack occurring, and weight the expected outcome accordingly. Table 7.3 shows the estimated probability of pest attack for the Ely region of the UK (Mumford 1978).

TABLE 7.3. *The probability of potential yield losses due to pest attack in the Ely region UK (Source: Mumford, 1978).*

Potential yield loss	Probability
0%	0.6
15%	0.3
45%	0.1

With this information, the expected outcome of each strategy in this region can be assessed as follows:

Expected outcome of strategy 1
$$= £[(685 \times 0.6) + (631 \times 0.3) + (523 \times 0.1)]$$
$$= £[411 + 189.3 + 52.3]$$
$$= £652.6$$

Expected outcome of strategy 4
$$= £[(712 \times 0.6) + (633 \times 0.3) + (475 \times 0.1)]$$
$$= £[427.2 + 189.9 + 47.5]$$
$$= £664.6$$

In terms of expected value, strategy 4 is best, yielding an extra £12/ha/ annum, on average. But this assumes the farmer is risk-neutral, that he makes his decisions on the basis of the long-term performance of each strategy, and is unconcerned that in 1 year out of ten (on average), when heavy attack occurs, this 'best' strategy yields £48/ha less than strategy 1 (Table 7.2). As we have already seen, this is unlikely to be the case, and it is to account for different attitudes to risk that attempts have been made to express pay-off in terms of utility, involving an assessment of the decision maker's personal evaluation of different monetary outcomes (Webster 1977).

In practice, such sophistication is unlikely to be warranted, especially in view of the difficulties involved in assessing pay-off for a diversity of farm conditions, particularly where numerous options exist concerning the number and timing of pesticide application. Where endogenous pests are involved, the picture becomes even more confused. Indeed, since actions taken against endogenous pests can affect the level of attack in subsequent years, decision theory itself becomes inappropriate, and different approaches have to be adopted, such as computer simulation models. For instance, to analyse strategies for controlling the Australian cattle tick, *Boophilus microplus*, a modelling approach has been used to assess the performance of various strategies on tick population dynamics (Sutherst *et al.* 1979). While this model can be used to search for optimal strategies, the identification of robust strategies that perform 'well' across a broad range of biological and management parameters appears far more useful. The results of simulation runs with various control strategies are shown in Table 7.4, illustrating how a number of 'pay-off' features can be investigated.

TABLE 7.4. *Comparison of selected control strategies against the cattle tick in Australia (after Sutherst* et al. *1979). The optimal weeks to apply control measures (shown in parentheses – where week 0 is the start of spring) are determined by computer search. The combined costs of these strategies (including tick control and damage costs) are given in Australian dollars (A$)/head/annum for low (100 000 eggs/head in the initial population) and high (500 000 eggs) infestations. Unquantified performance criteria are assessed on a three-point scale, where + indicates a favourable performance, 0 a moderate and − an unfavourable performance.*

	British cattle			Zebu cattle
	Acaricide dipping		Rotation of cattle between 2 paddocks (*weeks 4, 12, 20, 26, 38, 46)	Acaricide dipping
	3 dippings per year (weeks 15, 18, 21)	5 dippings per year (weeks 6, 9, 12, 15, 18)		1 dipping per year (week 6)
Low infestation	5.09	4.58	4.35	1.63
High infestation	9.81	5.54	5.85	3.05
Effect on population in subsequent seasons	−	+	−	+
Robustness	−	0	−	+
Acaricide resistance costs	0	−	+	+
'Handling' costs	+	+	+	−

*Subject to the constraint that cattle are moved at least every 8 weeks in summer and 12 weeks in winter to prevent pasture damage.

7.3.2 Adaptive control

In contrast to prophylactic control, adaptive pest control is undertaken in response to information on current levels of pest attack. Typically, the pest is monitored in the field and the decision to apply pesticide is made on the basis of the estimated level of attack. To provide a decision rule for such situations, Stern *et al.* (1959) first introduced the concept of the 'economic threshold', which essentially is defined in terms of that population density at which the amount of injury caused 'will justify the cost of artificial control measures'. Although entomologists in particular have interpreted this in various ways, the implication has always been that the benefit of control should exceed its cost.

In practice, economic thresholds are usually determined in a pragmatic way, on the basis of empirical, trial and error, experience in the field. In many cases, this will undoubtedly be the most feasible approach to adopt. Nevertheless, there is also merit in taking a more rigorous, analytical approach to the problem. To illustrate how this might be achieved, let us consider the case of the sugar cane froghopper in Trinidad.

This pest, which is the major pest of sugar cane in Trinidad, is monitored in the crop every 3 days in the wet season, aerial spraying being carried out in response to the assessed level of attack. Since adults appear to cause most damage, by feeding on the leaves of cane, it can be argued that the extent of this damage depends on the number of adults in the crop and the number of days they spend feeding: that is, damage is a function of total adult-days. From spraying trials, that allow sugar yield to be related to different levels of adult-days, it was found that, for practical purposes, this relationship could be regarded as linear (Norton and Evans 1974).

With this information, the yield loss caused by a given number of adults present in the crop on a particular day is:

$$d \, a_t l \tag{7.1}$$

where d is the damage coefficient, expressed as the loss in sugar yield (tonnes/ha)/adult-day/100 stools, a_t is the number of adults/100 stools on day t, and l is the average life expectancy of adults a_t, in days. Converting this to monetary terms, the loss in revenue is:

$$pd \, a_t l \tag{7.2}$$

where p is the price of sugar tonne^{-1}.

To decide whether it is worth applying an insecticide, two further items of information are required – the proportion of adults killed, and the cost of purchasing and applying insecticide. Using this information, the point at which it is profitable to apply a non-residual spray – such as malathion – can be assessed. It is when:

$$pda_t lk \geqslant c \qquad (7.3)$$
$$\text{(benefit of control)} \quad \text{(cost of control)}$$

where k is the proportion of adults killed/spray, and c is the cost of applying non-residual insecticide/ha. That is, the economic threshold for control is where:

$$a_t = \frac{c}{pdkl} \qquad (7.3a)$$

Since d and k have been estimated (Norton and Evans 1974; Conway *et al.* 1975) and p and c can be readily obtained, the main difficulty lies in assessing average life expectancy – l (Conway *et al.* 1975). We can expect the value of l to vary from field to field and season to season, depending on climatic and natural enemy conditions. With the values for the other parameters shown, the effect of changes in l on the economic threshold are shown in Fig. 7.6.

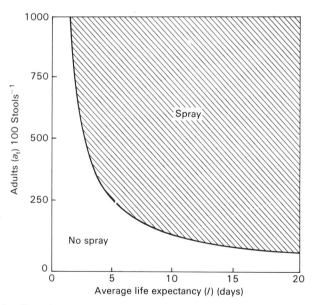

Fig. 7.6. The effect of average life expectancy (l) on the economic threshold for the sugar cane froghopper; where the cost of spraying $(c) = \$TT\ 7.04\ ha^{-1}$; the price of sugar $(p) = \$TT\ 200\ tonne^{-1}$, the loss in sugar yield $(d) = 3.228 \times 10^{-5}\ tonnes\ ha^{-1}\ adult\text{-}day^{-1}\ 100$ stools^{-1}, and the proportion of adults killed $(k) = 0.8$ spray^{-1}.

Clearly, it is only when initial estimates show l to be below 10 that it is necessary to consider greater precision. Again, where the insecticide to be used is residual, expressions (7.3) and (7.3a) need to be modified to consider the number of adults emerging during the period when the insecticide is still active. In practice, this could be assessed from estimates of nymphs that are monitored at the same time as adults.

While changes in other parameters can be accommodated within the context of expression (7.3a), a fundamental difficulty arises when the reproductive value of adults is considered. Since four generations of froghopper develop in the crop during the wet season, the threshold derived above is only valid for the 4th generation, and even then we have to assume there is no quantitative relationship between the 4th generation in one year and the 1st generation in the subsequent year. It is to account for this future effect that Conway *et al.* (1975) use dynamic programming to investigate how the best distribution of insecticide sprays over a particular season depends on the level of attack and the density-dependent relationship between generations.

7.3.3   *Prophylactic versus adaptive control*

The suggestion is frequently made (e.g. Royal Commission on Environmental Pollution 1979) that monitoring and forecasting schemes can reduce the use of pesticides by enabling farmers to change from prophylactic to adaptive control. Even where appropriate information is available, however, the extent to which farmers are willing to make this change will depend on a variety of other factors. To illustrate, consider the choice between two strategies – prophylactic (calendar) spraying and an adaptive, monitoring-and-spraying, programme. Hypothetical net revenue lines associated with these strategies, as well as the no-spraying strategy, are shown in Fig. 7.7 as a function of the level of pest attack.

In years when the pest is virtually absent, differences in net revenue reflect the cost of calendar spraying and of monitoring. As pest attack increases, the course taken by these net revenue lines depends on a number of factors, including the damage relationship, the economic threshold rule, and the effectiveness of the two spraying programmes in reducing pest attack. Having attempted to estimate the respective position of each curve, the likelihood of farmers changing from calendar spraying to a monitoring-and-spraying programme can then be assessed in the context of farmers' perceptions of the probability distribution of pest attack (for different farm categories and locations), their perceptions of the accuracy of monitored information, and their objectives and constraints.

7.3.4   *Pesticide resistance*

Apart from pollution, an equally important reason that is given for wishing to encourage farmers to reject prophylactic pest control is the problem of

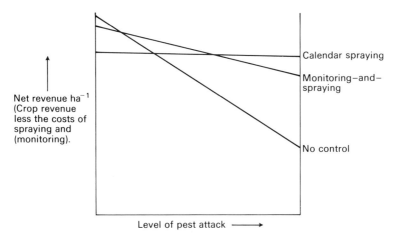

Fig. 7.7. Hypothetical net revenue curves for a prophylactic and a monitoring-and-spraying strategy at different levels of pest attack.

pesticide resistance. Clearly, in the long term, this hazard has to be evaluated. For any particular pest, the resistance problem can be viewed in terms of a resource (pesticide susceptibility) that can be depleted over time by the use of pesticides (Hueth and Regev 1974). Its importance rests on two conditions: the rate of depletion, and the 'exploration' and development of new pesticides.

Unlike most depletable resources, the depletion of pesticide susceptibility is not readily apparent until it is virtually complete. At this point, when pesticides are perceived by the farmer to have lost their effectiveness, corrective action is limited. Accordingly, the cost of resistance can arise from increased pest damage, or from increased control costs that are incurred either because application of the original pesticide is increased or because a substitute pesticide is used, which necessarily is less cost-effective.

To attempt to delay the development of resistance before this point, appropriate strategies have to be designed without the ability to test them in the field. It is here where a modelling approach to the resistance problem could be of considerable value, providing a means of understanding the complex interactions involved (Comins 1977a,b), and of investigating possible strategies for its delay (Curtis *et al.* 1978; Conway and Comins 1979). In the meantime, a more obvious means of delaying pesticide resistance is to reduce the use of pesticides. This may require the acceptance of higher pest damage or a reduction in the use of pesticides for 'cosmetic' purposes. Alternatively, forecasting and monitoring schemes that allow pesticides to be applied 'as needed', and particularly integrated control schemes (cf. Table 7.4) that lower the pest hazard can be equally effective.

The other half of the depletion equation concerns the development of new pesticides. In recent years, a marked decline in the introduction of new pesticides has been observed (Lewis 1977). To some extent, this decline can be attributed to the increased cost of environmental screening and the saturation of the major pesticide markets. Nevertheless, it may also be true that we are reaching a stage of 'diminishing returns' in searching for better pesticides, at least for specific pests. If this is the case, no longer can pesticides be regarded as a universal panacea. Their future necessarily will lie within the context of integrated control.

7.4 References

Bunting, A. H. (1972). *J. R. Soc. Arts* **120**, 227.
Carlson, G. A. (1970). *Am. J. Agric. Econ.* **52**, 216.
—— (1979). *Annu. Rev. Phytopathol.* **17**, 149.
Comins, H. N. (1977a). *J. Theor. Biol.* **64**, 177.
—— (1977b). *J. Theor. Biol.* **65**, 399.
Conway, G. R. and Comins, H. N. (1979). *Span* **21**.
—— Norton, G. A., Small, N. J., and King, A. B. S. (1975). *Study of agricultural systems* (ed. G. E. Dalton), p.193. Applied Science Publishers, London.
Curtis, C. F., Cook, L. M., and Wood, R. J. (1978). *Ecol. Entomol.* **3**, 273.
Gilmour, J. and Fawcett, R. H. (1973). *Proc. 7th Br. Insectic. Fungic. Conf.* **7**, 1.
Hueth, D. and Regev, U. (1974). *Am. J. Agric. Econ.* **56**, 543.
Kates, R. W. (1970). *Natural hazards in human ecological perspective: hypotheses and models*. Working paper No. 14. Natural Hazard Research, University of Toronto.
Lewis, C. J. (1977). In *Origins of pest, parasite, disease and weed problems* (eds J. M. Cherrett and G. R. Sagar), p.237. Blackwell Scientific Publications, Oxford.
Mumford, J. D. (1978). *Decision making in the control of sugar beet pests, particularly viruliferous aphids*. PhD. Thesis, University of London.
—— (1981). *J. Agric. Econ.* **32**, 31.
Norgaard, R. B. (1976). *Annu. Rev. Entomol.* **21**, 45.
Norman, D. W. (1974). *J. Dev. Stud.* **11**, 3.
Norton, G. A. (1975). *Meded. Fac. Landbouww. Rijks. Univ. Gent.* **40**, 219.
—— (1976). *Agro-Ecosystems* **3**, 27.
—— and Conway, G. R. (1977). In *Origins of pest, parasite, disease and weed problems* (eds J. M. Cherrett and G. R. Sagar), p.205. Blackwell Scientific Publications, Oxford.
—— and Evans, D. E. (1974). *Bull, Entomol. Res.* **63**, 619.
Royal Commission on Environmental Pollution (1979). *Agriculture and pollution*, Seventh Report (Cmnd. 7644) HMSO, London.
Slovic, P., Kunreuther, H., and White, G. F. (1974). In *Natural hazards: local, national, global* (ed. G. F. White), p.187. Oxford University Press.
Southwood, T. R. E. and Norton, G. A. (1973). In *Insects: studies in population management* (eds P. W. Geier, L. R. Clark, D. J. Anderson, and H. A. Nix), p.168. Ecol. Soc. Aust. (Memoirs 1), Canberra.
Stern, V. M., Smith, R. F., van den Bosch, R., and Hagen, K. S. (1959). *Hilgardia* **29**, 81.

Sutherst, R. W., Norton, G. A., Barlow, N. D., Conway, G. R., Birley, M., and Comins, H. N. (1979). *J. Appl. Ecol.* **16,** 359.

Tait, E. J. (1977). *Ann. Appl. Biol* **86,** 229.

Turpin, F. T. and Maxwell, J. D. (1976). *J. Econ. Entomol.* **69,** 359.

Webster, J. P. G. (1977). *J. Agric. Econ.* **28,** 243.

8
Environmental aspects

R. H. KIPS

8.1 Introduction

Pesticides are biocides purposely applied mainly to agroecosystems. Many of these chemicals will not be affected by and will produce no effects outside these particular systems. Others, owing mainly to their physical and chemical properties or to different fields of application, will affect and be affected by other ecosystems and sometimes attain global distribution.

Pesticide ecology thus presents two distinct but closely linked facets:

(a) The qualitative and quantitative study of the behaviour (metabolism and movement in space and time) of the pesticide under the influence of (changing) natural and Man-made biotic and abiotic factors in different (interrelated) ecosystems.

(b) The influence of the application of the pesticide on the various components, mainly biotic, of the ecosystems (biosphere).

In the field of environmental chemistry pesticides are classified as micropollutants, being present in very small quantities. They are referred to as xenobiotics because they are exogenous Man-made chemicals, foreign to the natural environment and show biological, mainly biocidal, activity. Since pesticides are applied as formulations consisting of a mixture of the technical-grade material and a number of adjuvants also of technical-grade quality, the impurities they contain have to be considered if their presence has any relevance in this context. Instances are known of impurities having a much higher mammalian toxicity than the pesticide compound itself e.g. tetrachlorodiphenyldioxine in the herbicide 2, 4, 5-T. Once applied the components of the pesticide formulation will undergo changes which should eventually lead to the mineralization of the active ingredients through physical, chemical, and biochemical processes. The example of metallic mercury, possibly derived from the application of arylmercury compounds in agroecosystems, which can be biologically converted to methyl mercury, a highly toxic compound which can be taken up by living organisms, shows that total mineralization, if it occurs, is not necessarily the end of the ecological impact of the pesticide. The complex nature of the problem is illustrated by the detailed study of the biogeochemical cycle of hexachlorobenzene (HCB) (Heinisch 1978). Although at one time it was generally accepted that only truly systemic or translocated pesticides could be taken up, transported and possibly accumulated in plant tissues, there is now ample evidence that many non-systemic pesticides find their way into the plant system (Heinisch 1978; Dejonckheere *et al* 1975a, 1976). Experiments with labelled compounds have conclusively shown that some of the radioactivity derived from the labelled pesticide applied to the soil is present in the plants grown on the treated soil as bound residues, not extractable by the usual analytical procedures (Kaufman *et al.* 1976).

This finding complicates the well-known problem of pesticide residues in treated agricultural produce, which are effectively kept under control in many countries (Dejonckheere 1979). Residues in feed and food have resulted in the presence of certain pesticide residues in animal and human adipose tissues. The changing pattern of pesticide usage and the phasing out of the persistent organochlorine insecticides in recent years is reflected in the analytical results obtained at different time intervals (Dejonckheere *et al.* 1977). The significance of this body load of xenobiotics, mainly pesticides, in terms of human health is not known but there is a general concensus that this form of contamination should be minimal.

8.2 Fate of the pesticide chemical in the environment

Efficiency of pesticide use is very low and varies between 0.03 and 60 % (Graham Bryce 1976). When applying pesticides, most of the chemical will directly or indirectly reach the soil. Some will drift or evaporate into the air during or after application, circulate into the atmosphere, and after more or less change in distribution, be returned to the soil or the water, which covers two-thirds of the earth surface. Run-off from treated land is another major source of pesticides in surface waters, and in addition large stretches of water may be treated purposely with pesticides for control of aquatic weeds, disease vectors, and undesirable fish.

Consideration of the fate of pesticides separately in the atmosphere, the lithosphere, the hydrosphere, and the biosphere has many advantages but is to some extent arbitrary. Soil contains air and water which have a vital influence on pesticide dynamics in the system; natural waters contain suspended solid particles and the bottom sediment is an important factor in aquatic ecosystems where, moreover, the pesticides may come in contact with air at the surface water – air interphase or in highly aerated waters, e.g. fast-flowing rivers; the atmosphere contains large amounts of solid particles and water. The biosphere is present in all three other systems.

Basically pesticides in the environment will undergo a fairly limited number of reactions through physical, physicochemical, chemical, or bio-chemical proccesses, qualitatively similar in all ecosystems: adsorption, desorption, ion-exchange, free-radical reactions, oxidation, reduction, hydrolysis, alkylation, dealkylation, decarboxylation, isomerization.

8.2.1   Soil

On reaching the soil, for most pesticides significant losses occur through volatilization soon after application. The remaining chemicals are more or less strongly adsorbed on the organic and mineral soil fractions, in practice on the organoclay complex. Adsorption retards the chemical and biological transformation and decomposition and may be so strong as to make the pesticides biologically and chemically unavailable. The dipyridilium cations of certain herbicides, for example, are so strongly adsorbed on clay minerals presenting a large negatively charged surface, that little or no desorption occurs and the compounds are inactive. Adsorption of organic molecules by the soil is influenced by the respective physicochemical properties and may vary widely under different natural conditions.

In the ion-exchange processes organic cations behave like metallic ones, whereby the ionic properties of the molecules are mainly influenced by the soil acidity. Protonation of amino or cyclic nitrogen atoms ionizes the triazine herbicide molecules, resulting in stronger adsorption with decreasing pH values (Bailey *et al.* 1968).

Photochemical and chemical reactions will produce changes in the pesticide molecules, and, although under practical conditions they cannot be separated from those induced by the soil biomass, the importance of these reactions is not to be underestimated. For most pesticides U.V. light between 200 and 400 nm is the most effective (Crosby 1970). 4 CPA dissolved in water and irradiated by sunlight underwent various reactions resulting in a dark, polymeric end product resembling humin-like substances.

The organic soil faction consists of 81 – 88% of humic substances (Calvet *et al.* 1977) which possess numerous functional groups and free radicals capable of inducing chemical breakdown reactions in pesticides, hydrolysis being especially important. Some of these reactions are catalysed by metallic ions, e.g. iron for DDT dehydrochlorination and copper for hydrolysis of organophosphorous compounds.

The soil organic matter is the site of and the main energy source for soil micro-organisms which are generally considered to be mainly responsible for pesticide degradation. Agricultural soil contains between 1 and 2 tons of living micro-organisms per hectare, calculated as dry matter (Calvet *et al.*1977), consisting mainly of bacteria, but also fungi, actinomycetes, algae, and protozoa, total and relative numbers and dominant species being influenced by the type of soil, seasonal variations, tillage, type of crop, fertilization (mainly nitrogen) and aeration. They are under most circumstances capable of degrading the few kilograms of pesticides applied each year to agricultural and horticultural soils.

Micro-organisms can use the pesticide as a carbon source, or, in the case of co-metabolism, another energy source is used which may or may not be structurally related to the pesticide molecule. All microbiological reactions are catalysed by enzymes which are either normally present in the organisms or are induced by the pesticide, and may also be present as free exo-enzymes in the soil. Such reactions can follow or precede non-enzymic reactions, a sequence which will finally lead to the disappearance of the pesticide either through mineralization or incorporation of part of the pesticide molecule into humic substances, apparently following first-order kinetics.

A pesticide is considered to be persistent if its half-life or disappearance time (DT50) is a year or more and if it builds up from one year to the next (Brown 1978). Apart from the chlorinated hydrocarbon insecticides few other pesticides persist for more than 12 months, half-lives being expressed rather in days, weeks, or months for most other insecticides, fungicides, and many herbicides. As the rate of disappearance is proportional to the concentration, total disappearance of persistent chlorinated pesticides or persistent break down products (chlorinated ring structures, polyaromats) of non-persistent pesticides is a long drawn out process. Edwards (1973) estimated that even if no more persistent chlorinated insecticides were to be used, there would still be residues in the soil for several decades and Heinisch (1978) cites a period of 50 – 100 years in the total environment for HCB, considering the many uncontrollable sources of input.

The part played by the soil macrofauna in direct pesticide breakdown is not clear, although indirect effects are obvious. Persistent insecticides may be concentrated, e.g. in earthworms as was noted by Barker as early as 1958 (Brown 1978) and affect links further up the food chain, contributing by their mobility in a small way to the removal of the pesticide from the treated field. How far this may also affect the fate of more or less persistent herbicides has not been fully investigated, interest being focused in most papers on the effect of the pesticide on the organisms.

The macroflora, crops and weeds, is capable of taking up a large number of pesticide chemicals and thus plays a part in their removal from the soil. Indeed it may be said that qualitatively most if not all pesticides are systemic, without, however, being systemically active. Lichtenstein (1969) proved this conclusively for organochlorine insecticides at very high application rates, but this is now known to occur for many compounds under normal conditions (Heinisch 1978; Dejonckheere *et al.* 1975a, 1976; Heinisch *et al.* 1976).

Plants are capable of metabolizing a number of pesticides and inducing some quite drastic chemical reactions to an extent unsuspected only a few years ago. Examples are the conversion of phosphorothionates to phosphates, thioethers to sulphoxide and sulphones, chlortriazines to hydroxy-triazines, urea herbicides to nitrobenzenes, and trifluralin to the corresponding acid β-oxidation of phenoxybutyric acid derivatives. Not all plant species can tackle all pesticides, and indeed this phenomenon is the basis of some of the few examples of biochemical selectivity.

Movement of pesticides in a given soil profile will depend mainly on adsorption and solubility of the compound. Most pesticides and their biologically active metabolites will only penetrate the deeper soil layers in very special circumstances, so that contamination of the ground-water-table or of drainage water is exceptional or minimal, although some particular cases of ground-water pollution and penetration in the deeper soil layers have been reported (Heinisch *et al.* 1976). Run-off will be influenced by the intensity of the rainfall and the slope of the land, although the amounts of pesticide removed from the soil in this way are relatively small. Removal of pesticides from the soil through wind erosion is also unlikely to be of great importance (Edwards 1973).

8.2.2   *Water*

Pesticides reach water from difference sources. Surface waters may be treated purposely with pesticides, mainly insecticides and herbicides but also with piscicides. Although these applications represent only a small fraction of the total pesticide input in the hydrosphere, they may be quite important locally.

A second source is drainage and surface run-off from treated soil, the importance of which will depend on a number of factors already cited. For most pesticides which are fairly insoluble and are more or less strongly adsorbed it is usually small (Edwards 1973).

Precipitation into water systems of pesticides that reach the atmosphere as vapour or in a particulate form is an important if not a major input. Some figures for organochlorine compounds in the Atlantic are quoted by Risebrough *et al.* (1968). Aquatic systems which adjoin agroecosystems where large quantities of pesticides are frequently used may be more heavily contaminated, e.g. endrin in the Mississipi River (Edwards 1977). Effluents from the pesticide industry, industries which use significant quantitites of pesticides, sewage plants, and cattle farms are obvious but local and mostly seasonal sources of pesticide contamination of water.

Pesticides when they reach water systems may be diluted through water flow, evaporation and uptake by the biota. Movement will occur in solution and with organisms containing the pesticides and their metabolites, but depends mainly on suspended material in the water currents and may be concentrated through physical – chemical processes and bioconcentration.

Compounds are broken down or transformed through chemical, biochemical, and photochemical reactions. Water is transparent to U.V. rays and photochemical reactions in water have been substantiated for many pesticides in numerous laboratory experiments. A chemical precipitation of pesticides may occur especially in sea water, possible resulting in accumulation of pesticides in river estuaries (Duke 1977).

Biochemical transformations are the result of enzyme reactions through the mediation of the living organisms, micro and macro, in the water ecosystem. With certain reservations it may be generalized that most (macro) marine species are less efficient in metabolizing pesticides than mammals (Hart 1977), and lipophilic compounds accumulate in many sea creatures (James *et al.* 1977). Bioaccumulation may be distinguished from bio-magnification, the latter referring solely to uptake in the foodweb although those processes cannot be separated in nature. Bioaccumulation is inversely related to the solubility of the pesticide. Most pesticides, whichever way they reach the water systems (ponds, lakes, rivers, streams, estuaries, oceans) are either already absorbed on solid particles or are quickly partitioned between the water, the suspended material, and the bottom sediment. Being mostly hydrophobic, their movement in the water environment is mainly by adsorption on solid particles (Hague *et al.* 1977) and they rapidly reach the bottom sediment, which is considered to be a source of possible future contamination of the water (Edwards 1973), although under the usually anaerobic conditions the breakdown of the persistent pesticides is greatly increased (Hill and McCarty 1967; Dejonckheere *et al.* 1975b). Because of this only the persistent pesticides (characterized by the great stability of the parent compound or the metabolites, good fat solubility, and favourable partition coefficient between octanol and water) appear to present a potential pollution problem.

A recent study in Canada clearly indicates that the pesticides in use today present few problems. Only atrazine and diazinon appeared in p.p.b. quantities in the agricultural watershed under investigation (Morley 1977). The significance, if any, of the presence of this type of biocide which does not bioconcentrate or magnify in the aquatic ecosystem has yet to be investigated.

Pesticide residues in the oceans appear to be barely measurable and it seems probable that the main source of contamination is precipitation from the atmosphere. Volatilization occurs and the residues in the sea water probably reach the ocean abyss and become unavailable. Concentration in oily surface films and plankton has also been suggested (Edwards 1977).

8.2.3 *Air*

There is general agreement that the main sources of pesticides in the atmosphere are spray drift and volatilization from treated crops, soil, and water.

Drift obviously occurs at the time of application and will depend on the method used, ULV treatment from the air being an extreme case. Particles with a diameter of less than 3 mm. take about 1 year to return to earth from a height of 1000 m, and when smaller than 1 μm. they mix with the air like gasses (Heinisch *et al.* 1976).

Volatilization is maximal soon after treatment, the pesticide being gradually adsorbed and bound (Ebeling 1973). Temperature and wind will influence the speed and intensity of the process. Volatilization of soil-applied pesticides is reduced by cultivation (Lichtenstein *et al.* 1964). Other possible routes for pesticides to reach the atmosphere are by wind erosion and burning of material containing pesticides, e.g. pesticide containers, incineration plants, and straw or stubble burning, the latter being unlikely to be of real significance (Wheatly 1973).

To what extent global transport of pesticides occurs via the atmosphere is not clear, and to what degree pesticides react under influence of radiation, oxygen, and water under natural circumstances is largely unknown. Heinisch *et al.* (1976) estimate that from 50 to 95% of the pesticides which reach the atmosphere are broken down mainly under the combined action of U.V. radiation and oxygen. They compare the atmosphere with a gigantic oven where the chemicals are destroyed and is thus mainly responsible for cleaning the biosphere. On the whole the ecological risks from the presence of pesticides in the air appear to be small, especially when viewed in the context of atmospheric pollution in general. However, this conclusion is highly conjectural and only further research and monitoring can provide more definite answers (Edwards 1973).

8.3   Effects of pesticides on the ecosystem biota

Pesticides are designed to produce an effect on target organisms such as insects, fungi and weeds, which generally results in killing a high proportion of the pest populations. Thus the introduction of a pesticide in an ecosystem of necessity will produce a drastic effect on the target biota and on the entire ecosystem.

However, since most chemicals used in crop protection are broad-spectrum biocides, other living organisms may be more or less affected by the impact of the pesticide on the ecosystem and it is these unintended, undersirable, and possibly harmful side effects that most investigations on pesticide ecology are concerned with. There are moreover the indirect effects to be considered which result from the more or less complete and more or less permanent disappearance of the population of certain species in the ecosystem, e.g. the effective destruction of the natural plant cover by the use of herbicides and of the food supply of insect parasites and predators; these are less well researched than the direct effects.

8.3.1   Insecticides

By far the greater part of all data available on the side effects of pesticides on the biota refer to the effects of persistent organochlorine insecticides mainly on birds and fish (Koeman 1979). The biological activity, the stability, and the lipophilic properties of these compounds make them very dangerous micro-pollutants. However, their effects on most non-target organisms studied, e.g. non-pest arthropods, soil and aquatic invertebrates, and terrestrial and aquatic vertebrates (except birds and fish) are, when applied at normal concentrations, usually short lived, difficult to interpret, or the casual relationship between pesticide treatment and the effect noted is unclear. The effects of the less persistent organophosphorous and carbamate compounds are even less noticeable, although their presence has been reported in slugs (Edwards 1973). Seed dressing of cereals with cyclodiene insecticides and methyl mercury fungicides resulted in high direct and secondary bird mortality, which could be directly correlated with these particular pest-control practices, banned in many countries in the late sixties.

Population trends for the period 1973 – 79 indicate a favourable reversal for a number of species which had been very seriously affected, particularly birds by the eggshell-thinning effect mainly of the DDT metabolite DDE. Clearly serious but reversible effects have been caused mainly on raptors by a few insecticides, but the majority do not represent a serious hazard (Koeman 1979).

Fish, like birds, are on a high trophic level in their environment. They take up insecticides directly from the water through the gills and this may be a more important way of entry than the food intake (Edwards 1973). Two insecticides, endrin and endosulfan, have been particularly involved in occurrences of large fish kills, either through accidental spills or by use under conditions where they could enter fish-containing water systems. However, although massive fish kills have been related to the use of endosulfan for tsetse fly control in Africa, repopulation took place fairly quickly (Koeman 1979). With the less-persistent insecticides such as the organophosphates, which have a much lower fish toxicity, the potential hazards are greatly reduced. The work of Dejoux, as cited by Koeman (1979), on the effect of temephos, used recently on a large scale in West Africa for the control of *Simulium damnosum* Theo, (the onchocerciasis vector) indicates that, although the numbers of many species of aquatic invertebrates decrease, the diversity of the biota remains constant. Fish apparently are not affected, either in diversity or in abundance, in spite of weekly treatments started in 1975, as indicated by the results of an ambitious and comprehensive ecological-monitoring programme (Levêque *et al.* 1977; Dejoux *et al.* 1979).

Although the persistent organochlorine insecticides have not produced any permanent drastic reductions in fish populations, certain sublethal side effects have been noted which call for continued watchfulness. Resistance to insecticides in fish (Ferguson *et al.* 1966) has some far-ranging ecological implications as many suffer sublethal effects, e.g. behavioural changes and reproductive failures. The latter are, however, difficult to relate solely to the presence of pesticides under natural conditions.

8.3.2. Fungicides

As this type of pesticide is designed to kill or inhibit the development of fungi, side effects are to be expected on non-target micro-organisms, particularly in the soil.

Soil fumigation is mainly used to control soil fungi (and nematodes) mostly under glass in intensive market gardening. The impact of such treatment, e.g. with methyl bromide, on the soil microflora is quite dramatic and leads to unacceptable bromide residues in certain crops such as lettuce and tomatoes (Van Wambeke *et al.* 1979). These transient fungicides can be contrasted with HCB which is possibly the most persistent fungicide chemical (Heinisch 1978).

Most fungicides produce a change in species composition due to differences in susceptibility of the various groups of organisms. Fumigation often results in increasing the bacterial population (Domsch 1959). Shifts in soil-borne fungus pathogen populations following treatment with selective systemic fungicides such as benomyl, resulting in the control of one disease but inducing another, is potentially extremely serious (Van der Hoeven and Bollen 1972). Advocating selectivity to minimize the impact of pesticides on non target organisms also presents its problems and may be more likely to induce the development of resistance (Oppenoorth 1972); *Botrytis cinerea* on cyclamen was highly resistant to benomyl only 1 year after the first treatment (Bollen and Scholten 1971).

Although the influence of most fungicides on the non-target soil micro-flora is substantial but short-lived (Brown 1978), their effect on the invertebrate fauna, especially predators and parasites, is important as they may have to be applied in systems of integrated insect control, and, although most fungicides are not insecticidal from a chemical control point of view, some may adversely affect insects (see Chaper 3).

Benomyl and other fungicides of this group are toxic to eathworms (Stringer and Wright 1973), another clear indication that in the field of pesticide ecology there is no room for complacency. The organic mercury fungicides, being converted in the environment to very toxic persistent and bioaccumulating methyl mercury compounds (Jernelöv 1969), coupled with their use as seed dressings, which was linked with high mortalities in seed-eating birds in Sweden, have been associated with the mercury pollution of the environment. However, the use of organic mercurials in agriculture, which is being phased out, introduced little of the toxic metal in the environment, no more than that added by rain or snow (Brown 1978).

8.3.3   Herbicides

Herbicides are applied to soil to control weeds which compete with crops for water, light, and nutrients, to water to control aquatic weeds, and to a lesser extent for treatment of weeds in road verges, industrial sites, and other non-agricultural land.

The direct ecological effects are the result of the very efficient way in which modern herbicides destroy or prevent the development of undesirable plant growth over ever-increasing stretches of the planet's surface. A striking example is the almost complete long term control of bracken made possible by application of asulam (Fryer 1977). Applied to the soil in agricultural ecosystems, the value of herbicides is measured by the degree to which all plant growth, except the crop, is eliminated. This leads to some apprehension about the survival of the wild flora and animal life depending on it (Koeman 1979). Moreover, some arthropods become pests through the complete disappearance of their normal food supply (Calvet *et al.* 1977).

The degree of susceptibility of the weed species towards the herbicides varies greatly, and intensive and regular use of these compounds leads to changes in the species composition of the wild flora and the dependent fauna. Massive application of herbicides in tropical forest ecosystems in Vietnam has led to such drastic changes in the fauna and flora, especially in the coastal mangrove swamps, that it has been estimated that substantial recovery may take more than a century (Westing 1977). This is a clear warning about the possible long-term consequences of the primary effects of herbicides, which have not received the attention they deserve, probably owing to the very low mammalian toxicity and the relative unimportance of the phylogenetic resistance problem connected with this type of compound. The ecological side effects on the soil microflora are on the whole rather slight and transient (Voets *et al.* 1977; Verstraete *et al.* 1979; Stryckers *et al.* 1978).

Some herbicides are active against plant pathogenic fungi, either through direct fungicidal action, by increasing the sensitivity of the host plant to the pathogen or by inducing changes in the biological equilibria resulting in stimulation or inhibition of the pathogen or its antagonists (Calvet *et al.* 1977).

There are conflicting reports on the influence of herbicide treatments on earthworms and soil arthropods. Obviously surface dwellers will be the most effected by any direct toxic effect (Van der Drift 1963). Possible explanations of the divergences noted are the difficulties in obtaining reliable and reproducible data on populations of most soil organisms, problems of interpreting the figures obtained, and relating laboratory results to the field situation.

The use of herbicides in aquatic systems may produce dramatic disruptions in biological communities similar to those resulting from traditional mechanical control of aquatic weeds, although direct toxicity to animal life is very low. However, the use of chemicals may induce permanent changes in the ecosystem, in some cases total inhibition of regrowth. These changes are not necessarily detrimental, as dense growth of certain weeds makes for impoverished biological systems (Robsen and Barret 1977). But the decomposition of the dead plants in the system produces drastic changes in oxygen supply, pH, and the light intensity at a given depth, resulting in profound changes in the aquatic biota.

8.4   References

Bailey, G. W., White, J.L., and Rothberg, T. (1968). *Soil sci. Am. Proc.* **32**, 222.

Bollen, G. J. and Scholten, G. (1971). *Neth. J. Plant Pathol.* **77**, 83.

Brown, A. W. A. (1978). *Ecology of pesticides.* John Wiley, New York.

Calvet, R. *et al.* (1977). *Les herbicides et le sol. A. C. T. A., Paris,*

Crosby, D. G. (1970). In *Pesticides in the soil, ecology, degradation and movement:* Symp. Proc. Michigan State University, East Lansing;

W. Dejonckheere (1979). *Residus van pesticiden. Overzicht en bespreking van resultaten van onderzoek naar residus van pesticiden in België gedurende de voorbije 10 jaar.* Laboratorium voor Fytofarmacie, Rijksuniversiteit, Ghent.

——Steurbaut, W., and Kips, R. H. (1975a). *Bull. Environ: Cont. Toxicol.* **13**, 720.

——Steurbaut, W. and Kips, R. H. (1976). *Pestic: Monit. J.* **10**, 68.

——Steurbaut, W., Verstraeten, R., and Kips, R. H. (1977). *Med. Fac. Landbouww. Rijksuniv. Gent.* **42/2**, 1839.

——Steurbaut, W., Willock, J., Kips, R. H., Voets, J. B., and Verstraete, W. (1975b). *Med. Fac. Landbouww. Rijksuniv. Gent* **40**, 1187.

Dejoux, C., Mensah, G., and Troubat, J. J. (1979). *ORSTOM. Rapport No. 27*, 55pp. Bouaké, Côte d'Ivoire.

Domsch, K. H. (1959). *Z. Pflanzenkrankh.* **66**, 17.

Duke, T. W. (1977). In *Pesticides in aquatic environments.* (ed. M. A. Q. Khan). Plenum Press, New York.

Ebeling, W. (1973). *Resid. Rev.* **3**, 35.

Edwards, C. A. (1973). *Persistent pesticides in the environment.* (2nd ed.). C. R. C. Press, Cleveland, Ohio.

——(1977) *Pesticides in aquatic environments* (ed. M. A. Q. Khan). Plenum Press, New York.

Ferguson, D. E., Ludke, J. L., and Murphy, G. G. (1966)*Trans. Am. Fish. Soc.* **95(4)**, 335.

Fryer, J. D. (1977) In *Ecological effects of pesticides.* (eds. F. H. Perring, and K. Mellanby), Academic Press, London.

Graham-Bryce, I.J. (1976). *Chem. Indust.* **1**, 545.

Haque, R., Kearney, P. C, and Freed, V. H. (1977). In *Pesticides in aquatic environments.* (ed. M. A. Q. Khan). Plenum Press, New York.

Hart, L. G. (1977). In *Pesticides in aquatic environments.*

Heinisch, E. (1978). *Biogeochemische Kreisläufe persistenter organischer Verbindungen.* Akademie Verlag-Berlin, DDR.

——Paucke, H., Nagel, H. D., and Hansen, D. (1976). *Agrochemikalien in der Umwelt.* V. E. B. Gustav Fischer Verlag, Jena.

Hill, D. W. and McCarty, P. C. (1967) *J. Water Poll. Cont. Fed.* **39**, 1259.

James, M. O., Fouts, J.R., and Bend, J. R. (1977). In *Pesticides in aquatic environments.* (ed. M. A. Q. Khan). Plenum Press, New York.

Jernölov, A. (1969). In *Chemical fallout*. (eds. M. W. Miller, and G. G. Berg, Springfield, Illinois.

Kaufman, D. D., Still, G. G., Paulson, G. D., and Bandal, S. K., (1976). *Bound and conjugated pesticide residues*. A. R. C. Symposium series no. 29, Washington.

Koeman, J. H. (1979). In *Advances in pesticide science. Part 1* (ed. H. Geissbühler). Pergamon Press, Oxford.

Levêque, C., Odei, M. and Pugh Thomas, M. (1977). In *Ecological effects of pesticides* (eds. F. H. Perring, and K. Mellanby, Academic Press, London.

Lichtenstein, E. L. (1969). *J. Agric. Food Chem.* **7**, 430.

——Myrdal, G. R. and Schultz, K. R. (1964). *J. Econ. Entomol.* **57(1)**, 133.

Morley, H. V. (1977). In *Pesticides in aquatic environments*. (ed. M. A. Q. Khan). Plenum Press, New York.

Oppenoorth, F. J. (1972). In *The future for insecticides. Needs and Prospects*. (eds. R. L. Metcalf, and J. J. McKelvey, John Wiley, New York.

Risebrough, R. W., Hugget, R. J., Griffin, J. J., and Goldberg, E.D. (1968). *Science*. **159**, 1233.

Robsen, T. O., and Barret, P. R. F. (1977). In *Ecological effects of pesticides*. (eds. F. H, Perring and K. Mellanby, Academic Press, London.

Stringer, A. and Wright, M. A. (1973). *Pestic. Sci.* **4**, 165.

Stryckers, J., Goddeeris, H., Van Himme, M., Bulcke, R. and Verstraete, W. (1978). *Med. Fac. Landbouww. Rijksuniv. Gent.* **43/2**, 1141.

Van Der Drift, A. (1963). *Neth. J. Plant Pathol.* **69**, 188.

Van Der Hoeven, E.P. and Bollen, G. J. (1972) *Acta Bot. Neerl.* **21**, 107.

Van Wambeke, E., Vanachter, A. and Van Assche, C. (1979). *Med. Fac. Landbouww. Rijksuniv. Gent.* **44/1**, 520.

Verstraete, W., Stryckers, J., Cadron, J., Van Himme, M., and Bulcke, R. (1979). *Med. Fac. Landbouww, Rijksuniv, Gent,* **44/1**, 699.

Voets, J. P., Angerosa Imas, M. O., Goddeeris, H., and Verstraete, W. (1977). *Acta Phytopathol. Acad. Sci. Hung.* **12(1-2)**, 31.

Westing, A.H. (1977). In *Ecological effects of pesticides*. (eds. F. H. Perring and K. Mellanby. Academic Press, London.

Wheatley, G. A. (1973). In *Environmental pollution by pesticides*. (ed. C. A. Edwards). Plenum Press, New York and London.

9
Safe use of pesticides

J. F. COPPLESTONE

9.1 Introduction

All humans are exposed to pesticides in some form. Accidental poisonings by pesticides arise not from inability to prevent exposure but from failing to limit it to an acceptable level. This level is well below that at which any adverse effect can be expected.

Exposures have to be evaluated according to the dosage which is likely to result. For a given organism, dosage of any chemical at a certain level and rate results in a response which may be either beneficial or antagonistic to the wellbeing of the organism. The dosage–response relationship is fundamental in toxicology and is the basis on which we try to understand the effects of the many chemicals that are absorbed daily into the human body (Casarett and Doull 1975; Hayes 1975a). Therefore, before discussion of the hazards to humans of pesticide exposure and the deduction from these as to how pesticides may be used safely, or without adverse effects, some general principles within the concept of the dosage–response relationship are set out briefly below.

In discussions of the safe use of pesticides, there is a tendency for some terms to be used in a rather confusing manner. Definitions are therefore given below. These may not be universally acceptable but it is hoped that they will at least clarify the points made.

9.2 Toxicity and hazard

Toxicity and hazard are words sometimes used synonymously. However, there is a clear distinction between them.

9.2.1 Toxicity

This is the capacity of a chemical compound or a mixture of compounds to cause harmful effect. It is directly related to dosage and is expressed in numerical terms, the most widely accepted index being the LD_{50} – the statistical estimate of the dose which kills 50% of a large population of test animals, usually rodents. The dose is recorded as milligrams of the compound per kilogram of the body weight of the test animal (mg/kg bw).

Since toxicity varies according to species and to the route of administration of the dose, no LD_{50} value is complete without the specification of the test animal and the route. The rat is the rodent most commonly used for comparison of toxicities and, for most compounds, it provides a reasonable correlation with toxicity in humans insofar as this can be estimated from accidental and suicidal cases of poisoning.

The routes of administration most commonly used in toxicity studies are oral, dermal, inhalational, intraperitoneal, and occasionally intracerebral.

Although at first it may seem that the number of deaths of the test animals is not (one hopes) a relevant index as far as human exposure is concerned, it is an 'all or none' recordable phenomenon. A single oral dose is given by intubation directly into the stomach; the compound may be applied to the shaven skin of the animal for a set period of time or injected into the peritoneal cavity. (For the methodology of testing, see WHO 1978a.) The effect produced may be acute or, with a few compounds such as organomercurys, chronic, since irreversible damage to tissues may occur in surviving animals.

To measure subacute toxicity, young animals are fed several levels of the test compound in their diet for about one-tenth of their lifetime (90 days in the case of rats), and their progress is compared to that of a control group receiving only the diet. Dosages are recorded in mg of the compound per kg of diet, and this is averaged over the period for the group to give a level of mg/kg diet/day. Individual dosages depend on the appetite of the animal and are therefore more variable than single doses, so larger groups of animals have to be used than for the acute toxicity tests. If substantial numbers of animals die, it is possible to calculate a 90-dose LD_{50}, but these tests are more commonly used to study the clinical and pathological progress of poisoning in the animals, and to calculate the 'no effect level' at which no effect in any animal can be detected.

Long-term feeding studies extending over the usual lifetime of the animal – 2 years in the case of rodents – are similarly carried out to measure any long-term effects such as carcinogenicity. As will be discussed later, the interpretation of the results of such studies is sometimes difficult since the levels fed are usually very high compared to potential human exposure, and there is a genetic tendency towards some types of carcinogenesis in many strains of rodent species.

For most pesticides, except the fumigants, inhalational toxicity is of less relevance than oral and dermal toxicity as most compounds are not highly volatile. Inhalational toxicity is assessed on a subacute multiple dose basis by exposing the animal to known concentrations of gas or aerosol for a given time, and is expressed as LC_{50}, or lethal concentration at which 50% of the test animals exposed to a stated concentration for a stated period of time die immediately, or within a certain period after exposure has ceased.

There are a number of other toxicological studies in animals to determine any effects on their reproductive capacity, changes in their genetic characters, and in their endocrine and immunological patterns. Unfortunately, with the exception of the lethality tests (LD_{50} and LC_{50}, etc.), there is no international concensus on the methodology or interpretation of these tests, and each study has to be considered on its methodological and statistical merits.

9.2.2   Hazard

This is the likelihood that, under a given set of circumstances, a compound may cause harmful effects. The circumstances include a variety of variables and so it is virtually impossible to quantify hazard in numerical terms. Instead, a continuous scale expressed in words from 'extreme' through 'high', 'moderate', and 'low' to 'slight' is usually used. (Sometimes the same words are used to qualify toxicity, usually on the basis of the oral or dermal LD_{50} in the rat. To avoid confusion, it is better to refer directly to the LD_{50} as low or high, a scale opposite to that used for hazard.) Hazard may also be described as acute or chronic according to the effect produced, although sometimes these words are used to describe the frequency of occurrence of the hazard. Since hazard can only be assessed in relation to effects on a particular organism in a certain situation, it should always be clearly stated who or what is at hazard. It is the hazards of pesticides to humans that are discussed below to the exclusion of environmental hazards which are dealt with in Chapter 8.

Although toxicity is only one of the variables concerned in the assessment of hazard, it has to be the starting point. The chemical class to which the compound belongs is associated with particular types of effect and, as indicated above, the toxicity of a particular compound is an index of its capacity to cause these effects. However, hazard also depends on dosage and this is determined by a number of independent variables, the main components being the concentration and formulation of the compound, the route of its entry into the body, the type of exposure or the environmental circumstances of use, and the variation in group characteristics. Each of these is discussed in detail below.

9.2.3   The concentration and formulation of the compound

In assessing hazard from the starting point of toxicity, it has to be clear as to what part or type of a formulation the toxicity indices actually relate. For example, the oral LD_{50} (rat) for pure malathion is 10 000–13 000 mg/kg bw (WHO 1978b); the same value for the technical product is usually in the region of 2000–3000 mg/kg bw. The difference is accounted for by the presence of impurities in the technical product (WHO 1979).

It is possible to calculate the LD_{50} of a formulation from that of the technical product using the formula:

$$\frac{LD_{50} \text{ technical product} \times 100}{\text{percentage of technical product in formulation}}$$

However, such calculations should only be made as a last resort if no information on direct tests of toxicity of the formulation is available. The formula neglects the influence of the toxicity of the diluent and the occasional influence of the diluent on the evolution of impurities under adverse conditions of storage. An example of the latter occurred in 1976 when the use of malathion 50% wdp (water dispersable powder) that had been stored in tropical conditions for some months resulted in a large number of cases of poisoning. The formulations concerned were found to contain higher than usual levels of isomalathion and other impurities, probably due to isomeriza-tion of the malathion by some of the 'inert' powder diluents. The result was that the LD_{50} indices for rats of these formulations were substantially reduced. It now seems clear that such a dramatic change in toxicity can only occur from this cause with powder formulations of some organophosphor-ous insecticides, but this had not been anticipated before the outbreak of poisoning actually occurred. Therefore it is essential that a watch should always be kept for the unusual, and that safety precautions should have a sufficient margin of safety and should be designed to prevent unforeseen fluctuations in toxicity from having serious effects.

Mixtures of active ingredients within formulations present other prob-lems. Here again a formula is available:

$$\frac{C_A}{T_A} + \frac{C_B}{T_B} + \frac{C_Z}{T_Z} = \frac{100}{T_m}$$

where C is the percentage concentration of constituent, A, B...Z in the mixture, T is the oral LD_{50} values of constituents, A, B, ...Z, and T_m is the oral LD_{50} value of the mixture. The formula can also be used for dermal toxicities provided that this information is available on the same species for all constituents. However, such calculations have the disadvantage that potentiating or protective interactions between the constituents are not taken into account.

An alternative is to consider the toxicity of the mixture to be that of the constituent of the lowest LD_{50} value, as if that constituent was present in the same concentration as the total concentration of all active constituents. This approach only works effectively if all the major constituents are of roughly the same toxicity. The only really satisfactory solution to the problem of mixtures is to determine the toxicity indices of each formulation.

The assessment of hazard must also take into account the possibility that a particular formulation might cause a contact dermatitis, or other disease of the skin or nails. This can be due either to the active ingredient, especially some carbamates and pyridyl-derived herbicides, or to the diluent, particu-larly oils and kerosenes. Although toxicity testing in animals usually includes some tests for irritancy, especially to the eye, these tests are not predictive for human allergic responses and only partly so for irritative responses. Information on these properties of pesticides usually accumulates through actual experience in use.

The physical state, solid or liquid, of a pesticide formulation plays a part in hazard evaluation as will be discussed in the section on exposure.

9.3   Routes of entry

There are four routes of entry by which pesticides can be absorbed into the body: the mouth (ingestion), the intact skin (dermal absorption), the lungs (inhalation), and through cuts, abrasions, and rashes of the skin (inoculation). Of these, the first two are by far the most important in the causation of poisoning.

9.3.1   Ingestion

This results in rapid absorption of a pesticide from the alimentary tract. It can occur in several ways, including the drinking of pesticide mixtures deliberately or accidentally, and the eating of food grossly contaminated with pesticide. Ingestion is a hazard more for the general public, especially children, than for the pesticide user.

9.3.2   Dermal absorption

This is the prime hazard for the operator, especially since most pesticides are applied as liquids or dusts. There is variability between compounds in their potentiality for dermal absorption: most insecticides of the organochlorine, organophosphorus, and carbamate groups are readily absorbed while some herbicides, including the phenoxyacetic acid derivatives, are hardly absorbed at all. For most compounds, dermal toxicity is lower than oral toxicity. Absorption is enhanced when the skin is warm and sweating, and is therefore more liable to occur in summer weather and the tropics.

9.3.3   Inhalation

Inhalation of gases, fumes, mists, or dusts is less common in pesticide application than is commonly thought, except in the case of gaseous fumigants and a few compounds with high vapour pressure such as chloropicrin. Although inhalation toxicity is usually as high or higher than oral toxicity, most particles of liquid or dust are of greater diameter than the 10 µm maximum that is a prerequisite for entry of the particle into the air cells of the lungs. Larger particles are trapped on the linings of the nose, pharynx, and trachea where a type of dermal absorption takes place, and is additive to that from other skin contamination. It has been shown and confirmed that in most agricultural spray applications of organophosphorous pesticides, inhalation accounts for under 1% of total absorption (Wolfe *et al.* 1967; Copplestone *et al.* 1976).

9.3.4 Inoculation

This is the least important route of entry and usually co-exists with some degree of dermal absorption. Its toxic potential can be high but it need not be a hazard if people with rashes are excluded from pesticide handling and if all wounds are protected with impervious dressings during work with pesticides.

9.4 Types of exposure

Hazard can be greatly influenced by the type of exposure since many factors may influence dosage. Among those which are discussed in detail are the physical state of the formulation, the manner of handling of the pesticide and the periodicity of exposure.

The physical state of the formulation is important. The most hazardous are the fumigant gases, and the special precautions required for handling these are usually both understood and followed. Most pesticide formulations are either liquids or solids and the former are the more hazardous at all stages of transportation, storage, and use. The chief danger in transportation and storage is the possibility of massive contamination of foodstuffs (often flour, sugar, or rice) due to spillage. Liquids are absorbed by the foodstuff and can also soak into floors making effective decontamination difficult. This is not an academic point. Many outbreaks of food-borne poisoning have occurred (Hayes 1975b) and frequently liquid formulations have been responsible.

In application, dermal absorption is enhanced by liquid formulations, which wet the skin and are often unnoticed. Granules are the least hazardous type of formulation provided that the active ingredient is dispersed uniformly through the granule and not coated on to an inert core.

The manner in which a pesticide is handled from its manufacture to its final degradation is the most important single factor in determining hazard. Over many years, wide experience has accumulated in the safe handling of pesticides, and acute accidental poisonings are usually the result of mishandling. The hazard of spillage in transportation and storage has already been mentioned. The commonest handling fault is the neglect of personal protective precautions. This is discussed in the section below on the control of hazards. Novel methods of application, adaptations of ULV equipment, hot and cold fogging etc., can all influence hazard but not always in an expected direction. Although the immediate operators may be well protected, other people not immediately connected with the application operation may be affected by drift or by entering or working in recently sprayed crops. Possible hazards of exposure should always be kept in view, both when a technique is first introduced or the equipment is new, and when familiarity has had its usual result and the equipment has become worn and less efficient. It is here that the adequacy of supervision and maintenance are important factors in controlling slow, almost imperceptible changes in hazard with time.

The periodicity of exposure is also important in the determination of hazard but its effect varies with the class of compound and its metabolism. A short-term exposure may produce an acute effect, and a case of poisoning may result; fortunately such events hardly ever happen if pesticides are properly used. Problems arise when the exposure is below the no-effect level on a daily basis. What then is the likelihood that an effect (acute or chronic) may be produced by daily exposure over a week, a month, a year, or a period of weeks or months repeated annually?

In the study of acute responses, it has been shown in animal experiments that the results of a 90-day feeding study can be as predictive as the results of a similar 2-year study (Weil 1963). Whether or not acute results occur depends on three factors:

(a) The speed of metabolism and excretion. Some of the newer pyrethroid insecticides are excreted almost completely by man within 24 hours of absorption.

(b) The potential of the compound for storage in the body, causing a cumulation of the compound until an equilibrium level is reached which is directly related to the level of intake. The best known examples of such compounds are DDT and other organochlorines most of which are stored in fatty tissues (WHO 1973). In occupational exposures, when storage levels are high, it can happen that a relatively small increase in daily exposure can produce an acute effect due to the raising of the equilibrium level. Also, the sudden mobilization of fat reserves for any reason, such as dieting, may produce acute symptoms even after exposure has recently ceased (Fitzhugh and Nelson 1947). However, no such event has been reported in the case of DDT.

(c) The potential of the effect of the compound to recover at a slower rate than that at which it can be induced by further exposure; this leads to accumulation of the effect. The best example of this is the effect produced by organophosphorous compounds which inhibit the enzyme cholinesterase in both plasma and red blood cells. Inhibition of plasma cholinesterase recovers more quickly than that in the red blood cells, although it is the level in the latter that is of more significance in the occurrence of symptoms of cholinesterase depression. After repeated consecutive exposures to some organophosphorus compounds, the red cell level may diminish by a small amount after each exposure until a level is reached at which symptoms occur and therefore exposure ceases. However, this is not the pattern with all compounds that inhibit cholinesterase. In the case of the carbamate compounds, the carbamoylated enzyme is very rapidly re-activated and therefore, although the level of cholinesterase may drop very rapidly and give rise to symptoms, its recovery is also so rapid when exposure ceases that no accumulation of effect occurs.

In the case of both accumulation of compound and effect, the resultant level after multiple exposure is the balance between storage or inhibition and excretion or re-activation. Therefore, the total multiple dose to produce a given effect is greater than the single dose required to produce the same effect.

Variations in group characteristics do not have as profound an effect on hazard as the variables discussed above. Characteristics that have been investigated include age, sex, race, and nutrition.

The importance of age in the mortality pattern is shown by the fact that in most developed countries about half the fatalities due to accidental poisoning are young children. In most cases, this is because of uncontrolled hazards in the storage of poisons and in the disposal of empty containers. It has also been shown that for the organophosphorous compound parathion, children are more susceptible than adults on an equal-dosage basis (Kanagaratnam *et al.* 1960).

Sex is probably less important although there is a preponderance of males among poisoning cases both in adults and children. This probably represents differences in exposure rather than a true sex difference, the excess in the number of male children possibly being due to their tendency to associate more with men in places where they might be exposed. Women tend to have lower levels of storage of DDT and its metabolites than men but the differences are small and may reflect differences in exposure levels.

In the United States (McLeod 1970) and some other countries, analysis of the race of cases of poisoning has shown up some apparent differences with non-whites preponderating, but this has been attributed partly to a high rate of occupational exposure among non-whites, and partly to problems of literacy and the language of labels. It has often been observed in other occupational fields that non-whites seem to be less prone than whites to allergic skin disorders.

It will be seen that the effects of age on hazard are due to exposure and that the effects of sex and race are rather speculative.

The effect of the nutritional status may be more significant. Only extreme malnutrition has produced observable changes in the toxicity of pesticides in animals and the changes are not constant for all pesticides. For example, complete deprivation of protein in rats increased the toxicity of captan by 2100%, of endosulfan by 20%, and of diazinon by 7% (Boyd *et al.* 1970). For severe protein deficiency (3.5% in the diet instead of 26%) the figures were 26, 4, and 2 respectively. It seems unlikely that humans who are employed and receive a wage, suffer from such degrees of deficiency. However, it has to be remembered that in many countries the average weight of males is well below the average 70 kg assumed in the developed World, and this difference is often considered to be mostly nutritional rather than genetic. Hence for the same dose or exposure, the dosage of the 50 kg man on a mg/kg bw basis is higher by 40%; this increases his hazard considerably, particularly in occupational exposures.

It has only been possible here to give a brief outline of the principal factors that influence hazard. No account has been taken of interactions between pesticides on successive exposures or between pesticides and drugs, and other environmental toxicants. Such relationships are very complex and not enough work has yet been carried out to allow general principles to be formulated. There is no evidence, however, that they constitute a major contribution to hazard.

9.5   The classification of hazards

Many attempts have been made to produce a classification of hazards to simplify the work of those that have to control them. It is generally agreed that toxicity of a pesticide as expressed in the oral and dermal LD_{50} values in the rat, is the starting point for the classification of hazard in the sense that this is the *acute* risk to health. Many classifications stop at the point of making empirical divisions in a ranked list of technical products, labelling these divisions 'high', 'moderate', 'low', etc.

It is true that it is virtually impossible to formulate any kind of comparison of hazard which takes in all or most of its components. There are theoretical ways of doing this but the more these are used, the more arbitrary and less useful the classification becomes. There are, however, two factors which can be taken into account in addition to the toxicity of the technical compound: the first is the strength of the formulation as expressed by percentage of the active ingredient, and the second is the physical state of the formulation to which humans may be exposed. As mentioned above, it is by far preferable that the toxicity of each formulation should be directly determined.

The classification now most widely used which incorporates these considerations is the *WHO Recommended Classification of Pesticides by Hazard* (WHO 1975, 1984). This suffers from the disadvantages of any classification which seeks to divide up a continuum but it is based on sound toxicological principles and is of use as a guide and as a start to the assessment of the hazard of particular pesticide applications.

9.6   The control of pesticides

The control of pesticides has three main objectives: first, to ensure that human exposure is not such as will cause adverse effects under any circumstances; second, to minimize as far as possible the exposure of the general public to any kind of pesticide, taking also into consideration the benefits to the public of its use; and third, to ensure the safety as far as possible of other animals and non-target organisms. The latter is outside the scope of this chapter but is dealt with elsewhere. As in the solution of other public health problems, effective control depends on the erection of a number of barriers so that failure of a single control measure should not result in complete loss of control.

The first barrier is *legislation,* aimed at limiting the use of pesticides of high hazard to circumstances for which there is no acceptably effective alternative, and the distribution of these pesticides to those who are trained their use. Standards have also been developed for the transportation and storage of pesticides. These require the complete physical separation of pesticides from food, and include a prohibition on the use of pesticide containers for food or drinking water at any time.

Restriction of distribution is essential if the public are to be protected from unscrupulous formulators and merchants whose activities in some developing countries without pesticide control have led to many cases of poisoning due to the inappropriate use of toxic pesticides; an example is the use of endrin to treat head lice. Distribution control infers a registration procedure. In some countries, the mechanisms of registration are so complex that registration has become an end in itself rather than being seen in its perspective as a prerequisite to control. For effective legislative control, in addition to registration of all compounds and formulations in use, an inspectorial system with access to chemical analytical facilities is essential.

The second barrier is *the protection of pesticide handlers,* whether manufacturers, formulators, baggers, mixers, or applicators. The aim is to reduce hazard by reducing exposure without unduly hampering the effective application of the pesticide. The protection of handlers can be approached in several ways.

The use of some type of *protective clothing* is essential. The emphasis to be laid on it depends on the toxicity of the pesticide formulation in the form in which it is handled. There is a certain minimum that must be used under all circumstances – a long-sleeved overall or equivalent local dress, and adequate footwear; these are designed to limit skin contamination, even by pesticides of slight hazard. In many other types of pesticide application, and always for the handling of concentrates, gloves are essential. Barrier creams give a false sense of security and should not be used. If the hazard of the formulation being applied is moderate or higher, head protection is needed, and for this group, eye protection is provided more comfortably by a visor than by goggles. Chemical respirators are usually only needed for compounds of high hazard that are applied in a form that might be inhaled. Dust masks or even cloths over the mouth and nose can be used with compounds of lower hazard with the caution that these must be frequently changed as they can enhance dermal absorption on the face if they are allowed to become wet. Aprons are needed if concentrates are being handled or formulations of high hazard are being applied. The same clothing needed for application is also needed during the washing and cleaning of application equipment.

None of these precautions are effective unless they are coupled with *protective hygiene*. This includes both the washing of working and protective clothing, and the personal hygiene of the man exposed, all designed to minimize as far as possible contact between the pesticide and the skin. As long as contaminated clothing is worn, so long does exposure and dosage continue, and therefore all working clothing should be washed regularly, daily if pesticides of other than slight hazard are being used. In the washing of protective clothing, also needed on a daily basis, special attention must be paid to the inside of gloves, boots, hats, aprons, and respirators.

Personal hygiene is probably the most important single protective measure. If properly carried out, dermal exposure is very considerably reduced. This infers regular washing of all exposed skin at each work interval and showering at the end of the day. Obviously, not only must each man be trained in these procedures, he must also be provided with facilities, including soap to more readily remove skin contamination. Washing is particularly important in tropical areas, not only because skin absorption may be enhanced when the skin is hot, but because climatic conditions often result in the wearing of working and protective clothing that is only marginally appropriate to the hazard of the pesticide being used.

There are other useful types of protection aimed at lowering exposure. Protection by distance is simply increasing the space between the man and the pesticide. It is more useful in industry than in pesticide application, although the principle seems sometimes to be neglected by the designers of application equipment. It has to be replaced by protection by technique such as training a man to avoid contamination by his own spray. To see an agricultural sprayman walking into the sprayed crop, or spraying to one side against the wind, is not as uncommon as it ought to be. Finally, there is protection by time. This means either limiting the amount of pesticide (often measured in pump charges) to which the man is exposed on a daily basis, or the number of hours worked per day, or the area treated.

The third barrier is aimed at the protection of the general public from smaller exposures to pesticides. It includes both the exclusion of all persons from treated areas where significant exposure might occur, and the observance of minimum intervals between the treatment and marketing of any edible crop.

Apart from legislative measures, the efficiency of control depends on the education of all those liable to be exposed to pesticides. The first line in education is the provision of adequate labels on all pesticide formulations, setting out as a minimum, in a form and language that can be understood, the approved name, the method of use, the precautions to be taken in use, the symptoms and treatment of poisoning, and the name and address of the supplier.

More than this is needed, for many people will not, do not, or cannot read the label. No person should handle any pesticide of more than slight hazard without proper training in its use. This infers that someone must be responsible for training. Too little training is carried out on a systematic basis, particularly for farmers.

To use statistics to indicate that no adverse effects are produced by pesticide exposure is analogous to saying there are no fish in the sea because they are not usually seen. Too often, doctors are not familiar with early signs of pesticide poisoning or do not notify cases. The control of hazards must include the education of doctors and the assurance that appropriate anti-dotes are available wherever they might be needed.

9.7   Monitoring of exposure

Monitoring of exposure to pesticides is an important exercise for some purposes, particularly for an understanding of the toxicology of a pesticide in humans. Animal experimentation and the determination of toxicological indices can only provide guidelines, particularly since species differences in response are not constant. It is by a study of the epidemiology of human exposure and of poisoning cases that we can assess how applicable are the animal indices to man.

Monitoring is an epidemiological tool to estimate exposure or absorption and dosage. It is most profitably carried out on those most exposed. These are to be found in occupational groups if certain difficulties can be overcome. Industrial groups may be also exposed to manufacturing intermediates or to mixtures of impurities with higher toxicity, while applicators are frequently exposed to a variety of pesticides, and it is not easy to isolate exposure to a single chemical to measure its absorption. Nevertheless, suitable groups can be found, and it is important that they should be, since it is improbable that any effect not found in those most exposed will appear in lesser exposed groups.

Methods of monitoring vary according to the class of pesticide. Some-times the level of the pesticide itself in body storage or transport can be measured; this is the method used for monitoring most organochlorine pesticides. Where the pesticide has a readily measurable dose-related effect, this may be measured, the best example of this being the cholinesterase depression caused by organophosphorus pesticides. Field kits are available for the measurement of the plasma, red blood cell, or whole-blood cholines-terase. The two latter are the most useful for monitoring purposes. For these and other pesticides, including some of the newer pyrethroids, excretion of the compound or its metabolites can be assayed, usually in urine.

When none of these methods are suitable or available, it is possible to use externally applied exposure pads to provide an index of potential absorption. The pads have an advantage that, while sophisticated laboratory facilities are still needed for their analysis, these need not be near the place of sampling, as in general it is easier to transport the pads than biological specimens. They have the disadvantage that the calculation of results entails a degree of extrapolation.

The publication of the results of surveys of exposure is important, even if the findings are negative. Only in this way can the 'no effect' levels of human exposure be found, and the wearing of protective clothing and other protective measures be encouraged by reference to facts rather than to the more commonly used general admonitions which are sometimes neither credible nor possible.

Another purpose of monitoring is the surveillance of working groups to prevent the occurrence of accidental poisoning among those routinely exposed to pesticides of moderate or higher hazard. The most common application of this outside factories is the surveillance of spraymen applying some organophosphorus insecticides in malaria control campaigns and other public health work, where they may be exposed to an insecticide for several weeks or months on a 6-day weekly basis. The aim of surveillance is to prevent the appearance of any symptoms of poisoning by withdrawing the person from contact if the effect of absorption passes a predefined level.

9.8   Exposure of the general public

The general public are exposed to pesticides in several ways. Often consider-able amounts may be used in the home or garden. If good control is exercised, the compounds obtainable by the public should only be those of slight hazard, notwithstanding the efficacy of more hazardous compounds for household purposes. No matter how comprehensive labels may be, tragedies are liable to happen due to misuse of the pesticide, or carelessness in the storage or disposal of containers, Other local large exposures can occur due to food becoming grossly contaminated, to mishaps and spillages, or to gross negligence on the part of applicators.

More widespread exposure of the public occurs through the presence of residues in food, but this exposure is at a much lower level. No case of poisoning has been described which could be attributed to a residue in food produced in accordance with good agricultural practice. The setting of acceptable daily intakes of pesticides and other chemicals takes into account large safety factors applied to the 'no-effect' level in animals and, if these levels are not exceeded, it is inconceivable that acute effects might result or even that exposure through residues might aggravate the effect of other larger exposures. Finally, the general public may be subject to exposure to some compounds through water and air. Such exposures are of very low level and of no significance.

Anxiety among the general public related to pesticide exposure is focused more on the possibility of chronic than acute effects, and particularly of those that may result from low-level exposure for very long periods of time. The main concerns are with mutagenicity and carcinogenicity.

Mutagenicity is assessed by a battery of tests designed to demonstrate any mutation produced by a pesticide on bacteria with known mutagenic tendencies, and on animals during reproduction studies and after a large single dose during early pregnancy. Positive mutagenicity tests on bacteria are frequently, but by no means always, associated with an increased incidence of tumours in rodents fed the pesticide at the highest tolerated levels for a period of 2 years.

The significance of the results of such tests is a matter still open to debate. On the one hand, a demonstrated tendency for a chemical to produce cancers in animals must give rise to caution that human exposure to the chemical should be kept to a minimum. On the other hand, for the pesticides currently in use, the levels to which animals have been exposed for almost the whole of their natural lifetime are so far removed from credible human exposures that, if the dose–effect relationship is a valid concept at all, the risk of long-term effects to humans is very small indeed. The acceptability of the risk has to be compared to that of other everyday risks and to the benefit expected from the use of the pesticide.

If a perceptible long-term effect occurs among humans, it would first be expected to appear among those with the highest exposure. Repeated health surveys of manufacturing staffs and a few surveys of applicators have failed as yet to demonstrate any long-term effect which could not have been anticipated from acute and subacute studies on animals, and from knowledge of specific biochemical modes of action in mammalian systems.

9.9   Conclusion

The benefits conferred on mankind by the use of pesticides are not generally realized by people who have become accustomed to the result of their proper use. Nevertheless the use of pesticides confers certain responsibilities on the user to protect himself, those who work for or with him and, more remotely, those other members of the general public who might be directly or indirectly exposed through their activities. In many ways, the pesticide user is in a position analogous to that of a car driver. Just as the latter has to possess a minimum knowledge to be safe to himself and others, so the pesticide user also needs knowledge to appreciate and control hazards. There is no pesticide that is so toxic that it cannot be used safely by reducing the hazard by taking extreme precautions; but on the other hand, there is no pesticide that is so safe that it can be used without minimum precautions, as many have found to their own, or somebody else's cost.

9.10 References

Boyd, E. M., Dobos, I., and Krijnen, C. J. (1970). Arch. Environ. Hlth. **21,** 15.

Casarett, L. J. and Doull, J. (1975). *Toxicology* (1st edn), p.17. McMillan, New York.

Copplestone, J. F., Fakhri, Z. I., Miles, J. W., Mitchell, C. A., Osman, Y., and Wolfe, H. R. *Bull. Wld. Hlth. Org.* **54,** 217.

Fitzhugh, O. G. and Nelson, A. A. (1947). *J. Pharmacol. Exp. Ther.* **89,** 18.

Hayes, W. J., Jr. (1975a). *Toxicology of pesticides* (1st edn), p.39. Williams and Wilkins, Baltimore.

—— (1975b). *Toxicology of pesticides* (1st edn), p.323. Williams and Wilkins, Baltimore.

Kanagaratnam, K., Wong Hock Boon, and Tan Kwang Hoh (1960). *Lancet* **i,** 538.

McLeod, A. R. (1970). *J. La. Med. Soc.* **122,** 337.

Weil, C. S. and McColister, D. D. (1963). *J. Agric. Food Chem.* **11,** 486.

WHO (1973). *WHO Tech. Rep. Ser.* **513,** 8.

—— (1975). *WHO Chron.* **29,** 397.

—— (1978a). *Principles and methods for evaluating the toxicity of chemicals, part I, environmental health criteria,* No. 6. WHO, Geneva.

—— (1978b). *WHO Tech. Rep. Ser.* **620,** 8.

—— (1979). *WHO Tech. Rep. Ser.* **634,** 9.

—— (1984). *Guidelines on the use of the WHO recommended classification of pesticides by hazard.* Unpublished document VBC/84.2. Pesticide Development and Safe Use Unit, Division of Vector Biology and Control, WHO, Geneva, Switzerland.

Wolfe, R., Durham, W. F., and Armstrong, J. F. (1967). *Arch. Environ. Hlth* **14,** 622.

10
Field and plantation crop pest control

L. BRADER, E. J. BUYCKX, J. C. DAVIES and W. REED

10.1 Introduction

It is clearly impossible in a single chapter to deal adequately with the problems of pesticide application in field and plantation crops, a subject which needs a book of its own. Therefore a few crops have been selected to illustrate in detail particular areas and aspects of the numerous problems which arise.

Cotton is one of the world's most important crops; usually it cannot be produced economically without adequate plant protection. Consequently it is the crop on which more pesticide is used than any other and which has therefore given rise to more problems – pesticide resistance, secondary pests, residues and environmental pollution – and hence has demanded and generated more research, for example with respect to pesticide application and integrated pest management, than any other.

The crops have been chosen from four categories – cereals, legumes, oil seeds and vegetables. These are important and representative crops, often grown by small-scale farmers in less-developed countries and have been selected to illustrate practical problems encountered in obtaining adequate chemical control of some of their important pests.

But not all crops and their associated pest complexes present problems. It must be emphasized that on several field crops the answers to the questions of whether to spray, and if so with what, are mostly well known. On many crops, especially non-irrigated, the economic thresholds (see Chapter 7) are clear-cut, for plant compensation for earlier damage often cannot occur, either because of the nature of the plant or because of the limitation of growth by seasonal change, particularly in non-tropical regions where the onset of winter limits growth.

But a simplistic approach to pesticide use in pest management on field crops is to be avoided at all costs. Although pesticides are undoubtedly invaluable to the many farmers who have the resources and capability to use them beneficially, it is essential that the alternative and/or complementary elements of pest management are fully considered and utilised in order to avoid or minimize undesirable side effects such as resistance and residues in food; growers should not be encouraged to reach for the pesticide can every time that an insect, disease, or weed is seen in a crop! Pest management should already start before the field is prepared for sowing. Choice of the appropriate cultivation methods, the right seed, clean and treated, and the optimum sowing date (in co-operation with other farmers in the area) can all help to reduce subsequent pest problems and pesticide use.

Pesticide use on field crops should be in response to counts of pests in the crop which exceed the 'threshold values'. Due attention should also be given to the conservation of beneficial insects, including the natural enemies of the pests, so broad-spectrum persistent pesticides should be avoided. Ideally, that pesticide and means of application should be chosen that will not only ensure adequate control of the target pests, but also the safety of the other fauna, including man. But in putting these broad principles into practice, other difficulties appear, determined by a wide range of factors. They include the biology of the insect, the particular ecological niche that the pest occupies, crop ecology, agronomy, and the physical problems of ensuring that the insecticide is sufficiently effective to ensure the speedy destruction of the insect before economic damage occurs. Clearly, in instances where insects act as vectors of disease, particularly virus, rapidity of kill is important; in other instances slow action and residual effect is acceptable or even preferable.

All these points have been discussed in principle in previous chapters of this book; the main objective of this chapter is to illustrate the various aspects to be considered when applying these principles in practical pest control in field crops, especially in relation to small-farmer practice.

10.2 Cotton cultivation

The cotton varieties in cultivation at present belong to four species of the genus *Gossypium: G. herbaceum, G. arboreum, G. barbadense,* and *G. hirsutum.* The first two species are grown mainly in Asia on a rather limited acreage, but by far the largest number of commercial varieties belong to *G. hirsutum* and include the Upland cottons. These varieties produce medium-staple cotton with a fibre length of 25–28 mm. *G. barbadense* includes the long-staple varieties with a fibre length of more than 28 mm. The latter are mostly grown in Egypt and Sudan; the Tanguis cotton of Peru and the Sea Island cotton of the West Indies also belong to this species. Sea Island cotton was also grown in southern United States until the beginning of this century, when the newly introduced bollweevil, *Anthonomus grandis,* made its production economically impossible. 'Moco' cotton, which is grown as a perennial on about 1 million ha in northeastern Brazil, is a mixture of *G. hirsutum* and *G. barbadense.*

The prototypes of the present day annual varieties were perennial shrubs, and these annual varieties possess two kinds of branches. The monopodia, or vegetative branches, are produced at the bottom of the main stem while the sympodia, or fruiting branches, occupy the rest of the main stem. Typical *hirsutum* varieties will produce on average two basal monopodial branches which may be followed by twenty or more fruiting bodies. It is typical of these cottons that they produce many more fruiting bodies (flower buds or 'squares', followed by flowers and bolls) than the plant can carry even when perfectly protected against pest attack. The excess fruiting bodies fall from the cotton plant; an example of this so-called physiological shedding is given in Fig. 10.1. It is well known that heavy shedding of fruiting bodies occurs after white bloom and before the bolls are 10 days old. Even under optimum growing conditions over 50% of the total number of white flowers produced

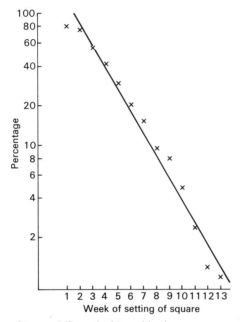

Fig. 10.1. Percentage of 'squares' (flower buds) resulting in mature cotton bolls in relation to the week of setting of the square during the development of the cotton plant. Average of observations of 50 cotton plants each year during 1967, 1968, and 1969 at Bebedjia (South Chad); *G. hirsutum* variety P 14, well protected against insect attack.

will never become fruits; the period of heavy shedding begins with the blooming peak and increases gradually as the growing season advances. About 20% of fruit parts are shed as squares. Besides its overproduction of fruit points, the cotton plant has the additional characteristic of compensatory growth and fruit production, which may manifest itself in particular when early set fruiting bodies are lost by pest attack but growing conditions continue to be favourable in late season. This fruiting pattern is of considerable significance for the protection of the plant against pests which attack the fruit, i.e. bollworms, and bollweevils. Bearing in mind also that the earlier bolls are heavier and produce better quality cotton, it is easy to understand that effective protection during a rather short, early season can produce an optimum crop.

The oldest records of cotton textiles date back to about 3000 BC in the case of West Pakistan and 2500 BC with respect to cotton found in Peruvian archaeological excavations (Brown and Ware 1958). In Africa, cotton textile fragments have been found in the Upper Nile (Sudan) in excavations of the Meroitic civilization, which existed *circa* 500 BC to AD 500.

In Asia, India has the longest commercial cotton-growing tradition. Originally, only Old World species, *G. arboreum* and *G. herbaceum,* were grown, but in the early 19th century American Upland varieties were introduced. Cotton was introduced from India into China where it was put to use towards the 14th century, and in Western and Central Asia it has been cultivated since the beginning of this era (Brown and Ware 1958).

Modern commercial cotton growing in Africa dates from the early 19th century when *barbadense* cottons were developed in Egypt. They were based on crosses between annual Sea Island cottons, introduced from the southern United States, and a locally grown perennial *barbadense*. In the rest of Africa cotton growing was developed in the late 19th and early 20th centuries mainly on the basis of the introduction of Upland varieties. A major element of the adaptation process was the breeding of varieties which were resistant to bacterial blight (*Xanthomonas malvacearum*) and jassids (*Empoasca* spp.) (Pearson and Maxwell Darling 1958).

In the major cotton-producing countries in South America – Brazil and Peru – regular cotton use and cultivation dates back to the 16th century. Peru is known especially for its Tanguis cotton, while in Brazil about half of the acreage is given over to Moco cotton, a perennial tree cotton.

In the USA cotton was introduced in the early 17th century and was mostly short-staple cotton. In particular, the introduction of Mexican cotton in the early 19th century contributed tremendously to the expansion of the cultivation of Upland cotton. Sea Island cotton was introduced in 1785 from the West Indies but remained mostly limited to the coastal lowlands of the southern states. Sea Island cotton was called 'lowland' cotton and the other cotton become known as 'upland', which led to the name Upland cotton for the *hirsutum* varieties.

10.3   Development of cotton pest control

The cotton plant is subject to attack by many pests; the major group is formed by insects of which over a thousand species have been recorded (Hargreaves 1948). A few of these, to which some mite species should be added, are of such economic importance that without their control only low yields of seed cotton are obtained wherever the crop is grown. Consequently effective cotton protection becomes a prerequisite for optimum production. Where insects have been controlled successfully, yields have been raised from less than 500 kg/ha to over 1000 kg/ha in rain-fed cotton (Gower and Matthews 1971) and to over 3000 kg/ha under irrigation.

In view of its overall economic importance, considerable research efforts have been devoted to cotton from which many new developments in pest control have originated. This is the case particularly in the United States, which will therefore be taken as the major example to illustrate the development of cotton pest control.

By far the most damaging cotton pest in the USA is the bollweevil, *A. grandis,* and notwithstanding the various control efforts, average annual losses were over 9% during the period 1929–1953. Losses caused by other insects were in total much less except in the western cotton-producing states (Brown and Ware 1958). However, in the latest decennia the relative importance of other pests, mainly bollworms, has increased due to the elimination of natural enemies and the development of pesticide-resistant strains.

The bollweevil was first observed in 1892 and spread over the cotton belt in about 20 years, although it never appeared in damaging numbers in the irrigated areas of the southwest. It has been the largest single factor in determining changes in cotton growing in the United States; for example as a result of its presence a general shift in the cotton belt has occurred from more humid to less humid areas where the development of the insect is less favoured. Sea Island cotton suffered so severely from attack that its cultivation had to be discontinued.

The pink bollworm, *Pectinophora gossypiella,* was first noticed in Mexico in 1916 and consequently was found in cotton in the United States in 1917, since when it has spread across the south western states. Other insects of significant economic importance are the bollworm, *Heliothis zea,* the tobacco budworm, *H. virescens,* the cotton leafworm, *Alabama argillacea,* the cotton aphid, *Aphis gossypii,* the cotton fleahopper, *Pseudotomoscelis seriatus,* lygus bugs, *Lygus hesperus,* and a number of spider mites.

Before 1892 cotton in the United States suffered little damage from insects or spider mites, although occasionally some losses were caused by the bollworm, the cotton leafworm and the cotton aphid. Farmers used Paris green, London purple or lead or calcium arsenate against bollworms and nicotine against aphids in the rare cases that these insects were controlled (Reynolds *et al.* 1975). However this situation changed drastically after the spread of the bollweevil. At that time with the pesticides and equipment available it was not possible to achieve effective chemical control of the bollweevil, as a result of which a combination of control measures was used to reduce the pest level by cultural methods. This included in particular the use of early maturing or shorter-season varieties and at the end of the growing season the destruction of fallen cotton debris to reduce the number of overwintering bollweevils. Thus there was relatively little need for insecticide applications (Bottrell and Adkisson 1977).

However, by 1923 the effectiveness of calcium arsenate dust for the control of the bollweevil had been clearly demonstrated and in addition the techniques for the aerial application of these dusts had been developed. Therefore generalized chemical control of cotton pests has been in effect since 1923, but this approach was not without consequence to the overall cotton pest situation and led to the erosion of ecologically-oriented pest control in cotton (Bottrell and Adkisson 1977). While the cotton leafworm became less important, cotton aphids became a bigger problem and nicotine sulphate had to be used for control purposes.

Chemical control became even more effective between 1945–1955 with the development and use of the synthetic organic insecticides, first DDT, followed by BHC, toxaphene, and chlordane. Mixtures of BHC/DDT/ sulphur and toxaphene/DDT provided excellent all-purpose treatments (Bottrell and Adkisson 1977). In fact weekly spraying was recommended until the cotton bolls became too hard to be susceptible to attack by *H. zea* and bollweevil. This was often preceded by early-season spraying against thrips (*Frankliniella* spp.), aphids, and cotton fleahoppers. The overall result was very effective pest control, maximum cotton production, but also the gradual development of unexpected changes in the cotton pest situation.

In the mid-1950s organochlorine-resistant bollweevils were found in the mid-south and by the 1960s they were widespread. This was followed by the appearance of DDT- and endrin-resistant strains of the bollworm and the tobacco budworm. It was then necessary to use organophosphorus and carbamate insecticides that were most effective against the bollweevil, but less so against *Heliothis* (Bottrell and Adkisson 1977). Moreover these insecticides proved highly destructive to parasites and predators which led to increased *Heliothis* outbreaks; thus mixtures with organochlorines had to be applied frequently and costs increased accordingly. In the Lower Rio Grande Valley of Texas and northeastern Mexico the tobacco budworm also ultimately became resistant to organophosphorous insecticides. Over 15 applications per season were carried out without adequate control results and finally cotton farms were abandoned in northeastern Mexico.

The above overview concerns the cotton-growing states where the boll-weevil is the key pest, but similar shifts are found elsewhere as a consequence of intensive chemical control. For example, in southern California *L. hesperus* and the pink bollworm are the key pests, but outbreaks of non-target secondary pests such as the cabbage looper (*Trichoplusia ni*), beet armyworm (*Spodoptera exigua*) and bollworms may cause greater damage than the key pests (Bottrell and Adkisson 1977).

In developing countries, cotton is often one of the crops first planted in agricultural development schemes, particularly in newly irrigated areas. In most cases, pest control methods are adopted which are dependent solely on chemical pesticides (FAO/UNEP 1974) and the trends described above can then be expected to take place. For instance in Nicaragua, the intensive use of insecticides, over 25 applications per growing season, has led to a situation where *H. zea* has become the dominant cotton pest and the previous major pest, *A. argillacea,* has become of less importance. In addition, towards 1965, *Spodoptera* species were the second most important cotton pest while they were virtually absent from cotton before the widespread use of insecticides (Falcon and Smith 1973).

It may be concluded from these examples that 'premature and exclusive use of insecticides has been the major inducer of the cotton pest insect problems in many areas of the world' (Bottrell and Adkisson 1977). The pattern of such pesticide-induced changes in the cotton pest situation may be classified into a series of stages or phases (Falcon and Smith 1973). These are the following:

(a) Subsistence phase. The cotton crop is usually grown in non-irrigated condition, yields are low, and there is no organized programme for protecting the crop from pests. Pests are controlled mainly by natural enemies and cultural practices.

(b) Exploitation phase. Cultural practices are improved in order to obtain maximum yields; crop protection is exclusively based on the use of chemical pesticides, and at the beginning high yields are obtained.

(c) Crisis phase. After a number of years of heavy use of pesticides, the key pests become resistant to the insecticides commonly used. Other insecticides, more frequent applications, and higher dosages are required, and new pest species often occur in damaging numbers.

(d) Disaster phase. Increasing use of insecticides and the unsuccessful control of the losses caused raises production costs and reduces returns to a level where cotton production becomes uneconomical (examples of this are found in Peru, 1956, southern Texas/northeastern Mexico in the late 1960s, and Sudan Gezira late 1970s–early 1980s).

(e) Integrated control phase. A combination of control measures is gradually introduced, leading to a rather stable pest control situation where pesticides are only applied when pest populations are above the economic damage threshold and where the fullest use is made of natural control methods (cultural control, natural enemies, and pest-resistant varieties).

(f) Deterioration phase. Relaxation of the above integrated control approach may again lead to gradually increased use of chemical pesticides.

From the above it is evident that over the years control of cotton pests has constituted the basis from which new approaches to pest control in general have emerged. In the context of this publication it will not be possible to trace the development of the cotton pest situation in each major cotton-producing country, but in Table 10.1 a general idea is given of the presence of the cotton pest species most frequently encountered in the Old World (Europe, Asia, Africa, and Australia) and the New World (North, Central, and South America).

The overall importance of the pests listed in Table 10.1 is such that every effort should be made to reduce the development of crisis and disaster situations as described above. Signs of such developments can be detected in virtually all major cotton-growing areas and this is illustrated in Table 10.2 which gives an overview of the changes in the pest situation in a number of selected countries (FAO 1976).

10.3.1 Some details of cotton pests

Cotton is one of the crops suffering most from insect damage; this is clearly confirmed by a study of the world pesticides market in the early seventies which shows that 21% of the world total is used on this crop (Cramer 1975).

Cotton insects may be divided in two main groups, on the one hand the leaf-, flower- and fruit-eating pests, on the other the sucking pests, the latter often being vectors of diseases. In the first group, the most important are the noctuid caterpillars, the bollworms *Heliothis* spp., *Earias* spp., *Diparopsis* spp., the armyworms and leafworms, the pink bollworm, *P. gossypiella,* and the cotton bollweevil, *A grandis* (confined to Central America and the USA). They have a direct impact on boll production.

Among sucking insects, the main pests are jassids (*Empoasca* spp.), whiteflies (*Bemisia* spp.), aphids, *Lygus* bugs, and cotton stainers (*Dysdercus* spp.). By feeding on the sap and often injecting saliva toxic to the tissues, they may cause stunted growth and reduce the photosynthetic activity of leaves. While feeding, aphids and whiteflies excrete honeydew which falls on the lower leaves and open bolls. The leaves become sticky and fungi grow on the honeydew, giving a black sooty appearance and reducing their activity. The lint is also stained by the fungus growth and ginning can be difficult because of stickiness.

The occurrence of cotton insect pests and the damage they cause are closely related to certain factors such as the amount of rain, particularly at the beginning of the rainy season, the stage of plant development and the period of infestation. In rain-fed cotton, the crop is sown at the onset of the rains. Generally speaking, the seedlings have their second pair of true leaves 3 weeks after germination. The first flower buds will appear about 3 weeks later and the first flowers after another 2 weeks. The flowering period extends from 6 to 8 weeks. As already mentioned considerable shedding will occur if

TABLE 10.1 *World distribution of principal cotton insect pests*

Order and genus	Old World species (Europe, Asia, Africa, and Australia)	New World species (North, Central, and South America)
Coleoptera:		
Alcidodes	Several spp.	—
Anthonomus	Present, but not recorded on cotton	*grandis*
Euthinobothrus	—	Several spp. in South America
Podagrica	Several spp.	—
Sphenoptera	Several spp., including *gossypii*	—
Syagrus	Several spp.	—
Lepidoptera:		
Xanthodes	*graellsii*	—
Alabama	—	*argillacea*
Cosmophila	*flava*	—
Diparopsis	*castanea, watersi*	—
Earias	Several spp.	—
Heliothis	*armigera*	*virescens, zea*
Sacadodes	—	*pyralis*
Spodoptera	*exigua, littoralis*	*exigua, frugiperda, ornithogalli*
Sylepta	*derogata*	—
Hemiptera:		
Dysdercus	Several spp.	Several spp.
Empoasca	*facialis, devastans, terrae-reginae*	*fabae*
Helopeltis	*schoutedeni*	—
Horcias	—	*nobilellus*
Lygus	*vosseleri*	Several spp. in North America
Nezara	*viridula*	*viridula*
Oxycarenus	Several spp.	—
Acarina:		
Cecidophyes	*gossypii*	*gossypii*
Hemitarsonemus	*latus*	*latus*

Three important pests not listed above are *Aphis gossypii, Pectinophora gossypiella,* and *Tetranychus urticae* which are present in practically all cotton-growing areas of the World (Adapted from Pearson and Maxwell Darling 1958.)

climatic conditions are unfavourable (poor rains, high temperatures), if irrigation is either deficient or in excess, or if insect infestation is serious. Considering the development of the cotton plant, a distinction can be made between early-season, mid-season, and late-season pests (Ripper and George

TABLE 10.2 *Changes in cotton pest situation (1975)*

Bolivia		Formerly unimportant pests becoming dangerous to cotton production
Colombia		‡ *Trichloplusia ni* ‡ *Pseudoplusia* ‡ *Tetranychidae* sp. ‡ *Anthonomus grandis* ‡ *Pectinophora gossypiella* ‡ *Sacadodes pyralis* † *Heliothis* sp.
Egypt		† *Bemisia tabaci* † *Empoasca lybica* † *Nezara viridula* ‡ *Heliothis armigera*
El Salvador		* *Anthonomus grandis* † *Heliothis zea* ‡ *Spodoptera exigua*
Greece		‡ *Pectinophora gossypiella* * *Heliothis armigera*
India		‡ spider mites
Pakistan (Sind Province)		‡ spider mites
Sudan		† *Podagrica puncticollis* † *Podagrica pallida* † *Caliothrips sudanensis* * *Empoasca lybica* ‡ *Bemisia tabaci* ‡ *Aphis gossypii* ‡ *Heliothis armigera* ‡ *Diparopsis watersi* ‡ rats
Turkey	Mediterranean: ‡ *Bemisia tabaci* 　*(since 1974)* § *Tetranychus* spp. § *Heliothis armigera* § *Spodoptera littoralis* † *Pectinophora gossypiella*	Aegean: † *Bemisia tabaci* * *Tetranychus* spp. † *Empoasca* spp. § *Aphis gossypii* † *Pectinophora gossypiella*

†, slight increase; ‡, important increase; §, slight decrease; *, important decrease.

1965). The first insects damaging cotton are often thrips, aphids, and jassids, and heavy infestation can completely stunt young plants. In addition, in the Old World early-season attack may occur of the bollworms *Earias* (causing destruction of the terminal bud, growth retardment, and subsequent development of a forked main stem) and *Diparopsis*. Exceptionally *H. armigera* may cause similar damage.

Mid-season pests occur on cotton as the first buds are formed, during flowering and the beginning of boll setting. As soon as flowering has initiated, bollworm infestation (*Heliothis* spp., *Earias* spp., and *Diparopsis*) increases rapidly.

Pests of the late season when green bolls are maturing are cotton stainers and pink bollworm. Often an end-of-season buildup of *Earias* and *Diparopsis* occurs, and with the onset of dry weather, outbreaks of aphids and red spider mite.

Of course conditions vary considerably from the drier savannah areas with a rainy season of only a few months where cotton is often irrigated, such as in the Gezira scheme in Sudan, to humid tropics such as Thailand. Under each set of climatic conditions a particular pest complex will prevail and differences in feeding habits and period of infestation of the various pests render their control difficult. Cotton being usually grown as an annual crop and assuming good cultural control practices that eliminate all remaining plants and debris, the cotton agroecosystem starts anew every growing season. Sources of infestation are neighbouring natural vegetation, food crops, and pests diapausing in the soil. Through migration, pest infestations may increase very rapidly. Natural enemies will follow, but their populations build up more slowly and not always fast enough; this is particularly true in areas with a long dry season. As mentioned earlier, the majority of the bolls which are the heaviest and give the best grade of cotton are set during the first 4 weeks of flowering. Therefore, it is important that the first boll setting is not lost, otherwise a later maturing and later picking will result. Experience in the Sudan has shown that in comparison with an early harvested crop, late maturity and late picking result in a higher percentage of lower-grade cotton (Ripper and George, 1965). In order to get high yields, growers must have healthy plants producing as many young bolls as possible early in the season and protect them to maturity. Understandably with the availability of potent broad-spectrum synthetic insecticides, the idea of preventing the occurrence of damage becomes prevalent. In many cases the strategy often adopted on grounds of simplicity was application on a calendar basis, starting with flowering and repeated at regular intervals. Because insect infestations vary in time and in intensity with each cotton-growing season, this results in unnecessary insecticide applications. The major weakness of such cotton pest control programmes is that they do not take sufficiently into account how and when losses occur, the changes in insect pest populations, and the effect of toxic chemicals on the beneficial arthropod fauna in and around cotton fields.

The main problems arising from sole reliance on insecticides are the development of resistance and the destruction of natural enemies. An excellent insecticide for the control of a given cotton pest today may become practically useless within a few years, because of the emergence of a resistant strain of pest. Resistance is a 'decreased response of a population of an animal or plant species to a pesticide or a control agent as a result of their application' (FAO 1967). Populations become resistant because of the selection by insecticides of individuals with a genetic constitution that permits them to survive and reproduce. The most common reaction to resistance to pesticides has been to augment dosage and frequency of application, which in turn increases resistance. According to Waterhouse (1976) the two principles basic to the judicious use of pesticides are (i) to avoid, where practical, unnecessary selection pressure for resistance and (ii) to anticipate the resistance problem. The probability of emergence of resistance in a cotton insect pest population is a function of the frequency, intensity, persistence, and extent of insecticide application. It is therefore advisable to exert a selection pressure as low as possible by limiting insecticide use to occasions when it is economically justified and using methods of control other than chemical. Obviously starting applications as late as possible and reducing their number is the answer. When resistance starts building up, it is advisable to switch to an alternative insecticide, although cross-resistance, such as DDT-synthetic pyrethroids, may to a certain extent limit the choice of another compound. Surveying for possible development of resistance should be part of any chemical control pro-gramme.

Except for the bollweevil, the pink bollworm, and *Lygus* bugs, most insect pests of cotton have important natural enemies attacking them in the egg and larval stages. In many areas, heavy use of insecticides virtually eliminates natural enemies. Improper application and inadequate dosage may result in deposits sublethal for the pest whilst the more susceptible and more mobile natural enemies are killed. Also, the elimination early in the season of insect pests such as aphids, spider mites etc. will deprive predators of their prey and prevent their population buildup by the time mobile flying pests such as *Heliothis* or *Spodoptera* arrive and start laying eggs.

A certain number of factors which are interrelated have to be taken into account in chemical control of cotton insects (Tunstall and Matthews 1972; Morton 1979): the choice of the insecticide, the timing of insecticide application, the method of application and formulation, and the potential yield.

10.3.2 Choice of insecticide

It is only rarely that losses in yield are caused by one insect pest. Usually several species, composing the pest complex, are involved – one or two being the economically most important ones, the 'key' pests – and their populations build up in succession as early-season, mid-season, and late-season pests. It is often impossible to control the major components of the pest complex with one insecticide, and two or three may be required. Rather than applying them as a mixture, they should preferably be used successively, except for the late-season application when there is a mite problem and an acaricide is mixed with an insecticide.

Field testing of pesticides that appear most promising on the basis of the information provided by the manufacturer is needed to select those best suited to the given crop/pest situation. Usually with the relatively large number of products on the market, several insecticides will be found effective. These will be further selected on the basis of mammalian toxicity, cost, spectrum of activity or selectivity with respect to natural enemies and residual persistence in the field. In principle, the cheapest product will be chosen but other considerations may have a certain precedence on cost. In developing countries, much of the cotton is produced by small-scale farmers who most of the time treat their crop themselves using knapsack or hand-held application equipment, having no protective clothing which anyway is uncomfortable to wear in the tropics. Therefore compounds of low mammalian toxicity have to be adopted. When spraying is mechanized, under closely controlled conditions, or when the wearing of protective clothing is feasible, insecticides of higher mammalian toxicity can be used.

According to the type of application equipment used, wettable powder or emulsifiable concentrate formulations are chosen. While the latter mix more easily with water, wettable powders can be supplied to small farmers in sachets, each containing the amount required for one knapsack load. In this way, wastage and mistakes in dosage are minimized (Gower and Matthews 1971).

Residual persistence is the most important factor in the development of an insecticide-application strategy using contact and stomach poisons (Morton 1979); it should provide continuous protection and allow less-frequent spray applications in conditions characterized, for instance, by a near-continuous and irregular pattern of infestation by bollworms. A residual insecticide deposited on cotton is most effective when the target pest is mobile. This mobile stage, which is also the damaging one, is the larva, and the younger the larva, the less insecticide required to kill it. On hatching, the larva has to crawl from the egg to the buds over well exposed and therefore easily contaminated plant parts.

The rate of degradation of the chemical is important to know, as in combination with plant growth it will cause a reduction of the toxicity of the insecticidal deposit for the target species. Organophosphates and carbamates at the usual rates of application are generally effective for about 5–6 days, while it seems that pyrethroids are slightly more persistent (7–9 days).

Whilst a certain degree of persistence is desirable in order to keep the frequency of application at a reasonable level, two factors have to be borne in mind, the buildup of resistance to pesticides and the rapid growth of the cotton plant producing new leaves and buds.

A cotton plant growing under favourable conditions can increase very fast in height and leaf surface. In Swaziland (Lea *et al.* 1967), the cotton variety Albar 637 under irrigation grows nearly 30 mm per day at the peak of growth, 10 to 11 weeks after sowing. Between two sprays at 7-days interval, cotton plants would be about 190 mm taller, presenting several new leaves and a new terminal which would be free of insecticide deposits. On the other hand it is well known that the young leaves on the top of the plant are a preferred ovipositing site for *Heliothis* spp. Newly hatched larvae could walk over uncontaminated parts to the terminal and newly formed buds. Further, heavy oviposition by *Heliothis* spp. often occurs in the period of maximum cotton growth. Therefore, with contact and stomach poisons a spray interval of less than 7 days at peak growth may be necessary to prevent first-instar larvae escaping insecticidal treatment (Morton 1979).

In Tanzania, it was found that small leaves at the top increased an average of 35% in area per week over a 5-week period (Webley and Parish 1967). Similarly, medium-sized and large leaves increased by 10.2% and 3.3% respectively. Therefore, the insecticide density of a week-old deposit on a young leaf may be reduced by 26% because of growth alone. For new deposits on older leaves, the effect is of less consequence but considering the preference of *Heliothis* to oviposit on young growth, it may be of sufficient importance with residual insecticides to be taken into account in planning a treatment.

10.3.3   Timing of insecticide application

In order to determine when insecticides are needed, it is essential to understand the relationship between pest infestation levels and crop loss (see Chapters 3 and 7 also).

One bollworm on a cotton plant is of no economic importance but several thousands of them per hectare may cause a noticeable reduction in yield. It is therefore necessary to determine 'economic damage thresholds', that is, the maximum population of a given pest that can be tolerated at a particular time and place without a resultant economic crop loss, i.e. the level at which the cost of control is equal to the cost of the damage caused. First, the population dynamics of the major pests have to be determined on the basis of general biology and phenology, and natural regulation; also, the losses they cause need to be assessed. For an insecticidal programme knowledge is required of how the situation may evolve. It can be obtained through population measurement and prediction and the basic approach is to sample the developing crop and its surroundings in a systematic and standardized way.

There are several methods for loss assessment, but it is important to recognize that economic crop loss depends not only on the degree of pest attack but also on the plant reaction to the attack. In addition to considering pest abundance, age structure, and duration of attack, the stage of the plants attacked and especially the surplus organs must be taken into account. It is essential here to distinguish between damage to the plant, and loss. All damage does not mean reduction in production, e.g. at certain growth stages the cotton plant may lose 50% of its leaves without any effect on yield (Falcon and Smith 1973). Of particular importance in this connection is knowledge of the cotton plant physiology and of the periods of growth and fruit production mentioned earlier.

Many techniques and procedures have been developed to measure qualitative and quantitative changes in growing cotton and associated fauna and to anticipate future events. For major pests such as bollworm and jassids, direct counts of the pests themselves or estimates of the damage caused by them must be made on a per unit or plant basis. For some minor arthropod pests such as thrips, aphids, and spider mites, casual observation by a practised eye is sufficient to determine if and when chemical control is needed.

With the information provided by population measurement and prediction and estimates of cost of insecticide applications and value of crop unit, the time to treat can be determined.

In general, the pest infestation of the cotton field is more or less rapidly followed by the establishment of predators and parasites. In order to avoid their destruction at the onset and the buildup of resistance in the boll- and leaf-worm populations, it is advisable to start the insecticide application programme as late as possible. Therefore, against early-season pests, methods of control other than chemical should preferably be used. If an insecticide must be used, preference should be given to a selective chemical for aphids or a systemic compound applied in granular form on the roots.

In recent years, it has become more and more evident that insecticide use has to be carefully planned and that pest scouting must play an increasingly important role. Scouting provides information which allows good timing of insecticide application, thereby assisting in limiting their indiscriminate use and increasing the profitability of insect control to the farmer. Scouting can be the responsibility of plant protection services or of the farmer. In Malawi simple scouting aids such as a pegboard are used successfully by small-scale farmers (Beeden 1971). In certain cases it can be based on the collection and counting of squares and young bolls fallen on the ground as a result of bollworm attack (Brader and Atger 1972).

10.3.4   Application and formulation

In most cases, spraying is preferred to dusting as it is more efficient and less dependent on weather conditions. Sprays may be applied from the ground or by air. Ground spraying may be through hydraulic nozzles, fixed to horizontal booms passing over the cotton plants, or vertical booms for droplet distribution within the crop. Also air blast sprayers blowing over or downward into the crop and ULV rotary atomizers are used. Aerial application is by boom and nozzle or rotary atomizers. These various types of equipment give different spray patterns with a wide range of droplet sizes. The important point is the distribution of the spray in the plants. Two aspects need to be considered (Morton 1979): penetration to lower parts of the plant and underleaf coverage (see Chapters 4 and 5 of this volume).

For bollworm control, all types of application equipment when properly used give adequate deposits on the young top parts of the plant. However, young *Heliothis* larvae may escape an insecticide application because they are protected inside a new bud or terminal growth. In widespread infestations when over half of the terminals are attacked, this may result in yield loss. To control these *Heliothis* larvae, an insecticidal deposit is needed below the top of the plant, either from a recent application or a new spray. In this case also it may be advisable to adopt a short interval between sprays.

Where *Earias* and *Spodoptera* are important pests, an effective deposit is required on the lower parts of the plant as this is where their larvae develop. For sucking pests, such as jassids, aphids, whiteflies, and mites, the insecticide has to be deposited on the under-surface of the leaves where *Spodoptera* moths also lay their eggs.

The droplet size also is of importance; it has been shown that smaller droplets penetrate better than larger ones into the crop as the large droplets are captured by the top leaves. Thus, a ULV spray at 5.5 litres/ha and 150 µm v.m.d. gives a better deposit than boom and nozzles at 22 litres/h and 250 µm v.m.d. (Morton 1979).

In conclusion, for good bollworm control the insecticide has to be deposited not only on the top part of the cotton plant, but also below. Among the available spraying equipment, sprayers distributing droplets above the crop, particularly in aerial application, achieve poor penetration. Airblast machines can give good distribution on the lower parts, but there is little deposited on the underside of leaves. The vertical boom passing between the rows with nozzles directed upwards and horizontally backwards is obviously a most suitable type. As long as cotton plants have not formed a close canopy the use of vertical boom and nozzle equipment is easier and likely to cause little damage. In order to allow better insecticide spray penetration and passage of sprayers, consideration should be given to (1) wider row spacings, as well as to (2) planting cotton varieties which are short and have a reduced vegetative growth and are not as leafy as taller ones, and (3) to avoid too high dosages of nitrogen fertilizers.

Morton (1975) found that a ULV spray penetrated better in the lower part of cotton plants when wind blew parallel to the rows instead of across. He also showed that when the wind blows towards the sun – e.g. early and late in the day – a higher under-side leaf deposit is obtained because leaves turn their upper surfaces to the sun and expose the under-side to the drifting spray.

The above-mentioned factors of chemical degradation, plant growth, and spraying technique can be integrated to build models of treatment concerning the use of different spray equipment, dose rates, and frequency of application and to determine the active ingredient deposit on the plant with time.

Small-scale farmers in the tropics, having little capital to buy spraying equipment, have been limited to the purchase of small hydraulic sprayers, generally of the lever-operated knapsack type or the compression type. However, manual operation of most of these sprayers is very arduous. Water may have to be carried quite a distance from the nearest stream or borehole. As a result insufficient spray is applied and patchy distribution achieved. ULV application by a spinning-disc sprayer may have overcome problems of inadequate water supply but for many small-scale farmers the costs of special formulations for use with this type of sprayer is prohibitive and wettable powders in minimal volumes of water are used in Malawi as the water-ultra-low-volume (WULV) technique (Matthews 1981). One of the constraints to the use of ULV or WULV spinning discs is the cost and availability of the batteries. Also, owing to the small droplet size, the risk of loss of insecticide through drift is high. Recently developed methods of spraying of electrostatically-charged droplets or 'electrostatic spraying' presents a new low-energy technique of distributing an insecticide on the various parts of cotton suitable for the tropics (Matthews 1981). (See Chapter 4.)

10.3.5 *Potential yield*

The insecticide treatment is an integral part of cotton production, and must be considered together with the other agronomic practices. Its profitability will therefore depend very much on the crop potential and on the care taken to grow the cotton in a given area. Timely sowing and proper plant densities, efficient weed control in the first weeks of growth, correct fertilizer use, and water supply when irrigated, all contribute to the economic effectiveness of insect control. Variability in growth due to wrong sowing dates, bad germination resulting in poor stand, late weed control, excessive nitrogen resulting in tall, unproductive growth reduce the efficiency of insecticide treatment.

Due attention must be given to the economics of cotton production and of the insect control programme in each set of conditions, in order to give the farmer a good return for his efforts.

10.4 Cereals – Sorghum

10.4.1 Sorghum pests

Sorghum [*Sorghum bicolor* (L.) Moench] is the fourth most important cereal crop grown in the world and occupies 8.1% of the world cereal acreage (about 59 million ha) and produces 4.9% of the world cereal yield (about 66.5 million tonnes). In semi-arid areas it is an extremely important staple human food, particularly in Africa and India, since it has important drought-withstanding characteristics; but it is also extensively grown as a fodder and feed crop, particularly in the Americas. One of the main reasons for the low yields of sorghum characteristic of the tropics (500–700 kg/ha) is pest attack. Both quantity and quality of produce are severely affected by a range of insect pest species (Appert 1957; Seshu Reddy and Davies 1979; Teetes *et al.* 1980). Several major cereal pests are widely distributed in sorghum-growing areas; but the low cash value of the crop, combined with low yields, makes application of pesticides only marginally cost–effective in normal seasons. This is particularly the case on peasant farms where the crop is of crucial importance in human nutrition.

The most ubiquitous and common pest is undoubtedly the sorghum midge, *Contarinia sorghicola* Coq., which occurs in almost all sorghum-growing areas of the world, presumably because it has been widely dispersed as 'resting' larvae by seed exchange (Geering 1953; Harris 1961, 1976; Passlow 1965; Bowden 1965). It is found not only on cultivated sorghums, but also on wild races of the genus on which it multiplies early in the season. The common practice of growing early- and later-maturing sorghums together exacerbates the midge problem, as does the staggered sowing of the crop. The eggs of the pest are laid in the minute flowers on the panicles. The hatching larva destroys the young ovary, causing a 'blind' or seedless spikelet. Often no grain is produced on heavily damaged heads. Attack can usually be diagnosed by the presence of white exuviae at the tips of the damaged spikelets. By that stage, however, insecticidal control of the pest is pointless. The midges, which emerge from the heads, normally just after sunset, live for less than 24 hours and the whole life cycle lasts only about 2 weeks. The problems posed by this insect in any control strategy are therefore considerable. Control is particularly difficult with the traditionally grown sorghums which are frequently more than 2 m tall and flower later than most of the improved cultivars now being increasingly introduced.

Another widespread pest is the dipterous sorghum shootfly, *Atherigona soccata* Rond, which is common in Africa and S.E. Asia, but absent from the

New World and Australia. This pest attacks seedling sorghum. The larvae, that emerge from small white eggs laid on the under-sides of leaves during the first 1–6 weeks of growth, quickly penetrate between leaf sheaths in the young whorl and sever the growing central shoot. This causes a typical 'dead heart' symptom. Pupation takes place either within the stem or in the soil, and the whole life cycle takes only 3–4 weeks. For most of its life the insect is protected by the plant tissue. It is exposed for only a short time between emergence from the egg and tunnelling in the plant tissue, again presenting problems in planning effective control measures.

The third group of pests of sorghum are the lepidopterous stemborers; the key important genera vary from one sorghum-growing area to another. Important species include the Pyralid, *Chilo partellus* Swinhoe, and the Noctuids *Busseola fusca* Hmps, *Sesamia calamistis* Hmps, *S. poephaga* Tams and Bowden, and *Zeadiatraea grandiosella* (Dyar). *C. partellus* is the main stemborer present in Asia, and is also important in lowland East Africa (Ingram 1958; Nye 1960), but it is absent from West Africa where *Busseola fusca* is the main pest species (Tams and Bowden 1953; Jepson 1954; Harris 1962). This species is also important in the highland areas of East Africa (Ingram 1958). *Sesamia* spp. appear to be present in most sorghum-growing areas of Asia and Africa, but seldom attain major pest status. *Z. grandiosella,* though present in the Americas, does not cause severe losses.

The life cycles of these lepidopterous borers are basically similar, in that eggs are usually laid on young sorghum and the emerging larvae, after a short period of dispersal (during which they may or may not feed on leaf tissue), finally enter the main stem of the plant either through the whorl or between the leaf sheaths. Tunnelling of the stem occurs, and feeding continues until the larvae pupate, usually within the stem. Several species produce two or more generations in a season, and some enter diapause or aestivation (Harris 1963; Sheltes 1978). Carryover in these periods often occurs in the old dry sorghum stalks. Some species maintain themselves throughout the year on alternative hosts. The life cycle of the insects is such that they are very largely protected by host plant tissue for much of their life cycles.

10.4.2   Control of sorghum pests.

The brief biologies of the important pests given above serve to illustrate several problems regarding insecticidal pest control in sorghum. The major constraint is the low monetary return obtained by farmers, particularly peasant farmers, who grow sorghum as a subsistence crop, often with poor agronomical practices and, consequently, with low plant populations. This means that returns of cash outlay are small, and serious thought has to be given to assessing whether the application of insecticide is in fact justified.

In improving the control of midge, a wide range of pesticides has been screened, and several have been found to give adequate control. However, these products are expensive and difficult to apply (Harding 1965; Randolph *et al.* 1971; Huddlestone *et al.* 1972). Frequently, by the time midge damage is noticed it is too late to apply control because the ovaries have already been destroyed and no grain can be produced. Careful field observations have to be taken daily to detect adult midges at appropriate times, i.e. late in the evening and very early in the morning, otherwise severe attacks may go undetected. The midge cannot be controlled with contact insecticides easily, as the oviposition site is within the floret and the larva remains inside the ovary for most of its life. Since many local cultivars sown by farmers produce heads 2 m or more above ground level, depositing an adequate dosage of insecticide may be difficult because of air turbulence. However, well-directed deposits may be achieved by using such conventional ground machinery as motorized knapsack appliances – though caution has to be exercised when motorized blowers are used because there is serious risk of contamination of spray operators. Since best results with midge control can be expected with a systemic pesticide, the high mammalian toxicity of many of the effective insecticides is a drawback.

In large-scale farming, particularly where hybrids of uniform height are grown, the aerial application of spray at flowering time is both easy and relatively efficient. However, several applications may have to be made in quick succession at 3- to 4-day intervals to coincide with midge emergence. Timing of the first application is important to ensure maximum kill of early-generation adults; but since flowering is relatively uniform in such situations and sowing dates do not vary greatly within an area, excellent results are usually obtained. The situation on small-scale and peasant farms is often completely different because of widely ranging sowing dates and, hence, of flowering and maturity dates. It is clear, given the short life cycle of the midge, that populations can build up on early-flowering sorghums that devastate later-flowering and maturing cultivars. The net result of the introduction of improved high-yielding sorghum cultivars on a limited scale in small-scale farms has often been a reduction of overall sorghum produc-tion because the preferred traditional cultivars, which most peasant farmers retain on at least part of their holdings, produce little or no yield. Control of midge is therefore often best effected not by using insecticides, given the problems mentioned, but by persuading farmers to adopt the best agronomic practices (Jotwani *et al.* 1971).

With the shootfly *Atherigona*, the fact that attack occurs early in the growth of the crop, and that the damaging larvae spend most of their lives within the plant stem, means that the best control is obtained with an insecticide applied with the seed at sowing (Veda Moorthy *et al.* 1965; Barry 1972). Results with conventional contact insecticides have often been less than encouraging (Ingram 1960; Davies and Jowett 1966, 1970). Currently in India and elsewhere, carbofuran is widely recommended and, since this is a systemic pesticide, protection of the crop persists for up to 1 month after sowing, giving very satisfactory control (Sepswadi *et al.* 1971; Jotwani 1972; Meksongsee 1972). Problems are often encountered, however, when for any reason the soil becomes saturated by heavy rain, such as occurs in the rainy monsoon season in India. This may leach out the insecticide rapidly, thus reducing its uptake by the plant, or standing water may reduce crop growth considerably, thus extending the period during which the crop is vulnerable to infestation by the fly. It is probably true that, in some instances, very rapid growth of sorghum in favourable circumstances also results in concentrations of systemic insecticide in the plant tissue that are suboptimal for effective control.

Control of sorghum stemborers is obtained relatively simply by the application of contact insecticide to plants at the funnel stage. Dusts, granules, and liquid sprays have been used to good effect on improved cultivars (Barry and Andrews 1974; Jotwani 1972), but have been uneconomic on traditional cultivars (Ingram 1958; Harris 1962). The biology of the different borer species can have a very significant effect on the efficiency of contact insecticides, e.g. *Sesamia* spp. enter the stem directly and are therefore exposed for a very limited time and to a small area of treated plant surface.

The major difficulty with insecticidal control operations is undoubtedly one of timing. Scouting can give early warning of the presence of the young larvae in whorls since the damage to leaves by borer 'windowing' is often easy to detect as plants grow. In some species, either the egg masses laid beneath leaf sheaths or young larval groups browsing on the leaf epidermis can be readily located. These young larvae are very susceptible to insecticides, and the use of a residual-contact insecticide can give a good kill. Timing of insecticidal applications could be greatly assisted by the use of pheromone traps to gauge both the extent of survival of moth larvae over the non-cropping season and the probable date of emergence of the first brood. However, such techniques have as yet been used only to a limited extent.

There is no doubt that development of cultivars resistant to pests is a very practical and feasible method of control in all three pest situations discussed above.

10.5   Oil seeds – groundnut

10.5.1   Groundnut pests

The groundnut or peanut (*Arachis hypogaea*) is grown as an oil seed and a food crop and ranks second only to soyabean in world importance in this crop category. It is the most important oil seed crop in the developing world, and areas in excess of 100 000 ha are sown in 24 countries, 95% of them developing countries. The crop is attacked by an extensive range of insect pests from the planting stage through to harvest.

One of the principal pests in Africa is the groundnut aphid, *Aphis craccivora* Koch, which acts as a vector for rosette disease (Evans 1954; A'Brook 1964; Davies 1971). The alate aphids enter the crop within 10–14 days of emergence of the seedlings and usually introduce foci of rosette infection from old volunteer groundnuts or infected 'out of season' plants. Apterae developing within the crop build up in numbers quickly, and migration of these across the soil surface between plants, or to adjacent plants in the row, results in the dissemination of the virus throughout the crop (Storey and Ryland 1955; Hull 1964; Davies 1971). Crop loss caused by the disease can be extremely high, amounting to total failure in some seasons. It has been clearly shown that early sowing and the use of dense plant populations has a marked effect on disease incidence (Booker 1963; A'Book 1964; Davies 1976).

Another important group of pests on groundnut, particularly in India, are the 'white grubs', and they appear to be of increasing importance. The species involved vary from country to country and the pests tend to be somewhat polyphagous. The most widely distributed species in India are *Holotrichia consanguinea* Blanch and *H. serrata* F. In Rajasthan losses of 20–30% were caused by the former species on 5000 hectares (Rai *et al.* 1969). Damage is caused by the larvae which feed on the root hairs and small rootlets of groundnut plants. The plants may therefore be killed off completely, particularly at the seedling stage; or more mature plants may have their rooting systems so severely damaged that they are unable to withstand the periods of drought stress typical of many tropical groundnut-growing areas. The adult beetles fly actively, often emerging after the first rains, and congregate particularly on neem trees, *Azadiracta indica,* and babul trees, *Acacia arabica,* which they often defoliate. Eggs are laid in the soil, and survival is high if soil moisture conditions are good. The larvae are able to withstand considerable periods of drought, however, often by penetrating deeper into the soil or by constructing an earth cell for protection (Veeresh 1977).

10.5.2.   Control of groundnut pests.

Control of the two pests mentioned obviously presents very different problems for the economic entomologist. The market value of the groundnut crop, however, makes the use of insecticide a practical proposition.

Successful economic chemical control of aphid species with contact insecticides on a range of crops has been limited in tropical situations. The groundnut crop is low-growing, which makes spraying relatively easy. The dense canopy, however, and the fact that aphids tend to congregate at the base of plants, particularly on flowers and young developing hypanthia ('pegs') (Davies 1972), as well as on the topmost and youngest leaves, makes successful deposition difficult. As aphids are very susceptible to predatory insects, particularly Coccinellids, Syrphids and Chrysopids, misuse of contact insecticides can result in even larger populations of the pests, since predators easily penetrate the canopy and detect colonies of aphids. The development of systemic or partially systemic insecticides, particularly those with low mammalian toxicities (e.g. menazon and dimethoate), has been an important advance because the translocation of the active insecticidal principle to the aphid feeding sites is possible.

Normally, in developed countries, aphicides are applied in large volumes of water using tractor-drawn machinery. Such applications are not practical in many areas of the tropics where water availability is a severe problem. Successful control of groundnut aphid has been obtained in the tropics with volumes as low as 90 litres/ha; but even this amount of water is often difficult to obtain and transport (Davies 1975a,b). The use of controlled-droplet applicators (CDA), such as the ULVA sprayer, with specially formulated insecticides, e.g. dicrotophos and dimethoate, has been successful, and partially overcomes the problems encountered in more conventional pesticide application (Davies 1975a). Controlled-droplet application is particularly attractive for arid and semi-arid areas. In practice, it has been found in many tropical areas that CDA spraying has to be done in the early hours of the morning or late in the evening, since wind speeds in excess of the optimum are usually prevalent between about 10 a.m. and 4 p.m. Additionally, between these hours there is a marked convectional movement of air from the soil and crop surface which interferes with the impaction of droplets on the plants. CDA has an added advantage in that, if the wind direction is carefully observed, four or more rows of groundnuts can be effectively sprayed by the controlled 'drifting' of a systemic insecticide. On small peasant farms a disadvantage of CDA is that it is difficult to obtain the correct batteries and to ensure replacement of these when the disc speed drops. The cost of the specially formulated insecticides is also a factor; but their use can help to ensure that correct dosages are applied at the correct time because they are supplied in measured quantities for stated units of area.

Owing to the early arrival of alatae in the crop, and the rapid growth and ground cover achieved, it is necessary to spray up to four times in a season to ensure that dissemination of the virus within the crop does not occur. In very favourable circumstances as few as two sprays have given excellent control. However, surprisingly, use of menazon seed dressing was not effective in controlling the spread of the disease. This was possibly due to the fact that sufficiently high concentrations of the toxic principle in the plants were not achieved for long enough to prevent the multiplication of aphids later in the crop's growth.

Chemical control of white grubs is difficult to achieve since insecticide has to be applied to the soil (Desai and Patel 1965) and attack is often not detected until a considerable amount of damage has been done. It may often be impractical and uneconomic to treat a whole field of groundnuts, particularly as the distribution of larvae in the soil tends to be patchy (Rai *et al.* 1969). A range of pesticides have nevertheless been used including aldrin, BHC, ethyl parathion, diazanon, fenitrothion, and phorate. They have been applied as soil drenches, as granules, or as dusts (Yadava and Yadava 1973; Vora *et al.* 1978), but it is difficult to ensure adequate admixture of insecticides in the soil. Further, the larvae are capable of considerable and rapid movement through the soil profile, particularly in sandy and friable soils, making the attainment of satisfactory levels of control very difficult to achieve. However, treatment of large blocks of groundnut with phorate, broadcast and mulched into the soil, gave very low levels of attack in comparison with infestation-levels in the control plots.

Attempts have been made to apply persistent insecticides (including DDT and BHC) to the seed in view of the difficulties and expense of soil treatment. Patel *et al.* (1967) and Krishnamurthy Rao *et al.* (1972) obtained significantly reduced mortalities of groundnut plants when using carbaryl and methyl parathion. The practicalities of the method, however, require further study.

Experiments have been carried out to attempt to kill freshly emerged adult beetles when they settle on trees at field crop boundaries. DDT has shown promise in some instances (Rai *et al.* 1969). However, there are difficulties in obtaining adequate coverage of tall trees with conventional equipment. High-power blast sprayers, using large volumes of water, are not practical in tropical situations and the expense of maintenance and running costs of air-blast sprayers make them uneconomical.

10.6   Pulses – Pigeonpea

Pigeonpea, *Cajanus cajan* (L.) Millsp., is a leguminous shrub, most cultivars of which grow to a height of 2 m or more. These shrubs, although potentially perennial, are normally grown as annuals, the plants being cut at harvest time, then threshed, with the woody stems being used for thatching or for firewood. Pigeonpea is used as a green vegetable in many areas of the tropics but finds most use as dhal, the dried split peas, in India where over 90% of the world's recorded crop is grown. Dhal is cooked in a variety of dishes that form an important protein source for very many vegetarians. It is the second most important pulse crop in India with an estimated area of $2\frac{1}{2}$ million ha and a total production of 1.8 million tonnes, thus giving a low mean yield of 700 kg per hectare. This crop is also of major importance in other countries, particularly in the Caribbean, and of minor importance in many countries of Africa.

Pigeonpea presents a complex of problems when considering pesticide use on the small-farmers' plots on which it is normally grown; to deal with these we first have to look at several factors in the ecology of the plant and its pests.

Pigeonpea is a crop well adapted to the no- or low-cash input system of the small farmers of the semi-arid tropics. It grows very slowly in the early vegetative stages and so can be conveniently sown as an intercrop in faster growing and earlier maturing crops. In India about 80% of the pigeonpea crop is grown in this way, with sorghum, millet, and cotton as the most common companion crops (Reed *et al.* 1981). It fixes nitrogen through the rhizobia in the nodules on its roots and will grow well on relatively poor soils with no added nutrients. It is relatively tolerant of drought and will produce a crop on poor soils and in seasons where most other crops would fail.

The plants are attacked by many insects during the vegetative stage and may suffer defoliation and temporary stunting but such attacks generally appear to have little effect on the final growth and yield. The most damaging pests form a complex that feed on the flowers and bore into the pods, by far the most important of these in India being the podborer, *Heliothis armigera* (Hb.), and the podfly, *Melanagromyza obtusa* (Mall.) (Saxena 1978; Davies and Lateef 1978).

Pigeonpea is well adapted to coping with the pests of flowers and young pods. The plants produce a great excess of flowers buds, 80% of which are shed at the flower or young pod stage and yields are little affected by flower removal for up to 5 weeks. Thus, a moderate pest attack on the flowers and young pods will have little effect on yield. Unfortunately, *H. armigera* attacks do not always occur in moderation and it is not unusual to find crops that have been denuded of all fruiting forms with the larvae then feeding on the leaves. Even then the plants will grow on to compensate, with a second or even third flush of flowering and pod formation, provided there is sufficient soil moisture.

Although the average yields across India are reported to be about 700 kg of dried seed per hectare, the crop in the south of the country is generally reduced to yields well below this average by *H. armigera* and other pests. Much of the produce is consumed on the farm where it is grown but the surplus finds a ready market, albeit at variable and relatively low prices to the farmer. Over the last few years there has been a rapid increase in cereal production but no rise in the production of pulses so the price has risen to above 2 rupees (US$0.25) per kg, thus giving an average gross return of more than 1400 rupees (US$175) per hectare. Pesticide use may give substantial yield increases, up to 200% or more in some areas and years, but the compensatory nature of the crop makes the establishment of economic threshold levels for the pests inordinately difficult, if not impossible. If there is a moderate pest attack at the first flush of flowering then the excess of flower buds produced will compensate for this attack. A heavier attack, which destroys all or most of the first flush of flowering, may serve only to delay the crop, for the pest attack is not often sustained long enough to destroy the second flush. As yet, it is not possible to predict the progress of infestation by the polyphagous *H. armigera.* On some occasions it virtually disappears from the plants after one or two generations, thus allowing the compensatory crop to flourish. On other occasions it has been recorded as persisting for several months, destroying the first and subsequent flushes of flowers and pods. The theory, originating from studies on cotton crops in the USA, that *Heliothis* spp. are essentially man-made pests, giving real problems only as a consequence of the upset to the natural control elements caused by injudicious pesticide use, certainly does not hold good for much of the pigeonpea-growing areas of southern India. Here, *H. armigera* is a major pest of pigeonpea in most areas and years in spite of the fact that well over 90% of the crop receives no pesticide treatment.

An application of a pesticide that controls *H. armigera* and other elements of the lepidopteran podboring complex will generally result in substantially improved yields. *H. armigera* larvae usually feed with much of their bodies exposed outside the flowers and pods and a good coverage with an effective contact pesticide will give a high-percentage kill. Ideally the spray should be timed to coincide with the hatch of larvae from the eggs, for the larger the larvae the greater the dosage of pesticide needed. The most commonly recommended pesticide for use against *Heliothis* and the other lepidopteran podborers is endosulfan. However, in surveys of farmers' fields throughout India (Reed *et al.* 1981) it was found that of the small proportion of farmers that apply pesticide to this crop, almost all use DDT and/or BHC. The reason for this would appear to be largely a matter of cost and availability. DDT and BHC are often applied as dusts in areas where water for spraying and sprayers themselves are not readily available.

DDT, BHC, and endosulfan give relatively poor control of podfly, for the eggs are laid through the pod wall, and the larvae develop and pupate inside the pods where they are protected from contact pesticides. Unless systemic chemicals are used, the adult would appear to be the only stage susceptible to pesticide use. Monocrotophos is said to give the best control of podfly but this chemical is expensive and its mammalian toxicity is such that it should not be used without adequate safety precautions and protective clothing. Such precautions are seldom observed by small farmers and the use of protective clothing when spraying in the tropics is a rarity, even on research stations.

It has already been stressed that the compensatory powers of pigeonpea make economic thresholds very difficult to calculate, at least on fields where there is no rush to harvest the crop. In areas where *H. armigera* is known to enter the crop in most years, a couple of sprays, one at the flower bud stage and another about 2 weeks later will probably increase the crop from the first flower flush by a substantial margin. The only real problem is how to deliver the pesticide to the target.

Pigeonpea grows up to two metres tall and much of the effective flowering is at the top of the plant. This is the target for the *H. armigera* egg-laying moths and so also for the pesticide intended to control the young larvae. Most farmers who grow this crop have no pesticide-application equipment. Many resort to the use of muslin bags or punctured tins through which pesticide dust is shaken over the plants. If a hand-operated knapsack sprayer and sufficient water is available for spraying, then it is by no means easy to get an adequate coverage of plants where the target is above one's head. Most pigeonpea is grown as an intercrop and the companion crop is often harvested just before the pigeonpea flowers; in this case movement through the crop with a sprayer is no problem. In pure crops, however, the pigeonpea forms an almost impenetrable jungle by peak flowering time and it is a struggle to force one's way through the crop, let along give a reasonable coverage with any conventional hand sprayer. In some areas bullock carts, containing rocker-type sprayers, are driven through the crop, the men in the cart spraying to each side of the cart over as broad a swath as possible. The bullocks and cart wheels cause considerable damage to the crop but this technique well illustrates the extent to which farmers are willing to strive to protect this crop. Aerial spraying could be a satisfactory means of treating this crop but the economics of using such spraying on a low-value crop that is grown in relatively small, scattered fields would probably not be favourable. Controlled droplet application using hand-held battery-operated spinning-disc applicators would appear to have a good potential, but suitable formulations of the appropriate pesticides that could be used for this type of application are not yet available in India.

An alternative to finding a means of treating such a difficult crop with pesticide would be to change the crop itself so that pesticide use would become easier. Efforts are already being made in this direction by breeding plants that are more compact and by experimenting with closer spacing, so that the crop is no more than a metre tall when it is vulnerable to pest attack and has to be protected.

10.7 Vegetables – tomatoes

Tomatoes (*Lycopersicum esculentum* Mill) are familiar and popular vegetables throughout most temperate and tropical countries. The fruits are eaten raw and in numerous cooked dishes and are also preserved and processed in different forms. In the tropics the crop is subject to many pests and diseases and tomato growers frequently resort to pesticide use.

Among the many common pests of tomatoes, the root knot nematodes

(*Meloidogyne* spp.), thrips, and *Heliothis* spp. occur in most areas of the tropics. Each of these pests present different problems of pesticide use and so have to be considered separately. The crop itself serves to illustrate some general problems with the economic and safe use of pesticides.

Several nematodes, but particularly the *Meloidogyne* spp., build up to damaging populations in the soil if tomatoes and other susceptible crops are grown too frequently. The nematodes cause root 'knotting', general malformation and retarded growth of the plants which give poor or no yields. In addition there are plant diseases, including fusarium wilt, that are known to be associated with nematode infestations. The simple measure of using lengthy rotations, with non-host crops being grown in most seasons, is the obvious means of combating these pests but this is seldom practical. The nematodes can survive, if not thrive, on a wide range of hosts. The typical vegetable farmer in the tropics lives on a farm within reach of an urban market, where land is expensive. He will usually be farming a small area, often with some supplementary irrigation and he has to grow high-value crops to survive economically. Many of the high-value crops are hosts to *Meloidogyne* spp. The most attractive means of reducing losses to the root knot nematode is to grow resistant tomato plants; such selections do exist in some countries but few farmers have access to their seed. Soil sterilization that will control nematodes and most other pests and diseases, using heat or methyl bromide, is a familiar and widely utilized practice for the tomato growers using glasshouses for production in the developed temperate countries. In the tropics, however, the cost and practical problems of soil sterilization generally rule out this control method. The chemical pesticides that reduce nematode infestations all appear to have limitations. The injection of nematicides, either in liquid or granular form, into the soil is generally a difficult and expensive operation often requiring specialized equipment (Matthews 1979; Chapter 17). Some of the nematicides, including aldicarb, are extremely hazardous if mishandled and most farmers are unlikely to read the warning on the container label. Many successful vegetable farmers will use relatively unskilled labour for pesticide application and they will seldom provide protective clothing or ensure that safety precautions are followed.

Relatively few species of thrips have been recorded feeding on tomatoes (Lewis 1973). Some may cause some economic damage by feeding on leaves and flowers, but thrips are of greater concern where they transmit the tomato spotted wilt virus. This virus is known to be transmitted by a few species, but particularly by *Thrips tabaci* and *Frankliniella schultzei*. The hosts for these thrips and the virus include tobacco, lettuce, potatoes, and many other cultivated and wild plants. When young tomato plants are infected they are seldom killed but remain stunted, rosetted and will bear no crop. When older plants are infected the fruits are reduced in number and size, and are blemished, which reduces their market value.

Thrips are fairly easily controlled by a wide range of contact or systemic insecticides, but an immigrant thrip that is virus-infected takes only a few minutes to transmit the virus to its host (Sakimura 1960). So unless the immigrant thrips are killed before they feed upon the plants, pesticide application is of little use. A continuous coverage of the plants with a quick knockdown pesticide is generally not practical. The only practical means of protection is to prevent the immigration of infected thrips by eradicating, or using pesticide on, the alternative hosts of the vector and virus. This is not easy where thrips are carried on the winds from infested hosts which are a considerable distance from an individual's farm; in areas where several farmers are growing tomatoes, however, co-operative action can lead to considerable mutual benefit.

Of the *Heliothis* spp., the polyphagous *H. armigera* is a common pest of tomatoes throughout the tropics and sub-tropics of the Old World. The larvae feed upon fruits, many of which subsequently rot, but even if they do not, tunnelled fruit seldom have any marketable value. As the larvae often feed with much of their bodies exposed, and also move over the plant, they make a relatively easy target for contact insecticides. In some countries where pesticides have been widely used, particularly on cotton, the *Heliothis* larvae may be relatively resistant to many of the available contact pesticides. In most developing countries, however, this is not yet a problem, so here persistent chemicals including DDT are sprayed or dusted on the plants to protect them from this and other pests. A survey of the available literature in India (Lalitha and Prasad 1978) showed that tomatoes and other vegetables that are sold in markets are often grossly contaminated with pesticides. It has also been recorded (Agarwal 1978) that many people in India have a surprisingly high amount of DDT in their bodies, even though a relatively small amount of this pesticide has been used on a per hectare basis, when compared with the amounts used in other countries. It is probable that similar situations occur in many other developing countries, but in most cases there are no data. The monitoring services that are the watchdogs for the consumer societies of the developed countries would soon put a stop to such violations of the pesticide residue regulations in their countries. In some developing countries there are no regulations, but even where they exist there is seldom an effective monitoring system.

The answer to such a problem would be, of course, to discourage the use of persistent pesticides on vegetables. However, persistent pesticides such as DDT are still the most cost-effective chemicals available and few farmers will see any merit in substituting more costly chemicals which have relatively short-term effectiveness. Governments are unlikely to ban DDT while it is still an effective pest control agent on cash crops, including cotton, particularly since the evidence that this chemical is a danger to health is far from convincing.

Perhaps the greatest problem of pesticide use on tomatoes and other vegetables in most tropical countries, however, is the constraint induced by the violent fluctuations in prices that farmers obtain for their produce. Vegetable production is often seasonal, particularly where most is rain-fed, so there are gluts and scarcities in each year. The farmer who is fortunate or expert enough to produce his crop when there is a shortage in the market can earn a very handsome profit and so can well afford expensive inputs including pesticides. His neighbours, who are unfortunate enough to harvest their crops when there is a glut, will have to accept prices that may be no more than one-tenth of the scarcity price, if they can obtain any price at all. Prices can vary by 200% within a month. Such price variations play havoc with economic threshold calculations.

Fortunately, tomato canning and processing factories are being established in several developing countries where tomatoes are extensively and intensively produced. Such factories will provide a stabilizing effect upon the prices paid to farmers.

10.8   References

A'Brook, J. (1964). *Ann. Appl. Biol* **54**, 192.
Agarwal, H. C. (1978). In *Pesticide residues in the environment in India* (eds C. A. Edwards, G. K. Versh, and H. R. Krueger). UAS Tech. Series No. 32, University of Agricultural Sciences, Bangalore.
Appert, J. (1957). *Les parasites animaux des plantes cultivées au Sénégal et au Soudan*, 292 pp. Jouve, Paris.
Barry, B. D. (1972). *J. Econ. Entomol.* **65**, 1123.
——and Andrews, D. J. (1974). *J. Econ. Entomol.* **67(2),** 310
Beeden, P. (1971). *Agricultural development in the Lower Shire Valley in Malawi.*
Booker, R. H. (1963). *Ann. Appl. Biol.* **52**, 125.
Bottrell, D. G. and Adkisson, P. L. (1977). *Annu. Rev. Entomol.* **22,** 451.
Bowden, J. (1965). *Bull. Entomol. Res.* **56**, 169.
Brader, L. and Atger, P. (1972). *Med. Fac. Landbouw. Gent* **37,** 408.
Brown, H. B. and Ware, J. O. (1958). *Cotton* (3rd edn) McGraw-Hill, New York, Toronto and London.
Cramer, H. H. (1975). *Semaine d'étude agric. et hygiène des plantes.* Gembloux, Belgique.
Davies, J. C. (1971). *The bionomics and control of Aphis craccivora Koch (Hem, Aphididae), and their effects on rosette disease attack and yield of groundnuts, Arachis hypogaea L.* Thesis, University of East Africa.
—— (1972). *Bull. Entomol. Res.* **62,** 169.
—— (1975a). *Trop. Agric. (Trinidad)* **52(4),** 359.
—— (1975b). *PANS* **21(1),** 1.
—— (1976). *Ann. Appl. Biol.* **82,** 489.
—— and Jowett, D. (1966). *Nature (London)* **209,** 104.

—— and Jowett, D. (1970). *E. Afr. Agric. For. J.* **35,** 414.

—— and Lateef, S. S. (1978). In *Pests of grain legumes, ecology and control* (eds S. R. Singh, H. F. Van Emden, and T. Ajibola Taylor), Academic Press, London and New York.

Desai, M. T. and Patel, R. M. (1965). *Indian J. Entomol.* **27,** 89.

Evans, A. C. (1954). *Ann. Appl. Biol.* **41,** 189.

Falcon, L. A. and Smith, R. F. (1973). *Guidelines for integrated control of cotton insect pests.* FAO, AGPP: MISC/8 (revised 1974), Rome.

FAO, (1967). *Report of the first session of the FAO working party of experts on resistance of pests to pesticides.* FAO Meeting Rep. PL/1965/18, Rome.

—— (1976). *Report of an FAO/UNEP consultation on pest management systems for the control of pests of cotton.* FAO, AGP: 1976/M/3, Rome.

FAO/UNEP (1974). *Report on an ad hoc session of the FAO panel of experts on integrated pest control,* FAO, AGP; 1974/M/8, Rome.

Geering, Q. A. (1953). *Bull. Entomol. Res.* **44,** 363.

Gower, J. and Matthews, G. A. (1971). *Cott. Gr. Rev.* **48,** 2.

Harding, J. A. (1965). *Texas Agric. Exp. Stn. PR2352,* p.7.

Hargreaves, H. (1948). *List of recorded cotton insects of the world.* 50 pp. Commonwealth Institute of Entomology, London.

Harris, K. M. (1961). *Bull. Entomol. Res.* **52,** 129.

—— (1962). *Bull. Entomol. Res.* **53,** 139.

—— (1963). *Bull. Entomol. Res.* **54,** 643.

—— (1976). *Ann. Appl. Biol.* **84,** 114.

Huddleston, E. W., Ashdown, D., Maunder, B., Ward, C. R., Wilde, G., and Forehand, C. E. (1972). *J. Econ. Entomol.* **65,** 851.

Hull, R. (1964). *Nature (London)* **202,** 213.

Ingram, W. R. (1958). *Bull. Entomol. Res.* **49,** 367.

—— (1960). *E. Afr. Agric. J.* **25,** 184.

Jepson, W. F. (1954). *A critical review of the world literature on the lepidopterous stalk borers of tropical graminaceous crops,* 127 pp. London Commonwealth Institute of Entomology.

Jotwani, M. G. (1972). In *Control of the sorghum shootfly* (eds M. G. Jotwani and W. R. Young). Oxford & IBH Publishing Co., New Delhi.

—— Singh, S. P., and Chaudari, S. (1971). *Final Technical Report, Division of Entomology,* p.123. Indian Agricultural Research Institute, New Delhi.

Krishnamurty Rao, B. H., Narayana, K. L., and Narashima, Rao, B. (1972). In *Proc. 4th Symp. Soil Biol. Ecol.* (eds C. S. A. Edward and C. K. Veeresh), p.206. UAS, Bangalore.

Lalitha, P. and Prasad, V. G. (1978). In *Pesticide residues in the environment in India* (eds C. A. Edwards, G. K. Veeresh, and H. R. Krueger), UAS Tech. Series No. 32, University of Agricultural Sciences, Bangalore.

Lea, J. D., Catling, D., Jackson, P., Cornish-Bowden, M. E., and Gibbs, J. D. (1967). *A comparison of the application of cotton insecticides by ground equipment and a fixed wing aircraft.* Part 1, Misc. Rep. No. 46. Department of Agriculture, Swaziland.

Lewis, T. (1973). *Thrips, their biology, ecology and economic importance.* Academic Press, London and New York.

Matthews, G. A. (1979). *Pesticide application methods.* Longman, London and New York.

—— (1981). *Outl. Agric.* **10,** 7.

Meksongsee, B. (1972). In *Control of sorghum shootfly* (eds M. G. Jotwani and W. R. Young). Oxford & IBH Publishing Co., New Delhi.

Morton, N. (1975). *Rep. Div. Agric. Res. Swaziland 1973–74.* University of Botswana, Lesotho and Swaziland, Swaziland.
—— (1979). *Outl. Agric.* **10**, 2.
Nye, I. W. B. (1960). *The insect pests of graminaceous crops in East Africa.* Col. Res. Series No. 31, 48p., HMSO, London.
Passlow, T. (1965). *Queensl. J. Agric. Anim. Sci.* **22**, 150.
Patel, R. M., Patel, G. G., and Vyas, H. N. (1967). *Indian J. Entomol.* **29**, 170.
Pearson, E. O. and Maxwell Darling, R. C. (1958). *The insect pests of cotton in tropical Africa,* 355 pp. Empire Cotton Growing Corp. and Commonwealth Institute of Entomology, London.
Rai, B. K., Joshi, H. C., Rathore, Y. K., Dutta, S. M., and Shinde, V. K. R. (1969). *Indian J. Entomol.* **31(2)**, 132.
Randolph, N. M., Meisch, M. V., and Teetes, G. L. (1971). *J. Econ. Entomol.* **64**, 87.
Reed, W., Lateef, S. S., and Sithanantham, S. (1981). *Proceedings of the international pigeonpea workshop,* ICRISAT, Hyderabad.
Reynolds, H. T., Adkisson, P. L., and Smith, R. F. (1975). In *Introduction to insect pest management* (eds R. L. Metcalf, W. L. Luckmann), 587 pp. John Warley & Dond, New York.
Ripper, W. E. and George, L. (1965). *Cotton pests of the Sudan.* Blackwell, Oxford.
Sakimura, K. (1960). In *Biological transmission of disease agents* (ed. K. Maramorosch). Academic Press, New York and London.
Saxena, H. P. (1978). In *Pests of grain legumes, ecology and control* (eds S. R. Singh, F. F. Van Emden, and T. Ajibola Taylor). Academic Press, London and New York.
Sepswadi, P., Meksongsee, B., and Knapp, F. W. (1971). *J. Econ. Entomol.* **64**, 1509.
Seshu Reddy, K. V. and Davies, J. C. (1979). *Pests of sorghum and pearl millet and their parasites and predators recorded at ICRISAT centre, India, up to August 1979.* Dept. Prog. Rep., Cereals Ento. No. 2, ICRISAT. 23 pp.
Scheltes, P. (1978). *Ecological and physiological aspects of aestivation diapause in the larvae of two Pyrillid stalk borers in Kenya.* Published thesis, University of Wageningen.
Storey, H. H. and Ryland, A. K. (1955). *Ann. Appl. Biol.* **43**, 423.
Tams, W. H. T. and Bowden, J. (1953). *Bull. Entomol. Res.* **43**, 645.
Teetes, G. L., Young, W. R., and Jotwani, M. G. (1980). *Guidelines for integrated control of sorghum pests,* FAO, Rome.
Tunstall, J. P. and Matthews, G. A. (1972). *Cotton Technical Monograph,* No. 3, Ciba-Geigy, Basel.
Veda Moorthy, G., Thobbi, V. V., Mattai, B. H., and Young, W. R. (1965). *Indian J. Agric. Sci.* **35**, 14.
Veeresh, G. K. (1977). *Studies on the root grubs of Karnataka with special reference to bionomics and control of Holotrichia serrata F.* (Coleoptera: Monograph Series No. 2). University of Agricultural Sciences, Bangalore.
Vora, V. J., Bharodia, R. K., and Kapadia, M. N. (1978). *White Grubs Newsletter* **2**, 38.
Waterhouse, D. F. (1976). *Proceedings XV International Congress of Entomology.* Washington, D. C.
Webley, D. J. and Parish, R. H. (1967). *The determination of DDT deposited on cotton plants in an airspray trial in Tanzania.* Misc. Rep. No. 598, Tropical Pesticide Research Institute, Arusha, Tanzania.
Yadava, C. P. S. and Yadava, S. P. S. (1973). *Indian J. Entomol.* **35**, 329.

11
Forest pests and diseases

T. JONES

11.1 Introduction

Some of the most damaging and notorious pest and disease epidemics that have occurred in the last century have involved either natural or Man-made forests. Yet the development of special pesticide measures to deal with these problems, either from research or by adoption of established agricultural practices, has, until the last two decades, been extremely slow.

The reasons for this are clear and relate directly to the low returns expected from forest products in general, the fact that bulk returns only accrue at the end of extended rotations, and the practical difficulties of treating very extensive crops, often of a mixed nature, over topographically difficult terrain. The combination of these factors alone has often precluded any prospect in forestry of using methods for pest and disease survey and control that have been successfully developed for other perennial crops.

Practical difficulties are of course minimal at the nursery stage when the plants are small and restricted to limited and accessible plots, and they are not critical in young forests and plantations where the canopy remains open and access to individual trees is easy. But as trees mature, management practices, including survey, damage assessment, and treatment, become increasingly difficult because of the physical size and form of the plants and the resulting height above or depth below ground of the feeding and breeding sites of their pests and disease organisms.

Pest management problems on mature stands can be increased by location and species composition. Natural forests (and indeed some plantations which were established without due thought to subsequent operational require-ments) are often remotely situated with disorderly espacements on difficult terrain where ground access is restricted and even the use of aircraft is limited by logistics and/or dangerous topography. Tropical forests, with their rich species complex, have scattered distribution of individual species which, whilst restricting the spread of mono-specific pests and diseases, often renders control operations totally impractical.

It is because of these practical difficulties that most attempts to control pests and diseases of mature stands have been confined either to the coniferous forests of the Northern Hemisphere, where large and relatively pure stands of a single species occur in accessible areas, or to single species plantations. In such situations, some spectacular pest and disease management programmes have been instigated, but frequently, even when conditions are practically favourable for them, control measures are precluded on financial grounds.

In terms of investment per unit area, the returns from forests are generally lower than from other crops and furthermore the period between investment and return is generally very protracted. Crop rotations for fast-growing tree species under ideal conditions, e.g., eucalypts for general timber and pines for pulp in East Africa, can be as short at 10–15 years; but for normal sawmill size softwoods of acceptable wood quality, 40 years is more normal and for hardwoods rotations of twice as long are not unusual.

With such long rotations, it follows that it may take years before the cost of any management practices, including control measures, can be recouped, and by the time the crop is harvested, these costs will have been compounded. It has been estimated that at $8\frac{1}{2}\%$ compound interest, £1000 expended on a 10-year-old timber stand could only be offset by £26 100 worth of increased timber production when that stand is harvested at 40 years old (Way and Bevan 1977). With compound interest – 'the forester's nightmare' – looming so large the decision to apply control measures or not will clearly depend very much on the value placed on the product, which varies with timber species, end usage, time and market prices, and predictions.

The economics of control may even vary for the same pest on the same crop in the same region. In Canada, for example, where spruce budworm *Choristoneura fumiferana* Clemens threatens millions of hectares of spruce-fir forests, the control policies of New Brunswick, Quebec, and Ontario are very different. In New Brunswick, some 20% of the total economy is invested in spruce-fir industries, much of the forest land is owned by smallholders who derive their livelihood from forest products, and most of the six million hectares of forest is accessible; here it is economically beneficial to try to protect the whole forest estate. In Quebec, out of a standing timber volume of some 2300 million m^3, which produces an annual increment of some 45 million m^3, only some 2.7 million m^3 is exploited annually. Such an exploitation rate can clearly be maintained with a much smaller standing volume and hence the loss of trees especially in inaccessible areas is far less critical than in New Brunswick. During the 1960 budworm outbreak, protection of the whole estate was considered impossible and only restricted areas of high risk were given special protection. In Ontario, very large areas of the north eastern and southern parts of the Province are covered with poor-quality forest where control is uneconomical, and in this Province the budworm epidemic was allowed to run its course. Control was confined to areas of scenic value and Provincial parks (Prebble 1975). These cases demonstrate how a pest which can devastate the economy of one area may be of little consequence in another.

To the foregoing account of the practical and financial constraints to the development and use of control measures in forestry must be added mention of the forester's outlook and his past and changing position.

Unlike the farmer who is trained from the beginning in the management of artificial plant populations, the forester traditionally receives a background education which fits him for duty primarily as a conservator of natural forest communities and resources. As such, he tends to view forest pest and disease problems as being the result of mismanagement of the natural ecosystem. This may or may not be correct but either way it frequently leads to a reluctance to solve these problems by imposing measures which may further disturb the natural environment. The forester's training and commitment to conservation may make him disinclined to adopt the positive approach of the agriculturalist to the pest and disease problems of this crop

It is reasonable to conclude that together the above-mentioned constraints have been largely responsible for the limited amount or total lack of effective action against important pests and diseases over many years, and their restrictive influences are still very apparent today. But the general situation has improved notably since World War II and with the increased demand per capita for timber and forest products throughout the world the limitations of the forester to protect his crop have lessened.

Many natural forests, where control programmes were most difficult, have been improved in value by enrichment plantings and later by supplanting with monocultures and so the permissible expenditure on protection has increased and some of the practical difficulties of treating mixed stands have disappeared. Simultaneously, there has been a very rapid expansion of Man-made forests, which are commercially much more viable than many of the earlier plantations. These are largely based on a few fast-growing high-yielding exotics with shorter rotations linked to dependent industries and guaranteed markets, and are so planned and sited as to facilitate management operation and reduce their costs to the minimum. Pests and diseases have often been taken into account at an early stage, and where the products and benefits of the forests can be obtained more or less equally from a range of different species there has been scope for avoiding foreseeable pest and disease problems at the development stage of the crop by a suitable choice of species and site – an option which ceases to exist once the species pattern and scope of the Man-made forest is established.

What these developments have achieved, in essence, is the simplification of the crop and its environment, with obvious advantages to commercial and management interests. In doing so, they have caused certain pests and diseases to decline (something which is often overlooked) and allowed others to become predominant. Predictably, the development of new crops, especially exotics, and new systems has been followed by new pest and disease problems. But whilst monocultures have provided conditions favouring the emergence of these new problems, they have, by their very nature, also gone far to facilitate control. Many of these pests and diseases have called for speedy control, such as only pesticide application can provide. A broad spectrum of effective pesticides is already available to the forester, and the Man-made forests facilitate their use, but as this review may show, there is a need to develop efficient application techniques for forest conditions especially for fungicides.

In the following section, which deals briefly with specific pests and diseases and their control with chemical pesticides, it is convenient to consider the forest cycle in five stages, namely those of the nursery, young plantations, mature forests, logs, and sawn timber.

11.2 Nurseries

11.2.1 Seeds

Whilst a certain proportion of seeds are damaged and destroyed by polyphagous soil-dwelling insects in nursery beds, this is rarely a serious problem and where it occurs it is easily overcome by application of insecticidal seed dressings (usually dusts with or without stickers) or insecticidal treatment of the soil. But losses of seed due to moulds during dormancy and germination and pre-emergence damping-off can be very serious and fungicidal seed treatments are often needed and widely applied. Fungicide seed treatment in forestry is probably the earliest and most widespread example of the use of pesticides for the control of tree disease on a relatively large scale.

Until recently, wide-spectrum fungicides have been used as seed dressings to control moulds and damping off (Hocking 1975) and to reduce the risk of introducing more specialized pathogens on the surface of seeds for planting in exotic localities, e.g., *Diplodia pinea* (Desm.) Kickx and *Lophodermium* spp. (Noble and Richardson 1968). Despite the widely held view that forest pathogens are only rarely seed-borne, Mathur (personal communication 1979) has reported *Colletotrichum dematium* (Pers. ex Fr.) Grove, *Macrophomina phaseolina* (Tassi) Goid., pathogenic *Fusarium* spp. and *Phoma* spp. from *Pinus kesiya* and *P. merkusii,* and there is evidence that the cause of pitch canker of pines, *Fusarium moniliforme* v *subglutinans,* may be seed-borne on *P. elliottii* Englm.

More specialized internal seed-infesting pathogens (Miller and Bramlett 1979) similar to the cryophilic fungus *Geniculodendron pyriforme* Salt described by Salt (1974) may yet be found. Systemic fungicides may control these (Morelet 1974) but there is little information on this approach at present.

While fungicide seed dressings are of doubtful value in preventing seed loss in storage (Hocking 1975), they have been more effective in seed beds against saprophytic and weakly parasitic fungi (Urosevic 1961; Gibson 1957; Fisher 1941; Gravatt 1931).

In tests to prevent damping-off diseases, the value of seed dressings has varied (Sutherland *et al*. 1975). In those cases where the causal agents of the disease were predominantly seed-borne, some success was obtained, but little or no control was achieved where the pathogens were soil-borne. In other circumstances, damping-off control by seed dressings has been obtained only where the pathogens are endemic and attack starts early after sowing (Lock *et al*. 1975). These poor results are partially attributable to the relatively long pre-emergence periods for many germinating tree seeds when the protectant may be leached from their vicinity. It is likely also that the nature of the soil environment may directly affect the effectiveness of the fungicides (Johnson and Harvey 1975). Fungicide mixtures have sometimes proved more effective for seed protection than the individual components applied singly (Mattis and Badanov 1974; Wall 1976).

Some improvement in the performance of fungicide seed dressings in forest nurseries has been obtained by the use of methyl cellulose or latex stickers to 'pellet' seeds (Gibson and Hudson 1969; Hocking and Jaffer 1969; Berbee *et al*. 1953; Carlson and Belcher 1969; Vaartaja and Wilner 1956; Dias 1973) but on other occasions these have given disappointing results (Lamontagne and Wang 1976; Frisque 1975). There is reason to believe that pelleting may at times directly impair germination (Muhle and Hewicker 1976; Lock *et al*. 1975).

Fungicidal seed dressings must of course be used with care where the tree species require obligate mycorrhizal associates for establishment and the inoculum for this is partly at least seed-borne (Theodorou and Skinner 1976).

11.2.2 Seedlings

Many of the insects and fungi which attack seedlings in forest nurseries, as might be expected, are familiar agricultural pests because the habitat closely resembles that of an intensively managed agricultural crop. Both the shoots and roots of the soft succulent plants are liable to attack by insects which normally infest herbaceous vegetation.

Stem and foliar diseases are particularly serious threats to young seedlings, and the forest nursery with its concentrations of even-aged crops of single-tree species provides an ideal environment for the spread of pathogens and insects, some of which under other circumstances may have little or no significance. Chronic or severe defoliation or die-back of nursery stock can lead directly to high mortality or render seedlings unfit for planting out, and furthermore there is often a risk that diseases contracted in the nursery may be carried on infected stock into the field and continue to spread in that environment. There is therefore a good case and a real need to try and control these diseases at the nursery stage, and fungicides applied as foliar sprays with knapsack sprayers and mistblowers have proved very effective, whilst systemics injected into the seed bed have given variable results.

Of the insect pests of seedlings in nurseries, lepidopterous larvae such as *Agrotis, Euxoa* and *Noctua* emerge from the soil at night and feed on the root collar region of seedlings and similar damage is caused by mole-crickets. *Gryllotalpa gryllotalpa* L. is a pest in Southern Europe and *G. africana* P. de Beauv. throughout Africa, Asia, and Australia. In East and Central Africa, weevils of the genera *Systates, Opseodes,* and *Entypotrachelus* can cause serious defoliation, as do invasions of a variety of grasshoppers from surrounding farm crops.

These pests and many more are easily controlled by persistent insecticides. In the UK, BHC (benzene hexachloride) or DDT is recommended as a spray for control of cutworms (Blatchford 1976), and, as an alternative to organochlorine insecticides, heptaclor as a seed dressing has been effective against *Agrotis* and *Euxoa spp.* in Bulgaria (Khinkin and Nikolov 1974). In Canada, chlorpyrifos and leptophos proved to be suitable replacements for carbaryl, which is used as a standard treatment for cutworms attacking tobacco. Poisoned baits using a mixture of DDT, aldrin, or dieldrin with sugar and maize meal were effective against cutworms and grasshoppers in East and Southern Africa while in Bulgaria cutworms were controlled by a bait of trichlorphon, molasses, and bran (Khinkin and Nikolov 1974). The synthetic pyrethoids, bioethanomethrin, bioresmethrin, and cismethrin as poisoned baits gave good control of *Agrotis ipsolon (Hfn)* in France (Lhoste 1975).

The major subterranean pests which feed on roots are termites and the larvae of Melolonthidae. The latter have been controlled in UK by BHC dust worked into the top soil (Blatchford 1976), but in East Africa and Rhodesia resistance to this and other pesticides has made control difficult. Termites are dealt with in a later section.

Seedling pests are very effectively controlled by a wide range of insecticides applied either to the soil or in baits. These treatments not only prevent insect damage in the nursery but can also provide some degree of post-planting-out protection (Blatchford 1976).

Diseases which kill seedlings in forest nurseries can be controlled by pre-planting sterilization of nursery bed soil or soil treatment with fungicides after sowing. Vaartaja (1964) gives an excellent review of this subject. Stem and foliar diseases can be controlled by application of fungicides to the plants, normally as sprays or by injection of systemics into the soil.

Pre-planting treament of nursery beds involves application of highly toxic and broad-spectrum fumigants, such as methyl bromide, chloropicrin, formalin, allyl alcohol, vapam, dazomet, and dichloropropane/dichloropropene mixtures which are either volatile or decompose rapidly to harmless by-products. These have been very successful (Magnani 1970, 1972), and in North America fumigation has been widely adopted as routine practice (Foster 1959; Hodges 1962) not only because of the protection provided but also because of the bonus it provides in weed and nematode control and release of soil nitrogen. But this technique eliminates microbiological competitors as well as pathogens, and hence diseases which arise after treatment may spread rapidly through the crop.

Soil drenching with copper compounds, dithiocarbamates, daconil, and captan applied at initial stages of attack generally provide good control of soil pathogens. But there is the danger that routine drenching regimes may be applied uncritically and wastefully where cheaper cultural practices might achieve the same goal. Systemics, such as benomyl, gave disappointing results (Johnson and Harvey 1975) initially, but are now widely used for control of forest seed-bed diseases.

There is strong evidence that the use of all kinds of soil pesticides may be deleterious to mycorrhizal formation, particularly where these involve materials that are cumulative (Bakshi and Dobriyal 1970; Hong 1976; Iloba 1978a,b). Where this is a problem it is best to use organic compounds, such as captan, which decompose fairly rapidly in the soil.

Pesticides are widely used to control stem and foliage pests and diseases of forest nurseries. Copper fungicides were regularly used before World War II to control *Lophodermium* spp. *Scirrhia acicola* (Dearn.) Siggers on pines (Boyce 1948). More recently dithiocarbamates, captan, chlorothalonil, and systemic fungicides have been used increasingly in nurseries, and spray programmes have been improved to give maximum control at minimal cost. In Europe and North America *Lophodermium* needle cast of pines largely caused by *Lophodermium seditiosum,* is now successfully controlled by maneb or carbendazim sprays (Stephan and Millar 1975) in schedules which in North America can be reduced to two seasonal applications if properly timed (Merrill and Kistler 1976). Dothistroma blight (*Scirrhia pini* Funk) has been controlled similarly in New Zealand by copper sprays (Jancarik 1969), and *Cercospora* pine blight (*Cercoseptoria pini-densiflorae* (Hori et Nambu) Deighton) has for years been treated successfully in Japan by a cocktail of Bordeaux mixture and Uspulun (an organomercurial preparation) (Ito 1972). More recently, Ivory (1975) has shown that chlorothalonil, captafol, or the systemic fungicides based on benomyl or thiophanate methyl can be equally effective against the blight without the risks involved in using mercuric compounds. Systemic fungicides have been used successfully to control diseases of conifers and broad-leaved seedlings, e.g. cyclohexamide for *Didymascella thujina* (Durand) Maire on *Thuja* spp. (Fernandez Magan 1974), benomyl, oxycarboxin, and thiophanate methyl for Coryneum canker [*Seiridium cardinale* (Wagener) Sutton] of *Cupressus* spp. in Italy (Parrini *et al.* 1976), and euparen preparations for terminal crook of pines in New Zealand. After initial failures (Foil *et al.* 1967), systemic fungicides, in particular bayleton (triadimefon) and benomyl, have shown an excellent level of control of fusiform rust of pines in the south-eastern states of the USA

(Kelley 1979; Mexal and Snow 1978; Hare and Snow 1976). Bayleton proved to be effective when applied to the seed, the soil, or as a spray to the growing seedling.

Among diseases of broad-leaved trees, copper preparations, maneb, benomyl, and thiophanate methyl have been found effective in controlling poplar leaf infections caused by *Septoria* spp. and *Marssonia* spp. in North America (Carlson 1974) and a wide range of preparations headed by benodanil and copper compounds gave good control of *Melampsora larici-populina* in New Zealand when application to poplars were correctly timed (Spiers 1976). Bertus (1976) found that very heavy sprays of thiophanate methyl were needed to control attack of the polyphagous fungus *Cylindrocladium scoparium* on *Acacia, Eucalyptus,* and other hardwoods in Australia.

Evidence, particularly from North America (Peterson and Smith 1975) and elsewhere, shows that a wide range of stem and foliar diseases in forest nurseries have been successfully controlled by fungicide spray programmes. Provided that the life cycle and epidemiology of the pathogens is understood, the normal process of redistribution of surface-active fungicides appears to be adequate to give a high level of protection without resort to any special application techniques. But one exception at least exists. In the case of terminal crook of pines in New Zealand and its control by captan organomercurial preparations, high-pressure spray application has been needed to penetrate the needle tuft around the terminal bud where the fungus becomes established and sporulates (Gilmour 1965). However, the successful use of systemics to control this disease has removed the need for this spraying technique.

Systemics have given variable results for foliage disease control when indirectly applied. This has been successful in the case of fusiform rust of pine, and Neely (1975) reports satisfactory control of *Guignardia aesculi* (Pk.) Stew. on *Aesculus glabra* and *Gnomonia platani* Edg. on *Platanus occidentalis* by soil injections of benomyl, but this treatment did not control *Gnomonia leptostyla* (Fr.) Ces et de Not. on *Juglans regia*.

As with nursery pests and the use of insecticides, so with nursery diseases and the use of fungicides comes the danger of resistance, in the form of the emergence of tolerant strains of pathogens, especially with the increasing use of selective systemics. Although this has not yet happened in forest nurseries, it is desirable to have in reserve alternative fungicides to those in current use, in case of this emergency.

11.3   Saplings and young plantations

This phase extends from the planting out of nursery stock to the time when the crop closes its canopy. For pests it is a distinct ecological phase, but because many diseases continue to affect trees into later life, discussion is postponed to the later section on mature stands.

Site factors have an important influence on the pest and disease complex of young plantations. In grasslands trees have to contend with herbaceous and agricultural pests and diseases while those in logged areas are threatened by pests surviving in the debris of the earlier forest.

Transfer of nursery stock to the field may result in a decline of those pests and diseases which were favoured by the dense uniform crop conditions of the nursery. Many nursery diseases cease to spread after infected stock has been planted out either because the pathogen is specialized to attack juvenile foliage only [as in *Didymascella thujina* (Durand) Maire] or because espacement and environmental conditions of plantations do not favour spread. Foliage attack by Cercospora blight of pines [*Colletotrichum acutatum* Simmonds f.sp. *pinea* (Dingley and Gilmour)] disappears completely from infected plants under similar conditions (Dingley and Gilmour 1972). Exceptions include Dothistroma blight of pines (*S. pini*), Diplodia die-back [*Diplodia pinea* (Desm.) Kickx], *Cylindrocladium scoparium* Morgan, leaf and stem disease of *Eucalyptus* spp., *Cylindrocladium pteridis* Wolf, *Gremmeniella abietina* (Lagerb.) Morelet on pines, and Coryneum canker (*Seiridium cardinale* (Wagener) Sutton) of *Cupressus* spp. The risk of planting stock infested with pests is rare but the pattern is the same. *Pineus pini* on a variety of *Pinus* species in East and Central Africa has been a notable exception (Odera 1972).

As plantations mature, changes in the host and cultural practices, such as thinning and pruning, may alter the pattern of pest and disease attack.

Extension growth of saplings is rapid and insects of high migratory capacity, relatively short life-cycles, and high fecundity are particularly well adapted to exploit the foliage and external parts of the stem where they may occur in considerable numbers, causing significant loss of increment. Severe defoliation of 37 000 hectares of sitka spruce by *Elatobium abietium* Walk in the UK in 1971 is estimated to have delayed harvest by 2 years which represented a loss of £1.75 million (Carter 1975). *E. abietinum* and *Adelges abietis* Ratz which are pests of Christmas trees are controlled by malathion and HCH respectively, but the timing of sprays against *Adelges* is important because the adults and eggs with their protective wax covering are not vulnerable to contact insecticides. Spraying therefore is restricted to the November–February period when the less-well-protected nymphs are exposed. In Kenya, heavy infestations of *Pineus pini* Gmelin on exotic pines in 1969 were successfully controlled by BHC.

Where young trees of one species which are normally diffusely scattered are concentrated in planting lines or squares, movement of their pests is facilitated. The leaf-gall psyllid *Phytolyma lata* Walk, which attacks *Chlorophora excelsa* A. Chev., in West Africa has limited powers of dispersal in natural stands but in plantations it may spread along the lines and cause severe damage (Roberts 1969; Gibson and Jones 1977). Planting conditions are apparently important and it is claimed in Kenya that on good soil with adequate rainfall, growth is so fast that attacks of the pest are by-passed and the apical meristem is unaffected (Templer 1948). Shade reduces attack, but also tree growth. Control with DDT and BHC sprays and dusts is possible, but uneconomic. Similar spread along lines is seen in the case of *Ambypelta cocophaga* China attacking *Eucalyptus deglupta BL* on the Solomon Islands.

Shootborers may be particularly troublesome on young trees, often killing the leading shoot with resulting loss of extension growth, distortion, and multiple stems, and reduction in the value of the timber. Pyralids of the genus *Hypsipyla* are particularly serious pests in the tropical regions of America, Africa, and Asia, and they have prevented the establishment of valuable Meliaceae in many countries. Each *Hypsipyla* species is adapted to the Meliaceae of its own range, but may be unable to infest those from elsewhere (Roberts 1964, 1968; Grijpma and Ramalho 1969; Grijpma 1970). Greenhouse tests by Allan *et al.* (1970, 1974) using several systemic insecticides applied to the soil around young *Cedrela odorata* L. showed that carbofuran and methomyl were the most effective, but protection was short-lived. Incorporating the systemics into polymer blocks provided a slower release mechanism and protection of seedlings by carbofuran was extended to 1 year by this means.

Planting of cut-over forests may lead to damage by several Coleoptera, such as the bark beetle *Hylastes* (Scott and King 1974), the weevil *Hylobius,* and the longhorn *Oemida gahani* Dist. (Gardner and Evans 1953, 1957).

The larvae of *Hylastes* and *Hylobius* live in stumps, logs, or branches resulting from land clearance or felling operations. Damage by these insects can be avoided by delaying planting for 2–4 years until their food supply is exhausted, provided there are no fresh breeding sites from which the beetles could invade (Scott and King 1974), but this is rarely practical or economic. In Britain good protection is afforded for 1–2 years by dipping the upper parts of the plant (but not the fibrous roots) into BHC solution in water prior to planting out (Blatchford 1976), whilst in Norway, an aqueous solution of DDT proved effective (Bakke *et al.* 1972). In Swaziland spraying the root collar region only with BHC emulsion in the nursery beds 24 hours before planting out gave post-planting protection against *Hylastes angustatus* Herbst. for over 6 months, even under prolonged and heavy rainfall.

Oemida gahani Dist. is an important heartwood borer of *Juniperus procera* Hocst. ex Endl, and exotic *Cupressus* plantations in the Kenya highlands. It breeds naturally in living *Podocarpus* and *Juniperus,* and also in the deadwood and stumps of a wide range of other indigenous trees (Gardner and

Evans 1953, 1957). Control is possible by removal of stumps and debris, exclusion of game from plantations to reduce mechanical damage, more frequent pruning schedules to ensure smaller and more rapidly occluding scars and the treatment of these scars with tar or insecticides to prevent entry of adult beetles.

Silvicultural practices designed to remove undesirable species from the tropical forest have caused pest problems in subsequent planting pro- grammes. Weed trees killed slowly by hormone spray girdling, mechanical girdling, and poisoning with sodium arsenite provided an abundance of breeding sites for Scolytoidea. In Ghana *Xyleborus* spp. infested and killed plantings of *Khaya ivorensis* (Jones and Roberts 1959), and protection was achieved by dipping the planting stock in BHC emulsion. Weed tree removal with resulting opening of the canopy caused healthy young trees of *Albizia* spp. to suffer from insolation and become susceptible to Scolytoidea in Uganda where control with BHC sprays was possible but uneconomic. Line planted *Swietenia macrophylla* King in Fiji suffered heavy infestation from Scolytidae breeding up in poisoned weed trees (Roberts 1964).

Termites can be serious pests especially where trees are grown on marginal areas and are under stress. Damage most often occurs in the dry season and less frequently where growth conditions are favourable. However, Sands (1962) notes that *Macrotermes bellicosus* Smeathman will attack vigorous new growth of the indigenous savannah tree *Isoberlinia ndoka* in Nigeria. Species of *Eucalyptus* are the most frequently quoted examples of trees damaged by termites probably because they are widely used for firewood plantations, shelter belts, pole and construction timber production, and often planted in arid or exposed sites where stress is high. Losses of 100% are quoted for *Eucalyptus saligna* Sm in the Cameroun (Monnier 1961), 80–86% loss of *E. camaldulensis* Dehn in Nigeria (Sands 1960), and 50–100% loss in Uganda (Curry 1964). Lethal damage to *Eucalyptus* is commonly restricted to the first 5–6 years after which attacks generally cause debility and gummosis.

Control may be achieved by destroying the nests mechanically or by pouring in insecticides, such as DDT, BHC, chlordane, dieldrin, or aldrin, or by injecting with fumigants, in the case of those species which have centralized mounds, e.g.*Macrotermes* and *Odontotermes*. In general, mound fumigation is more a palliative than a control. Other soil-dwelling genera which have diffuse nests cannot be controlled in this way. Seedlings may be protected from these species by insecticidal barriers, chlorinated hydrocar- bons being particularly effective. Where seedlings are grown in polythene sleeves dieldrin or aldrin dust incorporated in potting soil will provide protection not only in the nursery but also for sufficient time after planting out to allow the seedlings to become well established. Potting soil poisoning is best for Macrotermitinae. If sleeves are not used, the insecticide is mixed with the soil in the planting hole (Sands 1962; Harris 1965). None of these methods is adequate for *Mastotermes* which requires uneconomic and phytotoxic levels of insecticides – it is virtually uncontrollable in young plantations.

With regard to diseases, as these will be discussed in greater detail later, it is necessary here to mention only special commercial crops, such as Christmas tree plantations and seed orchards. Foliar necrosis resulting from even a mild attack of needle disease can render Christmas trees unmarketable and protection by intensive spray schedules is a most important part of their management. Similar intensive fungicide programmes have been necessary to protect pine seed orchards in the south eastern USA where *Cronartium strobilinum* Hedgc & Hahn can be extremely destructive to cones. Programmes based on ferbam applied as a spray every 5 days from the time of emergence of female strobili until seed setting are necessary to ensure adequate yields (Matthews 1964, 1967).

The methods of application of pesticides employed in young plantations is determined by economics, the size and configuration of the trees, the target area of the plants, and the size of the area to be treated. Where the trees and the area of plantation are small, i.e. up to 3–4 ft high, hand-operated knapsack sprayers have proved economical and effective because in essence application is a matter of 'spot' or placement treatment. The same equipment is suitable in much older forests where the target area is confined to the lower part of the stem, i.e. the root collar, or where application is confined to banding the trees at a convenient height. Special adaptations may be employed to provide a circular spray pattern, such as the unclosed circle of nozzles, as were developed for knapsack spraying of coffee trees in Kenya, or the more conventional guard attachment on knapsack sprayers used to encircle the plant whilst applying the spray. Long lances are frequently used and tractor-drawn or vehicle-mounted boom-type sprayers have been used effectively in very young plantations, such as grassland plantations in East Africa where insecticides were combined with herbicides, and wastage of the former was offset by the saving of labour costs of traditional hand-weeding methods.

Where trees are taller and their growth more profuse and the pesticide is required on the foliage, motorized knapsack sprayers are commonly used to 'mist' the trees and for extensive areas tractor-drawn or mounted machines similar to those used in fruit orchards have proved effective where there is a reasonable but unclosed canopy. Here again wastage of material may be offset by saving in labour costs. Shoulder-borne or stretcher-borne fogging and dusting machines are less commonly used, but they have proved effective in controlling leaf disease in young rubber plantations in Malaysia and *Buzura* sp. in young pine plantations in Uganda, where the sloping site gave favourable conditions for controlling the direction of spread of the fog (Brown, personal communication).

High-pressure sprayers capable of projecting fine sprays some 30 m or so vertically have been developed in North America and Europe, as well as elevated nozzles borne on hydraulic lifts, and these have been much more successful than vehicle-mounted rotating tower sprayers, such as was tried against *Manowia* in Malawi (Johnstone and Harvey 1975). Vehicular sprayers are often suitable for plantation crops where the ground is kept smooth and clear but quite unsuitable in forest crops where the terrain is usually rough and pitted and aerial spraying using low-volume or ultra-low-volume techniques developed for agricultural crops and vector control have proved more generally suitable and effective. One outstanding success was the use of ground sprayers for experimental control operations against *Dothistroma* blight of *Pinus radiata* in Kenya by Gibson (1971). Hand-operated ultra-low-volume machines have yet to make an impact in forestry but seem to have some potential in forest plantations.

11.4 Mature stands or older forests

The insect pests of this environment fall into three convenient classes of foliage feeders, phloem feeders and xylem feeders, whilst diseases can be considered in the three categories of root diseases, vascular wilts, and stem and foliage diseases.

Foliage-feeding insects are amongst the most destructive of forest pests and nearly all the important ones are larvae of Lepidoptera or Hymenoptera, with high reproductive capacity, highly mobile winged adults and relatively large larvae which individually consume a large amount of foliage. Many of those which have adapted from their natural broad-leaved hosts to conifers have caused havoc through their behavioural habit on the exotic host of eating the base of the needles, so that one bite equates to the loss of one needle, which explains the green carpet of needles under conifers suffering defoliation by these pests. Populations are normally kept in check by a large complex of natural enemies and diseases, but when this natural control is diminished for any reason or fecundity increases, albeit temporarily, the pests' inherent capacity for increase and dispersal can lead to massive and widespread outbreaks. In one area of New Brunswick, the population of the spruce budworm *C. fumiferana* increased 4000 fold and in another 10 000 fold in six generations. As one female budworm produces an average of 170 eggs, these increases represent survival rates of 4.7% and 5.6% respectively per generation (Prebble 1975).

Population increases may continue for several generations until brought under control by density-related natural enemies, disease or depletion of the food supply.

The spruce budworm affects the structure of the forest. In New Brunswick, prior to outbreaks, the forest is a mosaic of mature and over-mature forest in which balsam fir predominates, interspersed with dense uniform immature forest. Budworm epidemics, triggered by several consecutive warm dry spring seasons, kill firs in the mature forest, which allows a dense growth of seedlings and severely thins immature stands. By the time of the next epidemic some 50–100 years later, the immature stands will be mature or over-mature and seedling stands will be immature and so the cycle is repeated (Baskerville 1975). Outbreaks last for 8 years and damage is cumulative, and, because the insect has a preference for mature trees, it is in direct competition with the forester (Prebble 1975).

Defoliators cause severe damage and serious loss of increment and even death of the trees, but, apart from this, their hosts may become susceptible to secondary pests. In Britain in 1963, 50 hectares of pine were completely defoliated by the pine-looper *Bupalus piniaria* L., but in addition consider-able damage was caused to the leafless but living trees by the scolytid *Tomicas piniperda* L., which normally is confined to logs and unable to infest living trees. This secondary threat was anticipated when *Bupalus* outbreaks occurred in 1969 and felling operations were so organized as to ensure that the beetle had an abundance of its normal host logs available to it to attract it away from the *Bupalus*-damaged trees. The looper was controlled in 1954 by aerial sprays of DDT and in 1969 by aerial application of tetrachlorovinphos (Bevan 1974).

Defoliators, above all forest pests, have been subject to chemical control and because the acreages involved have been so large and the terrain so unamenable to ground based application techniques, aerial spraying has been the commonest and most effective measure employed. Between 1945 and 1974, a cumulative total of around 12 million ha of forest were sprayed from the air in the USA and the majority of target pests were defoliators. Of this 82% of the effort was directed against two pests, the Western Budworm *Choristoneura occidentalis* Freeman and the Gipsy Moth *Porthretria dispar* L. (National Academy of Sciences 1975). DDT, which was the insecticide used for over 20 years, was replaced by carbaryl from 1967 onwards. But aerial spray programmes of forests in USA are modest compared to those in Canada directed against only one pest species – the spruce budworm. In New Brunswick alone, where control operations began in 1952, the budworm epidemic spread to 2.7 million hectares by 1956 despite annual applications of DDT. In 1957, 2.4 million hectares of forest were sprayed with DDT representing nearly 90% of the infested area and, as a result, the rate of infestation by the pest fell very sharply. However, between 1960 and 1967 it was still found necessary to spray between 0.3 and 1 million hectares of forest each year and in 1968 an upsurge of the pest began which spread to 5.7 million hectares per year and covered almost the whole Province. Spraying at this time was increased to cover between 1.4 and 2.7 million hectares each year, using fenitrothion or phosphamidon (Prebble 1975).

Yet these spray programmes against budworm did not achieve the same success as, for example, those against *Bupalus* in Britain. Insecticides reduce *Bupalus* populations to an endemic level where they can be maintained by natural enemies (Way and Bevan 1977). Spraying against spruce budworm is less successful and, in this case, defoliation and tree death play a vital role in the regulation of populations. Spraying protected the tree from complete defoliation and thus provided the pest with an adequate supply of food to sustain its high reproductive rate. So the pest was kept in a protracted semi-outbreak condition which necessitated the annual use of insecticides. It is clear that defoliators exert a considerable influence on the ecology of the forest and the decision to use pesticides or not against them should be made with due attention to the likely long-term consequences.

Phloem feeders belong mainly to the coleopterous families Scolytidae, Curculinidae, Buprestidae, and Cerambycidae. The pests in question generally feed at the bark/sapwood interface or within the bark or sapwood and to them the nutritive value of the cambial layer, which may be increased by mechanical or fire damage to trees or drought, is particularly attractive. Wind-thrown and storm-damaged plantations are the sites of the most spectacular and damaging outbreaks. In Norway and southern Sweden, *Ips typographus* L. and *Pityogenes chaleographus* L. attacked storm-damaged spruce in 1970–72 causing the death of 1 million trees in Sweden, amounting to 0.3 million m^3 and in October 1978, 2.2 million m^3 of dead standing timber in Norway was attributable, in roughly equal proportions, to beetle attack and drought. Major outbreaks of *I. typographus,* which were common in Central Europe in the second half of the 18th century, have been significantly reduced by better plantation practice, including the removal of breeding material, (i.e. exploitation debris) the restriction of logging during periods of beetle activity, and debarking of timber left on the ground in the forests and plantations. But epidemics still occur due to storm damage or defoliation, and similar outbreaks of *Dendroctonus frontalis* Zimmer in the USA and Central America, beginning in 1957, caused losses of some 40 million m^3 in the Southern States in 1973 and the death of over 1 million trees in the State of Georgia alone. An infestation of *D. frontalis,* first noticed in the Olancho mountain of north-eastern Honduras in 1962, spread through 3 million hectares of natural pine forest by April 1964 and was moving West at a rate of 125 000 ha per month. Fortunately the epidemic declined due to natural control.

Even in the USA, chemicals are seldom used for direct control in standing trees, but lindane sprays in oil have proved effective (National Academy of Sciences 1975), and post-infestation control of scolytid bark beetles can be achieved using BHC applied as a spray to the bark. More commonly, control is by prompt harvesting of infested trees or by the use of trap logs treated with BHC, but Vite (1972) has suggested that the pests in flight avoid such trap logs and disperse into the forest while predators and parasites continue to be attracted to poisoned logs and are killed on them, thus increasing the incidence of the pest. Later evidence (Coulson 1974) suggests that this pest incidence was no more than a natural population fluctuation. An interesting control technique was the use of the artificial aggregating pheromone frontalure applied to selected bait trees whilst surrounding trees were treated with a herbicide which caused their phloem to become over-saturated with water and so drowned any beetles without harming their predators. This technique concentrated epidemics into selected areas, and trap trees were then destroyed before the beetles could disperse (Vite 1972).

Xylem-feeders can be classified into three biological groups; some, such as certain species of termites, digest lignin with the aid of gut micro-organisms. Others reject wood as such and rely for their food on the starches and sugars present in the wood; others again feed solely on the 'ambrosia' fungi which grow on the walls of their tunnels. They are all primarily pests of logs, sawn timber, and dead wood, but some have become important pests of living trees and prevention of their attacks and/or post-infestation control is usually uneconomic and difficult.

Coptotermes elisae has been controlled by injecting dieldrin into the gallery system and by destroying colonies in nearby stumps by pouring on to them a mixture of ammonium nitrate and fuel oil (Gray and Buchter 1969). Some post-infestation control of cerambycids, such as *Phryneta,* has been achieved by applying drenching sprays of BHC to the bark of the infested area (Jones 1961). Infestation of bostrychids such as *Apate monachus* F. have been eliminated by plugs of BHC-soaked cottonwool inserted into the entrance of the tunnels (Jones 1959). Tree-injection, such as is now used against Dutch Elm Disease in UK, can be successful, but is seldom economic.

Apart from termites, some of which can be economically and practically controlled by pesticides, the xylem-feeders present an intractable problem. The methods of pesticide application available and used to date are not practical propositions in commercial forestry though they have their value in amenity forestry. The problems which these pests cause are usually more effectively solved by management practices which render the environment less favourable to the insects.

Diseases of older forests have also posed serious problems. In the case of root and soil-borne diseases, despite considerable efforts over the past 50 years to protect trees from fungal attack and to eliminate infection centres by chemical soil or root treatments, it is only recently that control of these pathogens has become a practical possibility. This success springs from the work of Rishbeth (1950) which demonstrated the importance of stump infection by air-borne spores of *Heterobasidion annosum* (Fr.) Bref. [*Fomes annosus* (Fr) Cooke] in the spread of this root pathogen in pine plantations. This led to the practice of treating stumps immediately after felling with creosote, sodium nitrite solution, or urea (Phillips and Greig 1970). Recent work by Rishbeth (1976, 1979) has shown that useful levels of control of the root pathogen *Armillariella mellea* (Vahl ex Fr.) Karst. [*Armillaria mellea* (Vahl ex Fr.) Kummer] in hardwood stump inoculum centres can be achieved by inoculating these stumps with fungal competitors or treating them with ammonium sulphamate to encourage competitors. Treatment of infective soil or inoculum centres (stumps and root residues) with fungitoxic fumigants which had been developed to control *Armillaria mellea* (Vahl. ex Fr.) Kummer on citrus and tea was until recently considered too expensive for forestry. But Pawsey and Rahman (1976) have shown that the coal-tar preparation Armillatox, or 2% formalin, can be applied as a drench around trees quite safely at dosages that are toxic to the rhizomorph, of *A. mellea*. Filip and Roth (1977) have shown that a wide range of preparations either injected or introduced into stumps by borings effectively eradicated *A. mellea,* but did not affect antagonists such as *Trichoderma* spp.

Vascular wilts are important problems in forestry, but to date fungicides have been used against only one, Dutch Elm Disease. Whilst this has been recognized as a major problem for over 50 years, it was not until it reached the USA and caused disastrous damage to the American Elm that control measures on a large scale based on eradication and insecticidal measures against the vector were undertaken. Later the systemic benomyl was found to be effective against the fungus and this was applied as a soil drench and foliar spray or injected into the trunk (Biehn 1973; Smalley *et al.* 1973; Gregory *et al.* 1971). Subsequently trunk injection designed to introduce a special formulation of benomyl (Lingnasan) into the vascular system of the current year's growth was widely used in integrated control schemes to control early infections. With the return of the fungus *Ceratocystis ulmi* (Buism.) C. Moreau to the UK in more virulent form in 1969 the same measures, including trunk injection, were adopted.

Fungicidal control of *C. ulmi* has been confined to amenity and urban forest situations and no attempt has been made to use pesticides for controlling this or any other vascular wilt disease in the natural forest until recently, as they were considered too expensive and hazardous to the crop.

Stem and foliar diseases are important and long-standing problems in forestry, yet it is only recently that programmes of control with fungicides have been initiated. Stem diseases have received very little attention and so diseases such as pink disease *Corticium salmonicolor* Berk. et Br., which is controlled on rubber, still afflicts *Eucalyptus* plantations in India (Seth *et al.* 1978) and *Albizia falcataria* in the Philippines (Eusebio *et al.* 1979). The cankers caused by *Cryphonectria cubensis* (Bruner) Hodges (*Diaporthe cubensis* Bruner) which has posed such a threat to eucalypts in Brazil is now managed entirely through the use of resistant cultivars, and no attempt has been made to develop pesticides to deal with the problem (Hodges 1980; Hodges and Reis 1974; Ferreira *et al.* 1977). Exploratory research has indicated fungicides for control of a few stem diseases, such as Diplodia die-back of pines, but application difficulties make this impracticable (Van der Westhuizen 1968).

While exploratory work on fungicidal control is in hand for valuable crops, such as poplar, the case of dothistroma blight of pines provides the only notable example of successful control of a foliage disease in the forest by fungicides.

This disease first appeared in East Africa in 1957, where it caused chronic defoliation and reduction in increment in pines, particularly *Pinus radiata* D. Don. Its cause, *Scirrhia pini* Funk and Parker [imperfect state *Dothistroma septosporum* (Doroguine) Morelet], had previously been little more than an academic curiosity, but in the East African highlands and later in New Zealand, parts of Chile, Australia, and Spain, it found a combination of susceptible hosts and favourable environments that allowed it to develop rapidly as a forest pathogen of major importance.

Pioneer work by Gibson (1965) in East Africa laid the foundation for development of a control programme, and subsequent experiments (Gibson *et al.* 1966; Hocking 1967) showed that copper fungicides could provide an unusually high level of control of dothistroma blight. This was substantiated by field trials of these fungicides applied by air, over a 5–6-year period where two applications per year of cuprous oxide or copper oxychloride to coincide with the seasonal flushes of the disease gave good protection to susceptible young *P. radiata* (Gibson 1971).

The difficult topography, the high altitude of the plantations, and the limitations of air-spray services precluded the application of these findings to large-scale operations in East Africa, but aerial spray operations in New Zealand showed that a dosage much less than that used in East Africa was enough to ensure normal growth; these schedules have been refined further during the past 10 years (Kershaw *et al.* 1979).

Parasitic higher plants, mistletoes, present special protection problems. Some attempts have been made to develop chemical control measures for use against these parasites by trunk injection or by placement on the seat of infection by long lance sprayers, but with little practical success. It is clear that the close similarity in physiology of the host and parasite makes the search for development of selective chemicals for reliable control particularly difficult.

11.5   Logs

These are susceptible to attack by the xylem-feeders mentioned earlier. When green, their high moisture content makes them unattractive to Bostrychidae, Lyctidae, some termites, and Buprestidae, but highly attractive to some Curculionidae, Cerambycoidea, and especially Scolytoidea. As the logs slowly dry out, attacks by the last three families decline and those of the other xylem-feeders increase.

Infestation by longhorn beetles, bark beetles and most weevils can be avoided by removing the bark, but this operation offers no protection against the biological group known as ambrosia beetles (i.e. Platypodidae and some Scolytidae). These may attack logs within an hour of the tree being felled and continue to do so whilst the moisture content of the logs exceeds 40%, that is for as long as 5–6 months in the tropics, by which time the beetle tunnels will have penetrated deeply and spread very extensively (Webb and Jones 1956; Jones 1959). In this situation, the log, and the subsequent sawn timber, is unmarketable, not only because of the 'pinholes' in it, but also because of unsightly staining caused by fungi which spread from the beetle tunnels into the surrounding wood.

Where they can be properly applied, chlorinated hydrocarbons, and especially BHC, dieldrin, and aldrin, can provide excellent and economic log protection [for as long as 20 weeks in Ghana (Webb and Jones 1956; Jones 1959)]. The pesticides are used in oil solutions or as water emulsions and applied by either hand-operated or motorized knapsack sprayers. Oil solutions can be less effective than emulsions on logs with their bark intact and it is thought that the oils cause a rapid release of attractants from the bark/sapwood interface, thus inviting heavier beetle attacks. Either oils or emulsions are equally effective on de-barked logs.

Where blue stain, which is caused by fungi similar to those associated with ambrosia beetles, is to be avoided, spraying freshly felled logs with pentachlorphenol has given satisfactory protection. However, pentachlorphenol formulations are regarded as highly toxic to mammals so other methods including biological agencies are being sought with similar levels of protective efficiency.

11.6 Timber

Freshly sawn green timber is susceptible to attack by ambrosia beetles whilst drier and fully seasoned timber is susceptible to Bostrychidae and Lyctidae – the powder post beetles. Timber in stack can be protected by spraying with BHC but more effective control can be achieved by treating the timber as it comes off the saw, either by passing it through a spray race or a dip of BHC. Where blue-stain is a problem pentachlorphenol added to the BHC race or dip solution provides excellent protection, but borax must be added to the mixture to act as a buffer and prevent settling out.

Borax itself is a good protectant and impregnation of timber with it can be achieved by natural diffusion. Pressure impregnation of timber with creosote or chrome–arsenate salts, depending on end use, can provide permanent protection against insects and fungi. Established infestations in timber in stack or in use can be eradicated by fumigation with methyl bromide. Currently, research on the use of pyrethroids for timber protection is under way in the UK. As already noted some of these preservatives have undesirably high mammalian toxicity and safer alternatives are being sought to protect timber in use against biological degradation of all kinds. The possible use of Scytalidin, a product of *Seytalidium* sp., is an example of an indirect biological control agent for this purpose.

11.7　Conclusions

The use of pesticides and the development of appropriate techniques for their application to forests has lagged far behind such practices and technologies in agriculture. There is also a serious lack of information on the impact of pests and pathogens on forest crops. This is largely due to the limited financial investment in the crop and the large area involved which makes monitoring of pests and diseases difficult. Even where aerial surveys are employed data on pest and disease losses in forests are approximate and qualitative. A few countries with major interests in forest resources have special pest and disease survey systems, but in most countries information is erratic and unreliable.

Chemical control of forest pests has developed rapidly in the last 20–30 years, especially in the pure or almost pure natural forests of the northern hemisphere and in the monospecific plantations which have been extensively established in many parts of the world. This has been made possible by the increasing demand for forest products and concurrent increase in their value. Chemical control of forest diseases is still largely confined to the nursery stage as is well illustrated in a recent review (National Academy of Sciences, 1975). This reports that in North America chemical control was used for 39 out of 50 forest pest problems but for only three out of 40 in the case of diseases.

There are two contributory factors to this state of affairs. Firstly, the Lepidoptera that cause many of the more serious losses are easily visible, their attacks are often acute, short-lived, and spectacular, and their populations and post-treatment mortalities are quantifiable. Infestations are frequently localized and amenable to intensive pesticide treatment, often with dramatic results. Diseases, on the other hand, are slower to develop and are often chronic in effect and the results of fungicide treatments are rarely dramatic. Secondly, the fact that insects are easily seen whilst fungi are 'invisible' creates a bias in the attention they receive from field officers.

As the successful control of dothistroma needle blight by fungicides in New Zealand has shown (Kershaw *et al.* 1979), there are clearly good possibilities for developing chemical control measures against shoot and foliage pathogens on a large scale. The recent developments in techniques for application of insecticides could be applicable for fungicides in due course. Meanwhile, little effort has been made to interrelate the disciplines and develop solutions to common problems such as insect pests and diseases of foliage.

The present range of insecticides and fungicides are generally adequate and it is the problems of formulation and application that need resolving.

Systemic fungicides have yet to be used on large areas but offer great promise, provided that the problems of application can be overcome. Other chemicals are being developed to provide disease control by alteration of the vascular anatomy of the host (Nair and Kuntz 1975; Nair *et al.* 1969) or regulate leaf fall regimes (Sripathi Rao 1970) which will call for similar application techniques if they are to be used effectively for forest protection.

11.8 References

Allan, G. G., Gara, R. I., and Wilkins, R. M. (1970). *Turrialba* **20**, 478.

Bakke, A., Christiansen, E., and Saether, T. (1972). *Proc. 6th Br. Insectic. Fungic. Conf. 1971*, 808.

Bakshi, B. K. and Dobriyal, D. (1970). *Indian For.* **96**, 701.

Baskerville, G. L. (1975). *For. Chron.* **51**, 157.

Bertus, A. L. (1976). *Phytopathol.* **85(1)**, 15.

Berbee, J. G., Berbee, F., and Brener, W. H. (1953). *Phytopathology* **43**, 466 (Abstr.).

Bevan, D. (1974). In *Biology in pest and disease control*, p.302. Blackwell Scientific Publications, Oxford.

Biehn, W. L. (1973). *Plant Dis. Rep.* **57**, 35.

Blatchford, O. N. (1976). *Entopathol. News (Supplement)*, Oct. 1976.

Boyce, J. S. (1948). *Forest pathology* (2nd edn), p.550. McGraw Hill, New York.

Carlson, L. W. (1974). *Can. Plant Dis. Surv.* **54**, 81.

—— Belcher, J. (1969). *Can. Plant Dis. Surv.* **40**, 38.

Carter, C. I. (1975). *For. Rec. London* **104**, 17.

Coulson, R. N. (1974). In *Southern pine beetle symposium, March 7–8, 1974 Memorial Student Center, College Station, Texas*. Texas Agricultural Experiment Station, 5pp.

Curry, S. J. (1964). In *Forest entomology in East Africa: FAO/IUFRO symposium on internationally dangerous forest diseases and insects, Oxford 1964*, Vol. 1. FAO/FORPEST 64: 4pp X 1pp.

Dias, R. A. (1973). *Silvic. Sao Paulo* **8**, 25.

Dingley, J. M. and Gilmour, J. W. (1972). *N. Z. J. For. Sci.* **2**, 192.

Eusebio, M. A., Quimio, M. J., Jr., and Ilagan, F. P. (1979). *Sylcatrop* **4**, 191.

Fernandez Magan, F. J. (1974). *Rev. Plant Pathol.* **54**, 2511 (Abstr.).

Ferreira, F. A., Reis, M. S., Alfenas, A. C., and Hodges, C. S. (1977). *Fitopatol. Brasil.* **2**, 225.

Filip, G. M. and Roth, L. F. (1977). *Can. J. For. Res.* **7**, 226.

Fisher, P. L. (1941). *J. Agric. Res.* **62**, 87.

Foil, R. R., Merrifield, R. G., and Hansbrough, T. (1967). *Plant Dis. Rep.* **51**, 223.

Foster, A. A. (1959). *Forest Pest Leaflet USDA Forestry Service*, No. 32, p.7.

Frisque, G. (1975). *Bi-monthly Res. Notes* **31(4)**, 31.

Gardner, J. C. M. and Evans, J. I. (1953). *E. Afr. Agric. J.* **18**, 176.

—— —— (1957). *E. Afr. Agric. J.* **22**, 224.
Gibson, I. A. S. (1957). *E. Afr. Agric. J.* **22**, 203.
—— (1965). *Rep. Agric. Vet. Chem. Sevenoaks* **6**, 39.
—— (1971). *E. Afr. Agric. For. J.* **36**, 247.
—— Hudson, J. C. (1969). *E. Afr. Agric. For. J.* **35**, 98.
—— Jones, T. (1977). In *Origins of pest, parasite, disease and weed problems* (eds J. M. Cherrett and G. R. Sagar), p.413. Blackwell Scientific Publications, Oxford.
—— Kennedy, P., and Dedan, J. K. (1966). *Commonw. For. Rev.* **45**, 67.
Gilmour, J. W. (1965). *New Zealand Forest Research Institute Leaflet No. 10*, 4pp.
Gray, B. and Buchter, J. (1969). *Commonw. For. Rev.* **48**, 201.
Gravatt, A. (1931). *J. Agric. Res.* **42**, 71.
Gregory, G. F., Jones, T. W., and McWain, P. (1971). *USDA Forest Service Research Paper North eastern Forest Experiment Station*, No. NE-232.
Grijpma, P. (1970). *Turrialba* **20**, 85.
—— Ramalho, R. (1969). *Turrialba* **19**, 531.
Hare, R. C. and Snow, G. A. (1976). *Plant Dis. Rep.* **60**, 530.
Harris, W. C. (1965). *PANS A*, **11**, 33.
Hocking, D. (1967). *Ann. Appl. Biol.* **59**, 363.
—— (1975). In *Forest nursery diseases in the United States USDA Forest Service Agriculture Handbook No. 470*, p.114.
—— Jaffer, A. A. (1969). *Commonw. For. Rev.* **48**, 355.
Hodges, C. S., Jr. (1962). *USDA Forest Service, South East Forest Experimental Station Paper 142*, 16pp.
Hodges, C. S. (1980). *Mycologia* **72**, 542.
—— Reis, M. S. (1974). *Brasil Florestal* **5**, 25.
Hong, L. T. (1976). *Malay. For.* **39**, 147.
Iloba, Ch. (1978a). *Beit. Trop. Landwirtsch. Veterinarmed.* **16**, 179.
—— (1978b). *Eur. J. For. Pathol.* **8**, 379.
Ito, K. (1972). *Bulletin of the Government Forest Experiment Station Tokyo*, No. 246, p.21.
Ivory, M. H. (1975). *Commonw. For. Rev.* **54**, 154.
Jancarik, V. (1969). *New Zealand Forest Research Institute Leaflet No. 24*, 4pp.
Johnson, D. W. and Harvey, R. D., Jr. (1975). *Tree Planter's Notes* **26**, 3.
Jones, T. (1959). *Tech. Bull. W. Afr. Timb. Borer Res. Unit. No. 1*, p.20.
—— (1961). *E. Afr. Agric. For. J.* **26**, 1.
—— Roberts, H. (1959). *Second Rep. W. Afr. Timb. Borer Res. Unit*, 1955–58.
Kelley, W. D. (1979). *Phytopathology* **69**, 528 (Abstr.).
Kershaw, D. J., Gadgil, R. D., Leggat, G. J., Ray, J. W., and van der Pas, J. B. (1979). *Handbook for the assessment and control of Dothistroma needle blight, New Zealand forest service*, 43pp.
Khinkin, S. and Nikolov, N. (1974). *Rasitelna Zashchita* **22**, 14.
Lamontagne, Y. and Wang, B. S. P. (1976). *Tree Planters Notes* **27**, 5–6, 22.
Lhoste, J. (1975). *C. R. Seanc. Acad. Agric. France* **61**, 695.
Lock, W., Sutherland, J. R., and Sluggett, L. J. (1975). *Tree Planters Notes* **26**, 16–18, 28.
Magnani, G. (1970). *Publ. Cent. Speriment. Agric. For.* **11**, 75.
—— (1972). *Pub. Cent. Speriment. Agric. For.* **11**, 307.
Matthews, F. R. (1964). *J. For.* **62**, 881.
—— (1967). *Can. Dep. For. Publ.* **118**, 225.
Mattis, G. Ya. and Badanov, A. P. (1974). *Rev. Plant Pathol.* **54**, 2503 (Abstr.).
Merrill, W. and Kistler, B. R. (1976). *Plant Dis. Rep.* **60**, 652.
Mexal, J. G. and Snow, G. A. (1978). USDA *Forest Service Research Note, Southern Forest Experiment Station*, No. 50–238, p4.

Miller, T. and Bramlett, D. L. (1979). In *Proceedings of IUFRO Seed Biology Symposium Starkville, Mississippi USA May 1978* (ed F. Bonnet), USDA Southern Forestry Experiment Station, Starkville, Mississippi.

Monnier, M. F. (1961). *Proc. 2nd Inter African For. Conf. Pointe Noire, 1958* **2**, 281.

Morelet, M. (1974). *Rev. Plant Pathol.* **54**, 4177 (Abstr.).

Muhle, O. and Hewicker, A. J. (1976). *Allgemeine Forest -und Jagdezeitung* **147**, 10.

Nair, V. M. G. and Kuntz, J. E. (1975). *2nd FAO/IUFRO World Technical Consultation on Forest Diseases and Insects, New Delhi, India.* FAO/IUFRO D1/ 75/12-40.

—— Wolter, K. E. and Kuntz, J. E. (1969). *Phytopathol.* **50**, 1042 (Abstr.).

National Academy of Sciences (1975). Vol. IV. *Forest pest control.* Washington.

Neely, D. (1975). *Plant Dis. Rep.* **59**, 300.

Noble, M. and Richardson, M. J. (1968). *An annotated list of seed-borne diseases. Phytopathological Paper No. 8* (2nd edn), 191pp.

Odera, J. A. (1972). *E. Afr. agric. For. J.* **37**, 308.

Parrini, C., Intini, M., and Panconesi, A. (1976). *Rev. Plant Pathol.* **55**, 5970 (Abstr.).

Pawsey, R. G. and Rahman, M. A. (1976). *J. Arboric.* **2**, 161.

Peterson, G. W. and Smith, R. S. (1975). In *Forest nursery diseases in the United States USDA Forest Service Agriculture Handbook No. 470*, 125pp.

Phillips, D. H. and Greig, B. J. W. (1970). *Ann. Appl. Biol.* **66**, 441.

Prebble, M. L. (1975). *Aerial control of forest insects in Canada*, Ottawa, Department of the Environment, pp.330.

Rishbeth, J. (1950). *Ann. Bot. (London)* **14**, 365.

—— (1976). *Ann. Appl. Biol.* **82**, 57.

—— (1979). *Eur. J. For. Pathol.* **9**, 331.

Roberts, H. (1964). *Forest insect conditions in West Africa: FAO/IUFRO Symposium on Internationally Dangerous Forest Diseases and Insects. Oxford 1964.* Vol. 1. FAO/FORPEST 64: 7pp.

—— (1968). *Commonw. For. Rev.* **47**, 225.

—— (1969). *Commonwealth Forestry Institute, Department of Forestry Paper. No. 44*, 1–206.

Salt, G. A. (1974). *Trans. Br. Mycol. Soc.* **63**, 339.

Sands, W. A. (1960). In *Termite research in West Africa* (ed. W. V. Harris), Department of Technical Cooperation, London.

—— (1962). *Bull. Entomol. Res.* **53**, 179.

Scott, T. M. and King, C. J. (1974). *Forestry Commission. Leaflet 58*, 2.

Seth, S. K., Bakshi, B. K., Reddy, M. A. R., and Sujan Singh (1978). *Eur. J. For. Pathol.* **8**, 200.

Smalley, E. B., Meyers, C. J., Johnson, R. N., Fluke, B. C., and Vieau, R. (1973). *Phytopathology* **63**, 1239.

Spiers, A. G. (1976). *N. Z. J. Exp. Agric.* **4**, 249.

Stephan, B. R. and Millar, C. S. (1975). *Mitteilugen der Bundesanstalt fur Forst -und Holzwirtschaft* No. 108, 97.

Sripathi, Rao, B. (1970). In *Crop protection in Malaysia* (eds R. L. Wastie and B. J. Weed), p.204. The Incorporated Society of Planters, Kuala Lumpur.

Sutherland, J. R., Lock, W., and Sluggett, L. J. (1975). *Report, Pacific Forest Research Centre, Canada*, No. BC-X-125, p.20.

Templer, J. T. (1948). *E Afr. Agric. J.* **13**, 210.

Theodorou, C. and Skinner, M. F. (1976). *Austr. For. Res.* **7**, 53.

Urosevic, B. (1961). *Proc. Int. Seed Testing Assoc.* **26**, 537.

Vaartaja, O. (1964). *Bot. Rev.* **30**, 1.

—— Wilner, J. (1956). *Can. J. Agric. Sci.* **36**, 14.

Van Der Westhuizen, G. C. A. (1968). *S. Afr. For. J.* **65**, 6.

Vite, J. P. (1972). *Tall Timbers Conference on Ecological Animal Control by Habitat Management, Tallahassee, 1971,* 155.

Wall, R. E. (1976). *Bi-monthly Res. Notes* **32,** 12.

Way, M. J. and Bevan, D. (1977). In *Ecological effects of pesticides* (eds F. H. Perring and K. Mellanby), Linnean Society Symposium Series No. 5, London.

Webb, W. E. and Jones, T. (1956). *Record of the 6th Annual Convention – British Wood Preserving Association,* 63.

12
Control of dipteran vectors

N. G. GRATZ

12.1 Introduction

The insect order Diptera or the two-winged flies contains some of the most important vectors of human disease; foremost among these are the mosquitoes. Table 12.1 presents in outline a list of the major mosquito-borne diseases and of the genera of their mosquito vectors; later in the chapter consideration will be given to other important dipteran vectors, among them the tsetse fly vectors of African trypanosomiasis, sandfly vectors of leishma-

TABLE 12.1. *Major mosquito-borne diseases, their distribution and vectors*

Disease	Distribution	Vector species
Malaria	In many countries between 45° north and 40° south latitude	All four species of human malaria are transmitted by various species of the genus *Anopheles*
	Africa north of the Sahara 67 million people, only limited malaria risk	e.g. *Anopheles labranchiae, An. sergenti, An. claviger*
	Africa south of the Sahara 342 million people 291 million people at risk	e.g. *An. gambiae, An. funestus* main vectors
	America north of Mexico malaria free	—
	Central America 113 million people 50 million people at risk	e.g. *An. albimanus*
	South America 224 million people 60 million people at risk	e.g. *An. nuneztovari, An. darlingi, An. albimanus* etc.
	Europe Turkey	e.g. *An. sacharovi*
	Asia west of India 181 million people 146 million people at risk	e.g. *An. stephensi, An. superpictus, An. sacharovi* etc.

Table 12.1—continued

Disease	Distribution	Vector species
	S.E. Asia 734 million people 693 million people at risk	e.g. *An. culicifacies, An. stephensi, An. minimus, An. sundaicus*
	East Asia and Oceania 535 million people 266 million people at risk	e.g. *An. culicifacies, An. sundaicus, An. aconitus, An. minimus, An. balabacensis, An. maculatus, An. sinensis*
Yellow fever	Africa south of the Sahara between 15° N and 10° S. Not reported from East African coast. Jungle yellow fever is enxootic in northern South America and the Amazon basin including Colombian Ilanos and eastern regions of Peru and Bolivia. Epidemics sporadic often with high mortality	No cases of man-to-man transmission by *Aedes aegypti* in urban areas in Americas for 40 years. Chief vector of jungle yellow fever *Haemagogus* species. In Africa *Aedes simpsoni, Ae. africanus, Aedes* of *furcifer-taylori* group, *Ae. luteocephalus.*
Dengue and Dengue haemorrhagic fever	Africa south of the Sahara, S.E. Asia, parts of China, Malaysia, Indonesia, and the Western Pacific, parts of Central and South America, countries of Caribbean	Mainly *Aedes aegypti – Ae. albopictus* incriminated in some areas of Western Pacific
Japanese encephalitis	Western Pacific and South-East Asia from Korea to India	*Culex tritasniorhynchus, Cx. vishui* complex
Venezuelan equine encephalitis	Only in Western Hemisphere from Peru to central Texas	*Aedes, Mansonia, Psoraphora*
Eastern equine encephalitis	From Brazil to southern Canada	*Aedes sollicitans, Ae. vexans*
Western equine encephalitis	Western and central USA and Canada and scattered foci in eastern South America	*Culex tarsalis, Culiseta melanora*
Ross River fever	Australia, S. Pacific	*Aedes vigilax* in Australia and possibly also *Ae. polynesiensis* and *Cx. annulivorstris*

| Filariasis – Bancroftian and Brughian | Africa south of the Sahara, Egypt, many parts of Asia including India, Sri Lanka, Burma, Thailand, Indonesia, and S. China and Pacific Islands; West Indies, Colombia, Venezuela and Panama in the Americas. A focus remains in Turkey. About 200 million people infested with Bancroftian filariasis and 50 million with Brughian | Bancroftian = *Culex quinquefasciatus, An. gambiae, An. funestus,* various *Aedes* spp. Brughian: *Mansonia* and some *Anopheles* |

niasis, and blackfly vectors of onchocerciasis. In each of the cases dealt with below the description of the prevalence and distribution of the disease will be followed by a discussion of the various methods of control which might be used and the advantages and disadvantages of each.

12.2 Malaria

Malaria is the most important of all the mosquito-borne diseases. Until the mid 1940s, it has been estimated that some 300 to 350 million cases of clinical malaria occurred annually (Gratz 1976); in the developed countries, the disease has disappeared largely due to a combination of factors, among them the ecological changes resulting from socioeconomic developments, widespread access to medical attention and care, and organized malaria eradication efforts. In some of the less developed and developing countries, the reduction or eradication of malaria was due to the global malaria eradication campaigns that were begun in 1957 and which took advantage of residual insecticides such as DDT and synthetic anti-malaria drugs which had been developed during World War II. By late 1974 the disease had been eradicated from 36 countries or territories containing 10.3% of the total population of originally malarious areas of the world (WHO 1974a). However, since 1967 there have been a number of adverse factors which have affected the progress of anti-malaria activities in Asia, Central and South America, and Turkey. The main factors impeding further progress, or which have led to increased transmission, are worldwide inflation which has greatly increased the cost to governments of the large staffs used in malaria control, resistance of many malaria vectors to insecticides currently in use, the appearance of parasite resistance to anti-malaria drugs, and finally, the high cost of alternative insecticides (WHO 1979b). Virtually no major malaria control programmes are being undertaken in tropical Africa and over 290 million people are considered to be at risk to malaria transmission in this area; it has been estimated that due to the direct and indirect effects of malaria, many hundreds of thousands of infants and children below the age of 14 die every year in Africa (WHO 1974a). Severe epidemics have recently occurred in India, involving almost 5 million cases in 1977, though mortality from the disease has been held to a low level. By 1980 this had again been brought down to 2.8 million cases. The control of malaria thus remains one of the highest priorities in the health programmes of virtually all the tropical developing countries.

12.2.1   Malaria vector control

A campaign to control the transmission of malaria might centre around the treatment and cure of the disease in humans so as to eliminate the infection in the reservoir, or reduction of transmission by control of the mosquito vector or utilization of a combination of the two. The present chapter will, however, deal only with the means used to control the mosquito vectors.

Human malaria is transmitted only by mosquitoes of the genus *Anopheles;* Table 12.1 lists some of the more important of the many different species of this genus that are vectors in different parts of the world. Vector control can be based on the prevention of mosquito breeding by environmental manipulation to eliminate or render unsuitable the aquatic larval habitats, the use of insecticides, the use of biological agents, or a combination of these.

12.2.2   Environmental control

Environmental control implies the alteration of the environment in a manner that will make it unsuitable for the development of the target insect. This may involve drainage of swamps where *Anopheles* mosquitoes breed, the channelling of streams to increase water flow and eliminate standing pools or sluggishly moving water, the alteration or fluctuation of water levels in reservoirs or other water impondments to strand larvae breeding around the margins, or perhaps elimination of water plants to prevent breeding of shade-loving species. All of these possible methods, however, require an accurate knowledge of the ecology of the target species. Environmental control has been applied with much success in many different control schemes since the beginning of the century, and a number of current control programmes are based on water management. Unfortunately though, certain major vector species such as *An. gambiae* in Africa, breed in a multitude of small or large seasonal collections of water that are not economically subject to environmental correction. A species such as *An. aconitus* which breeds in rice fields is only subject to environmental control if water levels can be manipulated in the enormous areas of rice fields in which it breeds in Indonesia and elsewhere in South-East Asia. Nevertheless, larval habitats of many important vector species can be controlled or eliminated by environmental methods or their breeding can be prevented altogether by more carefully managed irrigation programmes in the tropics. Where they can be applied, environmental methods can eliminate or greatly reduce the cost and possible pollution due to chemical pesticides.

12.2.3 *Chemical control of* Anopheles *vectors*

Control of the vector mosquito species and malaria transmission depends either on the use of larvicides to control the aquatic stages or, much more frequently, adulticides. Larvicides are utilized for *Anopheles* control when the larval habitats are sufficiently restricted in scope and readily accessible and where, for one reason or another, application of environmental methods of control is not feasible. Adulticides are usually applied as residual formulations to the interior of houses where the target species is known to rest indoors for a period of time long enough for it to acquire a lethal dose from the insecticide-treated surface. The most common formulation used in malaria control programmes is that of wettable powders or water-dispersible powders. These are technical-grade insecticides with an inert carrier and a wetting agent in carefully defined proportions. These enable the wettable powder to be mixed with water in the field and then be applied by hand-operated compression sprayers to those interior wall surfaces where the adult mosquitoes are likely to rest.

Up until the time of development of the synthetic organic pesticide in the mid-1940s, the chemical control of *Anopheles* adults, where it was attempted at all, depended on the frequent application of pyrethrum extract as space sprays in houses. This method was very costly in manpower and materials. The long persistence of the synthetic organic pesticides enabled them to be applied to the interior walls of houses, and for a considerable time thereafter, sometimes for as long as several months, any mosquito which rests on the treated surfaces will be killed by the persistent action of the pesticide.

The most broadly used residual insecticide has been DDT; at its peak the global malaria eradication programme utilized some 60 000 tons of technical grade DDT each year, and more than 30 000 tons a year of 70–75% technical grade DDT are still being used for malaria control. Its safety record has been excellent and there is little or no pollution of the environment as a result of this particular mode of application. Unfortunately, however, vector mosquito resistance to DDT has been developing and spreading; by 1975 a total of 42 species of anophelines was resistant, of which 41 species are resistant to dieldrin and 24 to DDT, 21 of the latter having developed double resistance (WHO 1976); by 1980 this number had grown to 51 species of anophelines resistant to one or more insecticides of which 34 were resistant to DDT (WHO 1980). Where resistance has developed to DDT a number of other compounds have been used.

Benzene hexachloride (BHC) and its γ-isomer, lindane, was first produced as an insecticide in England in 1942 and widely used in agriculture.

Lindane has been used in malaria control at a target dosage of $0.5 \, g/m^2$ as it does not have the unpleasant odour of HCH. While an effective and reasonably safe insecticide, resistance to this compound has also developed in many anopheline vectors in part due to cross-resistance to dieldrin (see below).

Dieldrin was, at one time, used in a number of malaria eradication and control programmes but due to its high mammalian toxicity and the extensive resistance which very quickly developed to this compound, its use is no longer recommended (WHO 1970).

With the spread of *Anopheles* resistance to the organochlorines growing attention was paid to the organophosphorus (OP) compounds. This group (see Chapter 2) all have a similar mode of action as inhibitors of the enzyme cholinesterase both in mammals and in arthropods. Widely used in agriculture, some of the OP compounds of lower mammalian toxicity and sufficient persistence on treated surfaces are now also in use in malaria-control campaigns for residual application.

The most widely used is malathion, which is now in operational use in malaria control programmes in Turkey, Iran, India, Sri Lanka, and a number of other countries. It is applied at a target dosage of $2 \, g/m^2$ and is more persistent on organic substrata such as thatch, bamboo, and wood than on the mud surfaces of hut walls. Its toxicity to mammals is low with an oral LD_{50} to rats of 885–2800 mg/kg. However, poorly formulated wettable powders may contain significant quantities of isomalathion as an impurity which considerably increases its toxic hazard to man. *Anopheles* resistance to this compound has appeared in Turkey, Iran, India, El Salvador, and elsewhere in Central America and necessitated a shift to alternative compounds. Nevertheless, relatively large quantities of this insecticide are likely to be continued to be used in malaria vector control campaigns where resistance to DDT precludes further use of that insecticide.

Although it has a somewhat higher mammalian toxicity than malathion, the insecticide fenitrothion can still be used for malaria vector control providing that reasonable measures are taken to protect the spraymen from undue exposure to the compound when it is being mixed and applied. It has, in addition, a greater degree of persistence on both mud and organic surfaces than malathion when also applied at a target dosage of $2 \, g/m^2$ as shown in Kenya (Fontaine *et al.* 1975) and Java, Indonesia (Joshi *et al.* 1977). A large-scale field trial in East Africa showed that fenitrothion was effective in halting transmission of malaria (Fontaine *et al.* 1976). Increasing interest and increasing use is now being made of fenitrothion for malaria vector control.

A number of other OP insecticides have also undergone large-scale trials to determine their efficacy against anophelines; the most promising of these have been chlorphoxim and pirimiphos-methyl. Chlorphoxim when tested in northern Nigeria was found to be safe to spraymen and inhabitants and provided effective anopheline control for up to 10 weeks on mud walls and 12 weeks on thatch roofs (Rishikesh *et al.* 1977). Pirimiphos-methyl when applied at 2 g/m^2 in the same geographical area proved to be as effective as fenitrothion (Rishikesh *et al.* 1977) and has undergone promising field trials in Iran and Indonesia.

The appearance of OP resistance in anophelines stimulated work on chemical groups which it was thought were unlikely to show cross-resistance, and thus several carbamate insecticides have also been evaluated in the laboratory and field. Among these propoxur has shown a particularly good performance but its cost at the present time would probably exclude it from use in large-scale malaria control campaigns. In 1981 a trial was carried out in a series of 13 hamlets in Central Java applying bendiocarb 80% wdp (water-dispersible powder) at a target dosage of 0.4 g/m^2 against a DDT-resistant population of *An. aconitus*. Good control was achieved for at least 8 weeks and the results warrant a larger-scale trial (Fleming *et al.* 1983).

Recently, even more interest has been shown in the synthetic pyrethroids some of which, unlike the natural pyrethrins, show a considerable degree of persistence on sprayed surfaces coupled with a very high insecticidal activity. Field evaluation of two of this group, deltamethrin and permethrin, in villages in Nigeria (Rishikesh *et al.* 1978) showed that water-dispersible powders of these compounds applied at a target dosage of 0.05 g/m^2 and 0.5 g/m^2 were effective and produced satisfactory reductions in adult anopheline mosquito density. Though expensive, the low concentrations required and their degree of persistence may make their use feasible in areas where broad-scale insecticide resistance excludes the use of other compounds.

In those areas of the tropics where malaria remains an important public health problem, insecticides will undoubtedly continue in use for the foreseeable future. As societies develop better medical care and improvements in the environment this may enable the quantities of insecticides to be reduced. This will require a greatly increased degree of understanding and co-operation on the part of the populations which are being protected. For the time being malaria continues to be a severe burden in terms of morbidity and mortality in many tropical countries to the extent that it also represents a severe impediment to rural economic development, especially when linked with other mosquito-borne diseases such as those described below.

12.3 Mosquito-borne filariasis

Two species of tissue-inhabiting filarial nematodes whose immature forms, the microfilariae, can be transmitted by mosquitoes may cause serious disease in humans; these are Bancroftian filariasis caused by *Wuchereria bancrofti* and Brugian filariasis caused by *Brugia malayi*. Brugian filariasis is found only in several endemic foci in the Western Pacific and South-East Asia regions, but Bancroftian filariasis has an almost worldwide distribution in the humid tropical and subtropical zones. It has been estimated that there are 200 million persons in the world infected by Bancroftian filariasis and perhaps another 50 million infected by Brugian. There is some evidence to suggest that filariasis has increased in both its prevalence and distribution in many parts of Africa and Asia and that the total population at risk has almost doubled in the past 20 years (WHO 1974b). The increase in urban areas of the tropics where Bancroftian filariasis is transmitted by the polluted water-breeding mosquito, *Culex quinquefasciatus,* has been particularly marked (Gratz 1973); this is due to the rapid expansion in urbanization which exceeds the ability of sanitation services to cope with the increased populations. In the few developed countries where the disease was once a problem, it has disappeared or been reduced to but a very few occasional cases.

Man is the only vertebrate host of Bancroftian filariasis. The nocturnal form of *W. bancrofti* is found in several foci in the Western Hemisphere and throughout wide areas of tropical Africa, India, Sri Lanka, Burma, and southern China where it is transmitted by night-biting mosquitoes. In the Polynesian area the disease is transmitted mainly by day-biting mosquitoes of the genus *Aedes.*

Brugian filariasis (sometimes called Malayan filariasis) has a nocturnal periodic form transmitted by *Mansonia* mosquitoes: a nocturnal subperiodic form also transmitted by *Mansonia* species occurs in a number of different wild and domestic animal species in Malaysia and is therefore a true zoonosis in that country. In some foci in Indonesia and Malaysia, certain *Anopheles* species are also vectors of this disease. The clinical symptoms of Brugian filariasis are generally similar to those of Bancroftian, though its control poses special problems. There have been only a few successful vector control programmes directed against Brugian filariasis and much more research remains to be done before transmission of this disease can be effectively or economically achieved by vector control.

12.3.1   The control of filariasis

Human filariasis may either be controlled by chemotherapy, and correct use of the drug diethyl carbamazine for the treatment of infected people will remove most of the microfilariae and destroy most of the adult worms. Unfortunately, however, there are often unpleasant side effects from use of this drug and the course of the treatment is very often stopped before the patient is freed from the disease, and, in the presence of vector populations, transmission can still continue. In some countries it is difficult to carry out surveys of large human populations so as to determine who is infected by the disease and thus must be treated. For these reasons and because of the difficulty of wide use of the chemotherapeutic agent under carefully controlled conditions, interruption of filariasis transmission must often rely on control of the vectors.

12.3.2 *The control of filariasis vectors*

Bancroftian filariasis is transmitted by *C. quinquefasciatus* in urban areas of South-East Aisa, East Africa, in some few foci in the Eastern Mediterranean, and in some coastal areas of North-East South America. The optimum method of controlling the breeding of this species which breeds mainly in sewage accumulations is to eliminate its larval habitats by the installation of adequate sanitary sewage facilities. Unfortunately some of the cities which need this type of correction most are among those least able to afford the great capital outlay involved. As a result, a great deal of reliance must continue to be placed on chemical larvicides with which to spray the larval habitats. Attempts have been made to control the adults of this species by residual spray applications to the interior of homes, but the species readily rests both indoors and outdoors and attempts to achieve control by residual sprays only have failed. Space sprays, no matter how effective as a temporary control, are inadvisable owing to the considerable flight range of this species and its high biotic potential. Thus, where chemical control measures are to be considered, they must be aimed at the aquatic stages. It must be kept in mind that the life of the adult worm in the human body probably considerably exceeds 10 years. Thus if the aim of a control programme is to interrupt transmission of disease, the level of mosquito population which is low enough as to be unlikely to permit transmission must be maintained for at least this period of time. While this is technically feasible, it would be expensive and difficult to ensure unless a highly motivated control group worked continually to suppress mosquito populations for this period. In any event, it must be admitted that the lower level to which the mosquito populations must be held is unknown and will relate to the intensity of the disease in the human reservoirs in any given area. Most urban areas in any event have active mosquito control programmes, most of which are targeted at *C. quinquefasciatus,* because of the intense nuisance caused by the biting of this species rather than because of its role as a vector of disease. Unfortunately in most tropical areas the degree of control obtained is low because of the shortage of trained supervisory staff especially at a professional level. Until recently, larvicidal oils were widely used for the control of this species; such oils have several limitations. One factor will increasingly be their steadily mounting cost, another their lack of persistence in water with a high degree of organic pollution. In addition, the cost of transporting the large quantities of oil required is also high; finally, the larvicidal performance of many oils used for this purpose is poor and those especially refined for larvicidal purposes may be too costly for the developing countries to use. Hence increasing resort is necessary to chemical larvicides. Little use is made of the organochlorines owing to the widespread resistance that quickly developed in culicine populations where DDT, dieldrin, or others of this group were applied. Widespread use is, however, made of organophosphorus larvicides in mosquito control programmes in all parts of the world. A WHO field research unit carried out an extensive series of studies on different OP compounds against larvae of *C. quinquefasciatus* breeding in highly polluted water habitats in Rangoon, Burma (Self and Tun 1970). Many different OP compounds, such as parathion, malathion, and dichlorvos, were found, but the lack of any degree of significant persistence of action in the multitude of polluted water habitats, such as sewage ditches, pit latrines, septic tanks, and sewage pools, meant that frequent and costly treatment would have been necessary. More attention was therefore paid to those compounds which provided at least a week of persistence in the slowly flowing sewage ditches and other larval habitats; the most effective of these was found to be fenthion and chlorpyrifos. Temephos, while highly effective against mosquito larvae, is less effective in highly polluted water.

There has, so far, been relatively little resistance developed to the OP compounds in mosquito populations in most urban areas though this is likely to occur. In rural areas the widespread use of agricultural insecticides which inevitably find their way into irrigation water has frequently resulted in a high level of resistance of mosquito larvae. Where it has not been possible to rely on environmental or biological control methods as an alternative, it has become necessary to seek chemical groups which, while being effective larvicides, show little or no cross-resistance to the organophosphorus larvicides. Two groups which have shown considerable effectiveness against mosquitoes in most larval habitats are the insect growth regulators and the synthetic pyrethroids. The latter group, which will be discussed more fully later, is unfortunately also effective against most non-target aquatic organisms and its use as a larvicide is therefore largely excluded.

The use of insect growth regulators (IGRs) in mosquito control has, however, been steadily growing. This group includes insect hormones, growth regulators, developmental inhibitors, and synthetic hormone mimics. Chamberlain (1975) defined the IGRs as materials which are natural biochemicals or exogenously applied chemicals that cause morphological and physiological changes during the growth or development of insects or other arthropods. In the case of mosquitoes, exposure to these compounds at the larval stage either causes death of the larvae before pupation or failure of the adults to emerge successfully from the pupae. Most developmental work with this group was carried out in relatively unpolluted water but field trials were recently carried out in highly polluted water habitats in Jakarta, Indonesia, against *C. quinquefasciatus* (Self *et al.* 1978). In these trials, two growth regulators, diflubenzuron and methoprene, were applied to polluted water in cement and earthen drains, and prevented adult emergence for 2 and

5 weeks respectively after a single application. The continuing development of slow-release formulations is likely to extend the period of persistence of these compounds. While resistance can develop in larval mosquito populations to this group, only restricted use has so far been made of them and they are likely to remain useful weapons against mosquito larvae for some time to come.

12.4 Arbovirus diseases and vectors

Unlike the protozoan diseases, such as malaria, and the helminthic diseases, such as filariasis, which may persist for long periods of time in the human body, the arthropod-borne viruses usually called 'arboviruses' will, where they cause disease at all, generally give rise to a relatively short acute disease whose effects may vary greatly from one virus species to another as well as from one person to another. There are more than 200 known arboviruses, most of which are transmitted by mosquitoes and of which about only a quarter will actually give rise to human disease. All are characterized by being transmitted from one vertebrate host to another by an arthropod vector which, in addition to mosquitoes, may be ticks or *Culicoides* midges or sandflies. Many will also have animal or avian hosts in nature other than man.

Probably the best known, though no longer the most important of the arboviruses in terms of morbidity and mortality, is yellow fever. This disease is still endemic in large areas of tropical Africa and America, and was once the cause of many deaths in urban epidemics in the Americas. Sizeable epidemics have recently occurred in rural areas of Africa and smaller ones in rural or jungle areas of South America. There is no cure for the disease but a highly effective vaccine prevents inoculated persons or populations from contracting it. In both Africa and the Americas the disease has been thought to be maintained in the wild by a mosquito-monkey-mosquito cycle. The recent finding that transovarial transmission occurs at least in some species of wild mosquito vectors indicates that the epidemiology of the disease in nature is more complex than previously thought. In urban areas of both the Americas and Africa the vector is *Aedes aegypti,* commonly known as the yellow fever mosquito. In Africa this species is both a commensal one breeding in such man-made containers as water jars and tyres, and sylvatic breeding in tree holes and leaf axils. In the African forest and savannah areas other mosquito species of the subgenus *Stegomyia* of *Aedes* are vectors. In the Americas, *Ae. aegypti* is associated virtually entirely with man-made containers in cities and villages and a number of other mosquito species transmit the disease in the jungle.

12.4.1 *Aedes aegypti control in the Americas*

The control of *Aedes aegypti* in urban areas of the Americas has been an important public health objective, and since 1947 this species has been the target of a hemisphere-wide eradication campaign. While the campaign has not proved successful in eradicating *Ae. aegypti,* the control efforts combined with widespread vaccination have eliminated urban yellow fever and no case has occurred in any city or town in the Americas since the last reported case in Trinidad in 1954. While officially still underway, the eradication efforts are being increasingly hampered both by rising labour costs and by the continuing spread of resistance to the organochlorine insecticides which occurs virtually in every country of the Caribbean and many of the countries of Central and South America where reinfestations have occurred. The species remains susceptible to malathion which is finding increasing use both as a residual spray and in the form of an ultra-low-volume formulation applied by ground or aerial spray equipment. Only a few attempts have been made to achieve focal control of jungle vector species by ULV applications and the method is not likely to be practical against these wild vectors. In urban areas some use is also made of 1% temephos sand granules applied to fresh-water containers which cannot be eliminated or routinely sealed.

12.4.2   Aedes control in Africa

Field trials in Africa have shown that adulticidal and larvicidal measures which have proven effective against *Ae. aegypti* in urban areas of the Americas or South-East Asia would also prove useful there but, few if any routine control campaigns are now maintained against *Ae. aegypti* in African cities or towns. A field trial of the aerial application of malathion ULV concentrate at high dosage rates (Brooks *et al.* 1970) showed that *Ae. simpsoni*, an important peridomestic vector in Eastern Africa, could be controlled in limited areas by this method, but, as in South America, this approach would probably not be practical other than on an emergency basis. Studies in and around African villages both in East and West Africa have shown that there are usually far fewer domestic water containers in these villages, and that larvae of *Ae. aegypti* breeding in such containers can be relatively easily controlled by the application of 1% temephos granules, at least in those villages where containers storing domestic water are not so small that they are continually being emptied and changed. Studies in some coastal cities of East Africa have shown a relative increase of the *Ae. aegypti* population as such peridomestic breeding sites as old tyres, wrecked cars, tin cans etc. proliferate; this increases the risk of urban epidemics of yellow fever.

12.4.3   Aedes control in South-East Asia

Fortunately for the inhabitants of the area, yellow fever has never invaded any area in Asia, especially as populations of *Ae. aegypti* and the closely related *Ae. albopictus,* both of which are avid man-biters, reach extremely high levels in many of the large cities of India and the Indo-China peninsula. Another serious arbovirus disease is, however, widespread; dengue transmission is virtually universal wherever *Ae. aegypti* is found in densities high enough to transmit the virus, and in many areas from the Philippines to India a very serious syndrome of the disease called dengue haemorrhagic fever (DHF) has also appeared. This disease first appeared in the Philippines in 1954 and is associated with haemorrhagic manifestations which are especially serious in children. Fatality rates of 10–70% have occurred in some areas especially when inadequate supportive treatment was given to the patients because there is no cure for the disease which must run its short but serious course. All in all, DHF has become one of the ten leading causes of hospitalization and death in at least seven countries of South-East Asia. In the absence of any cure and, for the time being, of any vaccine to prevent it, emphasis has therefore been placed on control of the most important vector, *Ae. aegypti,* as the only way of preventing the disease. *Ae. albopictus,* a closely related species of the subgenus *Stegomyia,* has also been incriminated in some areas but its role is less important than that of Ae. aegypti. As in South America, *Ae. aegypti* in South-East Asia is found virtually entirely in man-made containers, the most important of which are various types of large clay jars or metal drums storing water for domestic use. In addition, the species breeds in a multitude of other small containers including discarded tyres, water-filled ant traps on table legs, discarded tin cans etc. *Ae. albopictus* which can also breed in such types of containers will also breed in natural containers such as cut-off bamboo stumps, coconut shells etc. The most effective way of preventing *Ae. aegypti* breeding would be to dispose effectively of waste containers and to mosquito-proof all domestic water containers. Unfortunately, in practice, this is very difficult to achieve. Reliance must therefore be placed on larvicides or adulticides. Temephos (Abate) has been found to be a safe and effective insecticide for the control of mosquito larvae breeding in domestic water containers when used in the form of 1% impregnated sand granules. Field trials in a neighbourhood of Bangkok (Bang and Pant 1972) showed that by adding the temephos-impregnated sand granules to all water containers to achieve a target dosage of 1 mg/l, and repeat treatment of positive containers, control was obtained of larvae for up to 3 months, and adult densities of *Ae. aegypti* in the area under control also fell to low levels. Other larvicides such as jodfenphos and pirimiphos-methyl also provided effective, though shorter, periods of control (Mathis *et al.* 1974). Many countries do not, however, have vector control programmes well enough organized to carry out the type of large-scale and long-term treatment programme which would be necessary to effectively control *Ae. aegypti* populations by larval control alone. In addition the cost of such programmes would also be high both in terms of the labour and materials. Since reductions of adult mosquito densities must be the control target, residual applications of insecticide to the interior of homes was considered. This was rejected when it was found that 90% of the *Ae. aegypti* mosquito populations in Bangkok rested on objects which would not or could not be treated by a residual spray, such as clothes, pictures, and devotional objects on walls, bedspreads, mosquito nets, shoes etc. (Pant and Yasuno 1970). In any event the effect of any residual treatments would not be immediate which would be unacceptable during an epidemic. Because of this it was necessary to consider space treatments which would effect an immediate reduction in adult mosquito populations. Many countries have come to rely on the application of large-scale ultra-low-volume aerosol insecticide treatments to rapidly reduce adult vector mosquito populations when epidemic outbreaks of DHF occur.

The principle of aerosol control of adult mosquitoes is to produce a mist or fog of droplets which will kill the flying or resting adult mosquito upon contact with them. Each droplet should contain enough insecticide to kill a mosquito upon contact and still be small enough to impinge upon the individual mosquito; the droplets should be of an optimum size to ensure that they do not settle out of an air-borne phase too rapidly, thus falling on to surfaces where they will have no insecticidal effect and yet not so small that they rise and drift away from the target area (see Chapter 3).

Aerosols may be produced by mechanical, thermal, or gaseous energy atomizing devices which can be utilized for both indoor and outdoor space treatments (see Chapter 5). Very small droplets are emitted at relatively low velocities, and for outdoor treatments advantage has to be taken of the wind to ensure effective dispersal. Aerosol applications are usually made when temperature-inversion conditions exist to ensure that the droplets remain in the target area for the longest possible period of time. When ground wind speeds exceed 10 km/h or conditions of turbulence exist, outdoor aerosol emissions usually become too difficult to control.

The simplest and best known form of aerosol dispenser is, of course, the commercial aerosol container for household use in which the insecticide is either in solution or mixed with a liquefied gas at about one atmosphere pressure. While relatively effective in individual homes, their use cannot be considered for epidemic control.

Thermal fogging machines (see Chapter 5) can produce insecticidal fogs which have provided effective control of *Ae. aegypti* populations when the relatively high percentage of 4% malathion solution was used. These machines have a number of disadvantages including the necessity of providing large quantities of non-active diesel oil carrier which consists of 96–98% of the material actually applied. They are also noisy and produce thick fogs which may disrupt traffic. More interest has consequently been placed on the use of ultra-low-volume or non-thermal fogs for *Ae. aegypti* control both from the air and by ground application equipment. While aerial applications are both expensive and exacting if effective, they can nevertheless cover very large areas rapidly which is an important consideration for epidemic control.

One of the first aerial ULV applications against *Ae. aegypti* populations was described by Eliason *et al.* (1970) in Florida. Four large areas (518 ha each) in an urban location were treated with various insecticides applied by a twin-engine aircraft. Applications were made with the aircraft flying at a height of 45 m and a speed of 240 km/h. The results measured by oviposition rates showed that in an area where malathion was applied twice weekly at a rate of 219 ml/ha (3 US fl oz/acre), *Ae. aegypti* oviposition was totally interrupted for 10 weeks during the 11-week treatment period. A combination of temephos/malathion or weekly treatments with malathion gave similar results. Earlier studies of malathion aerial ULV applications during an epidemic of St Louis encephalitis in Corpus Christi, Texas (Gardner and Iverson 1968) showed that there were no toxicological hazards to the human populations living in areas which had been sprayed by aircraft with this dosage of malathion.

A large-scale aerial ULV application of malathion concentrate was carried out over the city of Nakhan Sawan in Thailand with an area of 18 km^2, the city being treated twice, 4 days apart at a target dosage of 438 ml/ha (6 oz/acre). The landing rate of *Ae. aegypti* was reduced by 95 and 99% respectively after each application and reductions ranged from 88 to 99% for a 10-day period after the spraying. There were significant reductions in the landing rates of other mosquito species as well as houseflies, and the authors concluded that the method would be an effective and rapid way of reducing *Ae. aegypti* population in case of urban epidemics of DHF (Lofgren *et al.* 1970).

During an epidemic of DHF in Menado, Indonesia, aerial applications of malathion were followed by a reduction in the number of hospitalized cases of the disease presumably as a result of the applications (Self *et al.* 1977).

While it is possible to achieve rapid and effective control of important mosquito vectors of disease in the case of epidemic outbreaks, it is of paramount importance that applications be carried out by highly skilled pilots, trained to carry out ULV applications at the proper speeds and heights. It is also necessary to have ready access to aircraft fitted with correct application equipment (see Chapter 6).

In some areas of the world experienced ULV spray pilots and suitable aircraft may not be immediately available during an epidemic; also the use of a large twin-engine aircraft is expensive. Costs may range up to US$ 1000 or more an hour, including securing the aircraft if no suitable airplane with spray equipment is available in the area of the epidemic. Therefore a series of studies was carried out by WHO on ULV ground application equipment that might be more readily stockpiled or flown to the required site at the time of an emergency. The results of these trials have shown that it is possible to use vehicle or knapsack-carried ULV equipment to apply both malathion and fenitrothion concentrates into and around houses fairly rapidly. If large enough areas are covered and if target dosages are adequate, rapid reductions of *Ae. aegypti* populations can be achieved. If a sequence of such treatments is carried out, the ability of the mosquito population to reproduce itself can be so depleted that recovery to pretreatment levels can take many months (Pant *et al.* 1973).

Increasing use is being made of various types of ULV treatment in areas where DHF is important. However, the cost of the insecticide used can be high as is the cost of the application equipment and, wherever possible, ULV applications should be restricted for emergency use only.

12.4.4   *Japanese encephalitis*

Japanese B encephalitis, also caused, like dengue, by a group B virus, has been one of the most important arthropod-borne diseases in the Western Pacific and in parts of South-East Asia. While most infections with the virus are inapparent and virtually symptomless, a small percentage may cause severe encephalitis and death: fatality rates in clinically apparent cases may be 20–50%, and survivors are often left with permanent sequelae affecting the nervous system. The disease has been reported from eastern Siberia in the north through Korea to Indonesia in the south and as far west as India where it has recently been the cause of a severe epidemic in Bengal. The identity of the animal reservoir of the disease is still uncertain, but several animals, especially pigs, serve as amplifying hosts and produce a high level of viremia to infect the vector mosquito. The distribution of the disease is closely associated with that of the vector mosquito species of the *Culex vishnui* group which include *Culex tritaeniorhynchus, Cx. gelidus,* and *Cx. fuscocephalus.* Most of these species are associated with rice fields and their breeding covers vast areas of the countryside.

While the disease reached a high incidence in Korea, China (Taiwan), and Japan up to a few years ago, the heavily increased use of agricultural insecticides in these countries has apparently caused a significant decline in mosquito vector populations breeding in insecticide-treated rice fields. This, combined with the spreading use of vaccination for children, both in these countries and in the People's Republic of China, has resulted in a considerable fall in the incidence of the disease. As an example only 22 cases including 10 deaths were reported in Japan in 1972 whereas in the late 1940s before vaccination of horses and later pigs began, there were probably some 5000 cases a year. Insecticide resistance has begun to appear among populations of the vector and may result in the vector populations increasing again.

In some countries such as Thailand, India, and possibly Burma, the disease appears to be of growing importance and its control presents a serious problem, especially in areas where not enough agricultural insecticides are in use to have any suppressive effect on the breeding of the vector.

Some trials have been carried out in an effort to develop control methods that could be utilized in the case of epidemic outbreaks of Japanese encephalitis. Trials in Korea showed that ground applications of 95% ULV-grade fenitrothion by a LECO heavy-duty cold aerosol generator at 450 ml/ha in an area of 75 ha was able to achieve 75–90% reductions in adult populations of *Cx. tritaeniorhynchus*. Such reductions were however only maintained for a few days before adult mosquitoes infiltrated from surrounding areas or until new adults emerged since the treatment had no effect on the aquatic stages. Further trials were therefore carried out with aerial applications (Self *et al.* 1973) in 2 successive years over a 16 km² area utilizing a large fixed-wing aircraft. Malathion concentrate applied at 0.36 litre/ha gave insufficient control of the parous (possibly infective) females and no reduction in total numbers of adult *Cx. tritaeniorhynchus*. Fenitrothion concentrate applied at 0.45 litre/ha resulted in a 77–87% reduction in total numbers and an 87–98% reduction in parous females over a 4-day period. Such a method of control would, however, probably have to be limited to only the most severe foci of the disease in case of an epidemic.

12.4.5 *Venezuelan equine encephalitis (VEE)*

This disease has caused serious epizootics in South and Central America and as far north as Texas, killing large numbers of horses. The disease is maintained in a wild cycle by mosquito transmission between a large number of species of small mammals, and, occasionally, from these to equines or man by one of several species of mosquito. It may give rise to severe disease in man and equines. As an example, more than 1500 equines died in the south Texas epidemic which occurred in 1972 and there were 110 human cases, although no deaths.

Because of its rapid spread it is of concern both from the public health and veterinary viewpoint. As an example, when the disease first appeared in Texas in 1971, 1 300 000 equines were vaccinated, and 3 240 000 hectares were sprayed with insecticide (malathion) from the air in less than 2 months at a total cost of some $30 million. By comparison, only $35 million was spent yearly on all organized mosquito-control programmes in the USA at that time. Owing to the fact that this virus is found in a number of vertebrate hosts and can probably be transmitted by a considerable number of mosquito species in a wide variety of different ecological areas, it is difficult to propose any general vector control measures. Owing to the very rapid manner in which most epidemics of VEE have broken out and spread, larval control is unlikely to be effective; reliance will therefore have to be placed on ULV applications to control the adult mosquitoes circulating the virus. Two such large-scale ULV applications have been made both of which used malathion ULV concentrate, one in Ecuador in 1969 and the other in the southern USA in 1971; since in both cases the equine outbreak was either subsiding or an equine vaccination programme was being carried out, it is difficult to determine the effect of these sprayings on the spread of the disease other than to speculate that they probably contained the disease within the area of the outbreak (Chamberlain 1972).

12.4.6   Other mosquito-borne arboviruses

In addition to those described above, there are a substantial number of other arbovirus diseases with mosquito vectors. Some of these may occasionally give rise to severe local epidemics causing considerable morbidity and, in the case of Eastern, Western, and St Louis encephalitis in the Americas, mortality as well. Large epidemics have occurred in Asia and Africa of chikengunya and O'nyong-nyong fevers. There was an outbreak of Ross River virus in Fiji in early 1979 that may have caused many hundreds if not thousands of cases of illness (CDC 1979); the vectors are thought to be *Ae. vigilax* and *Cx. annulirostris*. Murray Valley encephalitis has been a problem in Australia for some time.

Rift Valley fever (RFV) is a mosquito-borne disease which was known to be endemic and enzootic in several countries of Africa. Human cases have generally been mild and occurred in small numbers. However, a serious outbreak occurred in South Africa in 1950–1951 in which 20 000 humans were infected, and over 100 000 sheep and cattle died. The virus was isolated from several different species of *Culex, Aedes, Anopheles,* and *Eretmapodites* mosquitoes in South Africa (McIntosh 1972). More recently a severe outbreak of this disease occurred in Egypt between October and December 1977 (WHO 1978). Along with a serious epizootic characterized by a high death and abortion rate in sheep and camels and a high abortion rate in cattle and buffaloes there were an estimated 18 000 human cases and some 598 deaths. There was another, smaller outbreak in 1978 (WHO 1979a). The mosquito *Cx. quinquefasciatus* was incriminated through the isolation of RVF virus from pools of this species. Attempts were made to control populations of this mosquito in the affected area by aerial applications of insecticide with uncertain effect.

12.5   Leishmaniasis

Like malaria, leishmaniasis is also a disease caused by protozoan parasites, in this case of the genus *Leishmania*. The vectors are sandflies of the family *Pychodidae*. The *Leishmania* species which may affect man may be classified into the visceral and cutaneous groups. Cutaneous leishmaniasis caused by *L. tropica* is very widespread in Africa, the Mediterranean basin, much of the Middle East and eastwards through the southern USSR, Iran, and the Indian subcontinent to the Indo-China peninsula and the Philippines. The cutaneous ulcers of varying degrees of severity can be exceedingly common in those areas where the sandfly vectors, such as *Phlebotomus papatasi, Ph. sergenti,* and *Ph. caucasicus,* are not controlled. In the New World cutaneous leishmaniasis or mucocutaneous leishmaniasis is found in different forms throughout much of Central America and large areas of Colombia, Brazil, Peru, Paraguay, Uruguay, Bolivia, and into Argentina and Chile; subspecies of *L. brasiliensis* are ascribed as the causative agents and the reservoirs are wild rodents and occasionally dogs.

Visceral leishmaniasis caused by *L. donovani* is also widely known as kala-azar, infantile leishmaniasis and dum-dum fever to name a few of its synonyms. In Asia, the disease is found in India and China, in the Middle East in Egypt and Israel eastwards through Iraq and Iran, and in many of the countries of the Mediterranean basin including all those of North Africa and Portugal, Spain, southern France and Italy, as well as in Hungary and Romania and in the Caucasian and Central Asia republics of the USSR. In Africa south of the Sahara it has been found in the Sudan, Ethiopia, Somalia, Kenya, Guinea, Nigeria, and the United Republic of Cameroons. In the Americas, the disease has been found in Mexico, Guatemala, El Salvador, Colombia, Venezuela, Peru, Brazil, Bolivia, and Argentina. The vectors of visceral cutaneous leishmaniasis are also species of *Phlebotomus* sandflies. The reservoirs include man himself as well as dogs and other canines, cats, and a number of wild rodent species. Untreated cases of the disease have a high fatality rate; the therapeutic agents available, usually the antimonials, have a high toxicity and must be used with care under close clinical supervision.

In most countries in which visceral or cutaneous leishmaniasis are endemic, the diseases are not reportable; however, many surveys have been reported in the literature which show the magnitude of the problem and indicate the general morbidity. Recent studies indicate that the incidence of the various forms of leishmaniasis may be more serious than has been thought; Lainson and Shaw (1978) consider that of the protozoal parasites, leishmaniasis is probably second in importance only to malaria. Reports to WHO give reason to believe that there may be as many as 400 000 new cases every year, an unknown number of which give rise to chronic sequelae. Visceral leishmaniasis is especially serious and in a recent epidemic in Bihar province of northern India it has been estimated that there were perhaps some 100 000 cases and 6000 deaths in 1976–1977 (unpublished reports to WHO).

12.5.1 *The control of the sandfly vectors of leishmaniasis*

The sandfly vectors are extremely susceptible to insecticides and there has been no instance of insecticide resistance in this genus. Practically everywhere that residual insecticides, and especially DDT, have been applied for the control of *Anopheles* mosquitoes as part of the malaria eradication or control campaign there has been a virtual disappearance of the sandfly populations; however, after spraying is withdrawn following the suppression or eradication of malaria, sandfly populations recover, and transmission of leishmanisis begins anew often at high levels owing to the loss of immunity of unexposed populations as has been described above. In most areas where leishmaniasis is endemic, governments have not initiated separate sandfly control campaigns for the control of the transmission of cutaneous leishmaniasis or occasional cases of visceral leishmaniasis and, unfortunately, control campaigns are all too often only mounted when epidemic outbreaks occur.

Where special spraying campaigns have been undertaken to control *Phlebotomus*, DDT at 1 or 2 g/m^2 and gamma HCH at 0.25 g/m^2 have provided very good control, though lindane is much less persistent. In some areas, and especially in the USSR, fumigants have been introduced into the burrows of small animals surrounding settlements to kill both the sandfly vectors resting in the burrows and the rodent reservoirs of the disease. Area fogging would probably give temporary control of sandfly populations but more work must be done on this technique as well as the possible application of ULV sprays to control sandfly vectors.

12.6   Onchocerciasis

Another serious helminthic disease of man, onchocerciasis, is also widely known as 'river blindness'. The disease is caused by the presence of the filarial parasite *Onchocerca volvulus* in the skin, subcutaneous, and other tissues of man where it often produces fibrous nodules and frequently causes eye lesions which lead to blindness. The disease is transmitted from one human host to another by bites of female blackflies of the genus *Simulium*. These become infected when they engorge on blood or tissue fluids from the skin of an infected host. Some of the ingested microfilariae are digested along with the blood meal but a proportion succeed in penetrating the wall of the fly's stomach and find their way to the thoracic muscles where development of the larvae then takes place. Depending on the ambient temperature the larvae develop for a period of about 6 days passing through two moults. The infective-stage larvae then pass to the labium of the fly where they may gain entrance through the skin of another human host when the fly next takes a blood meal.

In the tissues of the human host the larvae develop into adult worms fairly rapidly. Usually an individual must receive many infective bites before one or several couples of adult worms can establish themselves in the human body. In about 9 months time the larvae can change into a fertile female worm; at that point the female begins to produce large numbers of microfilariae which infiltrate into the skin and tissues, about 2500 microfilariae are produced daily.

Onchoerciasis occurs throughout the greater part of tropical Africa in both the rain forest regions and the savannah belt stretching from Senegal across to the Sudan in the north and south to Angola in the west and Tanzania in the east. A focus of the disease occurs in Yemen. In tropical America its distribution is more limited and up to recently, confirmed foci were thought to be limited to Guatemala, Mexico, Colombia, and Venezuela. However, the presence of the disease has now been confirmed in the Amazonas state of Brazil (Moraes *et al.* 1973), and later findings have indicated that there are three separate foci of the disease in Brazil including the Federal Territory of Roraima as well (Rassi *et al.* 1976). More recent surveys have also shown that the distribution and prevalance of the disease in both Venezuela (Rassi *et al.* 1977) and Colombia (Ewert *et al.* 1979) are greater than was previously thought to be the case and the public health importance of the disease in these areas is thus also now known to be greater.

It is thought that throughout the world as many as 30 million people are infected by the disease (Choyce 1972), and further foci will probably yet be found. In many parts of West and equatorial Africa more than 50% of the inhabitants are infected with onchocerciasis, 30% have impaired vision, and 4–10% are blind (WHO 1966). Where the age-specific prevalence of blindness is studied in greater detail, it may be found that as much as 30% of the adult male population in small communities in hyperendemic areas is blind. Percentages of infection and blindness in some of the Sudanese and central African foci are also very high. In many areas of West Africa there appears to be a pattern of human population retreat from the river villages many of which, though they are in the most fertile land, are also highly endemic for river blindness and onchocerciasis is probably the primary reason for this retreat. However, in many areas of the West African savannah there is serious land pressure due to overpopulation of favourable farming areas, and occasionally resettlement attempts are made in areas of high onchocerciasis transmission and this may quickly lead to infection and vision impairment in the new inhabitants. Clearly onchocerciasis is of considerable economic importance particularly in those area of West Africa where the better land is restricted to the river valleys highly infested by the vector fly and the disease. Many of those fertile areas remain unsettled because of fear of river blindness and blindness due to onchocerciasis among adult males is a serious impediment in existing communities.

Of the 1300 or more species of blackflies only a small number are vectors of onchocerciasis. In the Americas *simulium onchraaceum, S. metallicum,* and *S. callidum* have been incriminated as disease vectors. In West Africa all the blackfly vectors are members of the *S. damnosum* complex of which at least seven forms are now known to occur in the Volta River basin. *S. damnosum* is also the vector in East Africa except for some foci in Kenya and Zaire where the vector is *S. neavei.*

The control of onchocerciasis can be achieved either by drug treatment of infected individuals or by preventing transmission through the control of the blackfly vectors.

Only two drugs can be considered as effective against the parasite *O. volvulus* and be used for the treatment of patients with onchocerciasis: the first of these is suramin which kills the adult worm and also has some effect on the microfilariae. However, reported reactions to this drug include fever, headache, muscle and joint pains, nausea and pruritus and occular reactions that are probably allergic in nature. The second drug is diethyl carbamazine (Hetrazan). It has little or no effect on the adult worm but rapidly kills the microfilariae; severe reactions are not uncommon side effects of its use. While both drugs can be recommended for individual use under the supervision of an experienced physician, the toxic and side effects they commonly produce make their use in mass drug treatment campaigns impracticable.

From the preceding, it can be seen that in view of the impracticability of mass drug treatment and in the total absence of any immunizing agent, the only feasible method of interrupting transmission of onchocerciasis is by control of the blackfly vector.

12.6.1. Blackfly control

The control of the blackfly vectors of onchocerciasis is restricted to the control of the larvae. Control of adult blackflies could be carried out especially by aerial sprays but the control would be of very short duration and the adult population would be replenished almost immediately, either by fresh emergence from pupae or by the invasion of adult blackflies from adjacent uncontrolled areas since the flight range of adults may be 100 km or more. At one time DDT was widely and successfully used to control blackfly larvae which breed in the swiftly running water of rivers or streams attached to the submerged portions of plants or rocks most frequently around rapids. The larvae, which are usually found up to depths of about 50 cm below the surface, filter food particles from the water flowing around them, and will also filter out the toxic particles of insecticide. Concern with the effect of DDT on non-target organisms and the eventual development of DDT resistance in some foci of West Africa resulted in the replacement of this larvicide by temephos because of the latter's very low mammalian toxicity and relatively moderate effect on non-target invertebrate fauna.

The largest *Simulium* control programme being undertaken today and indeed the largest vector control operation is that of the Onchocerciasis Control Programme being carried out in the region of the Volta River basin of West Africa. In this region surveys carried out by National Health Services indicate that in an area of 700 000 km² about 1 million persons are infected with onchocerciasis and that of these at least 70 000 people are blind or their sight is seriously impaired. Strong international and national support has been found to support a programme for the eradication of onchocerciasis in parts of the seven most severely affected countries of the Volta River basin:

Benin, Ghana, Ivory Coast, Mali, Niger, Togo, and most of Burkino Faso (formerly Upper Volta). The objective will be to obtain a very high level of vector control for a period of up to 20 years to ensure the natural disappearance of the adult worms once transmission has ceased. While a previous, much more restricted programme in the same area, conducted with the assistance of the Fond Européen de Développement and the OCCGE (Organisation de Coordination et de Co-operation pour la lutte contre les grandes Endémies), utilized weekly treatments with DDT by ground applications to all *S. damnosum*-infested watercourses, the Onchocerciasis Control Programme carries out weekly treatments with 20% Abate emulsion concentrate at a target dosage of 0.1 p.p.m. for 10 min to all areas where blackfly vector breeding is taking place. Inasmuch as at least 14 000 km of river systems are within the programme area and a large proportion must be treated each week for a good part of the year it would be virtually impossible to carry out this application by ground treatment; spray applications are therefore made by a number of fixed-wing aircraft and helicopters fitted with specially designed rapid-delivery systems to enable the desired quantity of insecticide to be delivered immediately upstream from the *S. damnosum* larval habitats. The estimated total cost of this 20-year programme, which is one of the largest vector-control operations ever undertaken, will be at least US $120 million.

The length of the programme is calculated on the knowledge that the adult *Onchocerca* worms live for around 15 or 16 years in the human body and vector populations must be suppressed for at least this period of time to ensure that transmission does not begin again.

While temephos remains the insecticide of choice for the time being, nevertheless an active research programme is being carried out to develop alternative compounds that are as effective and ecologically acceptable for use against blackfly populations in those parts of the programme area where resistance to temephos has developed. Such resistance first occurred on the lower Bandama river of the Ivory Coast in early 1980 (Guillet *et al.* 1980) in a population of the forest breeding species pair, *S. soubrense, S. sanctipauli*. In those areas where temephos resistance was detected it was necessary to shift to an alternative insecticide with no cross-resistance to temephos. Chlorphoxim was used during the rainy season, and *Bacillus thuringiensis israeliensis* (serotype H-14) was shown to be effective against blackfly larvae at least under the low water conditions of the rivers during the dry season, i.e. at water flows of up to 50 m^3/s.

Unfortunately by October 1981, resistance to chlorphoxim also appeared at the same site and in the same species pair where temephos resistance had first been detected. This resistance has proved to be far less stable than that to temephos and after a one season suspension of chlorphoxim treatment, susceptibility reverted to normal enabling the compound to again be used until resistance will presumably develop again. More recently a considerably improved formulation of *B.t.* H-14 has been developed which enables use to be made of this bacterial toxin up to river flows of $250 \, m^3/s$ before it is necessary to shift to chlorphoxim in temephos-resistant areas.

While research continues on alternative approaches, such as environmental measures and alteration of water levels, use must continue to be made of chemical larvicides in as selective a manner as possible to ensure the continuation of the programme, and an active screening programme is under way in a search for additional larvicides.

12.7   African trypanosomiasis

Like malaria, African trypanosomiasis is a protozoan disease with parasites affecting both man and animals. The disease in its various forms may be found throughout a vast area of the African continent south of the Sahara covering about 10 000 000 km^2. The disease, both in its human form, known as 'sleeping sickness', and its various animal forms is transmitted by the bite of infected tsetse flies of one or another species of the genus *Glossina; Trypanasoma vivox,* an animal trypanosome, is also transmitted by other biting flies. Human trypanosomiasis in West and Central Africa is of a more chronic type caused by *Trypanosomiasis gambiense* and in East Africa is of a more acute nature caused by *T. rhodesiense*. The two parasites themselves are morphologically indistinguishable and their behaviour is similar in mammalian hosts and tsetse fly vectors in all but the one respect of the greater virulence of *T. rhodesiense*. The *T. gambiense* form of the disease is transmitted by riverine species of tsetse, *G. palpalis* and *G. tachinoides*. These flies may sometimes live close to human habitation and the essential feature in the epidemiology of this form is the close contact between man and fly, there being thus no necessity for a wild-animal reservoir of the infection. Recent studies have indicated that *T. gambiense* may be found in certain reservoir hosts, particularly the pig.

In contrast, *T. rhodesiense* is usually transmitted by game-feeding tsetse flies, *G. morsitans, G. swynnertoni,* and *G. pallidipes,* which feed selectively on certain species of wild animals and on man only occasionally. The wild animals, which are themselves unaffected, are the reservoirs of *T. rhodesiense.* Some animals such as the warthog and bushbuck seem to serve as especially important reservoirs as they are an important part of the tsetse flies' food supply.

The trypanosomes undergo cyclical development in the various species of Glossina. The tsetse fly becomes infective 18 to 34 days after it has ingested an infected blood meal. The female flies are larviparous, producing single fully grown larvae at intervals. These larvae pupate almost immediately and puparia are found in loose soil, moss, under organic debris, etc. but always in the shade. Since the immature stages are so well protected, vector control measures can only be aimed at the adults, and these measures are described below.

To this day, the prevention and control of animal trypanosomiasis and human sleeping sickness presents a formidable challenge to health and veterinarian authorities in Africa. Although there are comparatively few human cases of sleeping sickness reported every year, the disease seems to be only smouldering and could break out again whenever surveillance for new cases is reduced.

While control measures aimed at either control of the tsetse fly vectors or by use of human therapy or chemoprophylaxis have held the disease in check, support for some of the organizations dealing with the disease has declined. Because of the rapidity with which outbreaks can occur and spread it is necessary for governments in endemic areas to maintain a close epidemiological surveillance on the disease to prevent an epidemic return.

Trypanosomiasis of domestic animals, mainly cattle, is more widely distributed than the infection in man. The presence of this disease renders unusable for livestock husbandry large areas of Africa which are otherwise suitable for this purpose. The drought in the Sahel savannah area of West Africa in the early 1970s resulted in the death of many cattle and made the problem of animal trypanosomiasis affecting the remainder even more acute; as people move into new land areas due to the shortage of suitable grazing land they and their herds are likely to be exposed to the risk of infection by trypanosomiasis.

While it has so far been possible to keep the resurgences of the disease under control, presently existing campaigns will have to be expanded to cope with the growth and spread of human populations in Africa and to provide these populations with animal protein. The fact that the parasite resistance can occur to any of the trypanocidal drugs is a matter of growing concern and makes the need for effective vector control programmes even more imperative. Immunization of animals against trypanosomiasis appears only as a remote possibility at present.

12.7.1 Tsetse fly control

As noted above, insecticidal control measures can only be targeted at adult tsetse flies; this is largely the case for other types of control as well. Control of tsetse fly has been achieved by a variety of methods including the large-scale clearing of vegetation in infested areas to make the microclimate unsuitable for the flies, and the destruction of wild game hosts of tsetse. Neither of these methods is any longer economically or ecologically acceptable (Jordan 1978).

Much success has, however, been achieved by the spraying of residual insecticides on vegetation where the flies are likely to rest both against the riverine species and those of the wooded savannahs. It appears to be the only method available for large-scale trypanosomiasis control programmes (Speilberger *et al.* 1977; Challier *et al.* 1978).

Only three insecticides have seen very large-scale use against tsetse, all of them chlorinated hydrocarbons: DDT, dieldrin, and endosulfan. These are applied as emulsion or suspensions. The best manner of applying the insecticide is dependent on a detailed knowledge of the ecology of the target species of tsetse including its flight movements, resting places, and seasonal variations in density.

Many different techniques and formulations have been utilized for applying both DDT and dieldrin and these, as well as the frequency of application, depend upon both climatic conditions and the type of application equipment among other factors. DDT and dieldrin at concentrations of from 1.5 to 5% have been widely used in East, Central, and West Africa, and, when applied by ground equipment, an effort is made to apply them only to those portions of the vegetation where the flies usually rest. Dieldrin, however, is more harmful to the environment though it has been used at very low doses. In Nigeria almost 125 000 km² have been freed from tsetse fly by insecticide application and a further 12 500 km² are being freed every year. In Zambia an area of 1600 km² was treated with five applications of endosulfan (or Thiodan) at ultra-low-volume (ULV) quantities at 3-week intervals at 3 kg of active ingredient/km² by aerial application and *G. m. morsitans* was eradicated from almost all of the treated area (Park *et al.* 1972). At these low dosages endosulfan also appears to be reasonably safe to non-target life.

In the current climate of concern for protection of the environment there is a need for alternative, but equally effective, insecticides for use in tsetse control campaigns (Jordan 1978). Hadaway (1972) evaluated several new compounds in laboratory comparison with the organochlorines on *G. austeni*. The synthetic pyrethroid resmethrin was the most toxic followed by fenthion, dieldrin, propoxur, and chlorfenvinfos. Recent field trials with some of these and an even newer compound, decamethrin, in West Africa have shown that the pyrethroids are highly effective against a number of different species of tsetse flies (Baldry *et al.* 1981).

The use of traps against tsetse flies has been considered for many years but interest in this method waxed and waned until the development of a biocontrol trap by Challier and Laveissière (1973) which has since been extensively used against many species ot tsetse in West and Central Africa. Improved versions called moniconical traps incorporating insecticide (decamerhrin)-impregnated traps have been successfully tested in the Congo (Lancien 1981). A very successful trial of percale screens (90 cm × 115 cm) impregnated with decamethrin at a rate of 100 mg per screen and placed along the forest gallery reduced *G. tachinoides* and *G. palpalis* populations by 98% after 19 days, and by the second month reached a peak reduction of 99.6

and 100% (Laveissière and Couret 1981). Another line of very encouraging development is the use of extracted host odours with traps to increase trap catches for the sampling and control of tsetse (Vale 1982). This technique, whether by traps alone or with insecticide strips and/or attractants, carries considerable promise for village participation in tsetse control programmes. While there has been little environmental damage up to the present resulting from chemical control programmes the use of insecticides combined with traps where feasible would reduce this danger even further.

12.8 References

Baldry, D. A. T., Everts, J., Roman, B., Boon von Ochssee, G. D., and Laveissière, C. (1981). *Trop. Pest. Manag.* **27(1)**, 83.
Bang, Y. H. and Pant, C. P. (1972). *Bull. Wld. Hlth. Org.* **46**, 416.
Brooks, G. D., Neri, P.,˙Gratz, N. G., and Weathers, P. B. (1970). *Bull. Wld. Hlth. Org.* **42**, 37.
CDC (1979), *MMWR* **28(27)**, 323.
Challier, A. and Laveissière, C. (1973). *Cah. O.R.S.T.O.M. Sér. Entomol. Méd. Parasitol.* **2**, 251.
—— Eyraud, M., and Laveissière, C. (1978). *Cah. O.R.S.T.O.M. Sér. Entomol. Méd. Parasitol.* **26(1)**, 5.
Chamberlain, R. W. (1972). *PAHO Sci. Pub. No. 243,* 390.
Chamberlain, W. F. (1975). *J. Med. Entomol.* **12(4)**, 395.
Choyce, D. P. (1972). *Isr. J. Med. Sci.* **8(8/9)**, 1143.
Eliason, D. A., Kilpatrick, J. W., and Babitt, M. F. (1970). *Mosq. News* **30(3)**, 430.
Ewert, A., Corredor, A., Lightner, L., and D'Alessandro, A. (1979). *Am. J. Trop. Med. Hyg.* **28(3)**, 486.
Fleming, G. A., Barodji, Shaw, R. F., Pradhan, G. D., and Bang, Y. H. (1983). *A village-scale trial of bendiocarb (OMS-1394) for control of the malaria vector Anopheles aconitis in Central Java, Indonesia.* WHO unpublished document WHO/VBC/83.875.
Fontaine, R. E., Joshi, G. P., and Pradhan, G. D. (1975). *Entomological evaluation of fenitrothion (OMS-43) as a residual spray for the control of An. gambiae and An. funestus in Kenya.* WHO unpublished document WHO/VBC/75.547.
—— Pull, J., Payne, D., Pradhan, G. D., Joshi, G. P., and Pearson, J. A. (1976). *Evaluation of fenitrothion for malaria control in a large-scale epidemiological trial, Kisumu, Kenya.* WHO unpublished document, WHO/VBC/76.645.
Gardner, A. L. and Iverson, R. E. (1968). *Arch. Environ. Hlth.* **16**, 823.
Gratz, N. G. (1973). *Crit. Rev. Environ. Control* **3(4)**, 455.
—— (1976). *Chem. Pflanzenschutz Schädlingsbekämpfungsmittel, Band* **3**, 85.
Guillet, P., Escaffre, H., Ouedraogo, M., Quillévéré, D. (1980). *Cah. O.R.S.T.O.M. Sér. Entomol. Méd. Parasitol.* **18(3)**, 291.
Hadaway, A. B. (1972). *Bull. Wld. Hlth. Org.* **46**, 353.
Jordan, A. M. (1978). *Nature (London)* **273**, 607.
Joshi, G. P., Self, L. S., Shaw, R. F., and Supalin (1977). *A village-scale trial of fenitrothion (OMS-43) for the control of Anopheles aconitus in the Semarang area of Central Java, Indonesia.* WHO unpublished document WHO/VBC/77.675.
Lainson, R. and Shaw, J. J. (1978). *Nature (London).* **273**, 595.
Lancien, J. (1981). *Cah. O.R.S.T.O.M. Sér. Entomol. Méd. Parasitol.* **19(4)**, 235.

Laveissière, C. and Couret, D. (1981). *Cah. O.R.S.T.O.M.* **19(4),** 271.

Lofgren, C. S., Ford, H. R., Tonn, R. J., and Jatanasen, S. (1970). *Bull. Wld. Hlth. Org.* **42,** 15.

Mathis, H. L., Prapaipuk Chattraphuti, and Poonyos Reilrengboonya (1974). *Persistence of two new insecticides compared with Abate in water jars in Bangkok, Thailand.* WHO unpublished document, WHO/VBC/70.235.

McIntosh, B. M. (1972). *US Afr. Vet. Assoc.* **43(4),** 391.

Moraes, M. A. P., Fraiha, H., and Chaves, G. M. (1973). *Bull. Pan Am. Hlth. Org.* **7(4).** 50.

Pant, C. P. and Yasuno, M. (1970). *Indoor resting sites of Aedes aegypti in Bangkok, Thailand.* WHO unpublished document, WHO/VBC/70.235.

—— Nelson, M. J., and Mathis, H. L. (1973). *Bull. Wld, Hlth. Org.* **48,** 495.

Park, P. O., Gledhill, J. A., Alsop, N., and Lee, C. W. (1972). *Bull. Entomol. Res.* **61,** 373.

Rassi, E. B., Lacerda, N., and Guaimaraes, J. A. (1976). *Bull. Pan Am. Hlth. Org.* **10(1),** 33.

—— Monzon, H., Castillo, M., Hernandez, I., Pérez, J. R., and Conuit, J. (1977). *Bull. Pan Am. Hlth. Org.* **11(1),** 41.

Rishikesh, N., Mathis, H. L., Remasamy, M., and King, J. S. (1977). *A field trial of chlorphoxim for the control of Anopheles gambiae and Anopheles funestus in Nigeria.* WHO unpublished document, WHO/VBC/77.651.

—— —— King, J. S., and Nambiar, R. V. (1978). *A field trial of pirimiphos-methyl for the control of Anopheles gambiae and Anopheles funestus in a village-scale trial in Nigeria.* WHO unpublished document, WHO/VBC/78.689.

Self, L. S. and Tun, M. M. (1970). *Bull. Wld. Hlth. Org.* **43,** 841.

—— Nelson, M. J., Pant, C. P., and Salim Usman (1978). *Mosq. News* **38(1),** 74.

—— —— Theos, B., and Wiseso, G. (1977). *J. Med. Assoc. Thailand,* **60,** 482.

—— Ree, H. I., Lofgren, C. S., Shim, J. C., Chow, C. Y., Shin, H. K., and Kim, K. H. (1973). *Bull. Wld. Hlth. Org.* **49,** 353.

Speilberger, V., Na'isa, B. K., and Abdurrahim, U. (1977). *Bull. Entomol. Res.* **67,** 589.

Vale, G. A. (1982). *Bull. Entomol. Res.* **72,** 93.

WHO (1966). *Tech. Rep. Ser. No. 335.*

—— (1970). *Tech. Rep. Ser. No.* 443.

—— (1974a). *Tech. Rep. Ser. No.* 549.

—— (1974b). *Tech. Rep. Ser. No.* 542.

—— (1976). *Tech. Rep. Ser. No.* 585.

—— (1978). *Wkly. Epidem. Rec.* **53,** 197.

—— (1979a). *Wkly. Epidem. Rec.* **38,** 293.

—— (1979b). *Wkly. Epidem. Rec.* **54,** 105.

—— (1980). *Tech. Rep. Ser. No.* 655.

13
Control of ectoparasitic arthropods

J. R. BUSVINE

13.1 The importance of ectoparasites

13.1.1 Harmful effects of bites

The main annoyance of ectoparasitic arthropods is the irritation caused by their bites, which is due to an allergic response to the saliva they inject. The severity of the reaction varies from one individual to another and also at different times in the same host. The irritation may cause loss of sleep, and the scratching induced often leads to secondary infections. In domesticated animals, restlessness leads to poor condition and, in cattle, loss of milk yield and reduced weight.

Rather more serious is the toxicosis which develops from the bites of some ticks. An unknown poison which they inject causes paralysis, which can prove fatal, though fortunately it abates if the ticks are removed.

It has been suggested that the regular loss of blood could cause anaemia. This might be possible with large ectoparasites such as ticks or triatomid bugs, especially with small animals or heavy infestations. It would seem unlikely, however, that this could happen with very small ectoparasites like fleas or lice, since even a thousand such would remove only about a millilitre of blood per day.

13.1.2 Disease transmission

Arthropod vectors are responsible for spreading many serious tropical diseases (see Chapter 12). Ectoparasites do not have the advantage of extensive mobility of the free-flying diptera discussed in that chapter, but they are more closely associated with the host for longer periods and the host is thus more certain to become infected. In this respect, they show various grades of involvement with the host. Lice and mange mites (though the latter are not vectors) live permanently on the skin. Then there are the fleas, bugs, and soft ticks, which infest the dwelling or the sleeping place of the host. The hard ticks and the harvest mites alternate between close attachment and living in open country, where contact with a new host is problematical. In each case, the life cycle must be studied carefully to decide which stage is most vulnerable to control measures.

13.2   Control strategy

The following points are important.

(a) Vector control relates more to public health than to curative medicine, but nevertheless, a general attack on the pathogen by drugs or immunization must be considered as an alternative (or complementary) to it.

(b) Vectors which transmit strictly human diseases must be distinguished from those which bring infections from animal reservoirs. Typhus and louse-borne relapsing fever are examples of the first; plague, tick-borne arboviruses, and Chagas' disease, of the latter. While total eradication of the human diseases may be feasible, the diseases with animal reservoirs present virtually insuperable difficulties in this respect.

(c) In recent decades, the most successful vector control has been through the use of pesticides. But these are now circumscribed by the drawbacks of pest resistance and problems with toxicity and environmental residues, which have led to stringent regulations in their usage. In both respects, ectoparasites tend to present slightly different problems from free-flying dipterous vectors. In regard to toxicity, those parasitic arthropods which live in close contact with the host can only be attacked by chemicals of low mammalian toxicity. This is less important for the control of pests harboured in the structure of dwellings, or those breeding in open country. On the other hand, the risk of environmental pollution is substantially less for the closely associated ectoparasites.

A further important point is that ectoparasites infesting human bodies or clothing, and those living in the walls of dwellings, tend to disappear with improvement in standards of hygiene. This ideal solution cannot well apply to pests whose breeding grounds are remote from human habitations.

13.3   Tactical problems with human ectoparasites

The various kinds of pest will be dealt with roughly in order of the closeness of their parasitic association with Man. On these grounds, the itch mite will be considered first, though it is by no means the most important.

13.3.1   Sarcoptes scabei and Scabies

The human itch mite belongs to the family Sarcoptidae and is related to the mange mites which attack domestic animals and birds (see p.   ). The female adults of the human parasite tunnel in the horny layer of the skin and lay their eggs along their burrows. The young hatch as six-legged larvae, which later moult to nymphs and finally to adults. The young stages also live and feed on the skin, sheltering apprently in hair follicles; the whole life cycle takes about 3 weeks.

The presence of mites in small numbers, in the early stage of infestation, causes no symptoms. Later, the infested person develops an allergy, which causes intense irritation, especially in bed at night. These effects constitute the disease known as scabies. One result of the scratching response to the irritation is that many mites are destroyed, so that at this stage new infestations are hard to establish, which constitutes a kind of semi-immunity.

Infections generally start by close contact with an infected person, usually by sleeping in the same bed. The disease is often considered a venereal complaint, but perhaps it is more commonly a family infection, transmitted from parents to children or between sibs. In the first half of this century, it was noticed that scabies became more prevalent in the course of the two World Wars, for unknown reasons. In 1941, on the basis of hospital admissions, it was estimated that the general rate in Britain was nearly 2% (Mellanby 1941). Later, examinations of thousands of children showed a substantial decline after the World War II (Busvine 1980). However, in more recent years, there has been a disturbing rise in Britain, where reported cases among scholchildren rose from about 4000 p.a. in 1963–5 to over 14 000 in 1970 (Anon. 1974). A similar trend was seen in Czechoslovakia (Palicka and Merka 1971) and subsequently in Denmark, Norway, and parts of USA (Orkin *et al.* 1977), also in New Zealand (Tonkin and Wynne-Jones 1979) and Turkey (Tüzün *et al.* 1980). It has been suggested that regular fluctuations in incidence occur because of the semi-immunity caused by infection, so that the disease subsides until a new generation of susceptibles arises. Rather more probable, perhaps, is the recent spread of sexual laxity and contempt for hygiene among some young people, reflected in the simultaneous rise in venereal disease in several countries.

Studies of the itch mite and scabies have been undertaken in response to public concern in periods of prevalence. Substantial advances were made during the World War II by Mellanby and colleagues (Mellanby 1943, 1944); and the recent resurgence stimulated experts to hold a symposium in minneapolis (Orkin *et al.* 1977).

An important consideration in dealing with scabies is sound and prompt diagnosis. The secondary skin lesions caused are not very easily distinguished from other diseases, though the characteristic distribution of the rash may give a clue. Certain diagnosis follows the discovery of a female mite, which can be pricked out of her burrow, commonly from the hand or wrist. Curative treatment is generally preceded by a bath, though this is not essential. A suitable medicament is then applied to the whole surface of the body, below the head. The acaricide may be in the form of a lotion, an emulsion, or an ointment. It can be applied by hand, or in the case of less viscous preparations, by paint brush. It is wise to treat all members of a family, in case of inapparent infections. About 30 g of medicament is used for an adult. Of the compounds which have been used to cure scabies, the oldest is sulphur (noted by van Helmont, 1577–1644; Busvine 1976). It is effective as

10% sulphur in ointment; but it is messy and liable to cause dermatitis if used repeatedly. Benzyl benzoate as a thick (30%) emulsion was recommended by Mellanby and his co-workers during World War II. It is effective, but causes temporary smarting. The most widely used remedy at present is 1% lindane (gamma HCH), but though apparently safe, it cannot be used in some countries because it is banned as a chlorinated compound. Indeed, since it is applied all over the body, its use on children or infants should be done under medical supervision. A few other compounds have been recommended, e.g. crotamiton, tetraethyl monosulphide, diethyl thianthrene, and, most recently, thiabendazole.

A single treatment of an effective acaricide should cure over 95% of cases and a second one should clear up the remainder.

13.3.2   Human lice

Nearly all the ectoparasites to be considered in this chapter attack Man only incidentally, and many prefer to feed on other animals. Human lice, however, like the itch mite, are closely adapted to Man and live permanently on the human body; they do not naturally infest any other animal. There are three kinds of these lice; head lice, body lice, and pubic lice, infesting respectively, the scalp hair, the underclothes next to the body and the body hair of the pubic region. The head louse and the body louse are closely related; but recent research seems to confirm the opinions of those who blieve that they are separate species: *Pediculus humanus L.* and *P. capitis* DeGeer. (Busvine 1978).

The biology of all three species is roughly similar. All depend on frequent meals of human blood, perhaps several times a day lice starve to death fairly soon in the absence of blood meals – pubic lice in a day or so, body lice after about a week, and head lice are intermediate. Eggs are laid on human hair or on fibres of clothing. There are three nymphal stages, largely resembling the adults in form and habits. The life cycle takes 2–3 weeks and the adults live about the same time.

Although human lice are biologically rather similar, they present distinctly different problems in public health, so that their control should be discussed separately. However, they have one common factor; insecticides used against them cannot be relied upon indefinitely, because resistance is developing in two of them towards many compounds safe enough to be used near the human body. Since the prevalence of lice depends on poor personal hygiene in all cases, the ultimate solution is to raise living standards and levels of education to eliminate this.

Body lice. Body lice are considered first because they are the vectors of two dangerous diseases, typhus and relapsing fever. In the past, these (especially typhus) have been responsible for appalling epidemics with many deaths; for example, 30 million cases of typhus in Eastern Europe after World War I (with about 3 million deaths). Epidemics of these diseases, however, can only occur when body lice are very common. Regular washing of the underwear has largely eliminated body lice from civilized communities, though they persist among a small proportion of vagrants etc. At present, typhus and relapsing fever are restricted to parts of the world with low standards of hygiene, notably Central Africa and South America. There is now a global total of only 10 000–20 000 cases a year with a few hundred deaths.

Treatment of these diseases has been greatly improved by the introduction of antibiotics: but nevertheless, when epidemics occur, a rapid reduction in general lousiness is essential to stop the spread of disease. This is achieved by applying powder insecticide to the underwear of as many people as possible. If the lice are not resistant to it, a dust containing 10% DDT is best. But DDT has been widely used as a general measure of louse control as well as in epidemics and, as a result, lice have become resistant to it in many areas. Methods of testing for resistance have been developed by WHO, which can provide test kits. Where lice are resistant to DDT, a dust containing 1% malathion may be used; but in a few places malathion resistance has been noted. Alternatives are 1% gamma BHC or permethrin. The insecticide dust is either applied by a sifter-type tin, or blown up sleeves and other openings in the clothing with a dusting gun. About 50 g is required to treat an adult. The dust applied in such treatments persists for a week or two in the underwear, since people chronically lousy are unlikely to change their clothing. This gives protection from reinfestation.

Under modern sanitary conditions, the usual problem of body louse control is to deal with the relatively small numbers of lousy people entering public assistance hostels, charitable institutions, or prison. In such situations, heavy dusting of the clothing is unsuitable, since the obvious signs are resented; furthermore, the lethal action may be slow, so that intoxicated lice may be seen crawling over treated people before dying. Since there is no danger of reinfestation, a rapid treatment giving no protection is adequate; for example, fumigation or heat treatment.

The easiest method is to fumigate the clothing with ethyl formate, in a metal bin or plastic bag (Busvine and Vasuvat 1966). A dose of 2 oz/ft² (2 ml per litre) is used for a short (1 hour) exposure, or quarter this dose for a 5 hour exposure. An alternative is heat treatment by hot air; but as this is a poor conductor with little specific heat, an exposure of 1 hour to 70°C is necessary. The large steam disinfestors used in hospitals to disinfect bedding can be used; but the very high temperatures required to kill bacteria are not necessary and the whole procedure is rather cumbrous for this purpose.

(b) Head lice. Although laboratory tests have shown that head lice are

capable of transmitting the rickettsiae of typhus, epidemics of the disease have never occurred in the absence of plentiful body lice. However, though head lice do not pose a threat of epidemic disease, they are nevertheless unpleasant pests, mainly affecting children. Apart from the disgusting nature of the parasite, the continual irritation it causes results in head scratching, often leading to impetigo. Even in advanced countries, head lice are disturbingly prevalent. Rather like the scabies mite, they were found to be unexpectedly common in Britain during World War II, but subsequently declined in prevalence. About 1970, however, there was a serious rise in incidence, with nearly half a million children affected (Maunder 1971); but a later estimate suggested a fall to about half that figure (Donaldson 1976). Similar evidence of a rise in head louse incidence in the early 1970s has been recorded in Denmark (Hallas *et al.* 1977), Canada (Hopper 1971), France (Lamizana and Mouchet 1976), West and East Germany (Weyer 9978; Dittmann and Eichler 1978), and parts of the USA (Orkin *et al.* 1977).

Insecticides for the treatment of head lice must be acceptable to the patients (and, since most are children, to their parents). Otherwise, there will efforts to avoid treatment and substances applied under protest will be washed off as soon as possible. In contrast, an effective application which is odourless and invisible, will be accepted and even used by other members of the family. Thus aqueous preparations, or alcohol-based lotions are generally most suitable. For many years, an emulsifiable concentrate of gamma HCH has been widely effective; it is diluted in water, to 0.2% active principle, before use. Other preparations containing DDT have also been used. Unfortunately, strains of lice resistant to these compounds have appeared in several countries (Canada, Denmark, England, France, Hungary, the Netherlands, South Africa, and the USA (WHO 1976). In such cases, a lotion containing 0.5% malathion has been found effective and another with 1% carbaryl. These lotions are for direct application to the hair; but there are also medicated shampoos. It seems rather likely that these would be less effective because of the short period of exposure of the lice; and there would be less residue to kill lice hatching from any nits which survived. With an effective treatment, a single application should cure about 95% of cases, and a further one is made about a week after an inspection for possible failures. Direct applications should be left on the hair to provide a residue to kill surviving nymphs, so that removing the nits may not be essential. However, they are unsightly and may confuse the nurse making a re-examination, as it is difficult to distinguish live from dead eggs. Therefore, nits should be removed after treatment, using a fine-toothed metal comb.

(c) Pubic lice, or crab lice. Pubic lice do not present a danger as disease vectors, but they cause irritation. Statistics on their prevalence are expecially hard to obtain, but observations at a venereal disease clinic in England

suggest that they may be on the increase. Among some 10 000 patients, the percentage infested rose from 0.6% in 1954 to 3.2% in 1966. The highest incidence was in girls aged 15 to 19 and in men over 20 (Fisher and Morton 1970). In Denmark, the Government Pest Control Laboratory received only one or two enquiries on these lice in 1968–69, but 10 in 2973 and 15 in 1974 (Hallas *et al.* 1977).

The pubic louse invades a new host during close contacts between people, usually in bed. Because of its location, it is likely that they are often spread during sexual intercourse. But sometimes they may be spread by fomites (e.g. stray lice on shared towels) since infants are sometimes infested, on the eyelashes. The eyelashes are suitable for the parasite because they are thick and widely spaced, like pubic hairs.

Treatment for pubic lice formerly involved shaving of the infested parts; but this is no longer necessary if powerful modern insecticides are used. Dusts containing 10% DDT, or 1% gamma HCH would be effective; alternatively, the liquid preparations used against head lice can be employed. Treatments should be repeated after a week to kill any lice which may have hatched from surviving nits. So far, there have been no reports or resistance in pubic lice.

13.3.3   Fleas

Fleas differ from lice in having a non-parasitic larval stage living in debris near the sleeping place of the host. The adults themselves spend only part of their time on the host's body, in search of a blood meal, and they may also be found in the breeding site. In addition, they vary in restriction to a particular host, but are generally much less specific than lice. Neither the fleas which act as disease vectors, nor those which are merely annoying, are strictly human parasites.

(a) Fleas and plague. Plague is primarily a disease of wild rodents, among which it persists endemically, sometimes flaring up into an epizootic which kills many of them. It is transmitted among them by their fleas; but since these are very loath to bite Man, the wild rodent disease is not directly dangerous. Plague epidemics begin with the interchange of fleas between wild rodents and semi-domestic rats on the periphery of towns. The disease then spreads among the urban rodents, to which it is highly lethal. One of the two common urban rats, the black rat, is liable to invade wooden dwellings, though it rarely manages to infest in modern buildings. This brings it into close contact with Man in primitive towns as it did in mediaeval European cities. The last step in a plague epidemic is the passage to Man by a flea which will bite both rats and humans. Pre-eminent as vectors are species of the genus *Xenopsylla,* especially *X. cheopis,* the tropical rat flea. The rat flea of temperate regions, *Nosopsyllus fasciatus,* is very reluctant to bite Man, while

the so-called human flea, *Pulex irritans,* rarely bites rats (though it readily infests pigs). In human cases of plague, the bacillus responsible, *Yersinia pestis,* becomes concentrated in the buboes, or plague swellings in the groin and armpit. Further transfer by fleas is exceptionally rare, though sometimes a virulent pneumonic form develops, which can be spread by droplet infection.

Plague epidemics in history are legendary, from the 6th century Justinian plague and the 13th century Black Death to the Great Plague of London in 1665. With modern methods of treatment and control, however, the disease is now limited to some parts of the tropics. In the past decade, the numbers of cases reported annually to WHO range from 1000 to 6000, with 100 to 200 deaths. Where epidemics still have to be quelled, both the black rats and their fleas must be controlled. The fleas are attacked first, as the traps and poison baits take longer to eliminate the rats. Meanwhile, the fleas are reduced by insecticidal dusts, which are blown into possible harbourages and scattered along rat runways, thus contaminating the rats and hence their fleas. DDT dust was once the main weapon, but the plague fleas have become resistant to it in many parts of the World. Alternative insecticides must be used; but there are signs of resistance to gamma HCH in many places and to some organophosphorous compounds in South Vietnam. The new pyrethroids (resmethrin, permethrin) are other possibilities.

(b) Fleas as nuisances. Annoyance from flea bites in civilized towns and cities in the temperature region is now nearly always due to cat or dog fleas, rather than to *Pulex irritans,* which is becoming rare in such localities (Cornwell 1974). In Denmark, 90% of 2294 fleas from houses were identified as cat or dog fleas and only 1% as *P. irritans.* Moreover, the prevalence of pet fleas in that country is growing, as shown by the rise of complaints from about 10 up to 1000 a year (Anon. 1977). The reason for these changes is probably that improved housekeeping (e.g. vacuum cleaners) destroys the breeding sites of *P. irritans,* while the pet's basket or kennel is often undisturbed.

To eradicate pet fleas, a powder insecticide should be sifted into the fur of the animal. A non-toxic compound is essential, especially for cats, which tend to lick themselves clean. Suitable substances are the natural pyrethrins (1%, or 0.2% if synergized). There are also suitable shampoos, which may be obtained at a pet shop; and recently, a collar containing a volatile organophosphorous compound has been introduced for animals which frequently become infested. Whichever treatment is used, the pet's sleeping place should be thoroughly cleaned and, if necessary, treated with insecticide (e.g. by an aerosol spray). Similar attention should be given to upholstered furniture frequented by the pet.

13.3.4 *Blood-sucking bugs.*

Two kinds of blood-sucking bugs are troublesome to man; the bed bug (*Cimex* spp) and the American cone-nose bugs (Reduviidae, Triatominae) which are vectors of a serious trypanosomal infection, Chagas' disease. The latter will be considered first.

(a) Cone-nose bugs. The Triatominae are rather large bugs, with adults about 25 mm long. Though the adults have functional wings, they do not fly a great deal. Colonies of all stages of the bugs infest nests of birds and burrows of wild animals, hiding in crevices by day and occasionally emerging to seek a blood meal at night. The general appearance and habits of the five nymphal stages resemble those of the adults. Development is slow, taking 6 months to a year or more; and all stages are capable of withstanding long periods (months) of starvation.

There are some species which also (or mainly) infest human dwellings, especially the badly constructed huts in rural areas of South America, full of crevices in which they can hide. Although cone-nose bugs occur as far north as the southern USA, disease transmission is limited to Central and South America. In any given area, only one or two species of bug are important as vectors. They transmit the pathogen (*Trypanosoma cruzi*) from one infected person to another; but, more important, there is a reservoir of infection in wild mammals (though not birds). The disease has an acute phase, lethal to a proportion of children; but its main danger is that it seriously weakens the heart, so that many chronic sufferers die of heart failure in middle life. No satisfactory curative or prophylactic drug is known, so that disease reduction depends on vector control. This is generally achieved by spraying infested dwellings once or twice a year with a residual insecticide. The spraying techniques are similar to those used against mosquitoes (i.e. wettable powders, dispersed in water and applied with a compression or other sprayer; see Chapter 12). For some years, gamma HCH has been found effective and inexpensive: but now some vector species are becoming resistant to it. Various organophosphorous and carbamate compounds may be used as alternatives, though they are more expensive. In the long run, the best solution is improved housing, free from the numerous harbourages in whcih the bugs shelter, and are accordingly easier to keep uninfested.

(b) Bed bugs. The common bed bug, *Cimex lectularius L.,* is cosmopolitan, while *C. hemipterus F.* is restricted to tropical countries; both present very similar problems. Though not incriminated as disease vectors, their bites are unpleasant and irritating and they emit a disagreeable smell. In urban areas, they are characteristic of low-grade ('slum') housing. In general, their habits resemble those of cone-nose bugs, hiding in crevices by day and seeking blood meals at night. Again, the five nymphal stages resemble the adults in

appearance and habits; even more so, since the adults are wingless. Since they cannot fly, invasion of new premises occurs mainly by movement of infested luggage or furniture. Though rarely or never found wild, bed bugs will readily feed on other animals or on birds and are sometimes troublesome in zoos and animal houses.

Control of bugs presented severe difficulties until the introduction of residual insecticides, especially DDT, which greatly reduced their incidence in temperate countries. But unfortunately they have become resistant to organochlorine insecticides almost everywhere and to organophosphorous compounds in a number of places. The pyrethroids are still effective, but this will probably not last indefinitely. Room spraying with these residual insecticides is still the best method of controlling heavy infestations, where resistance has not developed. Apart from this, the only measure is improved domestic hygiene. The problem is much more difficult in hot countries, where bugs breed more rapidly.

13.3.5 Ticks

Ticks and mites belong to the order Acari, the former being distinctive because of their relatively large size and leathery cuticles. Nearly all are blood-sucking parasites of mammals or birds. The body is rounded and, instead of a true head, there is a 'capitulum' composed of the rather prominent mouthparts. There are two distinct groups, differing in form, biology, and importance: the soft ticks (*Argasidae*) and the hard ticks (*Ixodidae*). Hard ticks have a 'scutum' or shield-like plate on the body. This covers the entire back of the male, but only the anterior part of that of the female, the remainder being flexible and greatly extended when the tick feeds. The scutum (often decorated with a pattern) and the rest of the body are relatively smooth and the capitulum is clearly visible from above. Soft ticks have no scutum and the sexes are rather similar in appearance. The cuticle is wrinkled and granulated and the capitulum is situated underneath the front of the body and usually not visible from above.

(a) Agasidae, or soft ticks. These are considered first, though less numerous and important, because they are more 'domestic' in that they tend to infest the habitat of the host. Like blood-sucking bugs, they hide in crevices and emerge occasionally to feed, relatively briefly (except some larvae) before returning to hide. They are very resistant to long periods of starvation and also to arid conditions.

Various species are found in the lairs or nests of wild animals or birds; and people camping near them may be bitten. One form, a strain or species of *Ornithodoros moubata,* is exceptional in being adapted to living in primitive human dwellings in Africa, usually hiding in crevices in the floor.

The bites of soft ticks are unpleasant, but not dangerous in themselves;

but in some places they can act as vectors of a spirochaete disease, relapsing fever. This can be transmitted from various animals to Man, in sporadic, isolated cases. Also, the hut-infesting form of *O. moubata* may transmit it among the human inhabitants.

Control of soft ticks affecting Man essentially relates to the hut-infesting form. It is naturally tolerant of most acaricides, though *gamma* HCH is fairly effective. The best control measure consists in improving living conditions; even raising the sleeping places from the floor protects people from most bites.

(b) Ixodidae, or hard ticks. Most hard ticks infest open country, especially grasslands or scrub frequented by their hosts. The first stage (a six-legged larva) tends to feed on smaller animals, while the later stages (nymphs and adults) prefer larger ones. All three stages may occasionally feed on people, though no species is closely associated with Man.

The females lay enormous batches of eggs on the ground and the young larvae climb up herbage, attempting to catch hold of a passing animal; some ticks pass the rest of their lives on this animal. In others, the larvae and nymphs feed on the same animal, but the nymphs drop to the ground to moult and the adults seek another host. There are thus one-host and two-host ticks; but most species are three-host ticks, feeding on different animals as larvae, nymphs, and adults.

Hard ticks feed slowly, for many days, and the females in particular take huge blood meals, perhaps 20 to 150 times their original weight. They can then withstand periods of starvation in the absence of a host, but they vary in tolerance of low humidity. The larvae of some species rely on moisture at the grass roots to avoid desiccation.

Human reactions to bites of hard ticks are augmented by the appearance of the creatures embedded in the skin. Care must be exercised in making them detach; pyrethroids, or even mineral oil, may make them loosen their hold. In some cases, the bites may cause a paralysis, which can be dangerous, but this ceases if the ticks are removed.

Hard ticks are vectors of two kinds of dangerous disease, both of them transferred from animal reservoirs. Tick typhus, due to *Rickettsia* sp., occurs in various parts of the World: Rocky Mountain spotted fever in the western USA; *fièvre boutonneuse* in the Mediterranean region; African and Siberian tick typhus. Various species of tick are involved, of the genera *Dermacentor, Rhipicephalus,* and *Haemaphysalis.* The other tick-borne human disease consists of various arboviruses, which cause haemorrhagic fever or encephalitis. Most of the important ones occur in the Old World, especially the USSR and Central Europe, spread by species of *Ixodes* and *Dermacentor.* Colorado tick fever is an American example.

Because the reservoirs are in wild animals and the ticks are associated with them, human infections occur only sporadically, among hunters, campers, and rural workers. Generally, radical control measures are not feasible, though some Russians have successfully used DDT to destroy *Ixodes* spp. in frequented sites. Apart from this, people at risk can use repellants on their clothing.

The rickettsial diseases have become less dangerous as a result of the introduction of antibiotics, but the viral infections are still rather serious.

13.3.6　*Miscellaneous mites harmful to Man*

(a)　Trombicula spp. Mites of the genus *Trombicula* are only parasitic in the larval stage, the nymphs and adults being free-living and predatory. The larvae, like larval ticks, climb up herbage and catch hold of a passing animal or bird and travel over the body until they find a suitable place to burrow in. Small wild animals are often attacked in the ears, or other sparsely haired parts of the body. People who lie down in infested grassland or scrub may also be attacked, usually round constrictions of the clothing (waistband or garter etc). The larvae remain embedded for a few days then drop to the ground to continue development.

These mites are known as harvest mites in Britain and chiggers in America. Their bites cause intense irritation; and, in South-East Asia, they may transmit a potentially dangerous disease, scrub typhus, from a reservoir in the wild animals. Cases occur only sporadically, as with those spread by hard ticks; and again radical control measures are not generally feasible, though people at risk can use repellants. The infection responds to treatment with antibiotics.

(b)　Blood-sucking mites. One or two mites have the habit, like certain bugs, of hiding in crevices by day and taking blood meals by night. The red mite of poultry is one example which will bite Man. The rat mite, *Ornithonyssus bacoti,* is another similar pest, which disappears with adequate rodent control.

13.4　Control problems with animal ectoparasites

A considerable range of parasitic arthropoda attacks domestic animals (Lapage 1962). As with the parasites of Man, they will be considered in order of their closeness of association with the host, rather than their actual importance.

Treatment methods for the larger domesticated animals (cattle and sheep) differ less than those for humans and the relevant dipping and spraying techniques will be considered at the end of this section. (p. 318).

13.4.1　*Mange mites*

(a)　General remarks. Various species of mite live parasitically in the skin or

in hair or feather follicles of farm or domestic animals. Some are largely restricted to particular parts of the body (head, ears, lower legs etc.) while others are variously disposed about the trunk. The location of the infected area is important in relation to control.

All these mites are strictly parasitic; but some feed on epidermal cells and can be cultured on detached skin, while others are largely dependent on blood serum. Though somewhat more tolerant of starvation than the human itch mite, they normally die in about a week if removed from the host. Infection generally starts by contact with an infected individual, but sometimes fomites may be involved (e.g. curry combs or blankets in the case of horses).

Although the signs and symptoms vary, all these mites cause irritation which induces restlessness and attempts to rub or scratch the infected area. Skin sores and scabs develop and patches of hair, wool, or feathers may be lost.

A characteristic of several of these mites is that they exist as different forms, which cannot be distinguished morphologically, but appear to be physiologically adapted to different hosts. Cross-infection to a strange host (including Man) may occur, but such infections generally die out. The exact status of these forms (varieties, subspecies, species) is not always clear; they are commonly named after the particular host, as *equi, canis, bovis, ovis,* etc.

(b) Important genera and species. One of the most curious of these mites is *Demodex (canis, ovis, bovis,* etc). *D. folliculorum* commonly occurs on Man, especially in the eyebrows, but without apparent harm. In animals, the various forms of *Demodex* cause thickening of the skin, loss of hair, and irritation. Demodetic mange is particularly harmful to dogs.

There are other mange mites so closely related to the human itch mite that they are assigned to the same species, though named as varieties of *Sarcoptes scabiei.* In animals, these mites prefer parts of the body without dense hair, such as the face and ears of sheep, the head and neck of equines and the sacral region and neck of cattle. They cause bare patches of skin, with wrinkling and dry crusts, and the animal suffers considerable irritation.

In the same family is *Psoroptes,* with forms attacking cattle and horses; but probably the most serious is *P. ovis,* which is the cause of sheep scab. The infestation occurs on all parts of the body covered with hair, with scabbing and production of a serous exudate, which mats the wool together. Sheep scab is one of the oldest known ectoparasitic diseases of sheep and it is still a serious nuisance. Vigorous measures have managed to eradicate it from Australia and New Zealand, but it continues in other parts of the World (Page 1969).

A related mange mite, *Chorioptes bovis,* has strains adapted to various animals. In horses, it attacks the fetlocks, causing foot mange or itchy leg. In cattle, it occurs mainly in the perianal region and in the pasterns, particularly of the hind legs, causing papules and later scabs.

Other types of skin mites include the ear mites (*Otodectes* spp.) attacking the ears of carnivorous animals, including dogs and cats. In another family is *Psorergates ovis,* the sheep itch mite, found on fine-wooled merino sheep in Australia, South Africa, and the USA. Its effects, however, are less drastic than the sheep scab mite.

Various ectoparasitic mites attack birds, including poultry; for example, *Cnemidocoptes,* which is related to *Sarcoptes. C. gallinae* burrows into the skin and along feather shafts of fowls, causing depluming itch. A similar disease, however, may be caused by quite a different mite *Megninia cubitalis.*

Cnemidocoptes mutans is responsible for scaley leg of fowls and turkeys. These mites pierce the skin of the lower legs, causing inflammation and an exudate, which hardens and distorts the scales.

(c) Control of mange mites.

(i) Diagnosis. The first step is sound diagnosis. Other skin conditions, such as dandruff, fungal infections, and louse infestations, must be excluded. Clues may be obtained from the pathogenic signs and the known parasites of the animal in question; then, if possible, the actual mites should be collected and identified.

(ii) Localized treatments. Mites which congregate in specific areas may be controlled by inunctions or partial immersion. Examples are demodetic mange in dogs, depluming itch and scaley leg in fowls, foot mange in horses, and ear mange in cats and dogs. First, the hair on the afflicted part should be clipped, the skin cleaned and the scabs softened with soapy water. With sarcoptic mange, if the lesions are extensive, or the animals in poor condition, portions of the body may be treated in turn.

Infestations of *Psoroptes* or *Psorergates* in sheep may be difficult to reach, so that thorough treatment involves dipping or showering (see p. 319). This may, however, be combined with local hand dressing of bad lesions.

(iii) Chemicals used. A considerable number of chemicals has been used at different times against parasitic mites, some of them rather toxic. Of the older materials, sulphur is perhaps the most reliable, applied as a greasy ointment for local applications. Among modern synthetic compounds, lindane (gamma BHC) is outstanding. It may be used at 0.02% in a dip, or up to 0.15% in a local application. Gamma BHC has been widely used against sheep scab and its effectiveness was largely responsible for its eradication from Britain in 1952, although the disease was reintroduced in 1972. Unfortunately, there are indications of resistance in the mites in some parts of the World. In searching for alternatives, it must be remembered that mites respond to different chemicals from insects. Perhaps some of the acaricides used for human scabies may be of value (see p. 304).

13.4.2   Lice

(a) Types of lice. There are two distinct types of lice affecting animals; the

anoplura or sucking lice, and the mallophage, or biting lice. The biology and habits of sucking lice are roughly indicated by the account of human lice given earlier (p.   ). Biting lice have mouthparts adapted for chewing and most of them subsist on epithelial debris of the skin or feathers of birds. Some forms, however, also take the blood of their hosts and have mouthparts adapted accordingly.

Sucking lice are all parasites of mammals. Biting lice are predominantly parasites of birds, but some species have become adapted to mammals. Members of both groups, but particularly the sucking forms, are specific to a particular host and will only accidentally infest other animals or birds. Furthermore, there is often specialization to a particular region of the body. Among sucking lice, the head and body species of human lice have been mentioned; also, on sheep, there are two species of the same genus of sucking lice (*Linognathus*) adapted to the face and legs, respectively.

Among the biting lice of birds, there tend to be thin agile forms on the body, where they need to escape preening, and more sluggish, plump ones on the safer regions of the head and neck (Rothschild and Clay 1952).

Biting lice, like sucking lice, spend their whole life cycle on the host, and invade new ones during close contacts. They are also ill-adapted to survival away from the host and soon die from starvation.

(b) Harmful effects of lice. The chief ill effects of lice are due to the irritation they cause, which may lead to small injuries when the animals bite or scratch themselves. Excessive licking by calves may lead to the formation of hair balls in the stomach. The restlessness and consequent loss of condition may render the animals more susceptible to various diseases.

Animal lice are not especially important as disease vectors, though the pig louse may transmit swine fever by passing from a dead or moribund pig to a healthy one. The biting louse of the dog occasionally acts as an intermediate host of the tapeworm *Dipylidium caninum*.

(c) Important species. Cattle may be infested with three common species of sucking lice; *Haematopinus eurysternus, Linognathus vituli*, or *Solenopotes capillatus*, also a biting louse, *Damalina bovis*.

Sheep are the hosts of three sucking lice; *L. ovillus* (on the face), *L. pedalis* (on the legs), and *L. africanus,* the African blue louse. Among the biting forms, *Damalina ovis* attacks sheep.

Pigs may be infested with the large *Haematopinus suis* and horses by *H. asini*. Dogs are attacked by *L. setosus* and also by the biting louse *Trichodectes canis.*

Poultry are hosts of a variety of biting lice, some of which are not highly specific and will also attack turkeys and pheasants. *Cuclotogaster heterographus* is the head louse of hens, found among the feathers of head and neck. *Lipeurus caponis* is the wing louse, occurring under the large wing feathers. *Goniodes gigas* is a large form, found on the body, and *Goniocotes gallinae* occurs in fluff at the base of feathers. Other species are *Menopon gallinae*, the shaft or feather louse, and *Menacanthus stramineus*, the body louse of fowls, found on the breast, thighs, and perianal region.

(d) Control. Identification of the parasites is not of supreme importance but it can easily be done by collecting the lice and examining them.

Poultry must be treated carefully to avoid harming them. Some relief may be obtained by treating their perches with a strong solution of gamma BHC in a volatile solvent, to give a fumigant effect (3 ml of 1.5% per 25 cm). Direct treatment of the birds can be effected by applying pinches of insecticidal dust to the base of the feathers, using 0.5% gamma BHC or carbaryl. Alternatively, dust baths containing insecticide can be provided to allow the birds to treat themselves. Spraying or dipping can be done in warm sunny weather, using 0.5% trichlorphon or 0.1% fenchlorphos or malathion.

Dogs may be treated with washes containing 0.01% gamma BHC or dusts containing 0.1% gamma BHC. Cats, however, are very sensitive to chlorinated insecticides and pyrethrin or synergized pyrethrin dusts are safer.

Cattle and sheep may be dipped, as described later (p. 319).

13.4.3   The sheep ked

Melophagus ovinus is a wingless, biting fly, which lives as an ectoparasite on sheep. Other members of the same family (*Hippoboscidae*) are half way to this dependent state, being winged, but spending much of their time resting on their hosts (horses, cattle, deer). The sheep ked spends its whole life cycle on the animal among the fleece, the adults taking meals of blood at intervals. The larvae are born fully developed and pupate at once.

Heavy infestations may cause a degree of anaemia and the intense irritation produces its usual effects. Control is achieved by dipping or showering, as described later.

13.4.4   Fleas

Fleas are not parasitic on ungulates, so that the larger farm animals are not involved. Pigs are often attacked by heavy infestations of the so-called human flea, *Pulex irritans*. Poultry may be seriously annoyed by the hen flea, *Ceratophyllus gallinae*. The stick-tight flea, *Echidnophaga gallinacea*, not only attacks poultry but occasionally infests other animals. The females of this flea attach to bare areas such as in fowls, the face, comb, and wattal, often in large numbers. Young birds are rather susceptible to them and may even be killed.

Stick-tight fleas present special control problems because of their close association with the host. Local inunctions of insecticidal lotion (taking care to avoid the eyes) may be effective. Poultry can be treated with 5% malathion dust. The hen houses must be cleaned and treated (especially for *C. gallinae*);

and this is facilitated by the provision of impervious floors and good lighting.

The dog and cat fleas (*Ctenocephalides canis* and *Ct. felis*) quite often bite Man and their control has been considered earlier.

13.4.5 *Argasidae, or soft ticks*

(a) Types involved. Of the two main genera of soft ticks, *Argas* attacks mainly birds, while *Ornithodorus* prefers mammals. The latter genus, however, is less important as a pest of farm animals. *Argas persicus* is a common pest in many warm countries, attacking fowls, geese, turkeys, pigeons, and various wild birds. *A. reflexus*, though primarily a parasite of pigeons, may also attack poultry.

Chicken houses are someimes so heavily infested with *Argas* sp. that the birds suffer loss of sleep and anaemia; and young birds may even die. Furthermore, *A. persicus* has been recorded as causing tick paralysis in ducks.

Argas ticks can act as vectors of various diseases of birds, as follows. Fowl spirochaetosis, due to *Borrelia anserina,* is almost cosmopolitan. The pathogen is passed through the egg of the tick to the next generation. A poultry piroplasm due to *Aegyptianella pullorum* occurs in Africa and another sporozoan disease due to *Anaplasma marginale* in the USA. Chicken cholera, or fowl plague, caused by *Pasteurella avicidia* can be transmitted if the birds eat infected ticks.

(b) Control. The poultry are removed from their houses and kept in wooden crates with plentiful crevices in which the ticks can hide as they drop off the birds. Meanwhile, the infested houses are cleaned and sprayed with a powerful acaricide, such as gamma BHC. After about 10 days, the birds can be returned to the houses. The infested crates can be sterilized, or if this is impractical, burnt.

13.4.6 *Ixodidae, or hard ticks*

(a) Harm caused by ticks. Hard ticks cause the greatest economic losses in livestock production in the world today. Though disease and tick paralysis are more spectacular, the insidious effects of heavy infestations are very serious. These include: (1) loss of blood (1–3 ml for each tick completing development), (2) constant irritation, which causes licking and scratching, (3) damage to hides, and (4) increased susceptibility to pathogens. It has been calculated that a daily infestation of 50 or more *Boophilus* cause an annual loss of 1.67 lb (0.65 kg) per tick, or nearly four tons of beef in a herd of 100 animals. Similarly, milk loss is estimated at 40 gallons per year for each cow (Wellcome 1976).

Hard ticks are mainly responsible for tick paralysis, certain species being particularly troublesome in different parts of the world. Thus, *Dermacentor andersoni* is the main culprit in the Rocky Mountain region of North America and *Ixodes holocyclus* in Eastern Australia. In Southern Africa, the species responsible are *Rhipicephalus* spp., *Hyalomma truncatum* (causing 'sweating sickness'), and *Ixodes rubicundus* ('Karoo paralysis').

In addition to toxicosis, ticks are vectors of several diseases due to different types of pathogen. Those caused by protozoa are, perhaps, most important, for example, the piroplasmoses or red water fever (because of the red urine passed by the cattle) due to *Babesia* spp. In Africa, a related group of pathogens (*Theileria* spp.) is responsible for East Coast fever and corridor fever; while another species causes Australian theilerosis. A third protozoan, *Anaplasma marginale,* causes gall sickness, or anaplasmosis, in many tropical or subtropical countries. Finally, a rickettsia-type organism, *Cowdria ruminantium,* is the pathogen of African heartwater disease. All these are diseases of cattle, transmitted by various species of *Amblyomm, Boophilus, Hyalomma, Ixodes,* or *Rhipicephalus.* To them must be added two diseases of sheep, louping ill and nairobi sheep disease.

(b) General principles of cattle tick control. Because of the vast areas grazed by cattle, it is seldom feasible to attack the hard ticks in the field, at least by ixodicides. There are, however, two ways of reducing the wild populations. In Australia, the single host tick *Boophilus microplus* is almost entirely dependent on cattle, so that removing the beasts from the pastures for 3 months ensures that most of the newly emerging larvae die from starvation. Also, where the grazing areas are not too extensive, regular grazing with frequently treated or immune animals may remove some ticks and reduce the load.

In general, however, the main defence against ticks is by treating the cattle. The ixodicidal treatments should be geared to the life cycle of the tick in question. With a single host tick, two treatments should be given within the normal life-span of the parasite. This is about 24 days with *Boophilus,* so that a 21-day cycle should give good control. However, moulting nymphs, which are very tolerant, may escape the first treatment, and to kill them, a second one at 12 days should give excellent control.

With two- and three-host ticks, treatment is adjusted to the relatively rapid feeding of all three stages. This takes 4–5 days in some *Rhipicephalus* and *Hyalomma* and 8–10 days in some *Amblyomma.* These ticks are troublesome in selecting protected sites of attachment, so that complete kills are difficult to achieve. With ixodicides which leave protective residues for about 3 days, good control may be obtained with a 7-day cycle; but in places where tick-borne disease presents a danger, a 3–4–3- or a 4–5–4-day cycle may be necessary.

(c) Treatment method for cattle ticks. Ixodicides are applied as aqueous emulsions or suspensions (from wettable powders) in three general ways: (1) by a dipping bath, (2) by a spray race, or (3) by hand application.

(i) Dipping baths have been in use for more than 75 years. Animals are led to them by a narrow corridor ('race') and made to jump into a rectangular tank, deep enough to ensure complete immersion. The tank is of strong construction, commonly of brick or cement, able to contain 15 500 litres (3000 gallons) for large herds, or 9000 litres (2000 gallons) for smaller ones. After immersion, the animal is held in a draining pen, where additional hand dressing may be done. Losses due to evaporation and liquid removed by the animals must be made good; but normally the dipping liquid remains serviceable for some months. To reduce soiling by mud on the animals' hooves, they can be passed through a foot bath on their way to the dip.

One modification of the dipping bath has been proposed for places where water is scarce or the farmers impoverished; this is the Machakos dip method, used in East Africa. A bath with a capacity of about 2250 litres (500 gallons) is used and each animal is yoked and submerged up to its flanks. The treatment is completed by scooping buckets of wash from the dip and pouring them over the beast. This enables the owner of a small herd to treat his animals without the need of a large dip.

(ii) Spray races have been in use since about 1960. They consist of a system of pipes with nozzles, fixed along a corridor formed of concrete or metal walls. The cattle are driven along the passage or race and the ixodicide is sprayed on them from all angles. The liquid is delivered at about 800 litres (180 gallons) per minute and the fall-off drains to a sump and is recycled. Power for the spray pump is derived from a stationary petrol engine (6–8 hp), an electric motor, or the power take-off from a tractor engine.

The spray race is less violent than the dipping bath, which can injure pregnant cows or those with large udders. It is more economical with spray liquid and more flexible in allowing a change of ixodicide when necessary, without waste of a large quantity. On the other hand, it requires more careful operators to maintain it and ensure that the sprays are properly adjusted and functional. When operating properly, a spray race can treat up to 500 head of cattle an hour.

(iii) Hand spray treatments are suitable for farmers with small herds to treat. The animals are restrained by wooden 'crushes' and treated by types of sprayer commonly used in agriculture. These include the ordinary stirrup-pump sprayer and a double arm rocker pump. In both cases, the liquid is forced along a hose to a nozzle by hydraulic pressure applied by hand-operated levers; motorized pumps can be used to supply two or more spray lances.

With conscientious operators, hand-operated spray treatments can be very efficient, if the spray is applied to the ears, crutch and under the tail. The method, however, is somewhat laborious and unsuitable for large herds.

(iv) Hand dressing is an additional procedure which may be highly desirable in areas of danger from two- and three-host ticks. Particular attachment sites chosen by the ticks are carefully treated with high concent-

rations of ixodicide, as a grease, an oil (applied by brush or swab) or a localized spray.

(d) Treatment methods for sheep ticks etc. The methods to be described are applicable for control of sheep scab and for blowfly protection as well as for tick control. Control of ectoparasites on sheep is hampered by the great thickness of their wool. This handicaps a spray race type of treatment, as it may not penetrate sufficiently. Dipping baths, on the same lines as those used for cattle, are effective and widely used. An alternative, which allows some of the economy and flexibility of the spray race, is the showering method. This involves a circular or rectangular pen, in which about 20 sheep are held and saturated from above by spray nozzles mounted on a stationary or rotating boom. Acaricide is pumped out at rates up to 400 litres (88 gallons) a minute; and the excess and run-off from the sheep is recycled. Experiments have shown that this method provides good penetration of the wool, though not quite as good as that obtained by dipping (Kirkwood *et al.* 1978).

(e) Compounds used as ixodicides. Arsenic was the first chemical to be used in tick control on a large scale but because of their dangerous nature, lack of residual effect, and because of the resistance of some ticks, arsenicals have been largely superseded. Next came the organochlorine compounds, especially gamma HCH and camphechlor, which are still used where tick resistance to them has not developed. Concentrations up to 0.05% of the former and 0.37% of the latter may be used in dips.

Because of organochlorine resistance, a series of organophosphorous and carbamate compounds have been introduced, especially against *Boophilus* in Australia. Once again, resistance is proving a problem and has eliminated many compounds in some areas. Accordingly, new types of chemical are being tried. These include: chlordimeform (which is used in combination with organophosphorous compounds), chlormethiuron, clenpyrin, and amitraz.

13.5 Acknowledgements

I am indebted to Drs M. D. Matthewson and K. W. Page for reading and criticizing this chapter and for making helpful suggestions.

13.6 References

Anon. (1974). *Health of the Schoolchild.* Report of the Department of Education and
      Science.
—— (1977). *Annu. Rep. Danish Govt. Pest Control Lab.*
Busvine, J. R. (1976). *Insects, hygiene and history.* 262 pp. Athlone Press, London.
—— (1978). *Syst. Entomol.* **3**, 1.
—— (1980). *Insects and hygiene* (3rd edn), Chapman & Hall, London.
—— and Vasuvat, C. (1966). *J. Hyg, Cambridge* **64**, 45.

Cornwell, P. B. (1974). *Int. Pest Control* **16(4)**, 17.
Dittmann, J. and Eichler, W. (1978). *Beil. Angew. Parasitol.* **19**, 1.
Donaldson, R. J. (1976). *R. Soc. Hlth. J.* **96**, 55.
Fisher, I. and Morton, R. S. (1970). *Brt. J. Ven. Dis.* **46**, 362.
Hallas, T., Mourier, H., and Winding, O. (1977). *Entomol. Medd.* **45**, 77.
Hopper, J. M. (1971). *Can. J. Publ. Hlth.* **62**, 254.
Kirkwood, A. C. *et al.* (1978). *Vet. Rec.* Jan. 21st, p.50.
Lamizana, M. T. and Mouchet, J. (1976). *Med. Malad. Infect.* **6**, 48.
Lapage, G. (1962). *Monnig's veterinary helminthology and entomology.* 600 pp. Ballière Tindall & Cox, London.
Maunder, J. W. (1971). *Comm. Med.* **126**, 145.
Mellanby, K. (1941). *Med. Off.* **66**, 141.
—— (1943). *Scabies.* 81 pp. Oxford University Press.
—— (1944). *Parasitology* **35**, 197.
Orkin, M., Macbach, H. I., Parish, J. C., and Schartzman, R. M. (eds) (1977). *Scabies and pediculosis.* 203 pp. Lippincott, Philadelphia and Toronto.
Page, K. W. (1969). In *Veterinary parasitology: Soc. Chem. Ind. Monogr. No.* 33.
Palicka, P. and Merka, V. (1971). *J. Hyg. Epidemiol. Microbiol. Immunol.* **15**, 457.
Rothschild, M. and Clay, T. (1952). *Fleas, flukes and cuckoos.* 304 pp. Collins, London.
Tonkin, S. L. and Wynne-Jones, N. (1979). *N. Z. Med. J.* **90**, 8.
Tüzün, Y., Kotogyan, A., Çenes, A., Izoğln, E., Baransü, O., Özarmagan, G., Cilara, A., Gürla, A., and Tat, A. L. (1980). *Int. J. Dermatol.* **19(1)**, 41.
Wellcome Research Organization (1976). *Cattle tick control.* 65 pp. Wellcome, Berkhampstead.
Weyer, F. (1978). *Z. Angew. Zool.* **65**, 1.
WHO (1976). *22nd Rep. Insectic. Cttee. Tech. Rep. Ser. No.* 585.

14
Control of migrant insect pests

R. J. V. JOYCE

14.1 Introduction

Certain principles of pest management emerge from a study of the Desert Locust, *Schistocerca gregaria* Forsk., an insect which is indisputably migratory. This chapter will show that the same principles of pest management apply also to other insects not generally called migratory but which are mobile and damage crops in fields other than those in which they were born.

Control, according to the Oxford English Dictionary (OED), is the power of directing, or a means of restraint or check. In the context of pests, control means to check and, in protecting a crop or other host against the ravages of a pest, it is a means of regulating the numbers in the fraction of a pest population which is placing at hazard a population of the species to be protected. If we designate this as the damaging fraction we must seek a means by which we can distinguish it from the potentially damaging one and from the harmless one.

Traditionally, in agronomy, the damaging fraction is regarded as a discrete population which has colonized a crop in which it pullulates, generating, in due course, the economic injury level of infestation – defined as the level at which the value of the yield lost equals the cost of the control measures. To maximize crop protection, control should commence at the economic threshold which is a level lower than that of economic injury and one from which the latter can be predicted, i.e. the economic threshold defines the level of infestation 'at which control measures should be applied to prevent an increasing pest population from reaching the economic injury level' (Stern *et al.* 1959).

This complete separation of the damaging fraction as a discrete population is clearly inapplicable in the case of the Desert Locust where the economic injury level is much surpassed by invasion from sources outside the crop. The damaging fraction of the population is an incidental part of the potentially damaging fraction. Crop protection should be aimed at identifying what part of the whole population is potentially damaging and discovering ways of regulating its numbers.

The Desert Locust provides an example of an animal the World distribution of which must be regarded from the point of control as a single population. Protecting crops against its ravages has to be aimed at regulating the World population of the species, the whole of which must be regarded as potentially damaging.

14.2   The evolution of Desert Locust control

14.2.1   The nature of the control problems

The interdependence of the worldwide population of Desert Locusts was first formally recognized by a panel of experts established by the Food and Agricultural Organization (FAO) under the chairmanship of Sir Boris Uvarov to advise the Director-General of FAO on long-term policy of Desert Locust control (FAO 1956). The panel concluded that there was evidence of swarms produced at one end of the distribution region having, within a couple of generations, a decisive influence on events at the other. Thus the population for the purpose of control or regulation was the entire one, occurring over 30 000 000 km² or one-fifth of the total land-surface of the world, administered by over 40 different governments. The seasonal movements of the population are illustrated in Fig. 14.1.

The first requirement for optimizing control was to have available a sequential series of near-synoptic pictures of the worldwide distribution of Desert Locusts, so that their occurrence at any one time could be compared with that on preceding and subsequent occasions, and inferences could be drawn regarding the development of infestations and their redistribution.

Only FAO had the capability and moral authority to make possible the extension and establishment of reporting and survey on this scale. Through its Desert Locust Control Committee, on which served representatives of all countries in the Desert Locust distribution region, uniform procedures of systematic reporting made it possible to direct attention to new areas where weather conditions, particularly winds and rain, or with forecast derived from knowledge of current infestations in possible source regions with past experience, led to the expectation of the presence of the species.

To promote and co-ordinate this work FAO set up a number of regional bodies in areas where Desert Locust populations through successive breeding seasons could be self-sustaining, although a proportion of the progeny might migrate into other regions. The regional bodies were:

(a)   FAO Commission for Desert Locust Control in the Near East (Bahrain, Egypt, Iraq, Jordan, Kuwait, Lebanon, Qatar, Saudi Arabia, Southern Yemen, Sudan, Syria, Turkey, North Yemen).

(b)   Desert Locust Control Organization for Eastern Africa (Ethiopia, Djibouti, Kenya, Somali Republic, Sudan, Tanzania, Uganda).

(c)   FAO Commission for Desert Locust Control in the Eastern Region of its Distribution Area in South-West Asia (Afghanistan, India, Iran, Pakistan).

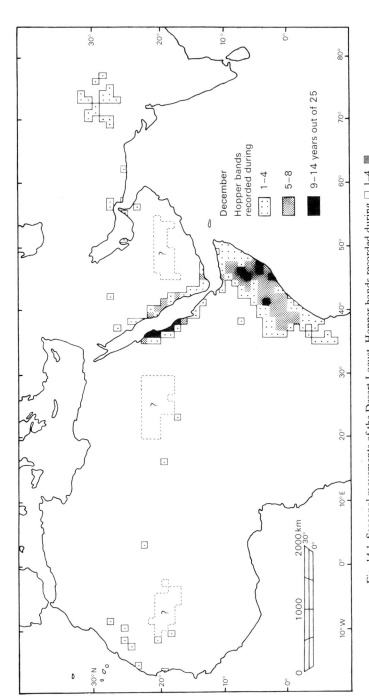

Fig. 14.1. Seasonal movements of the Desert Locust. Hopper bands recorded during □ 1–4, ▦ 5–8 and ■ 9–14 years out of 25.

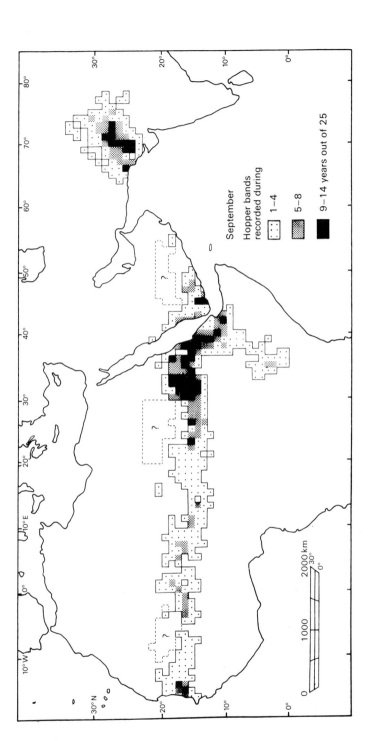

September

Hopper bands
recorded during

1–4

5–8

9–14 years out of 25

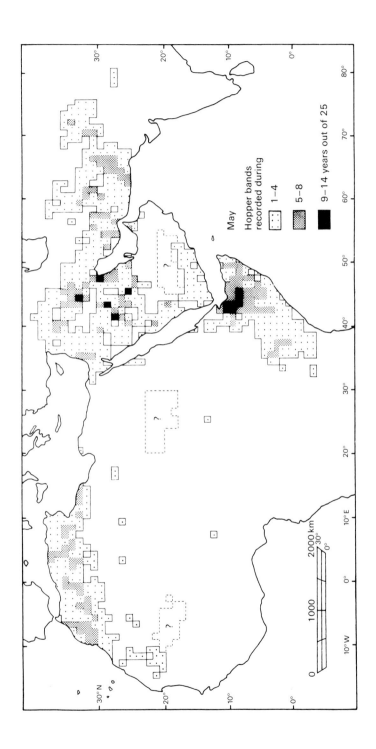

May

Hopper bands
recorded during

1–4

5–8

9–14 years out of 25

(d)   FAO Northwest African Desert Locust Research and Control Co-ordination Sub-Committee (Algeria, Libyan Arab Republic, Morocco, Tunisia).

(e)   Organisation commune de lutte antiacridienne et de lutte antiaviaire (OCLALAV) (Chad, Dahomey, Cameroun, Ivory Coast, Mali, Mauritania, Niger, Senegal, Upper Volta).

The role of FAO in the international and regional co-ordination of Desert Locust Control is described by Gurdas Singh (1972) and illustrated by Fig. 14.2.

Reports on Desert Locust infestation were compiled in a standardized format on which were stated the dates and co-ordinates of the infested areas and the estimated numbers, sizes, and stages of development of the juvenile (hopper) bands or adult swarms seen. All this information was transmitted monthly to the Desert Locust Information Service (DLIS) located until 1973 at the Anti-Locust Research Centre in London (later Centre for Overseas Pest Research) and now at the FAO Headquarters in Rome.

DLIS issues monthly bulletins which are circulated to all countries in the Desert Locust Distribution Region and these give a quasi-synoptic picture of the worldwide distribution of the entire Desert Locust population. From this picture forecasts are made, based on historical experience and current and expected winds and weather, of the probable redistribution which can be expected particularly during the coming month. The Service also provides special warnings of the imminence of significant development, sent by cable to threatened countries, based on current locust and weather situations.

14.2.2   Special features of Desert Locust survey

The special problem of Desert Locust control is that of scale. The total area liable to infestation during the monsoon rains is of the order of $5 \times 10^6 \, km^2$, and any one region may need the capability of searching for hopper bands over $500\,000 \, km^2$ of inhospitable desert during the 6–8 weeks available between oviposition by the parent swarms and evacuation of the area by adult progeny. Infestations in such source areas may be regarded as the potentially damaging fraction of the total population.

Surveys have greater value if they can be quantitative and of known reliability. A striking characteristic of hopper infestation is their clumpy or contagious distribution in which hopper bands are discrete entities within an infested area. The bands themselves similarly display a contagious distribution. The actual ground covered by marching bands may be from 0.1 to 4.0% of the infested area, and the numbers in hopper bands are such that each square kilometre of gross infested area can generate about 100–1000 kg of locust swarms (Joyce 1979).

Desert Locust swarms are protean in their structure and densities, reflecting changes in the structure of the air in which they fly and their

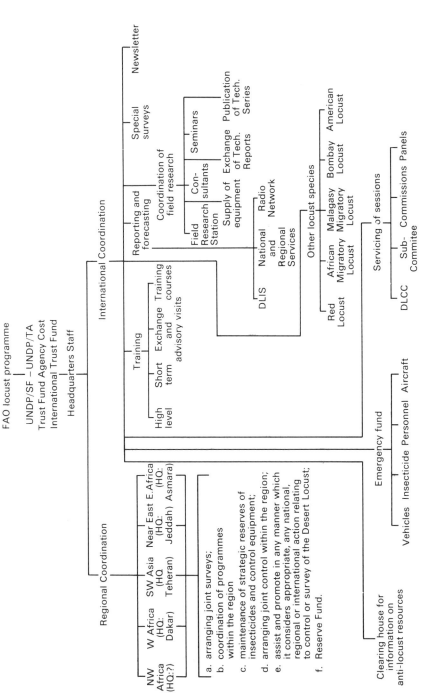

Fig. 14.2. The co-ordinating role of FAO in the research on and the control of the Desert Locust.

responses to it (Rainey 1958a). Accordingly, locust volume density may vary from 0.001 to 1.0 locust/m^3 (Gunn *et al.* 1948; Sayer 1959; Rainey 1958b; Ramana Murty *et al.* 1964; Waloff 1972.) Field assessment of control measures indicated an area density of 50 locusts/m or 50×10^6 locusts/km^2. This represents about 100 tons/km^2. Thus 1 km^2 of swarm could be produced from hopper bands distributed over 100–1000 km^2 of infested area according to whether the infestation rate was, say, 2.5 or 0.25%.

Survey for infested areas was dependent on scouts who were assumed to make 100% inspection of the areas liable to infestation. Such an assumption cannot be justified and the method failed to provide information on the fraction of the total infestation which remained undiscovered. A better method is a statistically valid sampling scheme constructed to provide a chosen level of security with respect to two probability risks, *viz.*

— discovery, namely, that the areas designated as infested above a stated tolerable level were in fact so infested – needing a high probability – and
— swarm escape, namely, that an area designated by the sampling scheme as uninfested in fact contained infestations in excess of the tolerable level – needing a low probability.

Such a sampling scheme requires the establishment of a level of infestation accepted as tolerable as well as quantification of these two risks. The values assigned to each of these parameters is in practice a function of the importance of the infested area as a source of swarms which damage valuable crops on the one hand, and the cost of mounting control operations in the area on the other. Thus the discovery and escape risks might for example be set at 80% and 10% for infestations in regions which produce swarms unlikely to invade crop areas, but at 98 and 5% respectively in Ethiopia and Somalia where swarms generated could invade the valuable crops of Kenya and Tanzania. The size and sample needed to provide these parameters can be calculated from historical knowledge of the distribution of hopper bands in the territory. A practical method of sampling was found to be achievable by aerial survey (Joyce 1981).

On the other hand, swarms present in crop areas may be regarded as the actually damaging fraction of the total population, and crop protection then requires 100% discovery rate. Since swarms are usually highly mobile, travelling in Kenya for instance at a rate of up to 100 km each day and not always in the same direction, 100% discovery would necessitate survey of the whole suspect area each day to locate and identify each swarm, as in the Souss Valley of Morocco (Rainey *et al.* 1979) and Somalia (Joyce 1962), a requirement which could be met only by aircraft.

The special features of survey for Desert Locust infestations are therefore: (1) it must be quasi-synoptic in the sense of providing information on the location of the total population, at any one time, so as to compare its distribution with that at other times (the duration of 'quasi' being determined by the length of time the population spends in a single locality – 4–6 weeks in the case of hoppers in an infested area, and 1 day in the case of swarms, as in Morocco); (2) it should be quantitative; (3) the reliability of the estimates needs to be selected in accordance with experienced judgement of the importance of the area in producing or containing potentially or actually damaging numbers.

14.2.3 Special features of Desert Locust control

The regulating or controlling of the potentially damaging locusts in the source areas requires the application of a regulator, which in practice is a synthetic insecticide, to infestations posing the most immediate threat, taking into account the probability of the area being a source of potentially damaging swarms. With regard to the actually damaging fraction of the population this has to be destroyed as quickly as possible over the entire range of its current distribution because the densities in any residual swarms far exceed the economic injury level.

Again the problem is one of scale, because areas measured in tens or hundreds of thousands of square kilometres may have to be treated in not more than 4–5 weeks against hopper bands in the source areas, and hundreds of thousands of tons of locusts destroyed as swarms invade crop areas.

A highly effective insecticide with, for example, median lethal toxicity of 5 µg per g of locust [the flying locust collecting about twice as much insecticide as a settled one experiencing a wind-speed equal to the flying speed of a locust (Rainey and Sayer 1953; Rainey 1958c; MacCuaig 1962a,b,c; MacCuaig and Yeates 1972)] can kill 50% of the individuals in a swarm if each collects 12.5 µg/g. Thus, to destroy a swarm, each 1 km^2 would in theory only have to be treated with $50 \times 10^6 \times 2.5\,g \times 10\,\mu g$ or 1.25 kg of active ingredient, if every toxic dose transmitted were collected by a locust and nothing else. In practice, this is impossible, but aerial spraying methods against locust swarms were developed by which over 5% of the median lethal doses delivered could be subsequently accounted for by locust corpses. The minimum application rate of a highly concentrated and effective locusticide such as Diazinon 85% or Fenitrothian 100% thus was 30–50 litres/km^2. In Kenya it was estimated that there was a 10% chance of swarms in excess of 100 km^2 being generated if only 2% of the potential breeding areas in Ethiopia and Somalia were heavily infested. Such swarms could be expected to move at times above 50 km per day (eating their own weight of food daily) (Rainey and Aspliden 1963) and would then be in range from a single aircraft base for not more than 5 hours. The control forces had to have the capability of delivering daily at least 1 ton of median lethal doses of insecticide if the crops of Kenya were to be 90% safe from damage. Techniques normally employed in crop spraying would require over 100 times this striking power, and moreover would be incapable of providing such a high probability of success.

Methods of spraying locust swarms from aircraft were pioneered by Gunn *et al.* (1948), and the Sawyer (1950) model of the dispersal of spray droplets and their collection by air-borne locusts was modified by Rainey and Sayer (1953) in the light of new knowledge of flight behaviour. These methods provided highly efficient control in large-scale practice (Joyce 1962) and are capable of improvement (MacCuaig and Yeates 1972). They are moreover suited for wider application (Rainey 1974) against other air-borne insect pests.

The development of methods of killing locust hoppers, from the use of poison baits to highly efficient treatment of desert vegetation with as little as 1–5 g of active ingredient of dieldrin per hectare, is summarized by Joyce (1974). The last concept, barrier spraying, first developed by Sayer (1959), is described in detail by Courshee (1965). It relies on the application of droplets of a size which are collected preferentially by the sparse desert vegetation rather than by other surfaces, such as the ground (Courshee 1959). The quantity of insecticide contained in each droplet and the number of droplets collected by the vegetation is such that the locust hopper, which consumes half its daily food whilst marching, will accumulate a toxic dose during its passage through one or more barriers of treated vegetation. Thus the locust's behaviour is exploited so that it finds the poison, rather than the control organization having to search for the locusts.

The dictates of scale in both swarm and hopper control demanded the development of target-specific methods of pesticide application. These are procedures which determine sequentially:
— the target stage, in relation to relevant knowledge of the ecology and behaviour of the target species,
— the target surface, selected in relation to the habits of the species and the route of entry of the chosen insecticide,
— the target dose, expressed as the quantity and distribution of the insecticide on the target surface,
— the total population of target surfaces which have to be hit in order to achieve the required biological goal,
— application techniques, which match the spatial and temporal dimensions of the population of targets, and which ensure that the greatest possible number of toxic doses released are collected by the target and the smallest by non-target surfaces,
— the operational and political capability of applying the required techniques on the required scale.
In the case of Desert Locust control in Eastern Africa, the last two requirements led to the replacement of baiting of locust hoppers from

ground-based teams by aerial operations, initially for budgeting reasons and subsequently justified by effectiveness (Joyce 1979).

The special features of Desert Locust control may thus be described as ones which require the application of an insecticide by target-specific methods to the whole of the pest population that exists at unacceptable levels as near-synchronously as possible (quasi-synchronous control).

The target-specific methods of application developed for locust control are amongst the most efficient ways in which insecticides have ever been employed. The 5% efficiency in accounting for the LD_{50} doses applied to locust swarms compares with, for example, 0.02% of y-BHC accounted for by Winteringham (1974) in the control of capsids on cocoa, and 0.03% of dimethoate in the control of aphids on field beans (Graham-Bryce 1976a,b). In hopper control 1–5 g of active ingredient per hectare compares with 200–400 g/ha commonly employed in grasshopper control in range land.

There can be little doubt that the application of these powerful measures terminated incipient upsurges of Desert Locusts, for instance in 1969, and have contributed to the prolonged recession of plagues since that date.

14.3   Basic principles for the control of migrant pests

14.3.1   Quasi-synoptic survey and quasi-synchronized control

These two features of locust control, quasi-synoptic survey and quasi-synchronized application of an insecticide by target-specific methods, are the principles of pest control which emerge from the dictates of a pest which is migratory par excellence.

The objective of quasi-synoptic survey is forecast, and the data can be understood only in the light of knowledge of the ecology and behaviour of the pest species. Of fundamental importance is flight behaviour, knowledge of which is essential for interpreting and forecasting changes in spatial distribution and for demarcating the population which has to be attacked to prevent damaging infestations appearing in crops to be protected.

The objective of quasi-synchronous control is to apply whatever regulator is chosen, say an insecticide, to the entire population demarcated by quasi-synoptic survey in a time not exceeding that spent in the same locality by the stage to be attacked. Thus, for Desert Locust hoppers the duration of 'quasi' is 4–6 weeks and, in the case of Desert Locust swarms, it can be the time the target swarms are available from a single aircraft base.

14.3.2   The origin of damaging populations

Of fundamental importance in constructing a control strategy is knowledge of the ecology of the pest species adequate to establish the source of the actually damaging fraction of the population. Many insect pests show a high degree of migratory activity (Southwood 1971). Some, like many species of aphids, such as *Myzus persicae* and *Brevicoryne brassicae,* and the rice planthopper *Nilaparvata lugens,* normally achieve pest status as a result of multiplication within the crop of a small number of immigrants. Control of such migrants in their source areas is either impractical, their host plants being too numerous and widespread, e.g. the cotton whitefly *Bemisia tabaci* with over 150 host plants in the Sudan Gezira (Gameel 1972), or too few and widespread, such as the cotton jassid *Empoasca lybica* (Joyce 1961), or inappropriate, the rate of multiplication in the crop being a function of factors other than the numbers of initial invaders. Joyce (1961) found, for instance, that there was a negative correlation between the numbers of immigrants of *E. lybica* and their subsequent rate of increase in the crop, both factors being a function of presowing rains which determined the amount of breeding on wild host plants on the one hand, and on the other, the nitrogen status of the food supply available in the cotton plant. Crop therapy, by chemicals or other means, clearly is the best approach to the control of such migrant pests, the objective being to maintain the intrinsic rate of increase below the control potential of natural mortality factors. The economic threshold can be defined only when the causes of excessive rates of increase are known.

Crop chemotherapy is often the best approach to control of those insects in which the damaging population arises through accumulation of immigrants in the crop. The density of such immigration is the measure of the economic threshold if this factor dominates the rate of multiplication to the economic injury level. The pink bollworm of cotton *Pectinophera gossypiella* is an example. Since the larvae inside the bolls are protected from insecticides, the most vulnerable stage is the immigrating moth.

Problems of control unique to migrant pests arise when immigration into crops is at a level which exceeds that for economic injury, as in the case of locusts and grasshoppers; the latter, however, in contrast to *Locusta* and *Nomadacris* have no discrete source areas. *Ailopus simulatrix,* which hibernates and aestivates in soil cracks in north-eastern Sudan, occurs in the clay plains at very low densities of the order of 10^3/ha (cf. Uvarov 1977), but crop loss is severe only when densities in the crop exceed about 10^5/ha (Joyce 1952). This 100-fold increase in density, though aided by accumulation in crops of choice, is largely a result of concentration in wind fields during evening flight. Some of the meteorological factors which generate such increased density of air-borne grasshoppers are described by Schaefer (1976). They include the front which in Sudan develops within the Intertropical Discontinuity (ITD), and which can give a 4-fold increase per hour (Rainey 1976), and cold outflows from convectional storms which can double aerial density in about $2\frac{1}{2}$ minutes (Schaefer 1976) (Fig. 14.3). Small eddies caused by turbulence or local surface variability, such as flooding, can provide local

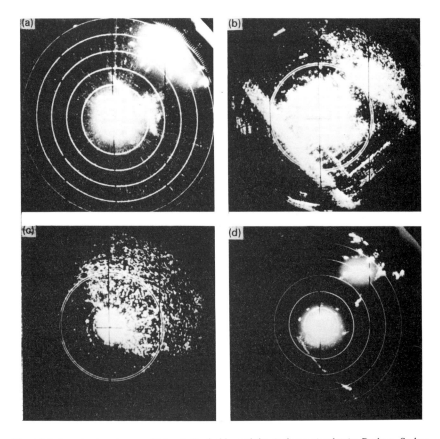

Fig. 14.3. Insects at a storm cold front. Probably mainly *Aiolopus simulatrix*. Radma, Sudan, 19th October 1971. (Figure 8.35: Schaefer, 1976, in *Insect flight,* p.182.) (A) 2013; range-rings 7.5 km apart and elevation 3°; rainstorm centred 35 km away to NE and cold outflow at 8 km, also approaching from NE, undercutting warm SW wind. (B) 2026; range-rings 450 m apart and elevation $1\frac{1}{2}$°; note leading edge of cold outflow approaching from NE and now 850 m away: canal bank also shown, running SE/NW to SW of radar site, and outlines of fields. (C) 2028; range-rings 450 m apart; looking up at 30° elevation at very sharply defined leading edge as it reaches the radar: (D) 2050; range rings 7.5 km apart and elevation $1\frac{1}{2}$°; cold outflow 13 km away, still receding to SW and visible to at least 45 km; storm collapsing (Schaefer 1976).

concentration (Pedgley *et al.* 1982). These can be expected to persist once the insects land, particularly if this landing is directed to particular vegetation as is said to be the case of the spruce budworm *Choristoneura fumiferana* (Greenbank *et al.* 1980). Such a concentration represents the actual damaging fraction of the total population, and it is this which quasi-synoptic survey has to measure and demarcate, and against which quasi-synchronized control must be directed.

14.3.3 *Opportunities for control*

Only when the origin of the damaging fraction of the total population has been determined can consideration be given to the relative merits of preventing its development by either reducing pest numbers in their source areas, or en route to, or after, their arrival in the crops to be protected.

The control of crop pests is customarily regarded as an agronomic input and its value determined by standard cost–benefit analysis in the same way as fertilizers or other agronomic practices.

The mobility of pest populations, however, makes this type of evaluation an over-simplification (Joyce and Roberts 1959). In contrast, the protection of crops against damage from a migratory pest such as the Desert Locust needs to be based on ecological concepts which identify that fraction of the total population which has to be regulated if damaging numbers are to be avoided, as well as to develop methods of regulation. Control serves an economic goal, but its value can only be estimated in terms based on a longer time and bigger spatial scale than are employed in cost–benefit analysis of standard crop-production procedures. It is normally undertaken by governments or other authorities which have the responsibility of working in scales of the required dimensions. An example is the Desert Locust Control Organization for Eastern Africa which can undertake control of pests and diseases which by their scale and nature are beyond the scope of individual plant protection organizations.

Chemical control of crop pests may be described usually as crop chemotherapy, that is, the target for the pesticide is the crop itself which is treated so as to render it lethal or otherwise unsuitable to the pest species. The control of migrant pests, on the other hand, directs attention to the need to develop methods by which insecticides can be transmitted most directly and with the minimum loss en route to the population of the pest species to be killed, whether this population is within or outside the crop, and on a scale determined by the population rather than by field or farm boundaries.

Knowledge of behaviour, particularly of flight, is essential for planning the necessary quasi-synoptic survey and interpreting the data collected. Similarly, it is also essential for the development of target-specific methods of pesticide application. Thus the development of efficient and economic methods of control for Desert Locust hopper bands required the integration of behavioural and chemical information. The effectively random orientation of groups of locusts in flying swarms (Waloff 1972), resulting in downwind displacement at a velocity lower than that of the wind, required the modification of the curtain method of spraying of the original Sawyer (1950) model; since the random flight together with the changing structure of the air in which swarms flew resulted in effective redistribution of individuals, control was achievable by releasing insecticide spray into the parts of the swarm temporarily the densest, without needing systematic traverses of the whole swarm.

Once released from the need to provide on all crop surfaces an even cover of insecticide, new and better methods of using these chemicals may be explored.

14.3.4   Control strategy

Strategy is concerned with when, where, and on what scale an activity is to be performed so as to achieve an objective at the least cost. Tactics are a means of executing a strategy. Strategy accordingly dictates tactics. In pest control, choice of the stage to be attacked is the basis of strategy and, from knowledge of its distribution, target-specific methods of pesticide application may be devised which match the temporal and spatial dimensions of these stages – the tactics of control.

In dealing with those migrant pests that require control before they enter the area at risk (Rainey 1974), we are in fact concerned with all pests causing damage in fields other than those in which they themselves were born when they invade in numbers above the economic injury level. The strategy for control must be based on the same principles as those which govern the control of pests which are migrant in the way defined by Kennedy (1961), namely, those with flight activity 'straightened out', meaning not diverted by appetitive stimuli. To a farmer the crop at risk is his own field into which pests from his neighbour's fields immigrate, and from which his pests emigrate to his neighbour's fields. Southwood (1971) points out that 'most arable crops are derived from ruderals, plants which are colonisers of bare ground. These are temporary habitats and thus pests of crops have from their evolutionary history high inherent rates of migratory activity'. Because their habitat is impermanent, crop pests are usually '*r*' strategists characterized by their rapid multiplication (MacArthur 1960). Moreover, their high degree of flight actively permits not only efficient exploitation of new and ephemeral opportunities but also, like diapause, an escape from unfavourable conditions. This constant flux of spatial redistribution within populations has been illustrated by Taylor and Taylor (1977) by analogy to the structure of a fern stele. To determine the scale of latitude and longitude demands detailed systematic mapping as undertaken for nearly 40 years for the Desert Locust and for about 20 years by the Rothamsted Insect Survey. There is evidence, however (Joyce 1975), that spatial redistribution of *B. tabaci* occurs over dimensions of latitude and longitude 1000 times greater than in Fig. 14.4.

Pest management must be based on a relevant definition of the population which is to be managed. A population is a group of individuals treated together for a specified purpose. In pest management a definition of population could be the numbers of individuals which have to be considered if regulation of the damaging numbers is to be achieved. In the case of Desert

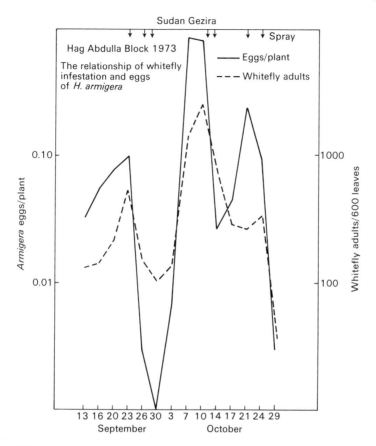

Fig. 14.4. The reinvasion of cotton-fields in the Sudan Gezira by *H. armigera* and *B. tabaci* after spraying, each peak associated with the passage of a storm flow, that on 21 October following the storm tracked by radar in Fig. 14.3. (Data from Russell-Smith, unpublished.)

Locusts this is the entire World population, in other cases only that fraction which is actually the damaging one. What constitutes a population in the sense required by pest management is a function of population dynamics and behaviour, especially the dispersal strategy of the pest species.

The flux of individuals within a population imposes the need to reconsider the implications of the terms 'economic threshold' and 'economic injury levels'. The latter arises from the former in any one field only when immigration ceases, i.e. when colonization is completed before the economic threshold is reached. Population in a given area increases not only when birth rate exceeds death rate, but also when birth rate plus immigration exceeds death rate plus emigration. It is sometimes assumed that immigration and emigration, as random processes, can be regarded as being in balance, dispersal being random resulting in 'scattering in all directions' (OED). Studies of insect flight, for example by Rainey (1976) and (Schaefer 1976), show conclusively that dispersal cannot be so conceived. The air in which dispersal occurs is structured, not only spatially but temporally, has velocity in the vertical as well as the horizontal direction, and transport of insects in it typically results in depletion of densities in one area and concentration in another. Thus the cotton jassid *Empoasca lybica* and the cotton whitefly *Bemisia tabaci* sometimes reach the economic injury level by massive immigration into cotton fields as a result of redistribution by wind fields and sometimes because birth rate has increased through repeated immigration. Russell-Smith (unpublished) drew attention to the number of occasions when large-scale immigration of whitefly occurred on the same storm fronts on which damaging levels of *H. armigera* were imported (Fig. 14.4).

This redistribution of individuals by immigration and emigration makes it necessary for any definition of economic threshold to include in it a statement of scale both in space and time, the same requirement as that dictated by migrant pests. Only when the spatial and temporal dimensions of the control problem have been clearly defined can tactics be devised which are tailored to these dimensions.

14.4   Migrant pest control in practice: some examples

14.4.1   Orthoptera and Hemiptera

The control of migrant pests involves locating and following the changing spatial distribution of the damaging populations. In contrast with the Desert Locust, potentially damaging population of which may occur almost anywhere in its distribution region, other locusts such as *Nomodacris septemfasciata,* the Red Locust, appear to have restricted and well-defined permanent breeding areas where damaging swarms first develop. Red Locusts which occur elsewhere constitute the harmless fraction of the total population. Control of such locusts is based on constant surveillance of these outbreak areas, and the imposition of suitable control measures when numbers reach a threatening level as a result of local pullulation. Such preventive control measures have also been instituted by the establishment of an internationally financed body to suppress outbreaks of the African Migratory Locust *Locusta migratoria migratorioides* in the vast plains of the middle reaches of the Niger river in Mali, and in recent years around Lake Chad, where the general environment is very similar to that of the reed-bed plains of the lower reaches of some Russian rivers which form the outbreak areas of the Asiatic Migratory Locust *Locusta m. migratoria* (Uvarov 1977). Such preventive control measures have been developed as a result of decades of study of the ecology and behaviour of the pest species.

On the other hand, the Australian Plague Locust *Chortoicetes terminifera* has no discrete outbreak areas but rather outbreak conditions which may be very widespread. According to Key (1954) *C. terminifera* was an insect of the grasslands, savanna woodlands, and other more open communities, and its present habitats are those where denser and taller grasses have been replaced by a mosaic of tall grasses, which serve as shelter, and short grasses, which provide food. Similarly, the Bombay Locust *Patanga succincta,* originally an insect of forest margins, became a serious pest in Thailand following the clearance of forest for crop production. In Sudan the sorghum crop in some years is severely damaged by *Aiolopus simulatrix,* a grasshopper which in favourable seasons can complete one generation in the young growth of the tall-grass areas, where *Sorghum purpureosericeum* is dominant, and a second generation in the short grass-area, to which it migrates on the southerly winds which bring the rain to these more northerly areas. It is this generation returning on the northerly winds which, when concentrated in wind fields, is the damaging population. A similar problem is presented by *Oedaleus senegalensis* in the Sahel of West Africa.

When the distribution of the potentially damaging population is so widespread and frequently unpredictable, survey for it may be impractical, and control, because of its low density, economically impossible. Control effort then has to be concentrated on destroying the actual damaging fraction of the total population en route to, or in the crop, as is done in the case of *C. terminifera.*

Destruction of migrant pests when they are concentrated during hibernation or aestivation is sometimes a favoured means of control. An example is *Eurygaster integriceps,* the Sunn pest of USSR, Turkey, and Iran. This species has a short active period of 2–3 months during spring and early summer when it breeds on and damages cereal crops. The rest of the year is spent inactive in mountain retreats, in aestivation during the hot dry summers and in hibernation during the cold winter months (Brown 1962). Destruction of any part of this potentially damaging population reduces the size of the actual damaging fraction, so that dead season control is a useful means of crop protection even if less than 100% complete. At the same time direct destruction of the damaging population in the crop is now regarded as a better means of crop protection. The Dura Andat *Agonoscelis versicolor,* on the other hand, hibernates and aestivates on trees in central Sudan. At the beginning of the rains the bugs migrate to weeds where they complete a generation. This new generation may then invade and damage sorghum fields. Control is said to be achieved by killing aestivating andat in their dry-season roosts. This is easily done by dislodging the sluggish bugs from the trees and arranging that they can return only by traversing a barrier of insecticide (e.g. BHC) dust (Joyce 1953). Control, however, will be effective only if a major part (say 90%) of the potentially damaging population is discovered and destroyed. This is because the size of the actually damaging population depends on the success of the breeding on weeds, itself a function of the rainfall which is extremely variable. Environments which are unstable in the sense of having a low degree of permanence and predictability favour animals with a high potential rate of reproduction, *r* strategists, which are able to exploit the rare favourable occasions for breeding even if their rapid rate of multiplication destroys their habitat. They do this without evolutionary penalty because their habitat is already doomed by its instability. Thus the breeding success of *A. versicolor* may vary one hundredfold so that loss of 90% of the aestivating population can be easily compensated by favourable rains.

Similarly, control of the Bombay Locust *Patanga succincta* is said to be achieved by the destruction of the dry season concentrations which accumulate in stands of tall perennial grass such as *Imperata* spp. These concentrations are very temporary because the populations of locusts are active and continually disturbed by burning of the grasses, so that it is doubtful whether dry-season control kills any but a small fraction of the population. Dry-season control does not obviate the need to discover and kill those locusts which actually invade and breed in the crop or in its vicinity, and this operation probably makes the major contribution to crop protection.

A control concept widely promoted in the past was often expressed in military terms – study the enemy and attack it at its weakest point. Though the concept remains true, the weakest point can only be determined by knowledge of the relevant features of the ecology and behaviour of the pest, and not selected simply because the pest presents itself as a favourable target for destruction. For example, during the summer of 1979 the numbers of wheat aphids *Metopolophium dirhodum* air-borne over East Anglia were estimated by Schaefer *et al.* (1979) at 10^{12}, weighing some 100 000 tons, flying at densities during several days in excess of $1/m^3$. These could have been killed by aerial spraying in a few hours of flying, with an expenditure on insecticides perhaps about 1000-fold less than that employed in standard crop spraying. Unfortunately our knowledge of the ecology of this species makes it unlikely that such destruction would have contributed to either control or crop protection because numbers in the following season are a function of factors governing survival over the winter.

In all these cases, however, any opportunity which exists to kill the actual damaging population without contaminating the crop needs to be explored and exploited.

14.4.2 Lepidoptera

(a) Stemborers and bollworms. Stemborers, for example, *Tryporyza oryzae,* lay their eggs on the flag leaf of rice, and the first-instar larvae burrow into and destroy the tiller within minutes of eclosion. Destruction of these larvae inside the plant is only of value if this decreases the numbers of the next generation, but if the crop is reinvaded or the next generation comes from a

neighbour's field, the control effort has been wasted. As in the case of pests normally classed as migrant, the insecticide has to be applied as near-synchronously as possible to the entire damaging population. The quasi-synoptic survey in this case was achieved in Java by light traps emptied each day (Joyce *et al.* 1970). Damage was largely prevented by killing the moth throughout the entire range of its distribution where its numbers exceed a tolerable level. The success of this strategy has been described by Singh and Sutyoso (1973) (Fig. 14.5). Best yields were obtained when areas in excess of 5000 ha were sprayed quasi-synchronously.

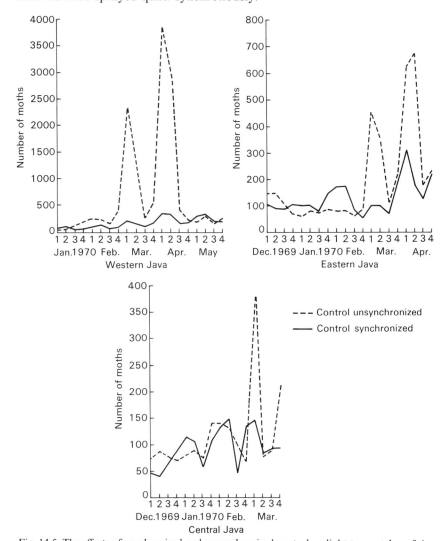

Fig. 14.5. The effects of synchronized and unsynchronized control on light-trap catches of rice-stemborer moths. Synchronous control: ULV aircraft spraying. Unsynchronized control; ground-spraying by individual farmers. Number of moths is mean *Tryporyza* count per night per trap, from weekly catches at ten traps. (From Singh and Sutyosa 1973.)

An example of a similar strategy is the control of the bollworm *Heliothis armigera* in the Sudan Gezira, where larvae destroy young fruiting points of cotton plants within minutes of eclosion. It was found that the economic injury level of larvae was invariably achieved as a result of invasion of the cotton crop by moths which had bred elsewhere, for example in groundnut or sorghum fields, even if the invasion were small (e.g. less than 50 females/ha). The cotton crop can be protected best by destroying adults before they oviposit, or at least by destroying the larvae immediately after eclosion. Topper (1980) found that moths evacuated the groundnut fields by flight each evening and equivalent numbers built up in the cotton where they oviposited and fed. The cotton crop was largely evacuated before dawn for the more protected day-time retreats offered by the groundnuts, where the moths accumulated as daylight suppressed their movement. Since eggs are normally laid one at a time, each oviposition usually alternating with some flight, the moths spent much of their evening hours air-borne. Schaefer (1976) showed that the plumes of adult *H armigera* which took off nightly from the groundnuts could become involved in massive redistribution as a result of rain storms and other meteorological phenomena. Haggis (1981a,b), using *H. armigera* eggs on cotton as an indication of the recent presence of adult moths, showed how patterns of egg laying occurred over areas which could be measured in hundreds or thousands of square kilometres, and these patterns changed from day to day often in association with recognized and measured changes in the wind fields.

Pest-control strategy dictated the moth of *H. armigera*, or at least the larva on eclosion on cotton, as the target stage. The behaviour of the target stage dictated that acceptable tactics should be those capable of treating tens of thousands of hectares of cotton within hundreds or thousands of square kilometres of land area during the time the adults were present in the cotton and before the eggs hatched, the incubation period of *H. armigera* in Sudan being 2–3 days. Similarly, quasi-synoptic survey required the development of methods which permitted economic sampling of these large areas on the same time scale. This was described by Russell-Smith (1975) and Joyce (1976).

Lawson *et al.* (1979) and Topper and Lawson (1984) directed both their surveys and insecticides at the adult stage of *H. armigera* which Lawson (1980) had shown were killed by three routes of the insecticide chosen (monochrotophos), namely, through contamination of the extra-floral secretions on which the moth fed, direct contact, and indirect contact, the moth accumulating a toxic dose from very small deposits (1 µg/cm^2 of leaf). This procedure made it possible to reduce the appliance of monochrotophos from 350 to 250 g a.i./ha. Moreover, as demonstrated by Dietrich *et al.* (1980), the adult Noctuids do not possess the powerful multi-functional oxidases which are possessed by their polyphagous larvae and are unlikely to develop tolerance to insecticides.

The yield benefits over areas of 20 000–50 000 ha of cotton from quasi-

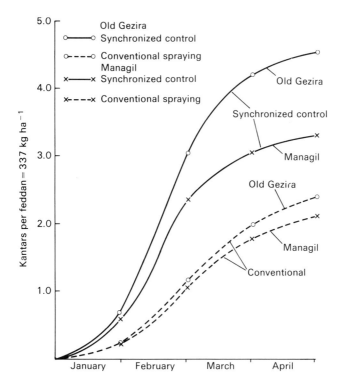

Fig. 14.6. Sudan Gezira: accumulated monthly yields of seed cotton (*Gossypium barbadense*) during the 1975/76 season from areas in which pest control was based on quasi-synoptic survey and quasi-synchronous control compared with conventional practice.

synoptic survey and quasi-synchronized control in the Sudan Gezira are illustrated in Fig. 14.6.

Armyworms. Notable amongst migrant pests are the armyworms, the larvae of the noctuid moths, *Spodoptera frugiperda* in the Americas, *Mythimna separata* in S.E. and E. Asia, and *Spodoptera exempta* in Africa. In relation to *M. separata* and other highly mobile pests, Barfield and Stimac (1979) point out that Integrated Pest Management strategies are designed to treat symptoms of pest problems rather than causes. This, they say, 'has been the result of a lack of focus on the fact that pest life system boundaries are dynamic and that crops in a productive system are coupled by flows of organisms among them.' The dominating role of population movements has been recognized in eastern and central Africa where *S. exempta* has been studied over a wide range of its distribution using a pattern of light traps established in some six to ten countries (Betts and Odiyo 1968; Betts *et al.* 1970; Brown *et al.* 1972; Betts 1976; Roome 1974), and the movement of populations in relation to prevailing wind systems inferred (Betts and Haggis 1970; Haggis 1971). This system led to the development of a useful method of forecasting armyworm outbreaks in eastern Africa (Betts *et al.* 1969; Brown *et al.* 1970; Odiyo 1972, 1974). Since 1974 pheromone traps have also been used (Campion *et al.* 1976).

The drift-spraying technique using insecticides in a carrier of low volatility and applied at ULV rates, first developed for locust control, proved to be a highly effective way of killing armyworm in pastures, a complete kill of larvae being obtained over a swathe of 400 m with an area dosage of 130 g of DDT/ha and a volume rate of about 500 ml/ha (Brown *et al.* 1970). This application is about one-tenth of that normally employed to kill armyworm by conventional spraying, though in tall dense crops, such as maize, complete kills were obtained only when application exceeded 400 g of DDT/ha in over 1.5 litres/ha (Joyce unpublished).

Killing larvae in crops may provide crop protection, but not pest control, in the sense of regulating pest numbers. For this, more information is required on the origins of the potentially damaging population, and research to this end is now in progress in Eastern Africa under a collaborative research programme between the Desert Locust Control Organization of Eastern Africa, the Kenya Agriculture Research Institute and the UK Tropical Development and Research Institute (formerly Centre for Overseas Pest Research) (Fig. 14.7). Using radar and a new optical recording device, Riley *et al.* (1981) showed that mass migratory flight took place at dusk, the moths climbing to heights up to 1000 m above ground level and flying downwind often at more than 35 km/h, achieving a minimum displacement of 6–9 km and sometimes at least 20 km in a single night. Two types of mechanisms leading to concentration were recorded, namely, topographically induced rotors and storm outflows (c.f. Schaefer 1976). Rose and Dewhurst (1979) found that freshly emerged moths, when ready to fly, moved into nearby trees and bushes and accumulated there, reaching peak numbers at midnight. After this time most leave, particularly big departures taking place just before dawn. While in the trees the moths avidly feed on sugars, a feature which provides a potential approach to control.

Marked migration of armyworm moths occur during the northward movement of the Intertropical convergence zone, so that infestations in Kenya derive from earlier ones in Tanzania and those in Ethiopia from earlier ones in Kenya (Brown *et al.* 1969; Tucker *et al.* 1982), and on to Yemen. Eradication of concentrations at one time should break the cycle and provide control of populations in eastern Africa. This depends on finding economic methods of quasi-synoptic survey and quasi-synchronous control. Meanwhile, any significant reduction in overall population in one of these countries will benefit other countries later; hence the importance of a body such as D.L.C.O.E.A. which is financed by the Governments of all the countries concerned and is able to operate across national boundaries.

(c) Spruce budworm. A well-documented example of the dependence of

Research objective

The development of a control strategy for African armyworm, on a regional basis.

Fig. 14.7. The process for the development of a control strategy for the African Armyworm, *Spodoptera exempta,* constructed by the Kenya Agricultural Research Institute and the Centre for Overseas Pest Research (after D. J. W. Rose, unpublished).

control on the understanding of migration is that of the Spruce budworm *Choristoneura fumiferana* in New Brunswick, Canada, the periodic outbreaks of which destroy millions of hectares of forest throughout eastern Canada and Maine, USA. Several millions of hectares of forest with damaging levels of larvae, in Quebec, New Brunswick, and Maine, have been sprayed annually now for several decades; 10 million acres were sprayed in New Brunswick alone in 1976 at a cost of $17 million (Baskerville 1976). Miller and Kettela (1975) record that, between 1952 and 1973, the percentage reduction in larval infestations following aerial spraying ranged from 76 to 99%. There was also a corresponding reduction in the number of adults which develop from survivors and which are the parents for the following season. Unfortunately, egg densities in sprayed areas are often equal to or greater than those in unsprayed areas; Fig. 14.8 shows the relationship of pupal counts to egg densities in sprayed and unsprayed areas. For every pupa in unsprayed areas there were some 30 egg masses compared with nearly 100

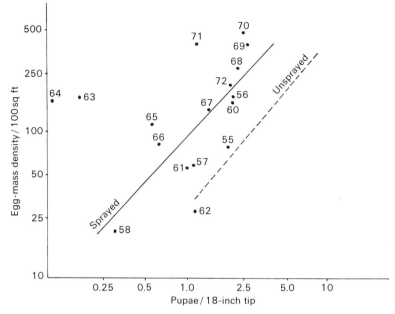

Fig. 14.8. The effect of aerial spraying of forests in New Brunswick, Canada, is to reduce the infestation by the Spruce budworm, *C. fumiferana* (expressed here as numbers of pupae per 18 in tip), and thus protect the foliage. This reduced foliage damage makes the stand more attractive to immigrant moths (here expressed by the egg-masses per 100 ft² of foliage), necessitating further spraying in the following season. (From Miller and Kettela, in Prebble, 1975.)

in sprayed areas. The spraying had been successful in protecting the foliage, but moths preferred to oviposit on this rather than on damaged foliage and immigrated to sprayed areas to do so.

The population dynamics of the Spruce budworm have probably been studied in greater depth than that of any other insect. Clark *et al.* (1979), in making their impressive model of the population dynamics of the species, employ the age structure and foliage condition of the host tree as key factors in determining increases and decreases in pest numbers. This model predicts, and experience has confirmed, that keeping the trees green by killing spruce budworm larvae prolongs the pest outbreak by removing from its environment the key factor, namely, reduced availability of optimum food, which alone can reduce its numbers. Current spruce budworm infestation of eastern North America is the most widespread of the century, including extensive damage in Newfoundland and Cape Breton far exceeding that of previous outbreaks (Ketella 1983). In this case crop protection is the antithesis of pest control, and a radical new approach is necessary to achieve the latter. A detailed study of the migratory strategy of the moth (Greenbank *et al.* 1980) suggests that such an approach is possible.

In New Brunswick the moths emerge from pupae, generally in the early afternoon, in late June in the south and mid August in the higher lands further north, and live for about 10 days. The females, which begin to emerge 2 days after the males, mate on their first night and remain at the site of emergence for the first 2 nights. During this time they deposit about 50% of their eggs (Greenbank 1957). Thereafter, both males and females engage in nightly flights, the females carrying with them 30–50% of their complement of eggs, that is, 50–90 eggs each. The major take-off is at sunset and the moths actively climb, initially at a rate of 0.6 m/s, and, once in steady winds, orientate themselves so that they are flying more or less downwind, reaching in fair weather an average cruising height of about 130 m above ground level where normally there is a temperature inversion, a height which can, however, be much exceeded in disturbed weather (e.g. a mean height of 700 m recorded by Dickerson *et al.* 1983).

At this cruising altitude their air speed is about 2 m/s in winds which varied from 5 to 22 m/s. The duration of cruising flights varied from 30 minutes to a maximum of 8 hours. Clark *et al.* (1979) estimated from radar data that a few moths land within 10 km from their exodus site, 25% fly less than 30 km and 25% more than 75 km in a single night. A small percentage may travel up to 240 km in a single flight (Greenbank *et al.* 1980). Later work indicates that these displacements are likely to be underestimates. The densities of air-borne moths at their cruising altitude reached a maximum of about 3 per 1000 m^3 of air (Schaefer 1976), a density greater than that of a swarm of Desert Locusts in daytime flight during fair weather in East Africa (Uvarov 1977), and provided a biomass of some 2 kg of moths per hectare. There were occasions when densities were increased 10-fold during the passage of sea-breeze fronts (Schaefer 1979).

The study, which continued over four seasons, showed that almost the entire population of Spruce budworm more than 2 days old of any one region was air-borne for at least 30 minutes and most for 1–2 hours following sunset during the 10-day period of emergence. Of these, 90% were concentrated in a layer about 100 m deep at the level of the temperature inversion, about 150 m a.g.l. (above ground level), at a density of more than 1 per 100 m^3, equating to about 10 000 moths/ha.

To detect the density of air-borne moths Schaefer (1979) (also in Greenbank *et al.* 1980) developed an air-borne radar system which was capable of measuring sequentially the number of moths at 32 different levels, starting 50 m below the aircraft and ending approximately at ground level. It then averaged 15 successive measurements at each level to produce smoothed profiles of density and orientation at each of the 32 levels at a rate of 8/s that is, at an air speed of 150 knots, a smoothed profile for each 10 m of track (Figs 14.9 and 14.10). The survey aircraft was also fitted with a collecting net designed to provide a soft landing for insects (Spillman 1979), so that the

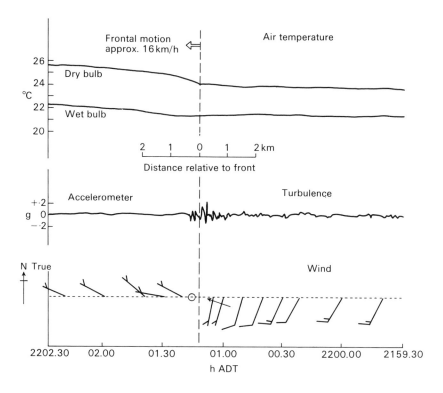

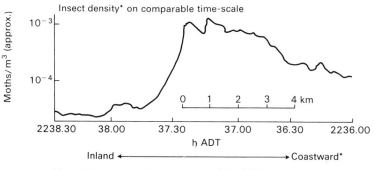

*Insect density on airborne radar at 280–310
asl; DC–3 traverse at 2236–38 hours ADT

Fig. 14.9. Concentration of Spruce budworm moths at a sea-breeze front in New Brunswick, Canada. Air temperature, turbulence, and wind velocity measured in a DC-3 aircraft whilst traversing a sea-breeze front in New Brunswick, Canada on 10 July 1976. The lower graph shows the absolute density of flying spruce-budworm moths recorded at the same time by an air-borne radar fitted to the DC-3. Although moths had been air-borne for only about 30 min, considerable accumulation had taken place in the front. The difference in density of air-borne moths above sprayed (inland) and unsprayed (coastward) forests is clearly visible. (From Greenbank *et al.* 1980.)

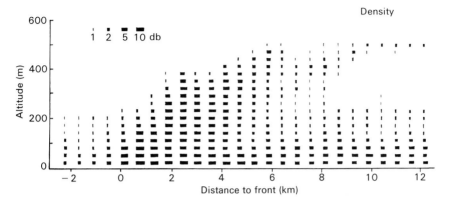

14.10. Distribution of relative moth density within the convergence zone of a sea-breeze front recorded by air-borne radar at 22:37 h, 10 July 1976. Density scale is moths/m³. (From Greenbank *et al.* 1980).

identity and sexual status of the insects responsible for the radar echoes could be confirmed.

Methods of killing economically such moths concentrated in both the horizontal and vertical planes were also devised. Spillman (1979) showed that moths flying in still air are 25 times more likely to be dosed to a given level if a spray cloud is dispersed in particles of 5 μm diameter than in ones of 40 μm diameter, and chemical analysis of moths sprayed in the field supported this conclusion (Uk 1976). Lawson (1976) demonstrated that under the relatively stable meteorological conditions typical of New Brunswick at the time of moth flight, the effective life of the cloud formed by a Micronair atomizer emitting at 6 ml/m was greater than 54 minutes, during which time the cloud had travelled 20 km from its source. At this distance concentrations were sufficiently high to provide a dose of 0.2 ng/0.77 cm², the equivalent area of a flying Spruce budworm moth. Lawson estimated that 60% of the emitted spray reached the ground after 50 minutes with a maximum deposit of 36% 19–34 minutes after emission, equivalent to 0.56 g a.i./ha, giving experimental confirmation of a spray dispersal model constructed earlier by Schaefer and Allsopp (unpublished).

The system thus developed permitted the possibility of protecting forests from invading moths with a suitable insecticide, such as a pyrethroid or even pyrethrin, at rates of ground contamination of the order of 1 g a.i./ha and achieves a rate of work at least two orders of magnitude faster than conventional spraying against larvae, a rate of work commensurate with the economic requirements dictated by the pest's biology, and leading, moreover, to zero contamination of the ground or the air.

It is clear that this extension of the original Sawyer model of curtain spraying has potentially wide applications to the control of those migrant pests which can be attacked en route to the crops, notably grasshoppers and armyworm moths. Aircraft fitted with Doppler navigation equipment used for precision wind-finding (Rainey and Joyce 1972) can locate and maintain their track within zones of wind convergence (Rainey 1976) and record, with Schaefer's downward-looking radar, the number and orientation of flying insects below it, identify them by capturing samples in the Spillman net and, if necessary, kill them by spraying with (for example) a photo-unstable insecticide (cf. Schaefer 1980).

14.4.3   Diptera

Perhaps the most impressive example of the importance of scale in the control of migrant pests is that of the WHO programme for eradication of river blindness in West Africa through control of the blackfly *Simulium damnosum* (see also Chapter 12). The vector is very widely distributed in West Africa and the original control programme covered some 654 000 km^2 in seven countries. After control had started in 1977 in the first phase of the project in the centre of this area, it was discovered that infected flies were reinvading the controlled area from the non-controlled areas surrounding it. Subsequent research on the migratory behaviour of the vector (Garms *et al.* 1979) showed that it was capable of long-distance travel, up to 500 km, and to cope with this the programme area had to be extended southwards in the Ivory Coast to cover the breeding sites of the reinvading blackflies. Investigations by Garms *et al.* (1979) showed that reinvasion was an annual occurrence beginning with the passage over the area of the ITCZ which heralded the wet season, and was characterized by the sudden and simultaneous appearance of immigrants over large areas. Despite the southern extension, major reinvasion still occurs from the west of the controlled area, so that studies are being made of the possible extension of the programme to include Guinea and Senegal. Magor and Rosenberg (1980) also showed that, on occasions, reinvasion occurred during the dry season from the north-east, possibly associated with the passage of Mediterranean depressions. They conclude that '*S. damnosum* females caught over a period of time at an invaded site are likely to have come from several sources and this factor should be taken into account when developing a control strategy.'

Quasi-synoptic survey and monitoring of control are attained by the use of human baits, two-men teams with each man working alternate hours between 07.00 and 18.00 h every day at some 600 sites, the catch being expressed as flies/man/day. This figure is subsequently transformed to an

Annual Biting Rate (Walsh *et al.* 1981), as a means of evaluating the success of the programme.

Control has been sought by concentrating attack on source areas which are the rapids in streams where larvae of *S. damnosum* live and develop from eggs to pupae in about 8 days. Each week over 200 potential breeding sites are visited and searched for infestations and each infested site is treated by the aerial application of a larvicide.

However, source areas evidently exist outside the programme area and for the present system to achieve control of the species, either the programme areas must be extended or the need faced for prolonging the programme. Multiple resistance to available insecticides has, however, already arisen but this dilemma would be removed if methods were found to kill adults as well as larvae. Since adult *Simulium* employ sugars as their flight fuel (Hocking 1953), an attractive possibility is the use of sugar baits at their emergence and/or oviposition sites.

14.5 References

Barfield, C. S. and Stimac, J. L. (1979). *Proceedings of the 9th International Congress of Plant Protection, Washington DC, .*

Baskerville, G. (ed.) (1976). *Report of task force for evaluation of budworm control.* 210 pp. New Brunswick Department of Natural Resources, N.B.

Betts, E. (1976). In *Insect flight* (ed. R. C. Rainey) p.113. Blackwells Scientific Publications, Oxford.

—— and Haggis, M. J. (1970). *Rec. Res. E. Afr. Agric. For. Res. Org.* **1969,** 109.

—— and Odiyo, P. (1968). *Rec. Res. E. Afr. Agric. For. Res. Org.* **1967,** 111.

—— Odiyo, P. O., and Rainey, R. C. (1969). *Rec. Res. E. Afr. Agric. For. Res. Org.* **1968,** 123.

—— Rainey, R. C., Brown, E. S., Mohamed, A. K. A., and Odiyo, P. (1970). *Rec. Res. E. Afr. Agric. For. Res. Org.* **1969,** 110.

Brown, E. S. (1962). *Bull. Entomol. Res.* **53,** 445.

—— Bettes, E., and Rainey, R. C. (1969). *Bull. Entomol. Res.* **58,** 661.

—— Stower, W. J., Yeates, N. M. D. B., and Rainey, R. C. (1970). *E. Afr. Agric. For. J.* **35(W),** 350.

—— Odiyo, P., Betts. E., Sondugaard, K. M. M., and Onyango, J. (1972). *Rec. Res. E. Afr. Agric. For. Res. Org.* **1971,** 199.

Campion, D. G., Odiyo, P. O., Murlis, A. M., and Nesbitt, B. F. (1976). *COPR Misc. Rep. No. 25.*

Clark, E. C., Jones, D. D., and Holling, C. S. (1979). *Ecol. Model.* **7,** 1.

Courshee, R. J. (1959). *Bull. Entomol. Res.* **50,** 355.

—— (1965). In *Operational research* (Final Report Vol. 1), p.206. Special Fund Desert Locust Project No. UNSF/DL/OP/5. FAO, Rome.

Dickerson, R. B. B., Haggis, M. J., and Rainey, R. C. (1983). *J. Clim. Appl. Met.* **22,** 278.

Dietrich, V., Lucktkemeier, N., and Voss, G. (1980). *J. Econ. Entomol.* **73(3),** 354.

Gameel, O. (1972). *Rev. Zool. Bot. Afr.,* **86,** 50.

FAO (1956). *Report of the panel of experts on long-term policy of Desert Locust control, London, April, 1956.* Rome, FAO.

Garms, R., Walsh, F. F., and Davies, J. B. (1979). *Tropenmed. Parasitol.* **30,** 345.

Graham-Bryce, I. J. (1976a). *Chem. Ind.,* 545.

—— (1976b). *Proc. 8th Br. Insectic. Fungic. Conf.* **3,** 901.

Greenbank, D. O. (1957). *Can. J. Zool.* **35,** 385.

—— Schaefer, G. W., and Rainey, R. C. (1980). *Mem. Entomol. Soc. Can.,* No. 110, 48pp.

Gunn, D. L., Graham, J. F., Jacques, E. C., Perry, F. C., Seymour, W. C., Telford, T. M., Ward, J., Wright, E. N., and Yeo, D. (1948). *Anti-Locust Bull. London* **4,** 121pp.

Gurdas Singh (1972). In *Proceedings of the International Study Conference on Current and Future Problems in Acridology* (eds C. F. Hemming and T. H. C. Taylor), p.475. COPR, London.

Haggis, M. J. (1971). *E. Afr. Agric. For. J.* **37,** 100.

—— (1981a). *Bull. Entomol. Res.* **71,** 183.

—— (1981b). *Workshop on* Heliothis *management (15-20 Nov. 1981),* ICRISAT, Patancheru, Andhra Pradesh, India.

Hocking, B. (1953). *Trans. R. Entomol. Soc. London* **104,** 223.

Joyce, R. J. V. (1952). *Anti-Locust Bull. London* **11,** 103pp.

—— (1953). In *Annual Report of the Research Division, Ministry of Agriculture, Sudan, 1950–1951.* McCorquodale, Khartoum.

—— (1961). *Bull. Entomol. Res.* **52,** 191.

—— (1962). *Report of the Desert Locust Survey 1955–1961.* E. Afri. Com. serv. Org. Nairobi, Kenya.

—— (1974). *Br. Crop Prot. Counc. Monogr. No.* 11, 29.

—— (1975). In *Seminar on the Strategy of Cotton Pest Control in the Sudan Gezira, Wad Medani, Sudan,* CIBA GEIGY, Basel.

—— (1976). *5th Int. Agric. Av. Congr., Nat. Agric. Cent., England* 1975.

—— (1979). *Philos. Trans. R. Soc. London, Ser. B 287,* 305.

—— (1981). In *Animal migration* (ed. D. J. Aidley), p.209. Cambridge University Press, Cambridge.

—— and Roberts, P. (1959). *Ann. Appl. Biol. 47,* 257.

—— Marmol. L. C., Lucken, J., Bals, E., and Quantick, R. (1970). *PANS* **16,** 309.

Kennedy, J. S. (1961). *Nature (London)* **189,** 785.

Kettela, E. G. (1983). *Spruce budworm in Eastern Canada and Maine 1967–1981: a cartographic history.* 23pp. CANUSA, Environment Canada.

Key, K. H. L. (1954). *The taxonomy, phases and distribution of the genera Chortoicetes Brunn and Austroicetes uv. (Orthoptera: Acrididae).* Canberra, Division of Entomology, CSIRO (441).

Lawson, T. J. (1976). *Spray Dispersal Trials.* New Brunswick Spruce Budworm Project 1976. AARU Progress Report 61/76, Cranfield, UK.

—— (1980). *The toxicity of Nuvaion 40 SCW and Curaion 375 ULV to adult Heliothis armigera and Spodoptera littoralis.* AARU Progress Report 103/80, Cranfield, UK.

—— Topper, C. P., and Hull, S. F. (1979). *Adult Heliothis armigera control trials in the Sudan Gezira.* AARU, Progress Report 88/79, Cranfield, UK.

MacArthur, R. (1960). *Am. Nat.* **94,** 25.

MacCuaig, R. D. (1962a). *J. Sci. Agric.* **9,** 677.

—— (1962b). *Bull. Entomol. Res.* **53,** 111.

—— (1962c). *Bull. Entomol. Res.* **53,** 597.

—— and Yeates, M. N. D. B. (1972). *Anti-Locust Bull. London* **49.**

Magor, I. J. and Rosenberg, L. J. (1980). *Bull. Entomol. Res.* **70,** 693.

Miller, C. A. and Kettela, E. G. (1975). In *Aerial control of forest insects in Canada* (Ed. M. L. Prebble), p.94. Department of Environment, Canada.

Odiyo, P. O. (1972). *Rec. Res. E. Afr. Agric. For. Res. Org.* **1971,** 200.

—— (1974). *Rec. Res. E. Afr. Agric. For. Res. Org.* **1974,** 133.

Pedgley, D. E., Reynolds, D. R., Riley, J. R., and Tucker, M. R. (1982). *Weather* **37,** 295.

Prebble, M. L. (1975). *Aerial control of forest insects in Canada,* Department of the Environment, Ottawa, Canada.

Rainey, R. C. (1958a). *Q.J.R. Met. Soc. London* **84,** 334.

—— (1958b). *J. Sci. Food Agric.* **10,** 677.

—— (1958c). *10th Int. Congr. Entomol. (1956)* **3,** 262.

—— (1974). *Br. Crop. Prot. Counc. Monogr. No,* 11 (1974), 20.

—— (1976). In *Insect flight* (ed. R. C. Rainey), p.75. Blackwell, Oxford.

—— and Aspliden, C. I. H. (1963). *Anti-Locust Mem.* **7,** 54.

—— and Joyce, R. J. V. (1972). *7th Int. Aerosp. Instr. Symp. Cranfield 1972* **8,** 1.

—— and Sayer, H. J. (1953). *Nature (London)* **172,** 224.

—— Betts, E., and Lumley, A. (1979). *Philos. Trans. R. Soc. London,* 67.

Ramana Murty, B. V., Roy, A. K., Biwas, K. R., and Khemani, L. T. (1964). *J. Sci. Ind. Res.* **23,** 289.

Riley, J. R., Reynolds, D. R., and Farmery, M. J. (1981). *COPR Misc. Rep. No. 54,* 43pp.

Roome, R. E. (1974). *J. Entomol. Soc. S. Afr.* **37,** 63.

Rose, D. J. W. and Dewhurst, C. F. (1979). *Entomol. Exp. Appl.* **26,** 346.

Russell-Smith, N. A. (1975). In *Seminar on the strategy for cotton pest control in the Sudan Gezira. Wad Medani, Sudan, Feb. 1975.* CIBA-GEIGY, Basel.

Sawyer, K. F. (1950). *Bull. Entomol. Res.* **41,** 439.

Sayer, H. J. (1959). *Bull. Entomol. Res.* **50,** 371.

Schaefer, G. W. (1976). In *Insect flight* (ed. R. C. Rainey), p.157. Blackwell, Oxford.

—— (1979). *Philos. Trans. R. Soc. London,* 459.

—— (1980). *Symp. R. Aer. Soc. Aer. J. June 1980,* 131.

—— Bent, G., and Cannon, R. (1979). *New Sci.,* 441.

Singh, S. R. and Sutyoso, Y. (1973). *J. Econ. Entomol.* **66,** 1107.

Southwood, T. R. E. (1971). *Trop. Sci.* **13(4),** 275.

Spillman, J. J. (1979). *Agric. Av.* **17,** 1.

Stern, V. M., Smith, R. F., van den Bosch, R., and Hagen, K. S. (1959). *Hilgardia* **29(2),** 81.

Taylor, L. R. and Taylor, R. A. J. (1977). *Nature (London)* **265,** 415.

Topper, C. P. (1980). Ph.D. thesis, Cranfield Institute of Technology, Bedford, UK.

—— Lawson, T. J. (1984). *J. Econ. Entomol.* (in the press).

Tucker, M. R., Mwandoto, S., and Pedgley, D. E. (1982). *Ecol. Entomol.* **7,** 463.

Uk, S. (1976). *Air-to-Air spray of spruce budworm moths in 1975. Chemical analysis of spray collection by the moth.* AARU Progress Report 49/76, Cranfield, UK.

Uvarov, B. P. (1977). *Grasshoppers and locusts* (Vol. II). 613pp. Centre for Overseas Pest Research, London.

Waloff, Z. (1972). *Bull Entomol. Res* **62,** 1.

Walsh, J. F., Davies, J. B., and Cliff, B. (1981). In *Blackflies* (ed. M. Laird), p.85. Academic Press, London.

Winteringham, F. P. W. (1974). *Proceedings of EPPO Conference on Integrated Control in Horticulture, Kiev 1974.*

15
Vertebrate pest control

J. H. GREAVES and P. J. JONES

Part 1. Rodent control

15.1 Introduction

15.1.1 Rodents as pests

The *Rodentia* are the largest mammalian Order, containing approximately 400 genera and over 6000 named forms (Ellerman 1949). Though the Order is dominated by the relatively unspecialized *Muridae* (rats, mice, and their allies), it is a very varied group and contains forms adapted to semi-aquatic, desert, arboreal, and completely subterranean habitats. Broadly speaking, about a dozen species are regarded as major pests in global or regional terms, while on a more local scale many times this number are of known economic importance. Among the major pests, three cosmopolitan species, the Norway rat, *Rattus norvegicus,* the roof rat, *Rattus rattus,* and the house mouse, *Mus musculus* are normally found in close association with Man as domestic pests and in food stores; they may also be present as field pests. Other important agricultural pests include *Mastomys natalensis* and *Arvicanthis niloticus* which are widespread in Africa, *Meriones* spp. in North Africa and South-Western Asia, *Bandicota bengalensis* in the Indian subcontinent, various closely related species of the *Rattus* group (e.g. *R. argentiventer, R. mindanensis,* and *R. tiomanicus*) in South-East Asia, and *Sigmodon* and *Holochilus* spp. in tropical America.

Almost every kind of crop, particularly cereals, may be attacked at any stage from sowing to harvest. Damage may be inflicted upon the roots, stems, leaves, buds, and flowers as well as the more obviously attractive parts such as seeds, fruits, or tubers. Rodents are among the classical pests of stored food at all levels from the farm to the consumer. In urban situations damage to consumer items such as clothing and furniture may be important as well as structural damage to buildings and electrical wiring. Rodents can cause great anxiety in the home by their noisy nocturnal activities and by inflicting bites. Disease is a serious hazard presented by rodents around human habitation; among the better known rodent-borne diseases are plague, murine typhus, leptospirosis, food poisoning, and tapeworm infections.

15.1.2   Pest situations

In agriculture, some of the most severe and long-standing rodent problems are associated with subsistence farming. Its pattern of small cultivated plots interspersed with patches of uncultivated land provides a varied and favourable rodent habitat. Paradoxically, however, agricultural improvements such as the introduction of a new crop, or of irrigation or land reclamation, sometimes exacerbate the problem by improving conditions for the rodent population.

Where agriculture has a strongly seasonal pattern, rodent numbers usually increase to a peak with the approach of harvest, and then decrease abruptly with the sudden post-harvest decline in the food supply. The small surviving population, often confined to a limited refuge habitat, constitutes the breeding nucleus from which the next season's upsurge develops. Variation from one year to another in the quality of the refuge habitat controls the size of the surviving population and may thus strongly influence the size of the population peak in the following season. When conditions are exceptionally favourable the potential may exist for a rodent plague to ensue (Taylor and Green 1976; Brei 1977); conversely, unfavourable conditions such as frost, drought, or flooding may keep rodent populations low.

In towns, rodents readily move between individual premises through the sewers as well as above ground, and frequently infest public facilities such as refuse dumps. Infestation is therefore a community problem requiring co-ordinated action. Consequently, where municipal organization is weak, rodent problems tend to be more severe.

15.2   The technical foundations of rodent control

15.2.1   Essential biological and technical information

Successful rodent control is based upon detailed knowledge of the ecology and economic status of the pest species, the merits of the available control techniques, and their appropriateness to the prevailing cultural and socioeconomic conditions.

Information on damage and the distribution of the pest species should be acquired by systematic surveys involving measurements of crop damage and the identification of trapped specimens. Ecological studies are required to answer specific questions relating to control, such as the prediction of damage, the identification of vulnerable points in the life history of the pests, and to establish any limitations imposed upon different control strategies by the ecology of the pest species. Typically, the population dynamics, feeding habits, daily and seasonal movements, and the habitat preferences of the pest species need investigation.

Rodenticidal formulations should first be tested in the laboratory to verify their toxicity and acceptability to the target species, and then rigorously field-tested in the different major environments where it is planned they should be recommended for use. Field trials should yield quantitative estimates of efficacy based on population measurements, in order that different formulations and methods of application may be compared objectively. A useful introduction to rodenticide evaluation techniques has been published by EPPO (1975). Finally, a rodent control technique, once established on a sound technical basis, should be tested in operational conditions by means of closely monitored pilot programmes. These will allow the basic technique to be modified to fit the needs of the ultimate user and will provide estimates of the costs and benefits to be expected in a full-scale programme and help to define training requirements.

15.2.2　*Non-chemical control techniques*

Though this chapter is primarily concerned with chemical control, it should be remembered that there is a role for non-chemical techniques. In agriculture it is important to reduce rodent habitat as much as possible by eliminating patches of uncultivated land, draining swamps, and consolidating small fields into larger cultivated tracts. Clean farming practices such as weed control, increased harvesting efficiency, early post-harvest ploughing and the burning of crop trash are also important. In towns, environmental hygiene is essential to successful rodent control. Rodents should be excluded where possible, for instance by the construction of rodent-proof food stores; other successful types of rodent barrier include wire mesh collars fitted to individual oil palms (Toovey 1953) and smooth sheet metal bands fitted to coconut palms (Montenegro 1962).

15.3　Poisons, baits, and formulations

During the present century poison baits have taken the leading place in rodent control, being, on the whole, more effective and easier to apply than the traditional trapping and hunting methods they have displaced, and more widely applicable than other rodenticidal formulations.

Rodenticides are usually classified as either acute (single-dose or quick-acting) or as chronic (multiple-dose or slow-acting) poisons, a distinction which though based primarily on the speed of rodenticidal action mainly concerns differences between the two groups in methods of application.

15.3.1　*Acute poisons*

Acute poisons typically produce symptoms within an hour of ingestion with death supervening shortly afterwards, usually within 24 hours. Different compounds vary considerably in both efficacy and safety. In general, the more effective acute poisons tend also to be the more hazardous, and restrictions are commonly imposed upon their use and availability. Acute poisons have two advantages over chronic poisons. Firstly, since the bait is highly toxic a smaller quantity is required and the labour of distributing it is less, resulting in a cost advantage. Secondly, acute poisons act quickly and numerous dead rodents may be found within 24 hours of application. The disadvantages of acute poisons are that they tend to be more hazardous and less efficacious than chronic poisons, and that surviving rodents tend to be poison- and bait-shy.

Upwards of a dozen compounds have been employed as acute poisons. Only zinc phosphide, at a concentration of about 2.5% in bait, is at all widely used; it is a broad-spectrum rodenticide widely regarded as the most acceptable compromise between cost, safety, and efficacy (Hood 1972). Crimidine at a concentration of 0.1–1.0% has a limited use against house mice in Europe but is little used elsewhere, possibly because it is regarded as too hazardous (Knudsen 1963; Dubois *et al.* 1948). Sodium fluoroacetate at 0.25% and fluoroacetamide at 0.5–2.0% are both very effective rodenticides but are exceedingly dangerous to Man and other animals; their use is therefore usually restricted by legislation to places where security can be assured, such as sewers, ships, and locked warehouses (Anon, 1965). Alphachloralose at 4.0% in bait is used only in temperate climates and only against house mice. It is a relatively safe compound since, should accidental poisoning occur, recovery usually takes place if the subject is kept warm. Its effect is unreliable at temperatures above 15°C (Cornwell and Bull 1967). Norbormide, at 0.5–2.0% is essentially selective to *Rattus norvegicus* and it is therefore a very safe rodenticide (Roszkowski *et al.* 1964). Owing to limited efficacy, its use is rare except where safety is an overriding consideration (Rennison *et al.* 1968).

Several other acute poisons, some with a long history of use as rodenticides, are still used occasionally but are generally regarded as outdated because of shortcomings in efficiency, safety, or humaneness. These are antu, arsenic trioxide, barium carbonate, endrin, potassium cyanide, pyrinuron, red squill, sodium arsenite, strychnine, thallium sulphate, and yellow phosphorus. Several useful reviews of acute poisons have been published (Gratz 1976; Gutteridge 1972; FAO 1979).

15.3.2 Chronic poisons

Chronic rodenticides exert their toxic effect by a cumulative action resulting from ingestion on several occasions over a period of days. This slow, cumulative action gives them two major advantages over acute poisons. Firstly, bait-shyness does not develop, since the symptoms of poisoning are very delayed and the rodent does not associate them with the poisoned bait. A lethal dose is usually consumed before the onset of symptoms, and even where this does not occur, the rodent normally persists in feeding on the bait until death. The second advantage is that the concentration of rodenticide in the bait can be kept to a level which is high enough to ensure adequate cumulative toxicity but low enough to reduce any acute toxicity hazard to non-target animals to a low level. Because of this combination of high efficacy and relative safety, chronic poisons are the rodenticides of choice in the majority of situations. All chronic poisons are blood anticoagulants with the single exception of calciferol.

Anti-coagulant rodenticides render the blood uncoagulable and the rodents die from haemorrhage, typically 1–3 weeks after starting to feed on the bait. Vitamin K_1 is a specific antidote in cases of accidental poisoning, and will restore the clotting ability of the blood within a few hours of administration, though it cannot of course reverse the effects of any haemorrhages that have already occurred.

Approximately a dozen anti-coagulants have been used as rodenticides (Table 15.1). Warfarin, one of the first, has been the most widely used compound since the early 1950s, but others, notably coumatetralyl, chlorophacinone, diphacinone, and coumachlor are also popular. Less commonly

TABLE 15.1. *Anti-coagulant rodenticides and their usual concentrations in bait*

Rodenticide	Concentration (%)
Bromadiolone	0.005
Brodifacoum	0.005
Chlorophacinone	0.005
Coumachlor	0.025
Coumafuryl	0.025
Coumatetralyl	0.0375
Difenacoum	0.005
Diphacinone	0.005
Pindone	0.025
Valone	0.025
Warfarin	0.025

encountered are coumafuryl, pindone, and valone. Three potent new anti-coagulants, difenacoum, bromadiolone, and brodifacoum have recently been introduced for use against anti-coagulant-resistant strains. Preliminary reports indicate that all three, especially brodifacoum, are superior to the older anti-coagulants for most purposes, though they are somewhat more hazardous to non-target animals (Lund 1984).

Calciferol is a recently introduced chronic poison that acts by producing hypervitaminosis D, thus causing a severe disturbance of calcium metabolism (Greaves *et al.* 1974). It acts relatively quickly, typically causing death within 1 week as compared with 1–3 weeks for anti-coagulants. The compound is used primarily for house mouse control, and its use against other species is not well established. Unlike the anti-coagulants, calciferol has no specific antidote and is therefore a potentially more hazardous material. Calciferol is included in bait at a concentration of 0.1%, but is unstable in moist baits.

15.3.3 Bait bases

Almost every kind of edible material has been used as a base for rodenticidal bait at some time. Ground, broken, cut, crushed, or milled cereals and small whole grains are the most commonly used materials and mixtures of these may be popular locally. Any good quality cereal may be suitable, such as wheat, barley, oats, millet, sorghum, or maize and may be chosen on the basis of local availability and cost. In environments where there is a strongly competing food source, as in food stores and in grain fields near harvest, difficulty may be experienced in attracting rodents to the bait and the palatability of different bait bases must be carefully evaluated. The attractiveness of cereal baits may be increased by the use of additives. Other bait bases used occasionally include processed cereals, peanut butter, chopped fruit or vegetables, ground meat or fish, and even tinned fish. Water is often a cheap and effective bait for use with soluble rodenticides in dry environments such as food stores and on ships.

(a) Additives

(i) *Colouring*. A warning dye is normally included in all rodenticides at a concentration sufficient to impart a distinctive colour to the bait so that it cannot be mistaken for human or animal food, and to act as an indicator to show when the bait has been uniformly mixed.

(ii) *Binders*. To maintain a uniform mix of the rodenticide in bait containing coarse particles such as whole grains, it is essential to add a 'binder' or 'sticker' such as edible vegetable oil, technical white oil, or water. Binders have various secondary effects on the rodenticidal efficiency of the formulation. Edible oils usually increase bait palatability marginally while water increases it significantly. However, the stability of the bait is reduced if the edible oil goes rancid, while water markedly accelerates fungal deterioration. Technical white oil has a slight preservative effect and may potentiate the action of anti-coagulant rodenticides somewhat by impairing intestinal absorption of vitamin K; however, it may also reduce palatability slightly. The use of a corn oil binder facilitates absorption of calciferol. The use of a binder increases the safety of bait-mixing operations by preventing the generation of dust, and in appropriate instances the rodenticide may be slurried with or dissolved in the binder (notably fluoroacetamide or sodium fluoroacetate in water) before mixing with the bait base. Oils are normally included in bait at concentrations between 2 and 5% while water is generally incorporated at a concentration close to the maximum absorptive capacity of the bait base (50–100%) by soaking for from 1 to 24 hours.

Anti-coagulant bait can be formulated as blocks or pellets using paraffin wax as a binder, for use in the warm, humid conditions of tropical agriculture and in sewers. The wax gives the bait partial protection against weathering and fungal and insect attack. The bait is mixed with melted wax, normally in a 6:4 ratio, and the mixture is moulded into blocks or into thick sheets which are cut into blocks before the mixture has set hard.

(iii) Attractants. Apart from the attractant properties of water and edible oils mentioned above, only sucrose (at a concentration of about 5%) is well established as improving bait palatability. Many other substances are supposed, with little or no objective basis, to have attractant properties. Thus, the value of various spices, essences, and artificial flavourings is questionable; some such as aniseed oil and coriander seed have been shown to have repellent effects (Drummond 1968). Generally it is a good and sufficient rule to rely upon the use of good-quality basic ingredients to produce a palatable bait.

(iv) Preservatives. To stop baits going mouldy in warm, moist conditions, paranitrophenol or dehydroacetic acid may be included at 0.25% and 0.1% respectively, with some loss of palatability (Larthe 1957). Though insect attack on bait is rarely of great concern, it may be noted that the rodenticides zinc phosphide, sodium fluoroacetate, fluoroacetamide, and the anti-coagulant pindone have insecticidal properties.

15.4   Rodenticide treatments

15.4.1   Timing

There are essentially two approaches to the timing of rodenticide treatments, the choice depending on local conditions and on the way in which infestation and damage develop.

(a) Preventive treatments (with an anti-coagulant rodenticide). The treatment may be done on a permanent basis in places continually subject to infestation or, depending on local conditions, throughout the growing season of a vulnerable crop. In some cases it may be restricted to the few weeks after sowing or before harvest, or from the first sign of flowering of cereal crops onwards. Preventive treatment can be planned well ahead and eliminates the need to monitor infestation; they are particularly effective in environments highly vulnerable to infestation and in controlling rats moving about in urban areas (Shaefer 1975; Drummond *et al.* 1972).

(b) Remedial treatment when significant damage begins. In public health the

presence of a single rodent in domestic or commercial premises presents an unacceptable health hazard. Remedial treatments in agriculture require a readiness for immediate action if substantial crop losses are to be avoided. Because remedial treatments tend to consume fewer resources and because visible economic damage is a strong spur to action, they are probably the most usual form of treatment.

15.4.2 Distribution of bait

Bait is usually distributed in small strategically sited heaps known as bait points or bait stations. For safety and efficiency the distribution and size of bait points should be closely matched to the distribution, numbers, and behaviour of the rodents. In general, the use of large numbers of small bait points ensures that every rodent has access to bait with minimal competition from conspecifics, whereas fewer, larger bait points are labour saving. Where infestation is limited to field edges it will be appropriate to limit bait distribution accordingly.

Bait may be brought into as close proximity with the rodents as possible by placing it inside burrow entrances and where there are signs of surface activity such as runways or feeding sites. This method will often be preferred where infestation is localized and the rodent signs easy to find. It has the advantage of intercepting rodents as they move from their nesting places to their feeding sites. Alternatively, the bait may be distributed in a predetermined grid pattern, typically between 30–150 bait points per hectare or at 3–10 m linear intervals. In individual cases experimental studies may show that a much lower baiting density is adequate, or that the broadcasting of bait is to be preferred. Grid baiting will be the preferred method in preventive treatments and where infestation is widely dispersed.

In special circumstances other methods of bait distribution may be called for such as the placing of bait in the crowns of coconut palms (Smith 1967), the fixing of bait containers in forest trees (Santini 1978), the use of floating rafts for baiting semi-aquatic species (Kuhn and Peloquin 1974), and the use of mechanical bait dispensers or of a special probe to locate and bait the burrows of more fossorial species such as pocket gophers (Canutt 1970).

15.4.3 Baiting with anti-coagulants

Since anti-coagulants are ineffective unless consumed for several consecutive days, the bait must be applied in such quantities that it remains available for consumption by the rodents until death. This will rarely be less than 2 weeks and may take up to 5 or 6 weeks, depending upon the efficiency of the baiting and the susceptibility of the target species. Adequate application may be ensured by replenishing the bait at intervals of 3–10 days, taking care that no bait point is allowed to be completely depleted between visits. Replenishment should continue until all bait consumption ceases. With experience, the need for bait replenishment can be minimized by increasing the amount of bait distributed initially, and in agricultural situations an acceptable level of control may be obtained with only two or, exceptionally, one bait application. Typical application rates are 1–5 kg/ha distributed in 20–300 g bait points.

15.4.4   Baiting with acute poisons

(a) Direct poisoning. Since acute poisons are effective as a single dose, the bait can be distributed in smaller quantities than with anti-coagulants. Theoretically, because of their high toxicity, dosages of acute poisons could be only one-tenth that of anti-coagulants; however, to secure adequate coverage it is often necessary to use as much as a quarter to a half the amount that would be required in an anti-coagulant treatment. A typical minimum rate of application would be 0.5 kg/ha distributed in 5–20 g bait points, though higher rates may often be required.

(b) Prebaiting with unpoisoned bait. The effectiveness of direct poisoning is often severely limited by the fact that consumption of poisoned bait is tentative or intermittent, with the result that the rodents experience early warning symptoms and fail to ingest a lethal dose. Readiness of the rodents to eat the poisoned bait can be markedly increased by first prebaiting, i.e. distributing unpoisoned bait for 3–10 days, and replacing it with poisoned bait when the prebait is being eaten freely. Observation of the amounts of prebait consumed enables the necessary distribution of poisoned bait to be gauged more accurately, with consequent gains in efficiency and safety. However, an acute poison treatment with prebaiting may cost as much as an anti-coagulant treatment and it may well be less effective.

15.4.5   Protection of baits

Bait points sometimes need protection from the elements and from interference by humans or livestock. This is particularly so with anti-coagulant baits, which are at greater risk through having to be kept in position for a minimum of several days. One of the simplest types of protection is to lay the bait in small dishes or trays, which by preventing soil contact, help to stop the bait going mouldy; it is preferable to use disposable cardboard or plastic trays, since reusable trays may become permanently infected with mould-producing organisms. A common form of protection is some type of container such as a box which may be lockable, with access holes for the rodents and constructed from waterproofed cardboard, wood, plastic, sheet metal or even concrete in accordance with local requirements. Less sophisticated substitutes such as drain pipes, bamboo tubes, or simple thatched shelters may be appropriate. In some cases bait can be adequately protected by concealing it under whatever suitable debris is most readily to hand. In other cases, particularly in agriculture, grain baits may be rendered weather-resistant if formulated as wax blocks or, more especially, sealed in plastic or waterproofed paper bags. Bagged and block baits are particularly easy to distribute and if the protection they afford is adequate the considerable trouble of providing specially constructed bait containers is avoided (Teshima 1976; Wood 1971).

15.4.6 *Safety precautions*

Rules for the safe use of rodenticides, as established under pesticide registration or other laws, are normally included in the label. Typically, such rules might specify secure storage of rodenticides, prevention of access to bait by children and non-target animals, labelling bait containers, wearing rubber or PVC gloves when mixing bait, precautions against inhaling toxic dust or ingestion of bait, and against accidental skin contact.

Small quantities of bait – up to 2 kg – are normally mixed by hand in a bucket with a long-handled all-metal spoon. Larger quantities should be mixed mechanically in a vessel reserved exclusively for the purpose, such as a seed-dressing drum, cement mixer, or industrial food-mixer. To prevent the spread of toxic dust during bait-mixing the vessel should be closed, unless the dust is suppressed by including a binder such as oil in the formulation.

When distributing bait, safety is best served by setting out the smallest total quantity that is consistent with the requirement for effectiveness, by similarly restricting the amount of bait placed at any one point, and by protecting or concealing the bait against interference. It is a good management practice and guards against a casual attitude to keep fairly detailed records of the placement and disposal of baits. The posting of warning signs may be advisable in some cases.

At the end of a treatment, potentially hazardous bait residues should be collected up. This is facilitated by the use of non-spill bait dishes or trays, which are in any case essential when liquid baits are used. Dead rodents may also be collected up, to avoid the risk of secondary poisoning of carrion-eaters, and destroyed together with bait residues by burial or incineration.

15.4.7 *Aerial baiting*

Aerial broadcast of bait from light planes or helicopters is occasionally used for rodent control in forests, rangeland, sugar cane, and, more exceptionally, on ordinary farmland. Both acute poisons and anti-coagulants have been applied with seeding or fertilizer attachments, using whole grains, oat groats, or small waxed pellet baits. Chopped carrot baits are distributed by air for rabbit control in New Zealand. Bait application rates are generally in the range of 2–10 kg/ha and aim to distribute 20–50 bait particles per square metre. Pank (1976) has noted that up to 50% of the active ingredient may become separated from the bait base during aerial application of conventional formulations and, after a carefully conducted study, recommended lecithin oil mixed into the bait as being a particularly efficient binder.

Areas up to 100 ha per hour or more may be treated by air, with very significant savings of time and labour (Podlishuk 1971). Results may, however, be disappointing unless the technique has been carefully developed on the basis of local experimental studies of efficacy and safety (Redhead 1971; Kerkwyk 1974). The success and safety of aerial baiting may depend critically upon the proficiency of the pilot in several technical aspects of the control technique, including the identification of signs of infestation. Careful organization and monitoring of the treatments may be required including preliminary and post-treatment surveys and may be the subject of strict legislative control (Schilling 1976).

15.5   Non-bait applications of rodenticides

Although poison baiting is generally found to be the simplest and most efficient way of controlling rodents, there is a limited range of situations, chiefly those in which it is difficult to induce the rodents to eat bait, where non-baiting techniques are useful.

15.5.1   *Rodenticidal dusts*

When rodents inadvertently come into contact with a rodenticide-containing dust, some that adheres to their feet, tail and fur is subsequently transferred to the mouth and ingested in the normal course of their self-grooming behaviour. Rodenticidal tracking dusts are particularly useful for dealing with infestations in food premises, and are also used in treating domestic and other premises, as well as rodent burrows outdoors (Giban 1958; Green 1971). Since very small amounts of the dust are ingested, the concentration of the rodenticide must be high, typically 20–40 times as great as would be used in a bait formulation. Several anti-coagulants are available as dusts, and the use of acute poisons as dusts is occasionally reported.

The usual method of application is to lay the dust in patches 2 mm thick in places where the rodents run, smoothing the tracks away at intervals of a few days and if necessary replenishing the dust until rodent activity ceases. Dusts may also be applied with a blower to burrows, ducts, or voids frequented by rodents, or set out in shallow trays or suitable boxes, or used in conjunction with a baiting technique. The advantages of rodenticidal dusts are that they by-pass the vagaries of rodent feeding behaviour and remain effective over a longer period of time than do baits. Their two main disadvantages, both due to the high concentration of toxicant, are high cost and potential hazard: great care must be taken to prevent accidental contamination of food and working surfaces and contact with non-target animals.

15.5.2 Burrow fumigation

Formulations of aluminium phosphide and of sodium or calcium cyanide, designed respectively to generate phosphine or hydrogen cyanide on contact with moisture, can be useful for treating burrow systems in damp well-consolidated soil (Fernando *et al.* 1967). Aluminium phosphide is available as 3.0 or 0.6 g tablets, generating respectively 1.0 or 0.2 g of phosphine, and cyanide as powders or granules. The fumigant is inserted well into every burrow and the entrances sealed firmly with soil. Dosage rates depend on the size of the burrow: for typical small-rodent burrows one 0.6 g aluminium phosphide tablet or one teaspoonful of cyanide powder or granules will often be found to be adequate. The fumigant is normally applied with a long-handled spoon, though a pump may be used to apply cyanide gassing powders. The method is unsuitable for use where a strong growth of vegetation makes location of the burrow entrances difficult or increases the porosity of the soil: the fumigant readily leaks away, sometimes creating a hazard, and rodents may bolt from unsealed entrances. For safety reasons burrow fumigants should never be used in the vicinity of buildings. Cyanide gassing powders are extremely hazardous to operators unless used with the greatest care.

15.5.3 Ground spraying

Formerly, organochlorine insecticides such as endrin and dieldrin were widely used as ground sprays for the control of microtine rodents in plantations and orchards. This use of persistent insecticides is now outdated, and has generally been replaced with baiting techniques. Latterly, the use of the anti-coagulant rodenticide, chlorophacinone, applied at a rate of 0.34 kg/ha has been developed for use against the pine vole *Pitymys pinetorum* in orchards in the USA (Byers *et al.* 1976). In this case the effectiveness of the treatment has been shown to depend upon its presence in the treated vegetation of plants utilized as food by the rodents (Horsfall *et al.* 1974). The use of rodenticides as ground sprays poses safety and wildlife hazards and therefore must be regarded as experimental at present.

15.6 Resistance to rodenticides

Genetic resistance to warfarin and other traditionally used anti-coagulant rodenticides has been a problem in the control of the three cosmopolitan species in Britain for several years, and has also been reported from Denmark and North America (Greaves 1971; Lund 1972; Jackson and Kaukeinen 1972). In *R. rattus* the resistance is believed to be polygenic (Greaves *et al.* 1976). In *R. norvegicus* resistance is usually due to a single dominant gene and involves an alteration in the metabolism of vitamin K

(Greaves *et al.* 1977). Three cases of monogenic resistance have been recorded in *M. musculus* (MacSwiney and Wallace 1978) though in this species a polygenic mode of inheritance has also been suspected (Rowe and Redfern 1965). Resistance to anti-coagulants seems liable to develop wherever these rodenticides are in regular use, though it must be emphasized that known resistant populations are at present very limited in distribution.

Both resistance and bait-shyness (which may be regarded as an acquired form of behavioural resistance) are sometimes blamed, on the basis of inadequate evidence, for the failure of control measures. Control failures most commonly result from inadequate distribution of bait, the use of an insufficiently attractive bait, the migration of rodents into the treated area, or the use of too low a concentration of the rodenticide. It is important to exclude these possibilities before suspecting resistance. Field studies by a trained investigator and comparative toxicity tests in the laboratory with rodents from the suspected-resistant population and from a known-susceptible population are required in order to confirm the presence of resistance (WHO 1970; Drummond and Rennison 1973).

Acute poisons, calciferol, and to a limited extent coumatetralyl, retain their effectiveness against anti-coagulant-resistant rodents. The three anti-coagulants, difenacoum, brodifacoum, and bromadiolone, have been developed specifically for the control of rodents resistant to other anti-coagulants.

Part 2. Bird control

15.7 Introduction

The status of birds as pests is distinguished by two important features of their natural history: their great mobility and their tendency to aggregate in vast flocks when feeding, roosting, or breeding. Flocking behaviour is characteristic of birds that exploit locally abundant but patchily distributed food sources such as scattered, cultivated fields. Many species are thus preadapted as agricultural pests and it is here that birds are most important economically. Other species may be pests when presented as large roosting assemblages, for example at airfields and in urban areas.

Unlike rodents, birds that become agricultural pests do not have reproductive rates sufficient to permit a rapid increase in population size within a single cropping season; they are usually common anyway and their populations are regulated by factors unconnected with the availability of the cultivated crop. Thus unlike rodents, which have limited powers of dispersal and whose populations crash once the crop is harvested, birds are easily able to congregate in an area for as long as it is worth exploiting and to disperse or migrate again afterwards. These features impose strict limitations on the various strategies applicable to bird damage control and on the techniques employed.

15.8 Types of bird damage

Relatively few species achieve notoriety as major crop pests, though more may be implicated on a small scale in local pest complexes. Crops may suffer bird damage at any stage of development; the seeds may be taken immediately after planting (e.g. by pigeons), at germination (e.g. pheasants and geese on various cereals), seedlings may be grazed (e.g. beet seedlings by skylarks *Alauda arvensis*), maturing grain is taken by many species of granivorous birds, and post-harvest losses may occur during storage or when used as cattle feed at feedlots (e.g. by starlings *Sturnus vulgaris*). Bud damage to fruit trees in late winter (e.g. by bullfinches *Pyrrhula pyrrhula*) as well as damage to maturing fruit (e.g. by starlings) may seriously reduce yields. Purely mechanical damage may be caused to crops by trampling (e.g. various waterfowl roosting in rice), and plantation trees may be broken by the weight of roosting birds or killed by the accumulation of their droppings. Birds roosting on airfields (e.g. gulls and waders) may pose a serious hazard to aircraft, airstrikes causing severe structural damage and occasionally loss of life. Bird aggregations may also pose a public health risk (e.g. the possible transmission of salmonellosis by gulls at reservoirs). Introduced exotic species may also be considered as pests where they compete wth endangered or desirable indigenous birds, or even prey directly upon them (e.g. introduced barn owls *Tyto alba* preying upon native Fairy Terns *Gygis alba* in the Seychelles).

15.9 Choice of control method

The choice of control method requires a careful appraisal of the bird problem, an appreciation of the ecological context in which the damage occurs, and a decision on the type of solution needed, e.g. a distinction between the permanent removal of birds necessary, say, at an airfield or food warehouse, and the temporary removal or deterrence of birds from a field crop for the period it is vulnerable. One of three basic strategies may be adopted to avoid or reduce damage by birds, and various control methods (tactics) may be employed to carry these out. Such tactics include the use of non-chemical control methods as well as pesticides. Non-chemical methods deserve mention here, partly because they are more numerous, partly because they frequently offer a preferable alternative, both economically and environmentally, and also because chemical methods are often more effective when used in conjunction with other techniques.

The three strategies and their associated control techniques are listed here in order of increasing environmental damage (see also review by Dyer and Ward 1977).

(1) The avoidance of bird damage by crop or habitat management, e.g. by a change of crop or crop husbandry practice, the complete enclosure of a valuable crop where it would be economically worthwhile to do so, the modification of airport habitats to be unattractive as roosts, and the provision of attractive wildlife refuges as alternative roosting or feeding sites.

(2) Deterring birds by planting 'bird-proof' varieties of grains, the use of visual deterrents such as bright flashing colours, model or live hawks, corpses or model conspecifics in unnatural postures to deter others from landing, the use of acoustic scaring devices broadcasting distress or alarm calls or loud 'white noise', gas cannons, or by shooting over the crop.

(3) The destruction of the pest species by trapping, nest destruction, shooting, explosives, or poisoning by baiting or aerial spraying with avicides, in order to reduce the population to a level where serious damage no longer occurs.

15.10   Problems in bird control

Certain problems almost always arise when bird-damage control is contemplated, whether the chosen method is chemical or otherwise. These considerations play an important role in the choice of strategy and the actual control method.

15.10.1   Public opinion

Unlike the control of insects and rodents, bird control is an emotive subject in many countries and public sentiment may impose severe constraints on it. Birds are normally pests to only a small part of the community, usually farmers, for only part of the year. To others, such as wildfowlers, some species afford legitimate quarry for hunting. To most, however, birds simply form an enjoyable part of the natural environment. As a result, protective legislation concerning bird populations severely restricts what is permissible in bird-damage control. Past uncontrolled use of pesticides resulted in widespread environmental pollution with particularly detrimental side-effects on bird populations, justifiably causing a public outcry. Public feeling still runs high where birds are at risk and particularly where they are the direct object of attack, even if they are acknowledged pests.

15.10.2   Population recovery

A popular opinion still persists that bird populations are somehow vulnerable and that any concerted effort could eliminate them if desired. Experience in the field has proved to be quite different, however, and notable large-scale failures to achieve population control have included the massive eradication campaigns using chemicals waged against Red-billed Queleas (*Quelea quelea*) in Africa and against blackbirds (*Agelaius* supp.) and starlings in the USA, as well as the government-sponsored shooting of Wood Pigeons (*Columba palumbus*) in Britain. In all these cases the artificial mortality imposed by the control campaigns merely replaced that occurring through natural population processes, in effect culling from the population some of the individuals that would soon have died naturally. Thus the remaining birds were assured of a better chance of survival and the overall annual mortality was increased little, if at all (Murton 1966; Jones 1980).

It is now realized that the relief from damage afforded by lethal control methods is only temporary and localized, depending largely on the rate of reinvasion. Many species are migratory so that reinvasion is certain during the same season or the next. Where damage is caused by birds on passage the period of relief obtained from a control operation may be only a matter of a few days (Jones 1975). In other circumstances the period may be long enough for the crop to be harvested, after which the birds can be ignored.

15.10.3 Mobility

Repellent chemicals and other disuasive techniques do not protect crops from damage if the birds are otherwise faced with starvation. Because of their superior mobility birds are able to sample a great many different feeding sites during the course of a single day, unlike insect and rodent pests. If they then settle in a particular crop to feed it may be because they prefer to feed there rather than on wild food or crops elsewhere, or because there is no food elsewhere and they have no choice. A repellent is effective in deterring birds only if it renders the crop a less attractive place to feed than *alternative* areas or foods nearby. Such alternatives may include natural feeding sites but often they are simply the neighbouring fields that are not protected, so that one farmer's gain is another's loss. In many cases where bird damage has been examined in an ecological context the reason for the damage has been found to be a temporary shortage of the birds' preferred wild food. As a result, even where all the crops were 'protected' by repellents, damage continued as before.

15.10.4 Technical constraints

The range of chemicals available for use as repellents or as acute avicides is small. Although many hundreds have been tested, few were found to satisfy environmental protection legislation requiring that the compounds were either harmless to other wildlife, livestock, and humans, or could be applied specifically against the bird pests concerned. In addition chemical companies have been reluctant to promote such chemicals that would invite adverse public reaction and at the same time they have been unwilling to invest in the research and development of specific chemicals that even worldwide would have small total sales. The chemicals available were often developed primarily to be widely used as insecticides and their avicidal properties are incidental.

15.11   Chemical application techniques

Much of the work on acute avicides and chemical repellents remains at the experimental stage and most governments impose strict conditions on their use. As a result few application techniques have become standard or gained widespread acceptance, but some generalizations can be made.

Chemical repellents directly protect field crops by making them unpalatable to birds. Application is therefore always on to the crops themselves, either by air or ground spraying. Other chemicals may be used as 'frightening agents'. These are actually acute poisons but death is preceded by a period when the poisoned birds show considerable distress, uttering distress calls and behaving erratically. Under ideal circumstances the remainder of the flock become so frightened that they leave the area altogether. Chemical frightening agents are not normally applied directly to the crop because they are extremely toxic and great care must be taken to destroy treated plants before harvest. Since the objective is to develop rapidly an aversion to the feeding site, frightening agents are best offered in baits.

In places where it is certain that only the target species will be exposed to the bait, the poisoned bait may be offered undiluted in troughs or on bird tables. Broadcast bait is normally offered diluted with untreated bait; whereas a single treated bait particle normally contains a lethal dose for small-bodied target species, any large non-target species, e.g. game birds or waterfowl, would need to eat a large amount of diluted bait to ingest a lethal dose. A typical dilution rate for bait where the objective is to kill every target bird is 1:10. Where the bait carries a frightening agent the dilution can be greater, e.g. 1:33 or even 1:99, since only a few birds need be affected to achieve the desired result. The effectiveness of frightening agents depends on a sufficient number of birds ingesting the bait in a short time, e.g. within $\frac{1}{2}$–1 h, and responding quickly enough for the remainder to be frightened away before substantial damage occurs. At low bird densities the treated bait may not be taken quickly enough at 1:99 dilutions; an increase in the amount of bait put out does not improve the situation since the chance of discovery of treated particles remains the same. Instead, an increase in the proportion of treated particles to a 1:33 dilution should increase the response rate threefold.

Surfactants and the acute avicides that are absorbed through the skin are sprayed directly on to the birds, either from the air or by ground spraying apparatus. The technique of aerial application varies considerably depending on the type of aircraft and equipment available, with no particular combination of outstanding advantage. For *Quelea* spraying in Africa fixed-wing aircraft have been single-engined planes capable of a slow speed (Piper Super Cub, Piper Pawnee, Cessna 180/185) and suitable for fitting with either boom-and-nozzle spray gear or Micronair windmill-driven rotary atomizers. Nozzle sizes are normally D8–D12 on fixed-wing planes and D6–D9 for helicopter application. In experimental conditions large droplets (180–250 µm) were much more effective at hitting flying *Quelea* than small droplets (30–75 µm) (Ward and Pope 1972). In ULV spraying, however, the objective has been to drift a fine fog of spray into a roost or colony; droplet sizes as small as 30 µm achieved high kills with helicopter application and 70 µm droplets were successful using a fixed-wing aircraft (La Grange and Jarvis 1977, unpublished report, Department of National Parks and Wildlife Management, Rhodesia). For normal application, droplets slightly larger than 100 µm are preferred since they are better at penetrating plumage, and spray drift is reduced, allowing the pilot to fly higher and so be less at risk of hitting obstacles and of excessive bird strikes. It is common for the windscreen of a fixed-wing plane to become opaque and the engine air intakes to become clogged with bird remains. Helicopter rotor blades are easily damaged and accumulate drag so that the necessary rotor speed can only be maintained with full power (see also Bauer 1966). The disadvantages of aerial spraying are that it is expensive, dangerous, and requires such a high degree of skill from the pilot that it is usually the competence of the pilot above all other factors that determines the success of an operation. For this reason some progress is now being made with ground-based systems, such as tractor-mounted mist-blowers, used upwind to drift spray into a control site.

15.12 Chemicals available

Suitable chemicals for bird-damage control are mentioned here regardless of any local legislation governing their application. The list is not exhaustive and is intended only as a guide to the principal compounds available. The dosages and application techniques are likely to differ in each new situation where the compounds are used.

15.12.1 Repellents and frightening agents

(a) Methiocarb (Mesurol) – 3,5-dimethyl-4-(methylthio)phenol methylcarbamate. This has become the best known and most widely used bird repellent, principally for the protection of cereals and fruit at all stages of maturity. It has been used as a water slurry or mixed with dry graphite powder at concentrations of 0.25–0.5% by weight to protect seeds and seedlings of maize, rice, and soya beans from blackbirds *Agelaius* spp.,

pheasants, and doves (West *et al.* 1969; Stickley and Guarino 1972; Crase and DeHaven 1976) and has reduced damage to ripening rice, sorghum, and wheat at application rates of 1.4–11.3 kg/ha.

Methiocarb at 0.12% in water gave significant protection to ripening cherry trees when sprayed at 2000 litres/ha or until the trees were dripping wet (Guarino 1972) and may indeed repel birds from the whole orchard (Stickley and Ingram 1973), but at 0.04% concentration it seems to be ineffective (Guarino *et al.* 1973). However, 0.5% methiocarb applied at a lower dosage of 475 litres/ha did not significantly reduce damage, though the reasons for this were not clear (Dolbeer *et al.* 1973). Some field workers have reported that Methiocarb causes marked changes in bird flight paths and foraging behaviour, but such effects appear to depend on the size of the treated plots being small. With larger treatment areas the repellency is markedly reduced, probably because the birds can less easily find untreated plots as alternative feeding areas.

(b) Avitrol (4-aminopyridine). This highly effective frightening agent was first shown to be successful in controlling bird damage by Goodhue and Baumgartner (1965a,b). In Redwing Blackbirds *Agelaius phoeniceus,* for instance, a lethal dose is necessary to produce distress behaviour, which typically begins 15 minutes after ingestion and persists for 20–30 minutes before death intervenes (Guarino and Schafter 1967). Distress behaviour exhibited by only 1% of a blackbird population may be sufficient to frighten away the remainder completely (DeGrazio *et al.* 1972).

Successful treatments included clearing magpies, *Pica pica,* by baiting with poultry pellets treated with 2% by weight of Avitrol (equivalent to 1.4 mg per pellet) diluted in the bait in a 1:9 ratio with untreated pellets (Guarino and Schafer 1967). Unusually in this instance, the distress calls of the affected magpies did not frighten other magpies but attracted them; the affected birds were killed by the newcomers, which then took the bait themselves.

Treatments include spraying partially husked ears of standing maize with a viscous solution of Avitrol, so that each kernel contained a 0.5 mg lethal dose for a blackbird (DeGrazio *et al.* 1971). The treated ears had to be burnt as they remain intensely toxic. Spraying a crop is therefore not possible and the normal application is by baiting.

Successful baits comprised cracked maize containing 3% Avitrol by weight diluted in a 1:29 ratio with untreated maize applied at 1.1 kg/ha, or diluted with 1:99 parts untreated maize at 5 kg/ha (DeGrazio *et al.* 1972; Stickley *et al.* 1972). A recommended commercial treatment consists of 3% Avitrol-treated cracked maize bait diluted 1:99 with untreated maize applied at 1.1 kg/ha by air. At low blackbird densities a dilution ratio of only 1:33 has been recommended (Dolbeer *et al.* 1976; see also Besser 1978). Although aerial application is often more convenient, better results may be obtained by hand broadcasting near to resting places favoured by the birds (Besser 1978). There appear to be no secondary hazards to predators consuming affected birds, and the mortality of non-target species is usually reported as negligible.

15.12.2 Acute poisons and narcotics

(a) Starlicide (3-chloro-p-toluidine hydrochloride). A non-violent toxic compound developed for the control of starlings by baiting. Death is slow (and therefore normally away from the baiting site) and non-violent so that bait avoidance and site aversion do not develop. There is no hazard to avian or mammalian predators eating affected birds and it has a low toxocity to some other non-target species, e.g. house sparrows *Passer domesticus* (DeCino *et al.* 1966). Starlicide is effective when bait acceptance is good but Avitrol would be recommended if acceptance is poor (West *et al.* 1967).

Bait is normally prepared by treating poultry pellets to contain 1% or 2% Starlicide (Besser *et al.* 1967; Guarino and Schafer 1967; West *et al.* 1967; West 1968). In animal feedlots the bait may be exposed undiluted in troughs protected from the livestock but broadcast bait is diluted 1 : 10 with an untreated mixture of 1 : 3 pellets/cracked maize. The inclusion of maize may attract non-target species that do not eat the bait pellets but which in turn attract starlings to the bait (West 1968). Other suitable baits for starlings include raisins and rolled barley but the latter also attracted non-target *Zenaidura* spp. doves (West *et al.* 1967). Broadcast pellets remained for only 2–3 days in feedlots before being trampled and crushed; in pasture they quickly dissolved in rain but remained for several weeks in dry weather (West 1968).

A closely-related compound, 3-chloro-*p*-toluidine, has been used as a perch toxicant for use in urban areas where baiting was impossible. It was successful as a 20% paste formulation applied in 3 mm thick beads on roosting ledges (Schafer *et al.* 1969).

(b) Fenthion {O,O-dimethyl O-[3-methyl-4-(methylthio)phenyl] phosphorothioate}. Fenthion has had its widest use in *Quelea* (weaver bird) control in Africa under the trade name Queletox. This replaced the similar organophosphorous poison parathion, which was eventually considered too dangerous for bird control and has since been banned for this use in many African countries. The objective in weaver bird control has always been to kill as many birds as possible per operation. Fenthion is sprayed directly on to the birds and the most efficient application method is aerial spraying of the often vast areas of nesting colonies or roosts, usually at dusk or at night when the maximum possible number of birds is present.

The spray liquid is normally prepared from the manufacturer's 60% a.i.

(active ingredient) emulsion diluted with commercial diesel fuel as a carrier. Diesel is very effective in penetrating bird plumage to carry the poison to the skin where it is absorbed and is preferred to other carriers since it is cheaper and universally available. Typical application rates from fixed-wing aircraft are 20–25 l/ha of 20% fenthion (equivalent to 4–5 kg/ha) (Pope and Jones 1978; Jazovane 1978), and by helicopter of 10–24% fenthion at 40–60 litres/ha (4–14.4 kg/ha) depending on the type of target (Bauer 1966). Typical application rates for tractor-mounted mistblowers against *Queleas* are of 30% fenthion at between 30 and 60 litres/ha (LaGrange and Jarvis 1977). Against the Spanish sparrow *Passer hispaniolensis* in Morocco 25–50% fenthion is effective when applied at 20 litres/ha (5–10 kg/ha).

It is probable that these high dosages could often be considerably reduced without altering the numbers of birds killed. In a series of spray trials on *Queleas* a dosage of 2.9 kg/ha produced high immediate kills (Jackson and Park 1973), but this dosage was more effective using 60% fenthion applied at 4 litres/ha, than when applied as 24% a.i. at 10 litres/ha. Even lower applications by helicopter of 37.5% fenthion at 4.7 litres/ha (1.8 kg/ha) and 60% a.i. at 1 litre/ha (0.6 kg/ha) were also successful (E. Dorow, personal communication). Experiments with very low dosages of fenthion on weaver birds, although not causing death from direct toxic effects, disturbed their feeding behaviour to such an extent that they died later apparently from starvation (Pope and Ward 1972). Conventional assessments of kills might therefore be seriously underestimated if a significant proportion of birds died later outside the sprayed area.

In the USA 11% fenthion has been successfully used in wick perches in areas where toxic baits are not permitted or where bait-shyness is evident. The perch is a hollow metal tube containing a permeable wick that contacts the birds' feet. From 10 to 50 seconds contact may be required for absorption of a lethal dose depending on the bird species and body size; an alternative poison used in such perches is 9.4% endrin (Jackson 1978).

(c) Phoxim (a-cyanobenzylideneaminodiethyl phosphorothionate). Phoxim has a lower mammalian toxicity than fenthion and is an effective alternative acute avicide. Experimental aerial application for *Quelea* control was successful at 42, 64, and 53% applied at 30 litres/ha using diesoline as solvent; it has not yet been used on a large scale as it is more expensive than fenthion (Pope and King 1973).

(d) Surfactant PA 14 (a linear alcohol ethoxylate). PA 14 is a powerful wetting agent that markedly reduces the water repellency of bird plumage. When followed by drenching (by water spray or rainfall) and low ambient temperature death occurs rapidly from hypothermia. PA 14 has potential for large-scale eradication of bird populations where the use of organophosphorous poisons is banned, such as against roosting blackbirds and starlings

in the USA. Successful applications were initially made at low concentration, 0.1–2.0% PA 14, and high volume, using a B26 aircraft capable of dropping its 3800 litre load within 0.4 ha. The surfactant spray was followed by several applications of water to drench the birds. However, the use of such a large aircraft was inefficient, and low-volume, high-concentration sprays are now used, one spray run applying 25% PA 14 at 750 litres/ha or three runs applying 8% solution. The highest mortalities are obtained on nights with temperatures experienced by the wet birds between $-5°C$ and $+1°C$, together with rainfall exceeding 10 mm. Accurate weather forecasting is therefore essential before control is attempted (LeFebvre and Seubert 1970). The technique is unlikely to be useful in warmer climates where night temperatures remain high.

(e) α-Chloralose. This narcotic is sometimes used for the control of urban bird species, usually where bird protection legislation specifically prohibits the use of acute poisons, as in Britain. As a narcotic α-chloralose has a safety margin between a hypnotic and a lethal dose, which is particularly important where non-target species may be at risk, since they can normally be picked up and allowed to recover. Thus in operations against feral pigeons and house sparrows the total numbers of protected species affected accidentally was small (0.11 and 0.47%) and of these 42 and 76% recovered (Thearle 1968).

A major disadvantage of α-chloralose is that it often takes 20–50 minutes before a pigeon is immobilized, during which time it may have flown some distance. With sparrows the drug acts more quickly and they do not often move far from the feeding place (Thearle 1968). When used to eliminate town-nesting populations of herring gulls *Larus argentatus,* baiting using a lethal dose (150 mg of α-chloralose + 50mg of Secanol) is carried out only at the nest when incubation has properly commenced or when the young have hatched. Only the target nesting birds are killed and death is almost certain to occur on the nest or very close by (S. Kearsey, personal communication). Nevertheless, to avoid health hazards and complaints from the public, all such operations require very large ground staffs to collect dead and immobilized birds as quickly as possible.

Where a particular method of control fails the remedy should not simply be to try larger doses or a more potent chemical. It is quite possible that that particular control strategy may not have been appropriate in the first place. The ecology of the pest species and the context in which damage occurs should be re-examined; often other strategies will be found to be worth pursuing (Dyer and Ward 1977).

15.13 References

Anon. (1965). *Use of fluoroacetamide and sodium fluoroacetate as rodenticides.* 4pp. HMSO, London.

Bauer, S. (1966). *Agric. Avi.* **8,** 90.
Besser, J. F. (1978). *Proc. 8th Vert. Pest Conf.* 51.
—— Royall, W. C., and DeGrazio, J. W. (1967). *J. Wildl. Mgmt.* **31,** 48.
Brei, W. (1977). *Know How* **4,** 25.
Byers, R. E., Young, R. S., and Neely, R. D. (1976). *Proc. 7th Vert. Pest Conf.* 242.
Canutt, P. R. (1970). *Proc. 4th Vert. Pest Conf.* 120.
Cornwell, P. B. and Bull, J. O. (1967). *Pest Control* **35,** 31.
Crase, F. T. and DeHaven, R. W. (1976). *Proc. 7th Vert. Pest Conf.* 46.
Decino, T. J., Cunningham, D. J., and Schafer, E. W. (1966). *J. Wildl. Mgmt.* **30,** 249.
DeGrazio, J. W., Besser, J. F., Decino, T. J. Guarino, J. L., and Starr, R. I. (1971). *J. Wildl. Mgmt.* **35,** 565.
—— —— —— —— Schafer, E. W. (1972). *J. Wildl. Mgmt.* **36,** 1316.
Dolbeer, R. A., Ingram, C. R., and Stickley, A. R. (1973). *Proc. 6th Bird Cont. Semin.* 28.
—— —— Seubert, J. L., Stickley, A. R., and Mitchell, R. T. (1976). *J. Wildl. Mgmt.* **40,** 564.
Drummond, D. C. (1968). *Proc. Int. Symp. Bionom. Cont. Rodents. Kanpur, India,* 101.
—— and Rennison, B. D. (1973). *Bull. Wld. Hlth. Org.* **48,** 239.
—— Taylor, E. J., Bond, M., and Greaves, J. H. (1972). *Urban rat control: an experimental study.* 36 pp. Association of Public Health Inspectors, UK.
Dubois, K. P., Cochran, K. W., and Thomson, J. F. (1948). *Proc. Soc. Exp. Biol. Med.* **67,** 169.
Dyer, M. I. and Ward, P. (1977). In *Granivorous birds in ecosystems* (eds J. Pinowski and S. C. Kendeigh), p.267. International Biological Programme 12: Cambridge University Press.
Eppo (1975). *Bull. Eur. Medit. Plant Prot. Org. (Special Issue)* **5(1),** 49pp.
Ellerman, J. R. (1949). *The families and genera of living rodents* (3 vols) British Museum, London.
FAO (1979). *Rodenticides: analyses, specifications and formulations for use in agriculture and public health.* Food and Agriculture Organization of the United Nations, Rome.
Fernando, H., Kawamoto, N., and Perera, N. (1967). *Plant Prot. Bull., FAO* **15,** 32.
Giban, J. (1958). *Report of the International Conference on Harmful Mammals and their Control,* 13.
Goodhue, L. D. and Baumgartner, F. M. (1965a). *J. Wildl. Mgmt.* **29,** 830.
—— —— (1965b). *Pest Control* **33,** 16 and 46.
Gratz, N. G. (1976). *Chemi. Pflanz. Schädlingsbekämpfungs.* **3,** 85.
Greaves, J. H. (1971). *Pestic. Sci.* **2,** 276.
—— Redfern, R., and Anasuya, B. (1976). *J. Stored Prod. Res.* **12,** 225.
—— —— King, R. E. (1974). *J. Hyg., Camb.* **73,** 341.
—— —— Ayres, P., and Gill, J. E. (1977). *Genet. Res.* **30,** 257.
Green, J. (1971). *Environ. Hlth.* **79,** 133.
Guarino, J. L. (1972). *Proc. 5th Vert. Pest Conf.* 108.
—— and Schafer, E. W. (1967). *US Department of the Interior, Fish and Wildlife Service, Special Report. Wildlife 104,* 5pp.
—— Stone, C. P., and Shake, W. F. (1973). *Proc. 6th Bird Cont. Semin.* 24.
Gutteridge, N. J. A. (1972). *Chem. Soc. Rev.* **1,** 381.
Hood, G. A. (1972). *Proc. 5th Vert. Pest Conf.* 85.
Horsfall, F., Webb, R. E., and Byers, R. E. (1974). *Proc. 6th Vert. Pest Conf.* 112.
Jackson, J. J. and Park, P. O. (1973). *Proc. 6th Bird Cont. Semin.* 53.

Jackson, W. B. (1978). *Proc. 8th Vert. Pest Conf.* 47.
—— and Kaukeinen, D. (1972). *Science* **176**, 1343.
Jazouane, B. (1978). *Eur. Medit. Plant Prot. Org. Publ. Ser. B* **84**, 214.
Jones, P. J. (1975). *Proc. 8th Br. Insectic. Fungic. Conf.* 883.
—— (1980). *Proc. 4th Pan-Afr. Ornithol. Congr.* (1976) 423.
Kerkwyk, R. E. (1974). *Cane Grow. Q. Bull.* **37**, 141.
Knudsen, E. (1963). *Acta Pharmacol. Toxicol.* **20**, 293.
Kuhn, L. W. and Peloquin, E. P. (1974). *Proc. 6th Vert. Pest Conf.* 101.
LaGrange, M. and Jarvis, M. J. F. (1977). *Quelea finch ground control manual (using Sprayquip mist-blower)*. 19pp. Department of National Parks and Wildlife Management, Rhodesia.
Larthe, Y. (1957). *Sanitarian* **65**, 276.
LeFebvre, P. W. and Seubert, J. L. (1970). *Proc. 4th Vert. Pest Conf.* 156.
Lund, M. (1972). *Bull. Wld. Hlth. Org.* **47**, 611.
—— (1984). *Acta Zool. Fenn.* (in the press).
MacSwiney, F. J. and Wallace, M. E. (1978). *J. Hyg., Camb.* **80**, 69.
Montenegro, A. (1962). *Coff. Cac. J.* **5**, 192.
Murton, R. K. (1966). *The Statistician* **16**, 183.
Pank, L. F. (1976). *Proc. 7th Vert. Pest Conf.* 196.
Podlishuk, A. (1971). *Prakt. Schädlingsbekampfer* **23**, 117.
Pope, G. G. and Jones, P. J. (1978). *Eur. Medit. Plant Prot. Org. Publ. Ser. B* **84**, 202.
—— King, W. J. (1973). *Centre for Overseas Pest Research, Miscellaneous Report 12*, London.
—— Ward, P. (1972). *Pestic. Sci.* **3**, 197.
Redhead, T. D. (1971). *Annu. Rep. Bur. Sugar Exp. Stns, Brisbane*, 47.
Rennison, B. D., Hammond, L. E., and Jones, G. L. (1968). *J. Hyg., Camb.* **66**, 147.
Roszkowski, A. P., Poos, G. I., and Mohrbacher, R. J. (1964). *Science* **144**, 412.
Rowe, F. P. and Redfern, R. (1965). *J. Hyg., Camb.* **63**, 417.
Santini, L. (1978). *Proc. 8th Vert. Pest Conf.* 78.
Schafer, E. W., West, R. R., and Cunningham, D. J. (1969). *Pest Cont.* **37**, 22.
Schilling, C. (1976). *Proc. 7th Vert. Pest Conf.* 110.
Shaefer, J. (1975). *All-India Rodent Control Seminar, Ahmedabad, India*, 344.
Smith, R. W. (1967). *Trop. Agric. Trinidad* **44**, 315.
Stickley, A. R. and Guarino, J. L. (1972). *J. Wildl. Mgmt.* **35**, 569.
—— Ingram, C. R. (1973). *Proc. 6th Bird Cont. Semin.* 41.
—— Mitchell, R. T., Heath, R. G., Ingram, C. R., and Bradley, E. L. (1972). *J. Wildl. Mgmt.* **36**, 1313.
Taylor, K. D. and Green, M. G. (1976). *J. Zool. London* **180**, 367.
Teshima, A. H. (1976). *Proc. 7th Vert. Pest Conf.* 121.
Thearle, R. J. P. (1968). In *The problems of birds as pests (Institute of Biology Symposia, 17)* (eds R. K. Murton and E. N. Wright). Academic Press, London and New York.
Toovey, F. W. (1953). *J. W. Afr. Inst. Oil Palm Res.* **1**, 68.
Ward, P. and Pope, G. G. (1972). *Pestic. Sci.* **3**, 709.
West, R. R. (1968). *J. Wildl. Mgmt.* **32**, 637.
—— Besser, J. F., and DeGrazio, J. W. (1967). *Proc. 3rd Vert. Pest Conf.* 89.
—— Brunton, R. B., and Cunningham, D. J. (1969). *J. Wildl. Mgmt.* **33**, 216.
WHO (1970). *Wld. Hlth. Org., Tech. Rep. Ser. No. 443*, 140.
Wood, B. J. (1971). *Pestic. Abstr. News Summ.* **17**, 180.

16
Control of mollusc vectors

J. M. JEWSBURY

16.1 Introduction

Schistosomiasis (bilharzia) is one of the most important parasitic infections of Man in the tropics. Estimates of the number of people infected range between 200 and 300 million; most of these live in Africa and the Far East but about 10 million live in South America (mostly in Brazil), and there are relatively small foci in the Middle East. The disease is caused by small worm-like parasites which live in the abdominal blood-vessels in Man. Infection is transmitted to certain fresh-water snails via human excreta contaminated with eggs of the parasite. Human infection is usually acquired during domestic, occupational or recreational contact with water containing infective stages (cercariae) produced by the snails. There are three important types of schistosome infection in Man, two causing damage principally to the intestine and liver, the third causing most damage to the urinary system. In Africa, the Middle East, and South America, intestinal schistosomiasis is caused by the worm *Schistosoma mansoni,* transmitted by *Biomphalaria* snails. In the Far East, a more severe form of intestinal infection is caused by *S. japonicum,* transmitted by *Oncomelania* snails. Urinary disease is caused by *S. haematobium,* transmitted by *Bulinus* snails in Africa and the Middle East. *Biomphalaria* and *Bulinus* snails are common in various types of fresh-water bodies (e.g. lakes, pools, borrow pits, and especially irrigation systems) and are fully aquatic; *Oncomelania* snails are also found in a variety of water bodies but are amphibious and not necessarily confined to water, being capable of surviving in moist conditions on land.

Schistosomiasis can be controlled by treating all infected people, by the provision and use of effective sanitation and safe water supplies and by snail control, principally by means of chemical control through the use of molluscicides but usually a combination of all three methods is used.

The snail hosts of schistosomes are extremely adaptable and this, together with great variety in the aqatic habitats in which they occur, makes it virtually impossible to lay down hard and fast rules for the application of molluscicides which will be suitable in all situations. This chapter attempts to establish broad principles and to describe the variety of problems and the various solutions which have been developed by different workers.

16.2 Mollusciciding

The part which mollusciciding can play in the overall strategy of schistoso-miasis control is an important one but it is only a part: experience has shown that mollusciciding on its own is unlikely to be successful in controlling the disease. It is usual to include therapy, and public health measures such as effective sanitation and safe water supplies, together with mollusciciding in a combined attack on transmission.

In any mollusciciding programme a sequence of events is followed which can best be summarized in a chart (Fig.16.1). A programme such as this would take perhaps 2–3 years to establish effectively. Any short cuts in the form of brief preliminary surveys, inefficient location of water bodies, inaccurate estimates of water volumes leading to under treatment, etc., are completely counterproductive and may render the whole project totally useless as a control measure. It cannot be emphasized too strongly that efficient, comprehensive preliminary investigations over as much as 2 years are essential if reduction in snails is to be achieved.

Costs are always important and this is especially true where budgets are restricted, but in the context of mollusciciding projects, two points should be borne in mind. Firstly, any mollusciciding programme implies the commit-ment of money for an indefinite period into the future. Experience has shown that it is virtually impossible to totally eradicate snails from a system, reinfestation usually recurring (perhaps within weeks or months) and requir-ing repeated treatment. Secondly, costs should include evaluations calculated on a basis of per head of population protected and per unit of water treated—it costs as much (and is as important) to effectively treat a body of water containing just a few snails as it does if it contains many snails, since the productive potential of snail hosts of schistosomiasis is very considerable.

Matters which should be considered before a mollusciciding programme is agreed in principle are discussed in detail by Farooq (1973), especially chapters 8 and 9. A general account of the biology of the snails transmitting *Schistosoma japonicum, S. mansoni,* and *S. haematobium* is given in WHO Monograph Series No. 50 (1965).

16.2.1 Preliminary investigations

The location and mapping of all water bodies in the proposed control area must be totally efficient if the project is to succeed. It is very easy to miss small water bodies (particularly if they are used for personal washing and are considered 'private') which may be important sources of infection. Human water-contact points often harbour snails (a variable proportion of which may be infected) but not all snail habitats are human contact points: however, some of these may act as sources of reinfestation for treated areas.

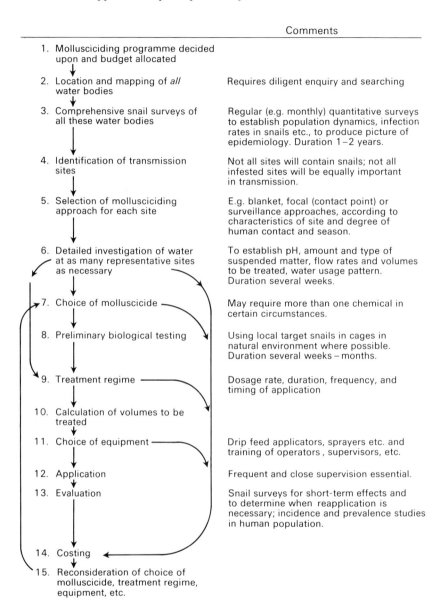

	Comments
1. Mollusciciding programme decided upon and budget allocated	
2. Location and mapping of *all* water bodies	Requires diligent enquiry and searching
3. Comprehensive snail surveys of all these water bodies	Regular (e.g. monthly) quantitative surveys to establish population dynamics, infection rates in snails etc., to produce picture of epidemiology. Duration 1–2 years.
4. Identification of transmission sites	Not all sites will contain snails; not all infested sites will be equally important in transmission.
5. Selection of mollusciciding approach for each site	E.g. blanket, focal (contact point) or surveillance approaches, according to characteristics of site and degree of human contact and season.
6. Detailed investigation of water at as many representative sites as necessary	To establish pH, amount and type of suspended matter, flow rates and volumes to be treated, water usage pattern. Duration several weeks.
7. Choice of molluscicide	May require more than one chemical in certain circumstances.
8. Preliminary biological testing	Using local target snails in cages in natural environment where possible. Duration several weeks – months.
9. Treatment regime	Dosage rate, duration, frequency, and timing of application
10. Calculation of volumes to be treated	
11. Choice of equipment	Drip feed applicators, sprayers etc. and training of operators , supervisors, etc.
12. Application	Frequent and close supervision essential.
13. Evaluation	Snail surveys for short-term effects and to determine when reapplication is necessary; incidence and prevalence studies in human population.
14. Costing	
15. Reconsideration of choice of molluscicide, treatment regime, equipment, etc.	

Fig. 16.1. Mollusciciding.

Different sites may be used for different purposes—e.g. washing, drinking, defaecation and urination, swimming—and may differ in transmission importance. It may be necessary to resurvey the area two or three times at different seasons in order to include small water bodies which may dry up, or be obscured by vegetation or become flooded in the wet season. Collections of snails should be made and referred to experts for identification. Quantitative snail surveys should be carried out on a regular basis to build up a picture of the species of snails involved and their population dynamics. Snail egg masses often occur and should also be recorded and identified as far as possible. It may be necessary to visit a site on a number of occasions if it dries up, since some snails aestivate successfully for several months in the dry season, and are difficult to find.

Snail survey techniques are simple and involve various methods for estimating snail numbers using some form of snail scoop (Fig.16.2). Alternative methods of carrying out snail surveys for both amphibious and aquatic snails using different types of equipment are discussed in WHO Monograph No. 50 (1965).

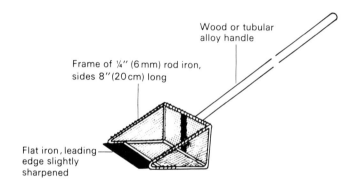

Fig. 16.2. Snail scoop.

A routine for quantifying results should be established early. The number of snails can be determined per scoop (or per so many scoops, e.g.50), per man-time period (e.g. man—ten minutes), or per unit of distance or area. The first of these is perhaps the more common, with a timed manual search being employed where snail densities are low. The actual method chosen is less important than the consistency and accuracy with which it is adopted, but statistically sound quantitative sampling of an infected area may be restricted or preluded by the variable nature of the habitats (WHO Monograph No. 50 (1965) 201–203; Yeo 1962).

Whether snails collected in surveys should be returned to the site from which they were taken after counting depends on the purpose of the survey. If population dynamics are not being studied it is not necessary to return them.

Particularly in the early stage of a survey representative samples of snails (possibly all snails collected) should be examined for schistosome infection by shedding or crushing; methods are described in WHO Monograph No. 50 (1965) 190–192. Shiff *et al.* (1979) describe variation in populations of infected snails with season. It should be borne in mind that failure to shed cercariae does not mean absence of infection, merely absence of mature infection or failure to stimulate cercarial emission. The advantage of the shedding method over crushing is that the snails generally remain alive and studies on population dynamics need not be unduly upset. Infection rates may be very low [of the order of 0.1%, for instance in St.Lucia (Jordan *et al.*1978), but infected snails are often more common around contact points (Chu and Vanderburg 1976).

16.2.2 General considerations

Two alternative approaches to mollusciciding are possible. These are blanket treatment whereby all water bodies are treated with an appropriate amount of molluscicide to achieve a satisfactory snail kill; and focal or contact point treatment where only those areas which actually harbour snails (plus an appropriate area surrounding the infested site) are treated. The former is particularly suited to the supply component of irrigation systems where the basic similarity and favourable environment provided by all parts of the system means that snail populations are often heavy and contiguous. Chemical can often also be carried to all parts of the system in the supply of irrigation water. Rivers and irrigation drainage systems are, however, efferent in character and each source must be treated to treat the whole. In the majority of rivers and streams and various types of natural or artificial lakes and dams snail populations may be smaller and relatively isolated from each other. This and the generally large total volumes of water relative to the snail populations make the treatment of the total water volume with chemical unrealistic and impractical for both financial and technical reasons, so that treatment of water only in the immediate vicinity of snails is practised.

Once an effective blanket control programme has been instituted and has run for one or two seasons with good control, so that a major impact has been made on the snail population, consideration can be given to a change from blanket treatment to a programme of snail surveillance with focal mollusciciding wherever and whenever snail reinvasion or resurgence occurs. The change to focal mollusciciding should not be made too soon, however, otherwise a breakdown in either snail monitoring or in effective treatment rapidly leads to a resumption in transmission. In some situations a blanket treatment is applied annually with regular monthly surveillance, focal mollusciciding taking place where and when necessary in the intervals between blanket applications.

A number of factors affect molluscicidal activity in water. Among the more important are water pH, the nature and quantity of suspended matter and aquatic vegetation. Some chemicals are effective in acid waters and relatively ineffective in alkaline waters while for other chemicals the reverse is true. Snails can occur in a wide variety of waters and the molluscicide chosen must be active in the pH range present. Values may change with season (e.g. from pH 5.6 to 8.9 in St.Lucia, and from pH 6.5 to 7.5 in the Lowveld of Zimbabwe). Suspended matter (silt and colloidal material) may be present in large quantities after heavy rain or annual floods (e.g. Nile) and has a relatively deleterious effect on some molluscicides. Aquatic vegetation is a widespread problem but may be more abundant at some seasons than others, e.g. in November–June in the Sudan. Vegetation tends to inhibit the action of molluscicides, as well as making water carriage of and penetration by the chemical difficult and providing (in some cases) a means by which snails can climb out of the water and escape the molluscicide. The distribution of infected and uninfected snails (and of human infection with *S. haematobium*) has been shown to be highly correlated to the distribution of the aquatic plant *Ceratophyllum* in Volta Lake (Klumpp and Chu 1980). Molluscicides are not all equally effective against all species of snails and a chemical must be chosen which is effective against the target species.

A knowledge of all these factors is necessary before a chemical can be provisionally selected. Specialist advice should be obtained and tests, e.g. for half-life of active ingredient, and to determine the optimum chemical concentration and contact time, should be carried out under actual conditions before a final decision is made.

16.2.3   Molluscicides available

Bayluscide®: niclosamide, clonitralide (Bayer AG Agro-chemicals Division, Leverkusen, West Germany): Available as a wettable powder containing 70% active ingredient and as an emulsifiable concentrate containing 25% of active ingredient (Bayluscide Technical Manual 1970 and undated). A locally produced version of Bayluscide is manufactured in Egypt and marketed locally under the name Mollutox.

Granular formulations have been described (Prentice 1970; Magendantz 1974; Upatham and Sturrock 1977; Prentice and Barnish 1980) but are 'home made' for use in 'difficult' habitats, for example deep water such as Lake Victoria where *Biomphalaria choanomphala* is a problem, in extensive marshy areas which are difficult to penetrate, and areas which are subject to periodical drying but are locally important in transmission.

Frescon®: *n*-tritylmorpholine, triphenmorph (Shell International Chemical Co.; London, UK): has been used fairly widely but is now no longer available.

(a) Other molluscicides. Molluscicide B2: Sodium 2,5-dichloro-4-boromophenol: used routinely in Japan since 1975 for control of *Oncomelania*.

Granular formulation, applied by hand to wetted soil at a rate of 5 g/m².

Chloroacetamide (C_2H_4ONCl): applied to 10 mg/litre to water containing *Oncomelania* in China; at concentrations above 20 mg/litre there is some toxicity to fish and rather more to animals and plants.

Tributyl tin oxide (TBTO): has the potential for being a very useful and effective molluscicide. However, definitive long-term toxicological studies on TBTO are lacking and there should therefore be reservations about its use in the field. Pellets have been produced consisting of 5–6% active ingredient by weight incorporated in a slow-release rubber matrix (Biomet SRM: M & T Chemical Co., Rahway, N.J., U.S.A.). Pellets apparently give prolonged control of aquatic snails when applied at a rate of 5–20 g/m² but the general environmental toxicity of the material is still unclear. This compound is in use in the Philippines.

Controlled-release glass containing copper ion: laboratory and small-scale field tests against a variety of snail species have shown that copper ion can be released in molluscicidal quantities from phosphate and borate glasses designed to dissolve in water at various rates over prolonged periods of time (e.g. months or years). The material is cheap and easy to manufacture and can be formulated in a variety of ways. It may offer an effective alternative to available molluscicides.

Nicotinanilide: laboratory tests on this candidate molluscicide have demonstrated its high activity against snails and their eggs in response to very low continuous doses and its high degree of target specificity. Further development of the product is in progress, including laboratory and field trials of slow release formulations.

(b) Plant molluscicides. A number of plant extracts having molluscicidal activity are known, but few have been adequately tested under field conditions, and long-term toxicological tests have not been carried out even on the most closely investigated of the group (endod, derived from the plant *Phytolacca dodecandra*).

Despite the successful demonstration of the practical use of endod in pilot field trials, the wide-scale use of the material would depend on well organized cultivation and harvesting of the plant and other procedures to ensure adequate supplies and standardization. These difficulties and the lack of detailed toxicological information on any of the potentially useful plant molluscicides severely restricts their use in the foreseeable future. McCullough *et al.* (1980) summarize the current status of molluscicides in general.

16.3   Control Techniques

16.3.1   *Preliminary field tests*

Snails can be exposed to different concentrations of chemicals and for different periods of time; the combination of these two variables is known as a concentration–time product. They are not directly related but in general an increase in one variable permits a decrease in the other. In practice one variable often has a practical limit set upon it (e.g. an available contact time) and the other variable has to be adjusted accordingly to achieve adequate snail kills.

Preliminary field testing of the chosen molluscicide(s) using target snails in cages in natural water bodies is virtually essential to confirm that all the variables have been properly evaluated. A range of concentration–time products around the proposed value is applied to a known number of snails (say, at least ten cages containing ten snails of each target species) plus appropriate untreated controls. Mortalities are determined after a period of at least 24 hours following the end of exposure. A dose and exposure combination is aimed at which will kill 99% or more of treated snails in the time period.

16.3.2 Timing of applications

Once the optimum dosage rate and duration of application have been determined (the concentration–time product) taking all the relevant factors into consideration, the timing and frequency of the application need to be considered. In general, molluscicides are more effective at higher temperatures than lower (Abdel Raheem *et al.*1980). From the knowledge of the population dynamics of the snails and of cycles of infection and cercarial emergence it is straightforward to deduce when the seasons of greatest risk to the human population occur. Snail populations fluctuate widely in response to a complex of environmental factors. Each situation has to be evaluated separately but in general snails multiply at the greatest rate immediately after the end of the rainy season, provided that temperatures are near optimal. Maximum transmission, on the other hand, may occur towards the end of the hot, dry season when water bodies are reduced in size and number, and the human population concentrates around a relatively small number of heavily infested and highly infective sites. Children are responsible for about 70% of transmission in many situations, and are usually found playing in water in the largest numbers at this time. There may be one or more additional minor peaks of snail breeding and of transmission but these should be evident from results of snail surveys. Molluscicide applications should therefore be timed to reduce snail populations as far as possible immediately before the transmission seasons and, where practical, also before the periods of maximum breeding. In the latter case, however, water volumes are often much larger, and more chemical is therefore required. Practical problems may arise such as more difficult access immediately after a rainy season when roads may be in bad repair, large areas of marginal land may be inundated, and vegetation may be growing rapidly.

The frequency of treatment depends on the rate of snail repopulation and this is determined by surveillance routines.Usually it is necessary to treat relatively frequently in the early stages of a programme with treatments becoming more spaced as the situation comes under increasing control.

Treatment schedules should be adapted to local circumstances. Frescon drip feed applications in the Gezira Irrigated Area, Sudan (where transmission is thought to occur throughout the year) were relatively ineffective except between September and December when water usage was sufficiently large to draw treated water into the ends of the system where most snails were found (Amin *et al.* 1976). In Middle Egypt, area-wide applications of Bayluscide are made in spring, summer, and autumn with complementary spraying of drains and water bodies. In Fayoum, Egypt, two annual area-wide applications are made of Bayluscide W.P. One application takes place in spring and the other in autumn; treatment is from a single point, supplemented by surveillance and focal mollusciciding during the rest of the year. In St. Lucia, West Indies, banana drains and streams are treated focally on a regular 4 week schedule throughout the year, since transmission is perennial (Sturrock *et al.* 1974). In Zimbabwe, Shiff *et al.* (1979) showed by snail surveys and cercarial-emergence studies that treatment in late winter (July) is effective in reducing the heavy transmission which would normally occur in late spring and early summer in the Highveld. Evans (personal communication), working in the extensive Lowveld irrigation systems, treats night-store dams annually by drip feed. Locally produced granular formulations of Bayluscide have been used successfully by Prentice (1970) and Magendantz (1974) to focally control *B. choanomphala* living in relatively deep water in Lake Victoria, East Africa, where the substrate (an uneven murram hard-pan) makes the application of conventional molluscicides uneconomical and technically difficult.

16.3.3   Choice of application equipment

There are two basic types of equipment for molluscicide application. These are first various types of constant-flow dispensers designed to release liquid suspension or emulsion at a constant rate for different periods of time according to design, and second, various types of power and hand-operated sprayers. The constant-flow dispensers are used mainly in applications to irrigation systems but can also be used to treat rivers and streams. Sprayers are used in all situations where dispensers are not appropriate (including drains in irrigation systems) and to 'top up' molluscicide levels in area where penetration of chemical from dispensers is incomplete.

(a) Constant-flow dispensers. Constructional details for a simple floating siphon are given in Figure 16.3. This type of dispenser is robust and can be made up from locally available materials in almost any workshop. It is used for applications of both wettable powder and emulsifiable concentrate

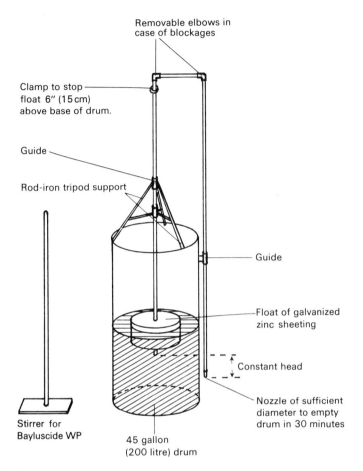

Removable elbows in
case of blockages

Clamp to stop
float 6" (15 cm)
above base of drum.

Guide

Rod-iron tripod support

Guide

Float of galvanized
zinc sheeting

Constant head

Nozzle of sufficient
diameter to empty
drum in 30 minutes

Stirrer for
Bayluscide WP

45 gallon
(200 litre) drum

Fig. 16.3. Floating siphon dispenser.

formulations of Bayluscide. The speed of emptying is controlled by the
nozzle size and the vertical distance between the nozzle and the liquid
surface. The nozzle should be of sufficient size to empty the full drum in half
an hour. The float should be airtight and of sufficient size to support the
weight of the siphon arm (which should run freely through the guides) but
still allow room for vigorous stirring of the drum contents while molluscicide
is being applied. Dispensers of this type are cheap to construct but are
somewhat bulky and awkward to transport.

Other types of constant-flow dispensers are described in the Frescon
technical manual published by the Shell International Chemical Co. (1974).

Several more dispensers should be available 'on site' than are actually required, in order to permit easy replacement of damaged or defective apparatus. Dispensers should stand on level ground when in use, and it is good practice to construct dispensing platforms where necessary, if dispensers are to be regularly used at any particular sites. Wherever possible the open types of dispenser should be covered to prevent leaves, dirt, etc, falling in and blocking the pipework. For the same reason any water added to the dispenser should first be passed through a simple gauze filter. To prevent siphoning, outlet hoses should be at least 0.75 in. (1.9 cm.) diameter and be placed above, not below, the canal surface. Flow rates should be accurately set by measuring with a stopwatch the time taken to fill a calibrated container such as a measuring cylinder. Where application rates are too high for one dispenser to supply accurately, two or more are set up, the chemical being divided equally among them.

(b) Power and hand-operated sprayers. Applications other than drip feeds tend to be relatively arbitrary in the concentration of molluscicide applied and rely heavily on the experience of the operator to achieve the desired concentrations. Equipment ranges from a large tank with a petrol-driven pump mounted on a trailer and requiring a tractor or similar vehicle for towing, through smaller petrol-driven pumps fed from a modified oil drum or similar container and which can be mounted in a small boat, to various types of hand-operated knapsack sprayers and even stirrup pumps. Since no precise concentration of molluscicide is aimed at, the principal requirements are that the pump provides sufficient pressure to give either a fine spray or a powerful jet with a long throw, suitable for penetrating reeds and other aquatic vegetation and for reaching some distance from the spray point. The pumps should be reliable and robust; a stock of minor spares should be carried in the field.

The actual equipment used in different projects depends on local conditions and availability. The Hudson X-pert hand sprayer (Hudson Manufacturing Co. Model No. 67322) (see Chapter 4) is used in St. Lucia (Sturrock *et al*.1974) with motorized knapsack sprayers (Kinkelder 60 sprayer) being used to force a jet of spray into the centres of marshes. Constant-flow dispensers made from 20-litre polythene bottles (based on a design described by Prentice 1971) are used to apply Bayluscide E.C. These are calibrated to empty in 1 hour and are robust and light enough to transport to even the most remote areas. Hudson sprayers have also been used in the Sudan to supplement drip-feed application and aerial spraying from suitably adapted light aircraft (Amin and Fenwick 1977). Aerial spraying has also been used by Barnish and Shiff (1970) in Zimbabwe, and by Barnish and Sturrock (1973) in St. Lucia.

A 'home-made' granular dispenser was used by Prentice (personal communication) to apply his granular formulation of Bayluscide to *B. choanom-*

phala sites in Lake Victoria. Granules were fed by gravity from a hopper on to a whirling disc, powered by a moped engine. The spreader was used in a boat and gave a swath width of about 5 m. Clarke and Evans (personal communication) used Finsbury, or Briggs and Stratton, 1-1/2 HP Hypro pumps in a small rowing boat to treat the edges of the lakes and night-store dams in Zimbabwe and also made extensive use of brass stirrup pumps fitted with 1 m. lance and solid jet nozzle in their mollusciciding programmes— the particularly powerful jet was effective in penetrating vegetation and distributing chemical below the water surface.

Liquid formulations of molluscicide are more suitable than powders for dispensing from power sprayers if the concentrate cannot be kept in suspension by stirring. Preweighed or dispensed charges of molluscicide powder of suitable size (e.g. sufficient to treat 100 m. of shoreline) are particularly useful for hand sprayers and obviate the need to weigh out in the field. Where water is added to the sprayer it should be filtered before addition. Difficulties can be experienced in obtaining adequate dispersion and mixing of chemical due to cold and warm layers of water, and to wind and currents. In these cases treatment may have to take place at different times of day or even at night, when problems may be less severe.

In summary, drip-feed dispensers are particularly suited to blanket treatments of rivers and particularly of irrigation systems; drains, the edges of lakes and rivers and small water bodies (e.g. borrow pits) are more easily and conveniently treated with sprayers.

16.3.4 *Estimating volumes to be treated and calculating amount of molluscicide*

Volumes to be treated can usually be calculated fairly precisely in irrigation systems and hence the amount of chemical required can be determined reasonably accurately. Under these conditions treatment can be operated at optimal efficiency. Calculation of volumes to be treated in the case of natural or artificial impoundments and of rivers and especially swamps is more difficult. Calculations are much easier where metric units are employed: the metric unit used to calculate large volumes is the cubic metre; the unit of flow or discharge is the cubic metre per second (cumec). The Imperial unit used to calculate large volumes is the cubic foot; the unit of flow or discharge is the cubic foot per second (cusec). The general principles to be followed are as follows:

(1) Determine the volume to be treated, preferably in metric units.
(2) Decide (by preliminary field experiment if necessary, bearing in mind the adverse effects of silt absorption, inactivation by ultraviolet light etc.) the best practical combination of contact time and chemical concentration in terms of active ingredient (in g/m^3, which is equal to mg/litre or p.p.m.).

(3) Multiply the concentration required by the volume to be treated to give the total amount of active ingredient to be applied.
(4) Multiply this value by the appropriate factor to allow for the proportion of active ingredient in the chosen formulation.

To overestimate chemical required is not economic but to underestimate and therefore to undertreat is totally wasteful of chemical, labour etc. Therefore, any approximations should increase rather than decrease the calculated amount of chemical, and a good rule of thumb in extensive and complex situations is to increase calculated dosages by 10% overall to allow for unexpected variations.

(a) Irrigation supply system. Calculation of volumes to be treated in irrigation systems is relatively straightforward in that the volume of water entering the system (or part of it) can usually be determined with some accuracy by reference to calibrated off-take gates, gauging weirs, or posts. The holding capacity (volume) of the system itself must also be taken into account. This is especially important if sufficient untreated water will not be drawn off (e.g. by irrigation demands) during mollusciciding to ensure replacement with treated water. Prolonged treatment (e.g. 7–10 days) with low concentrations of molluscicide involves the calculation of flow rates, not capacities, as water should be drawn into all parts of the system during the treatment period.

(b) Rivers, streams and irrigation drainage systems. Discharges are difficult to calculate because they progressively increase further down the system and chemical becomes more dilute so that booster applications may be necessary. Total volumes are not often calculated because of the cost of unnecessarily treating large volumes of water. Where necessary any relatively small flow rates can be determined by rectangular or 'V' notch weir or a Parshall flume. Alternatively, flow meters can be used or the time taken for a floating object to travel a known distance can be used to give an indication of the flow rate. In these last two cases, however, the average rate of flow is about 85% of the rate measured at the surface and appropriate adjustments to calculations are necessary.

(c) Lakes and dams. Treatment of total volume is unnecessary, costly, and impractical in most cases. However, calculation of volumes of night-storage dams in irrigation systems is often necessary. Volume estimates are the most difficult to calculate accurately, due partly to the frequently irregular shape and often (at best) imperfectly known depth and bottom configuration of the water body. The rule-of-thumb is to divide an irregular shape into components approximating to geometric shapes such as a wedge or pyramid. Where the volume is large and total treatment is not required (e.g. in a dam or river), treatment of a wedge-shaped strip around the shore is usual. An estimate of

volume can be arrived at either by accurately measuring or by pacing out the relevant distances and estimating the depth (or obtaining this information from the owner or operator).

16.3.5 Application by constant-flow dispensers

Application of molluscicide to an irrigation system is most successful when crop irrigation is at its height, ensuring a rapid distribution of chemical to all parts of the system. A suitable flat site should be chosen at the headworks of an irrigation system or at a major off-take feeding the section to be treated. The dispensers should feed into a turbulent section of the system so that thorough mixing of chemical and water takes place. It is particularly important that applications in irrigation systems are not interrupted by defective equipment since the routine of irrigation is usually suspended for the duration of treatment and crops can only stand withholding of water for a limited period without undergoing stress. Arrangement must be made some time in advance with the management so that flow rates and volumes in the dams (if any) can be adjusted for optimal treatment. Very low rates (e.g. 5ml/minute) of molluscicide discharge are difficult to control accurately and predilution with water may be required.

The general principle of treatment of a complex irrigation system containing night-storage dams, such as in the Lowveld of Zimbabwe, is to treat the supply system and night-storage dams by drip feed and to spray drains, swampy areas, and natural watercourses (Shiff *et al.*1973). The procedure for the supply system is to lower the water level in the dams to one-third capacity, and to set the control gates to complete filling in 24 hours. Chemical is applied at such a rate to treat the whole system (calculating on full capacity) within the first 12 hours, directing the treated water into each dam until it is two-thirds full at the end of 12 hours. Untreated water is then allowed to flow into the system for a further 12 hours to complete filling of the dams and mix chemical in them. The whole system is then held constant (without withdrawing treated water) for 24 hours (and longer if possible) to ensure sufficient contact time with snails to kill them. Any dams which are under-treated can be 'topped up' by spraying from a boat or by hand. Dams which are more than two-thirds full at the commencement of treatment will not receive adequate molluscicide by drip feed of the whole system and cannot be treated in this way. They must be treated by hand or boat spray methods, all the surface being covered with chemical. Main canals and off-takes, weirs, and gates are important sources of snails and breeding sites and require specially thorough treatment if snails are not to spread through the system.

Treatment of irrigation systems is in principle straightforward but in practice depends on highly skilled water management to ensure that appropriate amounts of treated water reach all parts of the system. On large systems involving several hundred dams of varying sizes with hundreds of kilometres of canals, weirs etc., management has to be of a very high order, and is the result of long practical experience. Difficulties may arise because of long travel times (beyond the useful life of the chemical) to distant parts of the system. Occasionally booster applications are made at different 'difficult' points in the system, half the chemical being supplied at the main application point at the same time as half is applied at the booster. Techniques for optimizing the siting of boosters have been described (Osgerby 1970). Difficulties may also be due to poor penetration of areas which do not have water drawn off at a sufficient rate to ensure flushing out with treated water. This type of problem has arisen in the Gezira project.

16.3.6   *Application by sprayer*

Application is relatively straightforward but depends on the skill and experience of the operator to ensure that sufficient chemical penetrates areas of long vegetation and relatively deeper areas of water. Mollusciciding of streams can often be done more effectively by two sprayers operating as a pair either on opposite banks if the water body is narrow or 'leap-frogging' along one bank. The entire width of the stream requires treatment. Spraying should be done from downstream towards the source, so that a long 'plug' of treated water is created.

A useful alternative to spraying small streams where vegetation is heavy and the terrain is suitable is Evans' dam-and-flush method. The water course is dammed temporarily at a suitable site (e.g. a bridge) using a steel plate. The water is impounded for several hours, heavily treated with molluscicide and then released suddenly. This flushes out areas which are difficult to treat by conventional spraying; it is also more economical with chemical than heavy conventional treatment methods.

Chemical analysis of treated waters is possible following any form of treatment but it is usually only carried out in the development stages of a project. Working concentrations are too low for easy accurate determination in the field, and sophisticated laboratory procedures are necessary.

16.4   Evaluation of treatments

Evaluation of the results of each application of molluscicide should follow within 7–10 days of treatment, using standardized quantitative snail-survey methods. A 100% reduction in snail population is unlikely to be achieved but less than 95–99% reduction requires detailed investigation to ascertain reasons for comparative (but important) failures. Additionally, cages of snails can be immersed in the water to be treated and mortalities assessed. The definitive means of evaluation of the programme is, however, not the reduction or elimination of snails but the reduction or cessation of transmission of schistosomes, and this can only be determined by careful evaluation of the human population (particularly using incidence rates which are compared with those determined at the beginning of the project). The determination of incidence rates is an important but major undertaking and the effect on snail populations provides a convenient interim indicator.

16.4.1 Costing

Costing is an important aspect of any health programme and some form of evaluation of cost-effectiveness should be applied to a mollusciciding programme, preferably right from the beginning. Items which need to be considered fall under the broad headings of staff, transport, capital equipment, and cost of chemicals etc. (Table 16.1). Detailed records need to be maintained, particularly of amounts and cost of molluscicide used, and of information which will allow the overall costs of application to be worked out. Calculations should be based on per head of population protected, per unit of water and per length of water course treated; these values are not necessarily directly related to each other so that for purposes of comparison with other projects all three should be available. Where a molluscicide programme is part of a wider schistosomiasis control project this will enable funds to be allocated to best advantage.

Mollusciciding is most cost-effective where the volume to be treated is small per person at risk. Therefore mollusciciding is well suited to arid areas where transmission is seasonal and confined to relatively small habitats. Mollusciciding *may* be unsuitable to control transmission in large lakes or rivers unless transmission is focal and focal control is adopted. Where population density is high and water volume per head therefore low, mollusciciding may be cost-effective even though actual volumes to be treated are large (e.g. in major irrigation schemes). Irrigation schemes where controlled water management is practised are well suited to cost-effective chemical control (Webbe and Duncan 1978).

Where snail infection rates are high (say 15% or more) a high level of miracidial contamination of the water is likely. Effective mollusciciding treatment will drastically reduce the number of snails (including infected ones) and will have an immediate effect on transmission. Where snail infection rates are low (say less than 1%), a low level of miracidial contamination is likely. Molluscicide treatment will not reduce the number of infected snails so dramatically, and treatment of infected patients is likely to have a more immediate effect on transmission than mollusciciding.

Different conditions, approaches, and scales of operation make comparison difficult between individual projects. As will be seen from Table 16.1, even the proportion of total cost accounted for by different components of the programme varies between projects; this is so even in the Kenya schemes where the operation was under the control of one organization.

TABLE 16.1. *Comparison of mollusciciding costs*

	Kenya			St Lucia	Zimbabwe				Cameroons	Tanzania	Sudan
					Highveld		Lowveld				
	Hola	Mwea	Ahero								
Period	1975	1967–74	1970–75	1970–73	1976	1977	1968–70	1978	1969–74	1968–70	1974
Chemical	35	44	59	44	43	47	27	53	28	74	83
Labour	58	37	28	42	35	23	61	41	48	22	8
Transport and equipment	7	19	13	14	15	21	12	5	24	4	9
Author and reference	Choudhry (Pers. comm., 1974, 1975)			Sturrock et al. (1974)	Shiff et al. (1979)		Shiff et al. (1973)	Evans (Pers. comm.)	Duke and Moore (1976)	Fenwick (1972)	Amin and Fenwick (1977)

The figures are the percentage of the total cost taken up by the component listed.
Costs of molluscicide, labour, and transport vary greatly from area to area; the costs of the two formulations of Bayluscide are almost equivalent in terms of active ingredient.

A more detailed breakdown of costs is given by Prentice *et al.* (1981) for the St Lucia project.

One point mentioned earlier is sufficiently important to justify repetition. Before a decision is taken to embark on a molluscicide programme it is imperative that those responsible for allocation of money should realize that such a project involves a commitment to make sufficient funds available for a period lasting perhaps many years. To embark on a programme and then to greatly reduce or withdraw financial support is largely a waste of investment since reinfestation and an upsurge in transmission rapidly follow reduction in efficiency of treatment. As a project continues some gain in efficiency and reduction in costs will probably occur. These changes should not be dictated by administrative authorities but should result from efficient management within the project.

In certain circumstances schistosomiasis control can be financially profitable as well as desirable from the public health point of view (Fenwick 1972). Paulini (1972, 1978) and Jobin (1979) have developed techniques for estimating costs and benefits of snail-control programmes, and Paulini (1974) describes methods for estimating the effects of various control measures (including mollusciciding) on schistosomiasis transmission.

16.5 Future role of molluscicides

While the future status of mollusciciding in schistosomiasis control will depend on the type of control strategy adopted, which will in turn be determined by the local epidemiological, ecological, and socioeconomic conditions, there is general agreement that judicious mollusciciding must remain among the methods of choice in any comprehensive schistosomiasis control programme. Moreover, in certain circumstances, the control of the snail hosts alone, whether by chemical or environmental means, can confer substantial protection, although it is seldom that any single control measure can be advocated without reservation.

In the future, population chemotherapy together with focal and appropriately timed mollusciciding are most likely to spearhead schistosomiasis-control programmes in the endemic foci that merit high priority (McCullough *et al.* 1980).

16.6 Acknowledgements

A number of workers actively engaged in snail-control programmes in various parts of the World have provided information on their projects, and I am grateful to the following for their help in various ways: Dr M. A. Amin and Dr A. Fenwick (Sudan), Mr A. C. Arnold (London), Mr A. W. Choudhry (Kenya), Dr B. O. L. Duke (formerly of the Cameroons), Dr Kua

Yuan Hua (China), Dr M. Ito (Japan), Dr D. B. Matovu (Tanzania), Dr H.-D. Matthaei (West Germany), Mr M. A. Prentice, Mr G. Barnish, and Dr J. D. Christie (St Lucia), Dr A. P. Warley (London), and Dr K. Yasuraoka (Japan). I am particularly grateful to Dr V. de V. Clarke and Mr A. C. Evans (Zimbabwe) for much detailed information. Permission to use information contained in the Field Officer's Manual used at the Blair Research Laboratory, Harare was kindly given by the Secretary for Health, Zimbabwe.

16.7   References

Abdel Raheem, K., El Gindy, H., and Al Hassan, M. J. (1980). *Hydrobiologia* **74**, 11.

Amin, M. A. and Fenwick, A. (1977). *Ann. Trop. Med. Parasitol.* **71**, 205.

—— —— Osgerby, J. M., Warley, A. P., and Wright, A. N. (1976). *Bull. Wd. Hlth. Org.* **54**, 573.

Barnish, G. and Shiff. C. J. (1970). *Rhodesia Agric. J.* **67**, 2.

—— and Sturrock, R. F. (1973). *Trans. R. Soc. Trop. Med. Hyg.* **67**, 610.

Bayer AG Agrochemicals Division (date unknown and 1970). *Bayluscide* (two technical information pamphlets). Bayer AG Leverkusen, West Germany.

Choudhry, A. W. (1974). *E. Afr. Med. J.* **51**, 600.

—— (1975). *E. Afr. Med. J.* **52**, 573.

Chu, K. Y. and Vanderburg, J. A. (1976). *Bull. Wld. Hlth. Org.* **54**, 411.

Duke, B. O. L. and Moore, P. J. (1976). *Tropenmed. Parasitol.* **27**, 505.

Farooq, M. (1973). In *Epidemiology and control of schistosomiasis (bilharziasis)* (ed. N. Ansari), 752 pp. Karger, Basle.

Fenwick, A. (1972). *Bull. Wld. Hlth. Org.* **47**, 573.

Jobin, W. R. (1979). *Am. J. Trop. Med. Hyg.* **28**, 142.

Jordan, P., Barnish, G., Bartholomew, R. K., Grist, E., and Christie, J. D. (1978). *Bull. Wld. Hlth. Org.* **56**, 139.

Klumpp, R. K. and Chu, K. Y. (1980). *Bull. Wld. Hlth. Org.* **58**, 791.

Magendantz, M. (1974). *Pflanzenschutz-Nachrichten Bayer* **27**, 46.

McCullough, F. S., Gayral, P., Duncan, J., and Christie, J. D. (1980). *Bull. Wld. Hlth. Org.* **58**, 681.

Osgerby, J. M. (1970). *Pestic. Sci.* **1**, 5.

Paulini, E. (1972) In *Schistosomiasis: proceedings of a symposium on the future of schistosomiasis control* (ed. M. Miller), pp. 135. Tulane University, New Orleans.

—— (1974). *On the problem of allocating funds for molluscides and drugs in schistosomiasis control.* WHO Cyclostyled Report WHO/SCHISTO/74.35.

—— (1978). *Proceedings of the International Conference on schistosomiasis, Cairo.* Vol. 1, p.393.

Prentice, M. A. (1970). OAU Symposium on Schistosomiasis, Addis Ababa (quoted by Magendantz, 1974).

—— (1971). *Pestic. Abstr. News Summ.* **17**, 64.

—— and Barnish, G. (1980). *Ann. Trop. Med. Parasitol.* **74**, 45.

—— Jordan, P., Bartholomew, R. K., and Grist, E. (1981). *Trans. R. Soc. Trop. Med. Hyg.* **75**, 789.

Shell International Chemical Co. Ltd. (1974). *Frescon – a molluscicide for the better control of schistosomiasis.* Shell International Chemical Co. Technical Manual, London.

Shiff, C. J., Clarke, V. de V., Evans, A. C., and Barnish, G. (1973). *Bull. Wld. Hlth. Org.* **48**, 299.

—— Coutts, W. C. C., Yiannakis, C., and Holmes, R. W. (1979). *Trans. R. Soc. Trop. Med. Hyg.* **73,** 375.

Sturrock, R. F., Barnish, G., and Upatham, E. S. (1974). *Int. J. Parasitol.* **4,** 231.

Upatham, E. S. and Sturrock, R. F. (1977). *Ann. Trop. Med. Parasitol.* **71,** 85.

Webbe, G. and Duncan, J. (1978). *Molluscicides: present and future roles in schistosomiasis control.* WHO Cyclostyled Report SCHISTO/WP/78.9.

WHO (1965). *Snail control in the prevention of bilharziaisis.* Monograph Series No. 50.

Yeo, D. (1962). *Bull. Wld. Hlth. Org.* **27,** 183.

17
Control of plant nematodes

A. G. WHITEHEAD and J. BRIDGE

17.1 Introduction

Nematodes are unsegmented, acoelomate worms (phylum: Nematoda), round in cross-section (hence 'round worms'). Plant-parasitic nematodes are small (adults 0.3–12 mm long) and mostly translucent, so they are rarely apparent in soil or plants. Most adults are vermiform, but the females of some genera are pear-, lemon-, or sac-shaped (e.g. *Meloidogyne, Globodera, Heterodera, Nacobbus*). Reproduction is mostly sexual, sometimes parthenogenetic. Eggs are laid singly in plant tissues or in soil, or in groups in a gelatinous matrix, the 'egg-sac' (e.g. *Meloidogyne, Nacobbus, Heterodera*). In *Globodera, Heterodera,* and other related genera the body wall of the female is tanned to form a protective envelope or 'cyst' around eggs which are not extruded. There are four juvenile stages separated by moults, for the cuticle, though flexible, cannot be stretched. The first moult may occur inside the egg shell, in which case the emerging juveniles are second stage. Most plant-parasitic nematodes have wide host ranges, but some reproduce only on plants of a single family. For example, the potato cyst-nematodes [*Globodera rostochiensis* (Woll.) and *G. pallida* (Stone)] reproduce only on certain Solanaceae and red-ring nematode [*Rhadinaphelenchus cocophilus* (Cobb)] has been isolated only from palms. Most plant-parasitic nematodes are obligate parasites of green plants; some may also feed on fungi. They are found in all soils that support plant growth and all cultivated plants are susceptible to attack by one or more species. Most of these nematodes attack roots and underground stems but some attack shoots.

The sedentary endoparasites have immobile females which, at least at first, are enclosed in plant tissue. The sedentary semi-endoparasites also have immobile females, whose anterior ends are embedded in plant tissue, the posterior part projecting into the soil.

Adult migratory endoparasites feed and reproduce inside plant tissues through and out of which they migrate. The root ectoparasites live wholly in soil, inserting their stylets into roots or underground stems to feed on host cells. Migratory endoparasites of shoots attack stems, leaves, and inflorescences. Identification of nematodes and the diseases they cause is often difficult and is best done by experienced nematologists.

Nematodes may occur in very large numbers in soil or plant tissue. For example, there may be 500 potato cyst-nematode eggs/g of soil (10^{12}/ha in the top 20 cm), 250 citrus nematode juveniles/g of soil and as many as 5×10^5 stem nematodes [*Ditylenchus dipsaci* (Kühn)], in one field bean (*Vicia faba* L.) plant. They can be found as deep in the soil as are the roots. They migrate slowly through soil but are dispersed from place to place in infested plants or seeds, in soil on vehicles, machinery and animals including Man and in flood and drainage water. Some nematodes are resistant to desiccation as eggs in 'cysts' or as juveniles in dried plant debris, both of which may be dispersed by wind.

17.2 Damage caused by nematodes

Nematodes injure plants in different ways. The mechanical and chemical effects of their feeding and migration into and through plant tissues may injure the plant directly. The damage often results in a smaller root system that exploits a smaller volume of soil, thereby impairing the plant's ability to extract enough moisture and nutrients, especially nitrogen, from the soil to maintain normal growth. In extreme cases the anchoring function of the roots is destroyed and the shoot may topple [e.g. banana toppling disease due to *Radopholus similis* (Cobb)]. Nematode attack may also increase the plant's susceptibility to bacterial and fungal diseases. Some root ectoparasitic nematodes (*Trichodorus, Paratrichodorus, Xiphinema,* and *Longidorus*) are also vectors of certain virus diseases, which they transmit when they probe plant cells to feed. Above-ground symptoms of nematode attack are often ill-defined. The plants may be stunted or unthrifty, and may wilt more readily than unattacked plants. The leaves may exhibit symptoms of nutrient deficiencies (especially nitrogen and magnesium), have scorched leaf tips and margins, and be bloated and twisted or show interveinal chlorosis. The inflorescence may be distorted or even aborted. As a result, yields are reduced, sometimes drastically.

Except for some nematode-transmitted virus diseases, damage to the attacked part varies with nematode numbers. The yield loss usually depends on the interaction between root size and the environmental stress to which the crop is subjected, and is greatest when soil moisture and nutrients are limiting. Damage to seedlings often affects crop yield more than damage to older plants. The yield (Y) of field crops can often be related to the initial number of nematodes (P_i) by a sigmoid curve (Fig. 17.1), defined by the equation:

$$Y = Y_{min.} + (1 - Y_{min.})cz^{P_i}$$

In this equation (Seinhorst 1965), Y is the observed fraction of maximum yield (1.0), $Y_{min.}$ is the minimum yield as a fraction of the maximum yield, P_i is the number of nematodes, c is a constant ranging from 1.05 to 1.15, which

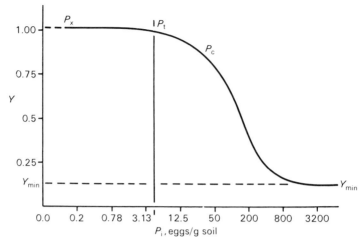

Fig. 17.1. Relationship between yield of field crops to number of nematodes; for explanation, see text.

represents 5–15% compensation by the plant for nematode injury and z is a constant, usually around 0.995, which represents the fraction of maximum yield left after one nematode has fed (Jones 1978). This equation was derived from pot experiments. In the field, the yield of unattacked plants cannot be determined reliably nor can reliable estimates of very small numbers of nematodes be obtained. The curve suggests that very small numbers of nematodes increase yields of lightly infested plants slightly. P_t is the threshold population above which crop loss occurs. Threshold values vary greatly. For virus-vector nematodes the threshold may be zero/g of soil because virus particles may be transmitted by a single nematode. For stem nematode (*Ditylenchus dipsaci*) on onions it may be as small as 0.001/g of soil. In Fig. 17.1, $P_t = 4.2$. As P_i increases beyond P_t, yield decreases to the minimum Y_{min}. If $Y_{min} = 0$, $Y = cz^{P_i}$. The relationship between nematode numbers (P_i) and yield loss is often difficult to determine in the field because a sufficiently wide range of P_i values may not occur in a soil, which is apparently uniform in all other respects. However, Brown (1969) studied the relationship between P_i and yield of potato tubers in soils infested with potato cyst-nematodes. Mostly his values of P_i were from around P_t to P_m and in a few cases he obtained values of $Y_{min.}$ and $Y_{max.}$ from which he estimated maximum yield loss at around 20 tons/ha. As the slopes of the regression lines vary considerably, maximum yield losses may be greater or smaller than this, depending on environmental stress suffered by the crop, on the cultivar grown, and on the virulence of the nematode population. Reliable, world-wide estimates of yield losses due to nematode attack are not available, but in the USA in 1970 they were estimated at about 10% (Anon. 1971). In many developing, especially tropical, countries losses may well exceed 10% but in W. Europe they are probably rather less. Of course, in any country, individual farms may be severely affected and others unaffected.

17.3 Aims of control

The first object of nematode control is to prevent injury to the crop plant, i.e. to reduce P_i to P_t or if possible to P_x, at which, yield may exceed that in uninfested soil. This entails killing or immobilizing a percentage (K) of the nematodes. Damage is prevented when:

$$K = \frac{100(P_i - P_t)}{P_i} \qquad (17.1)$$

In practice, it may not be necessary to reduce P_i to less than P_c, the first point of inflexion of the sigmoid curve.

The second object of nematode control is to prevent or limit increase in nematode numbers. Nematode increase is *density-dependent* but inversely proportional to nematode numbers (Fig. 17.2), so when P_i is reduced to P_t or

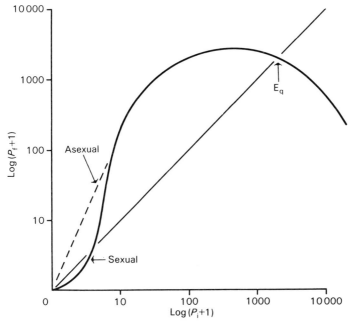

Fig. 17.2. Curve illustrating density-dependent increase of a nematode population in a crop; for explanation, see text.

P_x it approaches or reaches the maximum, because maximum crop growth provides the maximum number of feeding sites for the nematodes. If P_i is decreased further, *underpopulation* may occur in sexually reproducing species, i.e. nematode increase is less than expected because the adults are so few and far apart in soil or plant that successful matings are few. This does not occur in parthenogenetic nematodes, like *Meloidogyne* spp., large populations of which can develop from single juveniles.

Nematode increase is prevented (full control), i.e. P_f (nematodes after harvest) $= P_i$, when

$$K = 100(1 - \frac{1}{a}) \tag{17.2}$$

where a is the nematode increase for those nematodes that hatch. The actual multiplication is

$$\frac{P_f - P_u}{P_n}$$

where P_u is the number of viable nematodes, which did not hatch when the host plant was grown, and P_n is the number of nematodes attacking the plant after using a nematicide. P_u is difficult to determine, except when the eggs are retained in a cyst. In cool, temperate soils, maximum values of a may be about 10 for ectoparasitic nematodes like *Trichodorus, Paratrichodorus, Longidorus,* or *Xiphinema* and about 50 for several cyst-nematodes (*Heterodera, Globodera*). Nematodes with several generations on a host plant, such as stem nematode (*Ditylenchus dipsaci*) and root-knot nematodes (*Meloidogyne*), may increase 1000-fold or more. For such nematodes, K would have to equal or exceed 99.9%, in contrast to the ectoparasites (90%) or cyst-nematodes (98%) already mentioned. Similarly, K would have to approach or equal 100% to prevent transmission of certain virus diseases to plants by their nematode vectors. It is usually impractical, if not impossible, to achieve K values of this order.

Eqn (17.2) is empirical, for it ignores the fact that nematode increase is *density-dependent*. As a result, the kill needed to keep $P_f = P_i (K_{min})$ is rather less as P_i approaches the equilibrium population. Nor does it take account of eggs that fail to hatch, which form a fraction of the final population (P_f). Jones and Perry (*unpublished*) have produced equations applicable to nematicides used to control cyst-nematodes, which take account of these discrepancies. These have been derived from their mathematical model of the dynamics of cyst-nematode populations (Jones and Perry 1978). For fumigant nematicides

$$K_{min.} = 1 - \frac{1}{a(1 - C_p) - (a - 1)\Phi + C_p} \qquad (17.3)$$

where C_p is the proportion of eggs that do not hatch and are carried over into P_f, and $\Phi = EP_i^*$ (E = plant yield at the equilibrium population, expressed as a fraction of that in the absence of nematodes, $P_i^* = P_i$ expressed as a fraction of the equilibrium population. For nematostatic nematicides

$$K_{min.} = 1 - \frac{1}{a - (a - 1)\Phi} \qquad (17.4)$$

Eqn (17.3) applies for values of Φ greater than $[a(1 - C_p) + 2C_p]/[2(a - 1)]$. Similarly, eqn (17.4) applies when Φ is greater than $a/[2(a - 1)]$. When Φ is less than these values, i.e. in most field infestations, these equations can be simplified to:

$$K_{min.} = 1 - \frac{2}{a(1 - C_p) + 2C_p} \text{ (fumigants) or } 1 - \frac{2}{a} \text{ (nematostats)}$$

The population dynamics of other nematodes are less well understood, their population equilibria unknown, but eqn (17.2) gives a conservative estimate of K.

It is usually easier to prevent damage caused by nematodes than to control nematode increase fully. When P_i is small, a small value of K will prevent damage but have little effect on P_f if a is large. Conversely, when K is large, damage will be prevented even when P_i is great. Usually, values of K large enough to prevent damage but not large enough to prevent nematode increase can be obtained in commercial applications of nematicides. When increase is less than expected after growing a susceptible crop, *partial control* is achieved.

Nematicides may have to be applied at regular intervals to control nematode damage in perennials, or the increase of nematodes that multiply frequently on annuals.

17.4 Types of nematicide

Nematicides in commercial use belong to four groups of pesticides:
 (i) *Halogenated aliphatic hydrocarbons*—methyl bromide, ethylene dibromide (EDB), 1,3-dichloropropene (1,3-D) mixtures (e.g. D–D, Telone II), and 1,2-dibromo-3-chloropropane (DBCP)*.
 (ii) *Methyl isothiocyanate precursor compounds*—(metham-sodium,

*Now withdrawn from use for toxicological reasons.

dazomet) and *methyl isothiocyanate mixtures,* including solutions in xylol or in dichloropropene mixtures, with or without added chloropicrin.

(iii) *Organophosphates*—ethoprophos, fenamiphos, fensulphothion, thionazin.

(iv) *Carbamates*—carbofuran, aldicarb, and oxamyl.

Groups (i) and (ii) are soil fumigants, which kill nematodes and their eggs and many other soil animals. Groups (iii) and (iv) have comparatively little fumigant action and act as nematostats. Features of some widely used nematicides are given in Table 17.1.

17.4.1 Soil fumigants

Except for dazomet, which is formulated as fine granules ('prill'), soil fumigants are volatile liquids, which have the advantages of not requiring mixing with the soil and the ability to kill nematodes deep in the soil (Harrison *et al.* 1963; Taconis and Kuiper 1964; Whitehead *et al.* 1970). However, large amounts are needed to kill the majority of nematodes in soil, i.e. 100–600 (or more) kg/ha. The relationship between log dosage and probit mortality is linear (Peters 1954; Seinhorts 1973), probit mortality increasing by about one-half of one probit unit with each doubling of the dosage of fumigant. Large amounts of soil fumigants are phytotoxic and must therefore be applied to the soil several days, weeks, or months (according to the fumigant and the soil temperature) before planting. Dibromochloropropane, applied around the roots of woody perennials and some herbaceous perennials (e.g. banana), controls nematode damage without harming the plants.

Methyl bromide, metham-sodium, and dazomet also kill or inhibit soil fungi and may kill weed seeds or delay their germination. Dazomet had a harmful effect on mycorrhizae and on *Pinus* seedlings in inadequately aerated soils (Iyer and Wojahn 1976). Dichloropropene mixtures are usually weakly toxic to fungi and their residues sometimes harmful to wheat; occasionally they taint potatoes. Soil fumigants also inhibit or kill some soil bacteria. Surviving ammonifying bacteria, which convert organic nitrogen in soil to ammonium nitrogen, recolonize the soil more quickly after fumigation than the nitrifying bacteria, which convert ammonium nitrogen to nitrate nitrogen. Ammonium nitrogen, therefore, accumulates in the soil (Gasser and Peachey 1964). This may be an advantage to crops in sandy soils, from which nitrate but not ammonium nitrogen is rapidly leached from the surface by rain, but it may necessitate a reduction in fertilizer N to counteract excess uptake of N, which may adversely affect the quality of certain crops (e.g. sugar beet, tobacco leaf). Soil fumigants may also alter the availability of other soil nutrients, e.g. phosphorus, potassium, manganese, iron, aluminium, and silicon (Smith 1963; Singhal and Singh 1977).

TABLE 17.1. *Properties and application of some common nematicides*

	Common names	Methyl bromide	Dichloropropene (Telone) dichloropropane–dichloropropene mixture (D–D)	Ethylene dibromide	Dazomet	Fenamiphos	Ethoprophos	Aldicarb	Oxamyl	Carbofuran
	Trade names	Embafume C Dowfume MC-2 Dowfume MC-33 Brozone	D–D Vidden D Telone Telone II	Dowfume W-85, Soil Brom, Liro fumigant DB Fumephyt	Basamid Mylone	Nemacur, formerly Nemacur P	Prophos, Mocap	Temik	Vydate	Furadan Yaltox
Compound	Chemical name	Bromomethane	1,3-Dichloropropene	1,2-Dibromoethane (ethylene dibromide)	3,5-Dimethyltetrahydro-1,3,5-2H-thiadiazine-2-thione	Ethyl 4-methylthio-*m*-tolyl isopropylphosphoramidate	*O*-Ethyl *SS*-dipropyl phosphorodithioate	2-Methyl-2-(methylthio)propionaldehyde-*O*-methylcarbamoyl oxime	Dimethylamino-1-(methylthio) glyoxal *O*-methylcarbamoyl monoxime.	2,3-Dihydro-2,2-dimethylbenzofuran-7-yl methylcarbamate
	Formulation	+2% Chloropicrin	1,3-Dichloropropene and other chlorinated hydrocarbons	Concentration as required in solvent	Fine granules 98–99% a.i. (prill)	5, 10% granules	10% granules	5, 7½, 10% granules	10% granules 10.25% liquid	2,3,5,10% granules
	Boiling point (°C)	4	104–112	131–132	Principal active breakdown product is methyl isothiocyanate 119	—	—	Decomposes above 100°C	—	—
Active ingredient	Vapour pressure (mmHg)	1420 (20°C)	About 55 (20°C)	8 (20°C)	21 (20°C)	1×10^{-6} (30°C)	3.5×10^{-4} (26°C)	1×10^{-4} (25°C)	2.3×10^{-4} (25°C)	2×10^{-5} (33°C)
	% water solubility	1.75 (20°C)	0.1 (20°C)	0.3 (20°C)	0.8 (20°C)	0.07 (20°C)	0.075	0.6 (25°C)	28 (25°C)	0.07 (25°C)
Mammalian toxicity	Acute oral LD_{50} or LC_{50}	(Rat) 514 p.p.m.	(Rat) 140–500 mg/kg	(Male rat) 146 mg/kg	(Rat) 500 mg/kg	(Male rat) 15.3 mg/kg	(Rat) 62 mg/kg	(Rat) 0.9 mg/kg	(Rat) 5.4 mg/kg	(Rat) 8–14 mg/kg
	Suggested minimum soil temperature (°C)	5	7	16	10	—	—	—	—	—
Application technique	Dose	400–900 kg/ha	50–900 kg/ha	Up to 250 kg/ha	200–500 kg/ha	Up to 14 kg/ha/annum	Up to 25 kg/ha/annum	Up to 11 kg/ha	Up to 11 kg/ha	Up to 10 kg/ha/annum
	Typical methods of application	By pipeline direct from cylinder to soil under gas-tight cover	(i) Injected 15 cm deep in rows 25–30 cm apart (ii) Injected 15 cm deep in crop row positions	Injected 15 cm deep in rows 25–30 cm apart	Spread on soil surface and incorporated by rotary cultivation	Broadcast to soil around banana stems at regular intervals, or rotary-cultivated-in, or applied in seed furrows	Broadcast to soil around banana stems at regular intervals, or rotary-cultivated-in, or applied in seed furrows	Broadcast to soil and rotary-cultivated-in, or applied in seed furrows. Oxamyl also spread on soil around banana stems at regular intervals		Broadcast to soil around banana stems at regular intervals, or rotary-cultivated-in, or applied in seed furrows

Harvested crops may contain soil fumigant residues, e.g. bromine after fumigation with methyl bromide (Kempton and Maw 1972), ethylene dibromide (Brown *et al.* 1958) or DBCP (Newsome *et al.* 1977) and 3-chloroallyl methyl sulphide (Dekker 1972), 2,2-dichloropropane or 1,2,3-trichloropropane (Shepherd 1952) after fumigation with dichloropropene mixtures.

17.4.2 *Organophosphates and carbamates*

Much smaller amounts (0.5–10 kg a.i./ha) of organophosphates, carbofuran (a carbamate), or oxime *N*-methylcarbamates (especially aldicarb or oxamyl) than of soil fumigants are needed to control nematodes in soil. As such amounts are not usually toxic to plants, they can be applied at or just before planting, a big advantage, or they can be applied around the roots of established plants (e.g. banana).

Among the organophosphates, the diethyl phosphorothioates are more effective against potato cyst-nematode (*Globodera rostochiensis*) than dimethyl phosphates (Whitehead *et al.* 1972), and, although a number of organophosphates are effective nematicides in sandy soil, they often act poorly in soils with large organic fractions, on which they are strongly adsorbed. Nematodes in planting material (bulbs, corms etc.) may be controlled by dipping in a thionazin solution (Winfield 1973). Solutions of fensulphothion, fenamiphos, ethoprophos, and parathion have also been used for the same purpose. Parathion and thionazin have been used to control leaf nematode [*Aphelenchoides ritzemabosi* (Schwartz)] in chrysanthemum by repeated drenching of infested cuttings and plants (Bryden and Hodson 1957; Anon. 1972).

Like the organophosphates, carbofuran, though an effective nematicide in many soils, is of limited value in organic soils. The oxime *N*-methylcarbamates, especially aldicarb and oxamyl, are effective nematicides in a wide range of soils including very organic soils (Moss *et al.* 1975). They have 'half-lives' of a few weeks only; the half-life may be reduced if the soil is very alkaline or very warm.

All these compounds are highly toxic by ingestion, inhalation, or even by absorption through the skin, so for safe field use are best formulated as non-dusty granules.

(a) Translocation of organophosphates and oximecarbamates in plants. Applied to the soil, carbofuran, fenamiphos, thionazin, aldicarb, and oxamyl are absorbed by roots and translocated via the xylem to stems and leaves. This *acropetal translocation* may result in the control of nematodes and insects that attack the shoots.

Nematicides have been sought which, when sprayed on leaves, would be translocated to kill or repel nematodes feeding in or on the roots (*basipetal translocation*). Ideally, what is needed is an efficient nematicide which behaves in this way and is also non-toxic to plants or vertebrates. Peacock (1963) found that several phenyl tricyanopentanes applied to tomato leaves inhibited invasion of the roots by larvae of the root-knot nematode, *Meloidogyne incognita* (Kofoid & White), but did not affect those already established in the roots. Sprays of oxamyl or fenamiphos have been successful in some instances.

Spraying crops with oxamyl may lessen invasion of roots by nematodes (e.g. Potter and Marks 1971) but does not kill nematodes already developing in the roots. Juveniles of *G. rostochiensis* were able to develop in roots of potato plants sprayed with oxamyl (Harrison 1971), and dipping the shoots of susceptible, young potato plants in an aqueous solution of oxamyl (2000 p.p.m.) did not lessen nematode increase (Whitehead *et al.* 1973a). Gowen (1977) showed in pot experiments that oxamyl sprayed on the leaves of bananas controlled *Radopholos similis* and *Helicotylenchus multicinctus* (Cobb) in the roots.

Oxamyl sprayed six times per annum on to Valencia orange trees [*Citrus sinensis* (Osbeck)] budded on rough lemon rootstocks [*C. limon* (Burn. f.)] significantly reduced numbers of burrowing nematode (*Radopholus similis*) in the roots and significantly increased fruit yields in the second and third years of treatment (O'Bannon and Tomerlin 1977). Up to six sprays (each of 1, 2, or 4 kg of oxamyl/ha) at approximately 3-week intervals to potato foliage reduced the increase of *G. rostochiensis* on susceptible potatoes to about half, but more than 6 kg of oxamyl/ha left unacceptable residues in the tubers (Whitehead *et al.* 1979b). Oxamyl granules in the seedbed followed by one spray of 5.6 kg of oxamyl/ha to foliage of potatoes 8 weeks after planting controlled 'spraing' (caused by a virus transmitted by *Trichodorus* spp.) on two of three fields where it was tested, whereas the fumigant D–D controlled the disease at all three sites (Alphey 1978).

Basipetal translocation has also been demonstrated with fenamiphos (Homeyer 1971; O'Bannon and Tomerlin 1971). In Hawaii, almost 100% control of reniform nematode [*Rotylenchulus reniformis* (Linford & Oliveira)] resulted in the roots and in the soil around pineapple plants grown for 60 days from crowns dipped in fenamiphos solutions (1200 or more p.p.m.). Control after 120 days was almost as good (Zeck 1971). Although the tough leaves of perennials like pineapple and citrus are unharmed, fenamiphos sprays are toxic to the tender leaves of many annuals (e.g. okra, tobacco, tomato).

(b) Other effects of organophosphates and carbamates. These nematicides were first introduced for use as insecticides, and at current commercial rates for nematode control, which are greater than those generally recommended for insect control, they also act as powerful insecticides. Even 1 kg of

aldicarb/ha applied with sugar beet seed to control *Trichodorus* and *Longidorus,* which cause Docking disorder, also controls aphids and other seedling pests of sugar beet, including foliage pests, for many weeks by systemic action. Other invertebrates are also affected in soil treated with these nematicides. The organophosphate and non-fumigant carbamates may also affect soil fungi and bacteria.

(c) Residues of organophosphates and oximecarbamates. At dosages used to control nematodes there is usually less than 0.1 p.p.m. of organophosphates or oximecarbamates in harvested produce and such amounts are not considered harmful to Man or his livestock.

17.5   Mode of action of nematicides

Soil fumigants kill nematodes and their eggs. They diffuse through the pore system of soils into which they are injected, small amounts dissolving in the water films around soil particles and penetrating the cuticle of nematodes. Alkyl halide nematicides first induce a period of hyperactivity in nematodes but this gradually decreases and eventually paralysis ensues (Van Gundy *et al.* 1972). Even then the effect is not immediately fatal (Evans and Thomason 1971). Ethylene dibromide, for example, enters the body of *Aphelenchus avenae* Bastian about two and a half times as fast as water and has at first a narcotic effect, which reduces mobility, but this effect is reversible (Marks *et al.* 1968).

Organophosphates and oximecarbamates have a nematostatic effect at less than lethal concentration. In insects they inhibit acetylcholinesterases at nerve–cell junctions, so that acetylcholine accumulates there and blocks the passage of nervous impulses between nerve cells. This appears first to affect the sense organs, resulting in disorientation. The same seems to be true of nematodes. As the concentration of the nematicide increases within the nematode, movement becomes sluggish and may cease altogether. These effects are reversible until lethal concentration is reached within the nematode.

17.5.1   Effects on unhatched juveniles

In soil, hatching of potato cyst-nematode eggs was inhibited or retarded by aldicarb at 5 p.p.m. but not at 1 p.p.m. (Hague and Pain 1970). At 2 or 8 p.p.m., the organophosphates, CGA 12223 [*OO*-diethyl *O*-(1-isopropyl-5-chloro-1,2,4-triazolyl-(3)) phosphorothioate] and fenamiphos, markedly inhibited hatching, whereas oxamyl or carbofuran delayed hatching slightly (Hague 1975). Hatching of beet cyst-nematode eggs, in the presence of beet root diffusate, was stimulated by aldicarb at 0.01 or 0.05 p.p.m. (Steele and Hodges 1975). This is probably the result of hyperactivity, including induced stylet thrusting in the unhatched larva (Nelmes 1970). Aldicarb at 0.5 p.p.m. had no effect on hatching of *H. schachtii* (Hough and Thomason 1975) but as little as 1 p.p.m. in the soil retarded hatching of *H. schachtii* (Steudel 1972; Steele and Hodges 1975), and 5–500 p.p.m. inhibited hatching (Steele and Hodges 1975). Larger concentrations of aldicarb are needed in the soil water to inhibit hatching of root-knot nematode larvae (e.g. McLeod and Khair 1975) than to inhibit hatching of cyst-nematodes.

The inhibition of hatching by aldicarb is reversible, except at high concentrations, e.g. 20 p.p.m. for 12 weeks for potato cyst-nematode (Osborne 1973) and 48 p.p.m. for the root-knot nematode, *Meloidogyne javanica* (Hough and Thomason 1975). Organophosphate and oximecarbamate nematicides appear to inhibit hatching only when applied to soil in amounts far greater than customary in commercial practice.

17.5.2 *Effects on nematodes in soil and roots*

Many soil fumigants (Peachey *et al.* 1963), organophosphates and oximecarbamates (e.g. Whitehead *et al.* 1973c), greatly reduce the invasion of roots by cyst-nematode juveniles. This may be because they are killed or temporarily inactivated in the cysts or in the soil, or because there are toxic amounts of nematicide on or in the roots. Den Ouden (1971) found that the roots of potato plants grown in solutions of aldicarb, methomyl, thionazin, or fenamiphos repelled potato cyst-nematode juveniles before or soon after they had penetrated the roots, whereas the development of the nematode was not affected when roots were dosed 11 days after they had been invaded. As few as 0.1% or less of the initial numbers of (unhatched) juveniles may be found in potato roots 4 weeks after planting unrooted potato sprouts in samples of soil treated with large amounts of soil fumigants or small amounts of organophosphate or oximecarbamate nematicides (e.g. Whitehead *et al.* 1975a). *In vitro*, aldicarb inhibited locomotion of second stage juveniles of potato cyst-nematode but stimulated stylet thrusting. Aldicarb at 5 p.p.m. (Hague and Pain 1970) or 10 p.p.m. was lethal to the juveniles, but the effects of smaller concentrations could be reversed by transferring them to water (Nelmes 1970). However, Steele and Hodges (1975) found that development of beet cyst-nematode juveniles was affected if they were exposed to aldicarb solutions before being used to inoculate beet plants. Aldicarb is more toxic to the juveniles than aldicarb sulphoxide, which in turn is more toxic to them than aldicarb sulphone (Nelmes 1970). However, oxamyl, not its breakdown products is effective as a nematicide (Bromilow 1973). Clearly, nematostatic nematicides may affect nematodes at more than one point in their life cycles.

17.6 Resistance to nematicides

Resistance of insects to insecticides and tolerance of fungi to pyrimidine and benzimidazole fungicides is well known but as yet there is little evidence of resistance in nematodes to nematicides. However, repeated exposure of the fungus feeder *Aphelenchus avenae* to ethylene dibromide did result in the selection of strains more resistant to the fumigant (Castro and Thomason 1971). Bunt (1975) obtained no clear evidence of resistance developing in *Ditylenchus dipsaci* exposed for several generations to oxamyl or fenamiphos.

17.7   Application of nematicides

Nematode control is much affected by the choice and dosage of nematicide and by the method used to apply it to the plant or the soil. No one nematicide is suitable for controlling all nematodes. Similarly, the application strategy adopted will depend on the nematode to be controlled.

17.7.1   Treatment of seeds and setts

(a) Fumigation. An air-tight chamber is needed for fumigation of seeds. Stem nematodes in onion seed can be killed by fumigating the seeds with methyl bromide (Goodey 1945). Similarly, seeds of teazle and red clover (Goodey 1949), lucerne (Page *et al.* 1959), and shallot (Hague 1968) can be successfully fumigated. Powell (1975a) found that 1000 mg of methyl bromide/h/litre at 18°C reduced germination of onion seed by 11.5% but did not affect seedling vigour, whereas 2000 mg/h/litre severely reduced germination and affected seedling vigour. Vegetable seeds at 10 or 13% moisture are damaged less by methyl bromide fumigation than are seeds at 16% (Powell 1975b). Attempts to control the 'white tip' nematode, *Aphelenchoides besseyi* Christie, in rice seed at different moisture contents, with different amounts of methyl bromide and different exposure periods, without phytotoxicity, were unsuccessful (Todd and Atkins 1959).

(b) Dipping. Seeds, setts, or transplants may also be dipped in nematicidal solutions, thereby coating or infiltrating them with nematicide, to protect the young plants from attack or to kill nematodes in roots. The normal procedure is to fill a large tank or drum with the nematicidal solution into which is immersed a wire basket containing the seeds, setts, or transplants. For example, cotton plants were protected for 30–40 days from attack by root-knot nematodes when the seeds were steeped in DBCP before sowing (McBeth 1960). Similarly, some protection against the root-knot nematode, *Meloidogyne graminicola* (Golden & Birchfield), and other nematodes, was conferred on rice seedlings when the seeds were soaked for 12 or 24 hours in oxamyl, fensulphothion, or phorate at 500, 1000, or 2000 p.p.m. (Prasad and Rao 1976). Cysts of potato cyst-nematodes are destroyed when they are immersed in calcium or sodium hypochlorite solutions. Potato tubers submerged for 2 hours in sodium hypochlorite (1% available chlorine) were

thereby freed of cysts without harming the tubers (Wood and Foot 1975). Winfield (1978) concludes that stem nematode is controlled in tulips, which are grown for 1 year, by dipping the bulbs in an agitated solution of thionazin (2300 p.p.m.) at about 18°C for $2\frac{1}{2}$ hours, but that for narcissus, which are grown for at least 2 years, such treatment is inadequate.

Nematodes in the roots of transplants may be controlled, even eliminated, by immersing the roots in solutions of nematicides. For example, dipping the roots of citrus seedlings for 30 min in fenamiphos (250–600 p.p.m.) freed them of live *Radopholus similis* without harming the seedlings (O'Bannon and Taylor 1967).

17.7.2 Soil treatments

Liquid soil fumigants such as D–D, EDB, Telone, or metham-sodium are normally applied to soil by injection using either a hand-held fumigation gun or by tractor-mounted soil-injection equipment. The hand-operated fumigation gun can be used for small areas. It contains a small reservoir for the nematicide, a measured dose of which is injected into the soil when a plunger on top of the gun is pressed. In tractor-mounted equipment the nematicide is delivered by gravity or under pressure through tubes mounted behind chisel or blade tines or ploughs. Dosage is controlled by forward tractor speed, or, more accurately, by use of a land-wheel-operated pump. The depth at which the nematicide is applied is important and injection holes or furrows must be sealed with soil immediately to prevent rapid loss of the gases. Water-soluble or emulsifiable nematicides can also be applied directly in irrigation water although care has to be taken to control the dosage.

Granular nematicides can be applied to the soil by various methods. For small areas and spot treatments, simple 'pepper-pot' type applicators can be used, made from empty containers, e.g. coffee tins, by punching holes in the lid and fitting a handle. Metered, hand-held, shoulder-slung, and knap sack granule applicators are available which give a measured quantity of granules by gravity feed; motorized knapsack applicators are also manufactured. Better control of nematodes is achieved by incorporation of the granules into the soil but this is not always economic or necessary. Different tractor-mounted granule applicators are available which dispense the granules in a swath, bands, or in seed furrows. Details of the granule application equipment in current use are given by Matthews (1979).

(a) Seed-furrow and row treatments. Seedlings of annuals are usually damaged more by nematode attack than are older plants, so much of the damage may be prevented by applying an effective nematicide to the seed furrows during sowing or to the bands or ridges of soil in which the young plants are grown. The wider apart the crop rows are spaced the cheaper such treatments become. The seedlings of perennials may also be more sensitive to

attack than older plants, but the reverse is true for banana. Walter and Kelsheimer (1949) first showed the benefit of row treatments by fumigating the row positions in which tomatoes were then planted. The plants grew and yielded just as well and were just as well protected from early root-knot nematode attack as were plants grown where all the topsoil had been fumigated. Seed-furrow or row treatments have been used in controlling a range of nematodes and these are referred to below.

(b) Seedbed treatments. Soil in which nematode-susceptible seedlings are raised for subsequent transplanting in fields or plantations should be treated with an effective nematicide before the seeds are sown, especially if the beds are used repeatedly. Seedling losses in untreated seedbeds can be severe and poor crops often result from nematode-infested seedlings (e.g. tobacco). The cost of treating the seedbeds is small, relative to the value of the stock and the subsequent crop.

Seedbeds can be fumigated very effectively with methyl bromide under gas-tight polyethylene sheeting but seedlings may be raised more cheaply in pots of methyl-bromide-fumigated soil, e.g. tea (Kerr and Vythilingham 1967). DBCP, phorate, or thionazin applied to soil in pots controlled *Pratylenchus loosi* infesting young tea plants but did not eliminate them (Hutchinson 1963). The soil may also be fumigated with D–D, EDB, dazomet, or metham-sodium. Covering the soil with polyethylene sheeting improves the efficacy of D–D (Thorne and Schieber 1962) or Telone but did not improve the efficacy of dazomet (Whitehead *et al.* 1975a).

Covering the soil surface with clear polyethylene sheeting not only seals in the gas but in sunlight raises the soil temperature, in the tropics sometimes enough to kill nematodes near the soil surface (Whitehead, unpublished data).

The control of nematodes in tea and coffee nurseries was reviewed by Whitehead (1969). In tomato nursery beds, fenamiphos or carbofuran applied through an overhead sprinkler irrigation system partially controlled root-knot nematodes and markedly increased the output of marketable transplants (Johnson 1978).

(c) Planting-hole and localized topsoil treatments. Spot treatment with nematicides of the soil into which seedlings are transplanted is relatively inexpensive when the plants are wide apart, and it gives temporary control of nematode damage in the critical establishment period. Such treatments have been used to control damage to tobacco transplants and as an aid in the establishment of tea and coffee bushes. In S. India, Reddy (1976) found that oxamyl at 0.5, 1, or 2 kg a.i./ha, drenched into the soil around freshly transplanted tobacco seedlings, controlled *Meloidogyne incognita* and *Meloidogyne javanica* (Treub), and increased yield of cured tobacco leaf by over 50%. Soil fumigants, such as EDB, applied as spot treatments in tobacco-

planting stations can give good control of the root-knot nematode. This is still common practice in some tobacco-growing areas; maximum control is achieved by applying the fumigant 30–38 cm below the top of the soil ridge (Daulton 1967). Spot treatment using a hand-held injection gun is laborious and time-consuming when done correctly and, in large-scale tobacco farming, it can be uneconomic because of inefficient applications of the nematicide giving very poor nematode control (Bridge, unpublished).

Non-phytotoxic nematicides may also be applied to the soil around the stems of plantation crops. Good control of nematodes and profitable yield increases have been reported where bananas have been treated in this way. Treatments must be repeated at regular intervals to maintain adequate control (see below). Cyst-nematodes attacking potatoes may also be controlled by incorporating nematicides in the top 10–15 cm of the soil, which is then ridged up and into which potato setts are planted.

(d) Topsoil treatments. Effective chemical control of nematodes throughout the topsoil is expensive and often difficult to achieve, for the volume of soil to be treated is very great (2.5×10^3 m^3/ha in the top 25 cm of the soil). Volatile nematicides like D–D, Telone, or methyl isothiocyanate and its precursors can be applied under the plough sole during ploughing, through blade coulters mounted 20–30 cm apart on a tractor-drawn toolbar, or by under-soil spraying behind A-blade shares. Such materials have the advantage that they diffuse through the topsoil, but the disadvantage that they often escape so rapidly from the soil that nematodes near the surface are not killed unless the surface is 'sealed' by rolling, sprinkling with water, or both. The soil crumbler and powered flat roller of the Rumptstad fumigant injector, which runs a little faster than the forward speed of the tractor, makes a particularly flat and compact soil surface behind the under-soil fumigant sprayer, thereby increasing the kill of nematodes near the soil surface.

Non-volatile nematicides, such as organophophates and oximecarbamates, must usually be mixed well with the top 15 cm of the soil to control nematode populations effectively. This can be achieved by rotary cultivation (rotavation) but it is slow, for the rotor speed must be fast in relation to forward tractor speed to achieve good mixing and the hood must be lowered to shatter the clods. Rotary cultivators with L-shaped tines may 'smear' the soil in wet conditions, especially if it contains an appreciable amount of clay or silt; spike tines are more suitable on such soils. Rotary cultivators mix weathered surface soil with unweathered soil from below, which is undesirable for crops grown in level seedbeds and on the heavier soils. These problems can be overcome by blowing the granules into vertical bands 12.5 or 25 cm apart in the top 12–15 cm of the soil and then mixing them laterally by rotary harrowing as for example with the Lely 'Roterra' (Whitehead *et al.* 1981). Granules applied to the soil surface and harrowed in are not mixed uniformly with the soil, 60–70% of the granules usually remaining in the top 5 cm of the soil (Whitehead *et al.* 1975b). Such overall soil treatments even when effective can only be justified for valuable field crops or in glasshouses as a cheaper alternative to steam sterilization of the soil.

In established perennial crops, nematicides can be injected around the roots, applied as granules to the soil surface, from which the toxicant is leached into the soil by rain or irrigation water, or applied in irrigation water. Raski *et al.* (1976) have reviewed the chemical control of nematodes attacking grape vines in California. Very good initial control of root-knot nematodes and of the dagger nematode, *Xiphinema index* (Thorne & Allen), and thereby the 'fan-leaf' virus it transmits, is obtained by thorough fumigation of the soil after it has been deeply cultivated and before the vines are planted. Very large amounts of 1,3-dichloropropene or methyl bromide are needed to do this and careful soil preparation is important (McKenry and Thomason 1976a,b).

(e) Treatment of irrigation water. Overman (1975) controlled *M. incognita* in tomatoes by applying nematicides through a trickle irrigation system. In flood-irrigated clay soil in Texas, DBCP at 56 kg a.i./ha applied once in 15 cm of water reduced numbers of *Tylenchulus semipenetrans* on sour orange rootstocks (*Citrus aurantium* L.) bearing grapefruit [*C. paradisi* (Macfad.)] for 18 months–2 years and increased fruit yield in the first two seasons after treatment. Re-treatment after 2 years gave large increases in yield the following harvest. In contrast, oxamyl at 2.8 kg a.i./ha applied to the foliage twice a year reduced nematode numbers by 50% but increased yield greatly only in one of the two seasons in which it was tested (Timmer 1977). Chemical control of nematodes in citrus and banana are discussed more fully below.

17.7.3   Plant injection

Viglierchio *et al.* (1977) found that nematicide solutions injected under pressure into tree trunks or vine stems reduced numbers of *Pratylenchus* in the roots. In this way, numbers of *P. vulnus* (Allen & Jensen) in grape roots were significantly reduced by carbofuran, oxamyl, fenamiphos, sulphocarb, and DBCP, all but the last giving some control of *P. penetrans* in apples also.

17.8   Factors affecting control by nematicides

Many factors affect the control of nematodes by nematicides, especially fumigant nematicides. The most important are the distribution of the nematodes in the soil, nematicide concentration × time product (CTP), soil type, soil pore space, soil moisture, and soil temperature.

17.8.1 *Nematode distribution in soil*

The distribution of nematodes in the soil influences the choice of nematicide treatment and the control that can be achieved. The root-knot nematode, *Meloidogyne incognita*, has been found to depths of 120 cm in glasshouses where it is out of reach of normal steam or chemical soil sterilization techniques (Bird 1969). The needle nematodes, *Longidorus attenuatus* (Hooper) and *L. leptocephalus* (Hooper), are often more abundant below 30 cm deep in sandy soil but can be controlled there by injecting dichloropropene mixtures or chloropicrin into the soil in winter (Whitehead *et al.* 1970; Evans and Pandé 1972). Similarly, the dagger nematode, *Xiphinema diversicaudatum* (Micoletzky), may be numerous in soil to at least 60 cm deep, whereas *L. elongatus* (de Man) is mostly confined to the top 30 cm (D'Herde and van den Brande 1964). In soils heavily infested with potato cyst-nematodes in England, the nematodes were abundant in half of the soils to about 20 cm deep but in the other soils were as abundant 20–40 cm as 0–20 cm deep (Whitehead 1977). Such deep infestations of root-parasitic nematodes are difficult and expensive to control chemically, whereas the stem nematode, *Ditylenchus dipsaci,* being a shoot parasite, can be controlled by systemic nematicides mixed shallowly into the soil in the crop rows.

17.8.2 *Nematicide concentration × time product (CTP)*

Hague and Sood (1963) have shown that, except at extremes of time and concentration, the number of *G. rostochiensis* juveniles killed by methyl bromide (and probably other soil fumigants) is related to CTP.

At the same CTP, fewer nematodes were killed by large concentrations of methyl bromide for short periods than small concentrations for long periods (Hague *et al.* 1964). The actual CTP achieved is not necessarily the same as the applied CTP. In soil, the actual CTP can be greatly affected by porosity and by the loss of gas at the soil surface. The smaller the porosity of the soil, the longer it took for a dose of ethylene dibromide to kill *G. rostochiensis,* i.e. to reach the required CTP throughout the soil bulk (Call and Hague 1962). Where gas is escaping from the soil surface, it is the area under the actual CTP curve which relates to kill, not the apparent CTP. As organophosphate and oximecarbamate nematicides are degraded in time to substances not toxic to nematodes, it is important to know what CTPs are needed for nematode control with these compounds. Hague (1975) showed that for *Aphelenchus avenae* feeding *in vitro* on the fungus *Botrytis cinerea* long exposure to very small concentrations of nematicides added to the nutrient medium was lethal to the nematodes. For fenamiphos and oxamyl, an LD_{50} of 0.07 p.p.m. was obtained, and for CGA 12223 or carbofuran the figure was even smaller (0.03 p.p.m.). In soil, it may be possible to maintain small

concentrations of these nematicides by using 'slow-release' granule forma-
tions, though this might lead to a residue problem in the harvested crops.

17.8.3   *Soil type, pore space, and moisture*

Fumigant gases are much adsorbed on the clay and organic matter fractions
of the soil and so are less efficient in soils containing much of these fractions,
(e.g. Whitehead *et al.* 1973d). Adsorption is reduced when the temperature of
the soil is raised and its moisture content is decreased. Lewis and Mai (1963)
found that 1,3-dichloropropene mixture was more effective against *Ditylenchus dipsaci* in organic soils kept hot and dry than in those kept moist and
cold. Stark (1948) found that the greater the clay content of the soil the more
chloropicrin was adsorbed, whereas dry organic matter adsorbed little of the
gas. Organophosphate nematicides are also affected by soil composition.
Thionazin, for example, controlled *G. rostochiensis* better in peaty loam than
in silt loam (Whitehead *et al.* 1973c), whereas fenamiphos controlled the
nematode better in sandy loam than in peaty loam or silt loam (Whitehead *et
al.* 1973e). Although in two soils, one a peaty loam and the other a silt loam,
fenamiphos fully controlled the nematodes (Whitehead *et al.* 1973c), later
experiments (Moss *et al.* 1975) suggested that fenamiphos was less effective
the greater the organic matter content of the soil. This was in agreement with
the work of Bromilow (1973), who showed that redistribution of non-volatile
compounds by leaching could be related to Q, i.e. the ratio of the chemical
concentration in the soil to the chemical concentration in the soil water.
Nematicides with large Q values [e.g. fenamiphos-130 (7% in water phase in
a typical mineral soil)] were less effective than those with small Q values [e.g.
aldicarb-10 (50% in water phase) or oxamyl-2 (83% in water phase)], for they
were more readily adsorbed on to soil organic matter and so were redistributed less by leaching. It is the polarity of the compound, i.e. the degree of
charge separation in the molecule, which determines the size of Q. The larger
the organic matter content of the soil and the greater the value of Q, the less
the nematicide will move and the greater the need to mix it thoroughly with
the soil. In finely structured soils, organophophate and oximecarbamate
nematicides will move less than in coarsely structured soils. Redistribution of
the nematicide will also be affected by the amount of water flowing through
the soil profile. As would be predicted from their small Q values, the
effectiveness of three oxime N-methylcarbamate nematicides (aldicarb, oxamyl, and Tirpate*) was little affected by the clay or organic matter content of
the soil (Moss *et al.* 1975).

Organophosphate and oximecarbamate nematicides may be degraded
more rapidly in soils with high pH. For example, Bromilow (unpublished
data) has found that oxamyl has a half-life of only 1 week in very chalky

---

*2,4-Dimethyl-2-formyl-1,3-dithiolane oxime N-methylcarbamate.

(alkaline) soils compared with a half-life of 2–3 weeks in soils of pH around 7.

Nematodes live in the water films surrounding soil particles and few of them can penetrate pores less than 30 µm in diameter. To fumigate a soil effectively (in terms of nematode control) the gas must therefore be able to penetrate pores 30 µm or more in diameter. As the gases diffuse slowly through water, compared with air, this means draining the soil until such pores are open, which they are at about 100 cm of water suction or pH 2.0. According to Wallace (1963) usable pore space is related to nematode diameter (d) by the equation:

$$h = 3000d^{-1}$$

where h = suction pressure of the soil in cm water. So, for nematodes of a body diameter of 20 µm, a suction pressure of 150 cm water (pF 2.176) would be needed to ensure that they were all reached by fumigant gases.

17.8.4 Soil temperature

Ethylene dibromide and DBCP have small vapour pressures, so they take much longer at low temperatures to reach concentrations lethal to nematodes and to escape from the soil into the atmosphere than the more volatile 1,3-dichloropropene or methyl isothiocyanate. Even in warm soils, DBCP is slow acting (Ichikawa *et al.* 1955). In general, soil temperature is less important for most fumigant nematicides than soil moisture. Provided that the soil remains well drained throughout the fumigation period, satisfactory fumigation can be achieved in sandy soils with 1,3-dichloropropene mixtures or methyl isothiocyanate at soil temperatures above about 5°C. In moist soils, the higher the temperature the more rapidly are organophosphate and oximecarbamate nematicides degraded.

17.9 Effective nematicide treatments

17.9.1 Ectoparasitic nematodes

Some root ectoparasitic nematodes, which multiply slowly on host plants, can be controlled fairly easily with nematicides. For example, trichodorids and *Longidorus* spp., which feed on the roots of sugar beet seedlings in England causing 'Docking disorder', can be controlled adequately either by injecting 45 litres of 'Telone' or 67 litres of D–D/ha 15 cm deep in the rows in which sugar beet seed is sown 1–10 days later or, more conveniently, by applying 0.6–1.2 kg of aldicarb a.i./ha as granules in the seed furrow during sowing (Whitehead *et al.* 1971; Cooke *et al.* 1974). These row treatments kill or paralyse the nematodes close to the seedlings affording adequate protection to the developing plants. *Trichodorus christiei* (Allen) is a very damaging pest in southern USA. As it can pass several generations in one growing season on a host plant, populations are difficult to control chemically, though nematicide treatments will prevent damage to seedlings. The incidence of 'spraing' in potatoes, caused by tobacco rattle virus and transmitted by *Trichodorus* and *Paratrichodorus* spp. was greatly reduced by treating the soil with D–D, dazomet or methomyl (Cooper and Thomas 1971) or by applying small amounts of aldicarb, ethoprophos or oxamyl in the seed furrows at planting time (Brown and Sykes 1973). French and Wilson (1976) found aldicarb and oxamyl more effective against spraing in potatoes than fenamiphos, carbofuran or phorate and topsoil treatments rather more effective than seed furrow treatments.

Large amounts of methyl bromide, chloropicrin, or D–D (450–900 kg or more/ha) injected into the soil may kill nearly all trichodorids, *Longidorus* and *Xiphinema* (e.g. Harrison *et al.* 1963; Whitehead *et al.* 1970). Such treatments can only be justified to protect valuable crops from serious injury or to protect them from nematode-transmitted viruses, such as arabis mosaic virus in strawberries or fanleaf virus in vines. Some nematode-transmitted viruses can only be controlled by killing all viruliferous nematodes, e.g. tobacco rattle virus in gladiolus (Seinhorst and van Hoof 1976) and mosaic virus-H in hops (Pitcher and McNamara 1973). The short persistence in soil of most non-fumigant nematicides and the reversible effects they have on nematodes when used at recommended rates makes them generally unsuitable for controlling the nematode vectors of virus diseases of plants.

17.9.2   Root-lesion nematodes

Root-lesion nematodes have been fully or partially controlled by fumigating infested soils (e.g. Peachey and Winslow 1962; Miller and Hawkins 1969). Good control of *Pratylenchus* spp. has also been obtained by treating the soil with several non-fumigant nematicides [e.g. Thompson and Willis (1970) and Miller and Kring (1970)].

Pratylenchus coffeae and *P. loosi* (Loof) are very damaging pests of coffee and tea plants, respectively. In the nurseries, they can be controlled by fumigating the soil with methyl bromide (Kerr and Vythilingham 1967), D–D, or dazomet (Thorne and Schieber 1962). In plantations, D–D or metham-sodium injected into the soil before planting out the seedlings can greatly assist establishment in nematode-infested land (e.g. Visser 1959), but control is only partial.

17.9.3   Rice root nematodes

The rice root nematodes, *Meloidogyne graminicola* and *Hirschmaniella* spp., can be very damaging pests in rice. *H. oryzae* in rice was well controlled for at least 45 or 60 days when 12.5 litres of thionazin or 11.5 litres of DBCP/ha were applied to the top 15–20 cm of the soil by drenching and irrigation (Samantaray and Das 1971).

17.9.4 Nematodes of banana and plantain

The banana burrowing nematode, *Radopholus similis,* is a major pest, worldwide, in bananas. Careful experiments by Luc and Vilardebo (1961) in the Ivory Coast demonstrated that DBCP was not phytotoxic when injected into the soil around banana stems (as were D–D and EDB); it controlled the nematode and increased the yields of bananas greatly in infested groves. The recommended treatment was 40 litres of DBCP/ha injected into the soil at planting, 25 litres about 4 months later, 15 litres at 9 months, and 15 litres/ha at 12-month intervals thereafter. In Cameroun, 6–10 g a.i. of fenamiphos or ethoprophos applied in three dressings as granules on the soil around each plant controlled the nematode well and was more reliable than DBCP (Melin and Vilardebo 1973). Similarly in the Ivory Coast, ethoprophos was superior to DBCP, three dressings each of 5 g a.i. per annum applied in 1 m² around each stem giving the best control of *R. similis* and the best yield increases (Guerout and Pinon 1973). In St. Lucia, Windward Islands, West Indies, Gowen (1974, 1976) also found granules of fenamiphos, oxamyl, carbofuran, and ethoprophos more effective than DBCP in improving yields of bananas affected by toppling disease. He applied the granules in the planting holes and at 2 months and every 4 months thereafter in a circle of radius 30 cm (0.28 m²) around each stem. In Puerto Rico, fenamiphos, fensulphothion, ethoprophos, aldicarb, carbofuran, and oxamyl applied as granules to the soil surface and DBCP (e.c.) injected into the soil prevented yield loss in plantains (*Musa acuminata* Colla × *M. balbisiana* Colla, AAB) and extended the useful life of plantations infested with *R. similis* (Román *et al.* 1975, 1977).

Current treatments used to control *R. similis* (Anon. 1977) will also control the other important nematode pests of banana, namely *Helicotylenchus multicinctus, Pratylenchus coffeae,* and *Meloidogyne* spp.

17.9.5 Nematodes of citrus

The two most important nematode pests of citrus are *R. similis,* which initiates 'spreading decline', and *Tylenchulus semipenetrans,* which induces 'slow decline'. 'Spreading decline' may not be wholly due to *R. similis* (Feldmesser *et al.* 1959). Both nematodes have been found on citrus roots down to 4 m deep in the soil (e.g. Suit and Du Charme 1953). Long-term control of these nematodes therefore calls for very deep fumigation of the soil before planting and retreatment of the soil at intervals after establishment of the trees.

(a) Radopholus similis. This is very important in Florida. It has been

eliminated from roots 12 ft (3.7 m) deep by fumigating the soil with 200 US gallons of D–D/acre. Following soil fumigation using 500 gallons/acre, all citrus roots were killed down to 15 ft deep (4.5 m) (Collins and Feldman 1965). To halt or slow down the advance of spreading decline in Lakeland fine sand in Florida, the 'pull and treat' method is used. This entails the destruction of all trees in an area of 16 m greater in all directions than the infested area, along with the removal of as many roots as possible. The soil is then fumigated with 60 US gallons of D–D/acre (650 litres/ha), bare-fallowed for 2 years and replanted, preferably with trees grafted on to tolerant root-stocks (Poucher *et al.* 1967). Three applications of 4 US gallons of DBCP/acre metered into the flood irrigation water at monthly intervals gave excellent control of *R. similis* (McBeth 1960).

(b) Tylenchulus semipenetrans. McBeth (1960) reported 99.9% control of *T. semipenetrans* attacking citrus trees 13 months after a rapid 10–15 cm flood irrigation using 23.3 litres technical or 93.2 litres 25% e.c. of DBCP/ha, carefully drip-fed into the flood water as it passed through an agitation basin. This was subsequently confirmed by O'Bannon and Reynolds (1967) who found that over 99% of the nematodes attacking 30-year-old citrus trees were killed by 37 or 56 litres of 75% (w/w) e.c. DBCP/ha. The nematodes were killed to a depth of 1–1.5 m, depending on soil type. Trees subsequently became reinfested and, after 5 years, damaged by the nematodes. Retreatment with DBCP was, however, successful. In Israel, DBCP at 56 litres/ha applied in irrigation basins around orange and grapefruit trees controlled the nematode and increased yield and size of fruit. Bromine residues reached a maximum of 4.7 p.p.m. (11.5 p.p.m. in the peel) (Minz *et al.* 1961).

17.9.6   Cyst-nematodes

The control of cyst-nematodes by nematicides was reviewed by Whitehead (1973). They can be fully or partially controlled by treating infested soil with large amounts (100–1500 kg/ha) of soil fumigants, smaller amounts (10–100 kg a.i./ha) of organophosphate nematicides or small amounts (2–11 kg a.i./ha) of several oxime *N*-methylcarbamates or of carbofuran. In glasshouses, D–D alternating in successive years with steam sterilization was more profitable and controlled potato cyst-nematodes as well as steaming every year (Murdoch and Murdoch 1963). Treating the soil with methyl bromide or with oxamyl or dazomet and Di-Trapex CP or Telone fully controlled *G. pallida* and allowed good crops of tomatoes to be grown in heavily infested soil under glass (Whitehead *et al.* 1975a, 1979a).

Heterodera goettingiana (Liebs.), the pea cyst-nematode, was fully controlled by incorporating aldicarb or oxamyl in the seed bed by rotary cultivation before sowing peas (e.g. Whitehead *et al.* 1979e). Such treatments might well control the soybean cyst-nematode, *H. glycines* (Ichinohe), a serious pest in Japan, USA, Korea, Manchuria, and Egypt. Similarly, *G. rostochiensis* and *G. pallida* are fully or partially controlled by incorporating aldicarb or oxamyl into the soil before planting susceptible potatoes (Whitehead, 1974; Moss *et al.* 1975, 1976). In heavily infested soil, *H. schachtii*, the beet cyst-nematode, was fully controlled by the incorporation of either of the above nematicides into the topsoil before sugar beet seeds were sown (Whitehead 1974), but in other less infested soils control was only partial (Whitehead *et al.*, 1979d). Placing these nematicides in the seed furrows at sowing time will not control *G. rostochiensis*, *G. pallida*, or *H. schachtii* on susceptible crops grown in widely spaced rows because the nematicides are not translocated to roots growing beyond the treated soil. Some success has been recorded with seed-furrow applications in closely spaced crop rows. Brown (1973) found that 2.2 kg of aldicarb a.i./ha drilled with wheat seed increased yield as much and controlled *H. avenae* Filipjev as well as did 9 kg a.i./ha incorporated in the top 10 cm of the soil. Ethylene dibromide applied to the seed furrows at 2.5–20 kg a.i./ha lessened invasion of wheat roots by *H. avenae* in Australia and increased wheat yields (Gurner *et al.* 1980). Similar amounts of ethylene dibromide or dichloropropene similarly applied, did not improve the yield of susceptible oats in sandy loam lightly infested with *H. avenae* in England, probably because of phytotoxicity, whereas 1.5 kg of oxamyl/ha in the seed furrows greatly increased oat grain yield (Whitehead *et al.* 1982). Oxamyl drilled with pea seed controlled *H. goettingiana* as well in one experiment as did oxamyl mixed in the seed bed before sowing, but in other experiments such seed-furrow treatments were less successful (Whitehead *et al.* 1979e).

17.9.7 Stem nematodes

The stem nematode, *Ditylenchus dipsaci,* is often difficult to control with nematicides because it can multiply up to 1000-fold on a susceptible plant, so it can only be fully controlled if up to 99.9% of the population is killed or immobilized. The rice stem nematode, *D. augustus* (Butler), is even more difficult to control chemically because it attacks rice plants in flooded paddy fields.

Small amounts of aldicarb, oxamyl, fenamiphos, or thionazin applied to the crop rows can protect seedlings of sensitive crops, like onions, from injury and greatly increase yields (e.g. Winfield *et al.* 1971; Whitehead *et al.* 1973b). Because stem nematodes affect only the shoots, before and after emergence from the soil, treating onion rows with a small amount of an oximecarbamate nematicide may be more effective, as well as more convenient, than incorporating the same amount throughout the topsoil. However, as surviving nematodes may multiply rapidly on the crop later in the growing season, when temperatures decrease and moisture returns, a second application of systemic nematicide may be needed to control this second damaging attack. Damage to summer-sown onions has been prevented in this way, but only very small amounts of aldicarb can be applied as the second dose, the following spring, without leaving unacceptable aldicarb residues in the harvested bulbs (Whitehead *et al.* 1979c). Repeated applications of systemic nematicides or the use of more persistent nematicides is acceptable for ornamental crops.

17.9.8   Root-knot nematodes

Root-knot nematodes are difficult to control fully for, like stem nematodes, they may multiply 1000-fold or more on a susceptible plant. They are usually easier to control in annual than in perennial crops. Many experiments have been made on the chemical control of these nematodes, mostly with the annuals, tobacco, tomato, potato, and cotton, and the perennials, peach, grapevine, and pineapple. Full control has been reported with methyl bromide (Kissler *et al.* 1973), chloropicrin (Nikitina and Mishkina 1954), D–D (e.g. Raj and Nirula 1971), and ethylene dibromide (e.g. Paddick and Meagher 1973) and partial control with DBCP and methyl isothiocyanate nematicides.

The infested roots of a previous crop should be removed from the soil, or allowed to rot, before treating the soil with DBCP or ethylene dibromide (Lear and Raski 1962), because these fumigants do not kill the nematodes inside the roots. D–D and methyl bromide penetrate unrotted root galls better than does chloropicrin (Stark and Lear 1947). Methyl bromide applied under gas-tight polyethylene sheeting is very effective for seedbeds (Daulton 1956; Abrego and Holdeman 1961), and, applied by blade coulters followed by polyethylene sheeting or rolling and sprinkling with water, has been effective in California (e.g. Thomason 1959). Very good control of root-knot nematodes was obtained with large amounts of 1,3-dichloropropene or methyl bromide injected deeply into the soil (Raski *et al.* 1973). Rolling the soil after incorporating dazomet granules did not improve control (Burgis and Overman 1956). DBCP was more effective when injected into the soil than when applied in the seed furrows or drenched into the soil (Birat 1965). In sandy soil, 11.2–14.0 litres, (2.5–3.1 gallons) of DBCP/ha was more effective when injected 15–20 cm deep than when injected 5, 10, or 25–30 cm deep (Gilpatrick *et al.*, 1956). *M. incognita* on cotton was controlled better by topsoil fumigation than by fumigating the rows in which the seeds were later sown (Raski and Allen 1953). Repeatedly flushing hydroponic gravel beds with 2% formaldehyde prevented injury to tomatoes by *M. incognita* (Schuster and Wagner 1972).

Some non-fumigant nematicides, especially oxime *N*-methylcarbamates, have given good protection against root-knot nematodes. Full control has been reported with oxamyl (e.g. Hart and Maggenti 1971), and partial to full control with aldicarb (Brodie and Good 1973), fenamiphos, and ethoprophos (e.g. Johnson and Harmon, 1974). Fensulphothion, fenthion, or trichloronate prevented galling of tomato roots by *M. incognita* at 12.5 p.p.m. in the soil, DBCP did so at 6 p.p.m. but D–D was not nematicidal below 50 p.p.m. (Jarnevic and Coffee 1965). Brodie and Good (1973) found aldicarb, ethoprophos, or fenamiphos controlled *M. incognita* on tobacco better than did a number of soil fumigants. Aldicarb was more effective when incorporated at the same dosage in the top 15–20 cm than in the top 5–10 cm of the soil, whereas ethoprophos or fenamiphos was more effective when applied to the top 5–10 cm of the soil. Carbofuran apparently worked better when applied in the seed furrows than when incorporated in the seed bed.

17.10 References

Abrego, L. and Holdeman, Q. L. (1961). *Inst. Salvadoren invest. cafe, Santa Tecla, El Salvador, Bol. inf. supl.* **8**, 1.

Alphey, T. J. W. (1978). *Ann. Appl. Biol.* **88**, 75.

Anon. (1971). *J. Nematol. Spec. Publ. No. 1*, 7.

—— (1972). *Bull. Minist. Agric. Fish. Fd.* 201 (2nd Edn), 46 pp. (HMSO).

—— (1977). *Pest control in bananas. Pans Manual No. 1*, p. 80. Centre for Overseas Pest Research, London.

Birat, R. B. S. (1965). *Indian Phytopathol.* **18**, 82.

Bird, G. W. (1969). *Can. J. Plant Sci.* **49**, 90.

Brodie, B. B. and Good, J. M. (1973). *J. Nematol.* **5**, 14.

Bromilow, R. H. (1973). *Ann. Appl. Biol.* **75**, 473.

Brown, A. L., Jurinak, J. J., and Martin, P. E. (1958). *Soil Sci.* **86**, 136.

Brown, E. B. (1969). *Ann. Appl. Biol.* **63**, 493.

—— Sykes, G. B. (1973). *Ann. Appl. Biol.* **75**, 462.

Brown, R. H. (1973). *Austr. J. Exp. Agric. Anim. Husb.* **13**, 587.

Bryden, J. W. and Hodson, W. E. H. (1957). *Plant Pathol.* **6**, 20.

Bunt, J. A. (1975). *Meded. Landbouwhogesch. Wageningen,* **75**, 1.

Burgis, D. S. and Overman, A. J. (1956). *Proc. Fla. Hort. Soc.* **69**, 207.

Call, F. and Hague, N. G. M. (1962). *Nematologica* **7**, 186.

Castro, C. E. and Thomason, I. J. (1971). In *Plant parasitic nematodes* (eds B. M. Zuckerman, W. F. Mai and R. A. Rhode), p. 289. Academic Press, London and New York.

Collins, R. J. and Feldman, A. W. (1965). *Phytopathology* **55**, 1103.

Cooke, D. A., Dunning, R. A., and Winder, G. H. (1974). *Ann. Appl. Biol.* **76**, 289.

Cooper, J. I. and Thomas, P. R. (1971). *Ann. Appl. Biol.* **69**, 23.

Daulton, R. A. C. (1956). *Rhodesian Tobacco,* **13**, 4.

—— (1967). *Down to Earth* **22**, 20.

Dekker, W. H. (1972). *Meded. Fac. LandbWetensch. Rijksuniv. Gent* **37**, 865.

D'Herde, J. and van den Brande, J. (1964). *Meded. LandbHogesch. Opzoeksstations Gent.* **29**, 788.

Evans, A. A. F. and Thomason, I. J. (1971). *Nematologica* **17**, 243.

Evans, K. and Pandé, L. U. (1972). *Rep. Rothamsted Exp. Stn. 1971*, Part 1, 175.

Feldmesser, J., Rebois, R. V., and Taylor, A. L. (1959). *Plant Dis. Rep.* **43**, 261.

French, N. and Wilson, W. R. (1976). *Plant Pathol.* **25**, 167.

Gasser, J. K. R. and Peachey, J. E. (1964). *J. Sci. Fd. Agric.* **15**, 142.

Gilpatrick, J. D., Ichikawa, S. T., Turner, M., and McBeth, C. W. (1956). *Phytopathology* **46**, 529.

Goodey, J. B. (1949). *J. Helminthol.* **23,** 171.
Goodey, T. (1945). *J. Helminthol.* **21,** 45.
Gowen, S. R. (1974). *PANS* **20,** 400.
—— (1976). *Br. Crop Prot. Counc. Monogr.* No. 18, 140.
—— (1977). *J. Nematol.* **9,** 158.
Guerout, R. and Pinon, A. (1973). *Fruits* **28,** 751.
Gurner, P. S., Dubé, A. J., and Fisher, J. M. (1980). *Nematologica* **26,** 448.
Hague, N. G. M. (1968). *Plant Pathol.* **17,** 127.
—— (1975). *Proc. 8th Br. Insectic. Fungic. Conf.,* **3,** 837.
—— Pain, B. F. (1970). *Plant Pathol.* **19,** 69.
—— Sood, U. (1963). *Plant Pathol.* **12,** 88.
—— Lubatti, O. F., and Page, A. B. P. (1964). *Hort. Res.* **3,** 84.
Harrison, B. D., Peachey, J. E., and Winslow, R. D. (1963). *Ann. Appl. Biol.* **52,** 243.
Harrison, M. B. (1971). *J. Nematol.* **3,** 311.
Hart, W. H. and Maggenti, A. R. (1971). *Plant Dis. Rep.* **55,** 89.
Homeyer, B. (1971). *Pflanzenschutz-Nachrichten Bayer* **24,** 48.
Hough, A. and Thomason, I. J. (1975). *J. Nematol.* **7,** 221.
Hutchinson, M. T. (1963). *Rep. Tea Res. Inst. Ceylon, 1962,* Part II, 70.
Ichikawa, S. T., Gilpatrick, J. D., and McBeth, C. W. (1955). *Phytopathology* **45,** 576.
Iyer, J. G. and Wojahn, K. E. (1976). *Plant and Soil* **45,** 263.
Jarnevic, N. B., and Coffee, E. G. (1965). *Plant Dis. Rep.* **49,** 603.
Johnson, A. W. (1978). *Plant Dis. Rep.* **62,** 48.
—— Harmon, S. A. (1974). *Plant Dis. Rep.* **58,** 749.
Jones, F. G. W. (1978). *Univ. Leeds Rev.* **21,** 89.
—— Perry, J. N. (1978). *J. Appl. Ecol.* **15,** 349.
Kempton, R. J. and Maw, G. A. (1972). *Ann. Appl. Biol.* **72,** 71.
Kerr, A. and Vythilingham, M. K. (1967). *Tea Q.* **38,** 22.
Kissler, J. J., Lider, J. V., Raabe, R. D., Raski, D. J., Schmitt, R. V., and Hurlimann,
    J. H. (1973). *Plant Dis. Rep.* **57,** 115.
Lear, B. and Raski, D. J. (1962). *Phytopathology* **52,** 1309.
Lewis, G. D. and Mai, W. F. (1963). *Plant Dis. Rep.* **47,** 1097.
Luc, M. and Vilardebo, A. (1961). *Fruits* **16,** 205.
McBeth, C. W. (1960). In *1st Inter-African Plant Nematology Conference Proceedings,*
    CCTA/CSA Publ. No. 86, 19.
McKenry, M. V. and Thomason, I. J. (1976a). *Pestic. Sci.* **7,** 521.
—— —— (1976b). *Pestic. Sci.* **7,** 535.
McLeod, R. W. and Khair, G. T. (1975). *Ann. Appl. Biol.* **79,** 329.
Marks, C. F., Thomason, I. J., and Castro, C. E. (1968). *Exp. Parasitol.* **22,** 321.
Matthews, G. A. (1979). *Pesticide application methods.* 334 pp. Longmans, London
    and New York.
Melin, Ph. and Vilardebo, A. (1973). *Fruits,* **28,** 3.
Miller, P. M. and Hawkins, A. (1969). *Am. Potato J.* **46,** 387.
—— Kring, J. B. (1970). *J. Econ. Entomol.* **63,** 186.
Minz, G., Oren, R., and Cohn, E. (1961). *Bull. Natl. Univ. Inst. Agric., Rehovot, No.*
    353, 10 pp.
Moss, S. R., Crump, D., and Whitehead, A. G. (1975). *Ann. Appl. Biol.* **81,** 359.
—— —— —— (1976). *Ann. Appl. Biol.* **84,** 355.
Murdoch, G. and Murdoch, S. M. (1963). *Exp. Hort. No. 8,* 27.
Nelmes, A. J. (1970). *J. Nematol.* **2,** 223.
Newsome, W. H., Iverson, F., Panopio, L. G., and Hierlihy, S. L. (1977). *J. Agric. Fd.*
    Chem. **25,** 684.

Nikitina, T. F. and Mishkina, L. P. (1954). *Trudi Problemnikh i Rematicheskikh Soveshchavi Akademiya Nauk SSSR* **3**, 118.

O'Bannon, J. H. and Reynolds, H. W. (1967). *Nematologica* **13**, 131.

—— Taylor, A. L. (1967). *Plant Dis. Rep.* **51**, 995.

—— Tomerlin, A. T. (1971). *Plant Dis. Rep.* **55**, 154.

—— Tomerlin, A. T. (1977). *Plant Dis. Rep.* **61**, 450.

Osborne, P. (1973). *Nematologica* **19**, 7.

Ouden, H. den (1971). *Meded. Rijksfac. LandbWetensch. Gent* **36**, 889.

Overman, A. J. (1975). *Proc. Soil Sci. Soc. Fla* **34**, 197.

Paddick, R. G. and Meagher, J. W. (1973). *Austr. J. Exp. Agric. Anim. Husb.* **13**, 108.

Page, A. B. P., Hague, N. G. M., Jakabsons, V., and Goldsmith, R. E. (1959). *J. Sci. Fd. Agric.* **10**, 461.

Peachey, J. E. and Winslow, R. D. (1962). *Nematologica* **8**, 75.

—— Rao, G. N., and Chapman, M. R. (1963). *Ann. Appl. Biol.* **52**, 19.

Peacock, F. C. (1963). *Nematologica* **9**, 581.

Peters, B. G. (1954). *Rep. Rothamsted Exp. Stn. 1953,* 97.

Pitcher, R. S. and McNamara, D. G. (1973). *Ann. Appl. Biol.* **75**, 468.

Potter, J. W. and Marks, C. F. (1971). *J. Nematol.* **3**, 325.

Poucher, C., Ford, H. W., Suit, R. F., and Du Charme, E. P. (1967). *Fla. Dep. Agric., Div. Plant Ind. Bull.* **7**, 63 pp.

Powell, D. F. (1975a). *Plant Pathol.* **24**, 237.

—— (1975b). *Ann. Appl. Biol.* **81**, 425.

Prasad, K. S. K. and Rao, Y. S. (1976). *Z. Pflanzenkrank. Pflanzensch.* **83**, 665.

Raj, B. T. and Nirula, K. K. (1971). *Indian Phytopathol.* **24**, 155.

Raski, D. J. and Allen, M. W. (1953). *Plant Dis. Rep.* **37**, 193.

—— Jones, N. O., Kissler, J. J., and Luvisi, D. A. (1976). *Calif. Agric. Jan. 1976,* 4.

—— Schmitt, R. V., Luvisi, D. A., and Kissler, J. J. (1973). *Plant Dis. Rep.* **57**, 619.

Reddy, D. D. R. (1976). *Plant Dis. Rep.* **60**, 430.

Román, J., Rivas, X., Rodriguez, J., and Oramas, D. (1975). *J. Agric. Univ. P. Rico* **59**, 36.

—— —— —— —— (1977). *J. Agric. Univ. P. Rico* **61**, 192.

Samantaray, K. C. and Das, S. N. (1971). *Proc. Indian Natl. Sci. Acad., Biol. Sci.* **37B**, 372.

Schuster, M. L. and Wagner, L. J. (1972). *Plant Dis. Rep.* **56**, 139.

Seinhorst, J. W. (1965). *Nematologica* **11**, 137.

—— (1973). *Neth. J. Plant Pathol.* **79**, 180.

—— and van Hoof, H. A. (1976). *Neth. J. Plant Pathol.* **82**, 215.

Shepherd, C. J. (1952). *Nature (London),* **170**, 1073.

Singhal, J. P. and Singh, C. P. (1977). *Indian J. Agric. Sci.* **43**, 280.

Smith, D. H. (1963). *Proc. Soil Sci. Soc. Am.* **27**, 538.

Stark, F. L. Jr. (1948). *Mem. Cornell Univ. Agric. Exp. Stn. No. 278,* 61 pp.

—— Lear, B. (1947). *Phytopathology* **37**, 698.

Steele, A. E. and Hodges, L. R. (1975). *J. Nematol.* **7**, 305.

Steudel, W. (1972). *Nematologica* **18**, 270.

Suit, R. F. and Du Charme, E. P. (1953). *Plant Dis. Rep.* **37**, 379.

Taconis, P. J. and Kuiper, K. (1964). *Versl. Meded. Plziektenk. Dienst Wageningen* **141**, 177.

Thomason, I. J. (1959). *Plant Dis. Rep.* **43**, 580.

Thompson, L. S. and Willis, C. B. (1970). *Can. J. Plant Sci.* **50**, 577.

Thorne, G. and Schieber, E. (1962). *Plant Dis. Rep.* **46**, 857.

Timmer, L. W. (1977). *J. Nematol.* **9**, 45.

Todd, E. H. and Atkins, J. G. (1959). *Phytopathology* **49**, 184.

Van Gundy, S. D., Munnecke, D., Bricker, J. and Minteer, R. (1972). *Phytopathology* **62**, 191.

Viglierchio, D. R., Maggenti, A. R., Schmitt, R. V., and Paxman, G. A. (1977). *J. Nematol.* **9**, 307.

Visser, T. (1959). *Tea Q.* **30**, 96.

Wallace, H. R. (1963). *The biology of plant parasitic nematodes.* 280 pp. Edward Arnold, London.

Walter, J. W. and Kelsheimer, E. G. (1949). *Market Grow. J.* **78**, 12.

Whitehead, A. G. (1969). *Tech. Commun. Commonw. Bur. Helminth. No. 40,* 238.

—— (1973). *Ann. Appl. Biol.* **75**, 439.

—— (1974). *Proc. 7th Br. Insectic. Fungic, Conf.* **3**, 955.

—— (1977). *Plant Pathol.* **26**, 85.

—— Dunning, R. A., and Cooke, D. A. (1971). *Rep. Rothamsted Exp. Stn. 1970, Part 2,* 219.

—— Fraser, J. E., and French, E. M. (1979a). *Ann. Appl. Biol.* **92**, 275.

—— —— Greet, D. N. (1970). *Ann. Appl. Biol.* **65**, 351.

—— —— Storey, G. (1972). *Ann. Appl. Biol.* **72**, 81.

—— Tite, D. J., and Bromilow, R. H. (1981). *Ann. Appl. Biol.* **97**, 311.

—— —— Fraser, J. E. (1973a). *Ann. Appl. Biol.* **73**, 325.

—— —— —— (1973b). *Rep. Rothamsted Exp. Stn. 1972, Part I.* 166.

—— Fraser, J. E., French, E. M., and Wright, S. M. (1975a). *Ann. Appl. Biol.* **80**, 75.

—— Tite, D. J., Fraser, J. E., and French, E. M. (1973c). *Ann. Appl. Biol.* **73**, 197.

—— —— —— —— (1973d). *Ann. Appl. Biol.* **75**, 257.

—— —— —— —— (1973e). *Ann. Appl. Biol.* **74**, 113.

—— —— —— —— (1979b). *Rep. Rothamsted Exp. Stn. 1978, Part 1,* 170.

—— —— —— (1979c). *Ann. Appl. Biol.* **93**, 213.

—— —— —— Nichols, A. J. F. (1982). *Rep. Rothamsted Exp. Stn. 1981, Part 1,* 159.

—— Bromilow, R. H., Lord, K. A., Moss, S. R., and Smith, J. (1975b). *Proc. 8th Br. Insectic. Fungic. Conf.,* 133.

—— Tite, D. J., Finch, P. H., Fraser, J. E., and French, E. M. (1979d). *Ann. Appl. Biol.* **92**, 73.

—— Bromilow, R. H., Tite, D. J., Finch, P. H., Fraser, J. E., and French, E. M. (1979e). *Ann. Appl. Biol.* **92**, 81.

Winfield, A. L. (1973). *Ann. Appl. Biol.* **75**, 454.

—— (1978). In *Plant nematology MAFF/ADAS GDI,* p. 296. HMSO, London.

—— Murdoch, G., and John, M. E. (1971). *Proc. 6th Br. Insectic, Fungic. Conf.* **1**, 141.

Wood, F. H. and Foot, M. A. (1975). *N.Z.J. Exp. Agric.* **3**, 349.

Zeck, W. M. (1971). *Pflanzenschutz-Nachrichten Bayer* **24**, 114.

18
Plant disease control

J. M. WALLER

18.1 Introduction

The aim of this chapter is to describe the various ways in which chemicals can be used to control plant diseases, firstly by considering the basic biological principles involved, then by examining the various methods of applying chemicals for plant disease control and lastly by analysing some of the secondary problems which have arisen from their use. Chemical and biological aspects of fungicide use have been dealt with in earlier chapters of this book and in a number of other texts devoted to the chemical control of plant diseases (e.g. Evans 1968; Torgeson 1967, 1969; Lukens 1971; Marsh 1977; Siegel and Sisler 1977); much has been written on the control of specific diseases both in the form of review articles and research papers. A comprehensive review covering the whole field of chemical control of diseases on all crops cannot be undertaken within the confines of this chapter, but selected references to relatively few examples and to specific review articles will be made when appropriate.

Probably the first example of the chemical control of a plant disease was the treatment of wheat seed with brine in the 18th century to control loose smut. Subsequently in the 19th century, copper sulphate solution was found to be more effective for this purpose and the beneficial effects of sulphur on plant diseases were also known. The application of chemicals to growing crops for controlling plant disease was stimulated by the fortuitous discovery of 'Bordeaux mixture' by Millardet in 1882. At this time also the parasitic nature of many plant diseases was becoming established so that the use of chemicals to kill the parasite became a rational solution to plant disease control.

However, crop diseases were controlled by cultural practices long before chemical treatments became available, and nowadays genetically controlled crop resistance is the method of choice for disease control in the World's major crops. Pesticides still tend to be regarded more as a last resort for disease control on staple crops; they are used most frequently on high-value perennial or horticultural crops, or where there is a special requirement for particularly healthy produce, or where disease pressure is likely to be severe. Nevertheless, recent advances in chemical control of plant diseases have greatly expanded the effectiveness and economy of fungicidal usage.

Chemical control is, therefore, only one of several methods used in plant-disease management. The decision to use chemical control depends on a whole series of interrelating factors. Besides the economic consideration of balancing projected increases in yield or quality against the cost of the chemicals and their application, the use of alternative disease control methods, their relative efficiencies under the prevailing environmental and disease conditions, how they can be integrated with other agricultural operations, and the possibility of combining several methods all need to be considered. 'Integrated' control, using a combination of different methods against a range of diseases under varying circumstances has long been used by plant pathologists.

The vast majority of chemicals used to control plant diseases are fungicides, but there are also bactericides and antibiotics, as well as insecticides and nematicides used for controlling the vectors of some diseases. In general terms these continue to be referred to as pesticides despite the fact that they are used to control pathogens.

Since the use of fungicidal chemicals to protect crops against disease was first adopted, there have been a series of important developments in fungicide chemistry and mode of action which have radically changed the capability of chemical disease control. Advances in the knowledge of the epidemiology of plant diseases have greatly improved the efficiency and economy with which fungicides can be used. Therefore, chemical disease control is now becoming an increasingly potent and more widely used weapon in the efforts to combat plant disease.

18.2   The basis of chemical control of plant diseases

18.2.1   Characteristics of pathogens

There are a number of important differences between pests and pathogens which make the chemical control of plant diseases somewhat different from the chemical control of pests. These need to be appreciated when considering plant disease control within the overall framework of pesticide usage.

Unlike pests, pathogens are plants and are thus physiologically very similar to their hosts. There are consequently fewer biochemical differences on which selective toxicity can be based. The majority of fungicides are toxic to fungal pathogens and not to their plant hosts by reason of selective absorption; even the so-called systemic fungicides move primarily in the apoplast of the host and not within the protoplasm of plant cells. The mechanisms of fungicidal action have been recently reviewed by Lyr (1977).

Pathogens are considerably less mobile than pests and spend most of their life cycle inside the tissues of the host plant. This makes the placement of many fungicides and the timing of their application very critical. Another major difference between pests and pathogens which affect control is the way in which they are monitored. Pests are usually much larger than pathogens and can be seen and counted before they cause damage, whereas most pathogens are microscopic organisms which cannot be easily detected until the diseases which they cause are apparent; by this time most are already rapidly increasing and both visible and latent damage has occurred. Thus, the economic threshold concept used in monitoring insect pest populations cannot be used in the same way to monitor disease levels. This difference fundamentally affects the timing and strategy of chemical application for plant disease control.

Plant pathogens consist of fungi, bacteria (*sensu lato*), and viruses. The majority of plant diseases are caused by various species of parasitic fungi some of which are very specialized obligate parasites; others are less-specialized facultative parasites able to live saprophytically in soil or on dead host remains in the absence of a suitable host. Plant pathogenic species are scattered throughout the taxonomic range of fungi, and fungicides vary in effectiveness between the different classes of fungi. Bacteria cause many plant diseases but few fungicides are bactericidal in action, copper-based fungicides being an important exception. Chemical control of bacterial diseases by currently available pesticides is, therefore, rather limited. An increasing number of plant diseases are now being found to be caused by viruses; many of these have previously been overlooked as they frequently do not cause eye-catching or spectacular symptoms (except perhaps to those familiar with them!), although they often cause dramatic losses in yield. As yet, direct chemical control of viruses is hardly a commercial feasibility, but control of their vectors through use of a suitable pesticide can prevent these diseases from spreading. A few plant diseases are caused by spiroplasmas and organisms resembling mycoplasmas, or rickettsias; control of these can be achieved by the systemic application of certain antibiotics or by control of their vectors.

Chemicals can be used to prevent plant diseases either by destroying sources of pathogenic inoculum, or by protecting seeds or growing plants from infection. They can also be used to cure diseased plants. Prophylaxis or disease prevention is the most widely used method and until the fairly recent development of systemic fungicides was the only method available for the vast majority of plant diseases. Therapy or curing of diseased plants depends upon using chemicals which can penetrate the plant tissues to eradicate existing infections without readily harming the host.

18.2.2 Eradication of the pathogen at source
Destruction of sources of pathogenic inoculum is a principal method for the control of many plant diseases and has long been achieved by standard agricultural practices such as crop rotation and clean cultivation. With intensive agricultural production and the growing of high-value crops, chemical methods of achieving this are often used because of operational

advantages and economic feasibility. Soil sterilization with formalin, methyl bromide, and a number of other compounds, some often considered primarily as nematicides (see Chapter 17), are often used to control soil-borne diseases such as damping off, collar, and root rots caused by *Pythium, Rhizoctonia,* and *Fusarium* spp. in nurseries, glasshouses, and seed beds. Herbicides and similar chemicals may be used to destroy alternative hosts of obligate parasites such as rusts and viruses which can survive in weeds and volunteer seedlings in the absence of a suitable crop host. Chemicals can also be used to hasten the death of perennial plant remains such as old stumps which may act as sources of soil-borne pathogens.

18.2.3   Protection from infection

Most fungicides and bactericides are applied to plants in order to protect them from infection. The chemical remains on the outside of the plant and bacterial cells or spores which come into contact with it are killed. Maintaining an effective coverage on a growing plant surface, particularly those parts which are susceptible to infection by the pathogen, presents a number of difficulties which will be discussed later. The use of chemicals to protect roots against diseases is much more difficult than protection of the aerial parts of plants, and depends upon preplanting application to seed, soil, or roots or postplanting soil drenches. Powdery mildews present a special case as these pathogens grow over the surfaces of plants – only the haustoria penetrate epidermal cells. These diseases can be controlled by the application of contact toxins such as sulphur which directly kill the exposed mycelium of the pathogen. Similarly, application to seed may directly kill the mycelium of fungal pathogens on the seed coat as well as protecting the developing seedling from infection. Chemicals which can eradicate existing infections as well as protect against new ones are partly therapeutic in action.

18.2.4   Chemotherapy of plant diseases

Several primarily protectant fungicides have a slight curative effect on some diseases. Apart from powdery mildews, there are several other diseases where the pathogen spends the early part of its life cycle growing close to the surface, often beneath the leaf cuticle, before penetrating deep into the tissues. Apple scab (*Venturia inaequalis*) and Sigatoka disease of bananas (*Mycosphaerella musicola*) are two such examples and some fungicides applied soon after infection can penetrate the cuticle and eradicate these incipient lesions. However, since the late 1960s, a range of fungicides which are systemic, or partially so, within the host plant have been developed and are now used increasingly for the therapeutic control of plant diseases as sprays, seed applications, or injections into the vascular system of the plant. In practical usage, systemic fungicides operate both as protectants and therapeutants, since when they are applied to crops being attacked by pathogens they will eradicate existing infection as well as protect against future invasion during the period that they are effective in the plant. The ways in which chemotherapeutants act on plant diseases have been reviewed by Dekker (1977) and Kaars Sijpesteijn (1977).

18.2.5 Other effects

There are a few ways in which chemicals may act that do not conveniently fit into these three groups. Certain chemicals, particularly those having growth-regulating properties, may alter the metabolism of the host and make it less susceptible to disease or more tolerant of its effects. They may also be used to alter the growth pattern of plants so that certain diseases may be avoided. Others may affect the ability of the pathogen to multiply in the host tissue or reduce its sporulating capacity. Some of the more promising chemicals for the control of virus diseases act in this way, but chemical control of their vectors (often insects) which inhibit transmission between plants is the only practical method at present. Some chemicals do not kill the pathogen outright but merely halt or change its growth in such a way that the host is able to resist infection.

18.3 The relation of chemical control to disease epidemiology

18.3.1 Diseases caused by fungal pathogens

The method and rate of disease development and spread in populations of crop plants is of fundamental importance in designing rational control measures. Complete elimination of pathogens by chemical methods is seldom possible so that some inoculum usually survives after treatment. The capacity of this surviving inoculum to cause disease is therefore critical, particularly as some of the most destructive disease epidemics can develop very rapidly from a small source of initial inoculum. Van der Plank (1963) divided plant disease epidemics into two basic types depending on whether or not there was multiplication and spread of disease during the cropping season which he terms 'compound interest' and 'simple interest' respectively by analogy with the increase of invested capital. With perennial crops the division between the two types is less clear-cut as diseases can increase and spread throughout the crop over successive seasons.

In 'simple interest' diseases, now usually called monocyclic diseases, there is no secondary spread of disease during the cropping season and the amount of disease which develops in the crop is directly proportional to the amount or infectivity of original inoculum. This may be influenced by various factors such as soil conditions, cultivar, multiple infections, and growth compensation in 'gappy' crops so that the relationship between the initial inoculum and subsequent yield loss is not necessarily linear. Nevertheless, any reduction in amount or infectivity of the initial inoculum by chemicals will have a corresponding effect on subsequent disease development. As any surviving inoculum will not multiply to cause an epidemic increase in disease, a single early application of the chemical to achieve partial eradication of the source of the pathogen is usually adequate. Examples of such diseases include the cereal smuts and many soil-borne root-infecting diseases such as *Fusarium* and *Verticillium* wilts, and replant diseases of perennial crops. Chemical application to seed or soil is an efficient and fairly straightforward way of controlling these.

Most diseases which are capable of causing substantial losses and which often require chemical control are of the 'compound interest' type where successive generations of the pathogen are produced during the cropping season, now usually called polycyclic diseases. These spread and increase the amount of disease initially in a logarithmic fashion. The amount of disease developing in these situations is governed primarily by the rate at which the pathogen multiplies and its infectivity. These determine the rate of epidemic progress (*r*), and the relationship of chemical control to factors which influence this are reviewed by Van der Plank (1967). The latent period of the disease (time from infection to production of new propagules), the growth of individual lesions and the duration and amount of secondary inoculum produced by them are the most important intrinsic factors influencing the rate of pathogen multiplication, whereas the infectivity of the inoculum is influenced primarily by environmental conditions, especially the occurrence of climatic conditions suitable for infection, and host resistance. Fungicides can affect both these aspects. Protectants decrease inoculum infectivity by reducing the amount of inoculum able to infect the plant, but many act as anti-sporulants as well and may reduce the amount and viability of spore production from diseased tissues; some can also prolong the latent period. Systemic fungicides are more efficient at reducing *r* as not only do they inhibit infection, but they further reduce the epidemic by eradicating existing infectious lesions. Generally it is easier to control diseases by chemical means when other conditions such as host resistance and climate also help to reduce the infection rate. In those diseases where the latent period is short, sporulation of old lesions prolonged, and dispersal and infection widespread, control by chemical protection is more difficult. Reduction in the amount of initial inoculum will only delay the epidemic and when the rate of epidemic increase is high this delay will be slight; devastating epidemics can therefore develop rapidly from small amounts of initial inoculum. However, where the rate of epidemic increase is slower, or very erratic, substantial gains can be achieved in delaying the onset of the epidemic by the reduction in the initial inoculum. These effects are shown in Fig. 18.1.

In practice, fungicidal application to protect crops from compound interest diseases, typified by many of the foliar diseases of cereals and vegetables, relies on both a delay in epidemic development and a reduction of

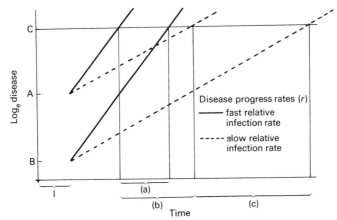

Fig. 18.1. The effect of relative infection rates and initial inoculum levels on disease control. A, Initial inoculum level; B, initial inoculum partially eradicated; C, critical level at which disease begins to reduce yield. I, latent period of disease (between infection and appearance). (a) Epidemic delay produced by reducing initial inoculum from A to B (fast disease progress); (b) delay produced by reducing infection rate; (c) delay produced by reducing initial inoculum level (slow disease progress). Disease progress rates (*r*) —, Fast relative infection rate; ----, slow relative infection rate.

r. Large (1952) demonstrated this quite clearly for the control of potato blight. Early spraying was able to delay epidemic development as well as to slow its rate so that significant destruction of potato haulm did not occur until most of the tubers were fully formed. Later spraying, when the epidemic had already started, was not so efficient, as slowing the rate of epidemic increase did not have such a large effect on yield. Berger (1977) has also shown that attempted protection later in disease epidemics is very much less efficient, and that any subsequent loss of disease control such as may be caused by a missed spray application or inefficient coverage results in a marked increase of disease that may obliterate any previous gains.

The development of suitable chemical control measures against coffee berry disease in Kenya (Griffiths *et al.* 1971) illustrates the importance of epidemiological factors. Early attempts to control this disease were based on reducing the amount of inoculum which survived between crops in the maturing bark of the twigs (Nutman and Roberts 1961). This was done by a programme of early-season sprays before the rainy season began and the crop flowered and produced berries. However, the much greater potency of inoculum produced by coffee berries which became diseased early in the rainy season resulted in an explosive epidemic developing during most seasons. Furthermore, changes in cropping and weather patterns resulted in overlapping crops so that the small amount of primary inoculum carried in the bark became epidemiologically relatively unimportant. Under these circumstances only a prolonged spraying schedule which protected the crop during its susceptible stages could control the disease (Fig. 18.2). It is noteworthy that little effect was achieved by early sprays and these could result in disease levels worse than where no spraying was undertaken at all – a point which will be referred to again later.

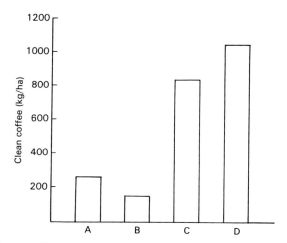

Fig. 18.2. Effect on coffee yields of spray timing to control coffee berry disease. (Modified from Griffiths *et al.* 1971). A, control—no spraying; B, early spray schedule; C, late spray schedule; D, full spray schedule.

18.3.2   Bacterial pathogens

Soil-borne diseases such as those caused by *Pseudomonas solanacearum* or *Agrobacterium tumefaciens* can be regarded as monocyclic diseases so that control of primary inoculum sources by methods such as soil sterilization can give adequate control. Bacterial pathogens which attack the aerial parts of plants generally produce polycyclic diseases which are restricted to suscept-ible stages of the host or by the occurrence of predisposing climatic conditions. Most of these also have a relatively slow rate of dispersal, and can be controlled by the reduction of primary inoculum sources. The application of a bactericide such as bronopol or copper preparations to cotton seed substantially reduces seed transmission of bacterial blight (*Xanthomonas campestris* pv. *malvacearum*) and this practice coupled with reasonable resistance to the disease in the cotton varieties currently grown in E. Africa has virtually eliminated losses caused by this pathogen (Ebbels 1976). Similarly halo blight of beans (*Pseudomonas syringae* pv. *phaseolicola*) can also be controlled by bactericidal seed dressing (Taylor and Dudley 1977). Antibiotics, particularly commercial preparations of steptomycin and tetracycline, are sometimes used on high-value crops such as tobacco and ornamentals; they have some eradicant effect, but are expensive, and insensitivity soon develops which can spread to other organisms in the environment.

Bacterial diseases of perennial crops, such as those caused by *Pseudomonas syringae,* where eradication of the primary source is more difficult, and bacterial vascular diseases, such as those caused by *Pseudomonas solanacearum,* still present major challenges to chemical control.

18.3.3 Virus diseases

Soil-borne virus diseases are mainly monocyclic diseases so that eradication of the primary source should give adequate control. For practical purposes, this means destruction of the vectors which carry the virus. Chemical control of nematode-transmitted soil-borne viruses (Lamberti 1981) seems to be more successful than the control of soil-borne viruses transmitted by fungi, but even so very efficient control of the vector, often hampered by practical problems which limit the efficiency of soil fumigation in field situations, is needed to get adequate control of the virus (Harrison 1977).

The success of preventing the spread of insect transmitted viruses by vector control is primarily determined by the way in which the virus is carried by the vector. Insecticidal control of the vector often has only a limited effect on the spread of non-persistent or stylet-borne viruses. These can be spread rapidly by very small populations of winged vectors and they have usually transmitted the virus by the time they have absorbed enough chemical to be killed. By contrast the spread of persistent or circulative viruses, in which there is often an incubation phase before the vector becomes infectious and where it may remain infectious for a long period, can be successfully prevented by these measures. This was well illustrated in the control of potato viruses by Birt *et al.* (1960). Potato leaf roll virus is persistent and was greatly reduced by controlling the potato aphid vector, yet incidence of the non-persistent virus Y was little affected, even though it was transmitted by the same vector. Although appropriately timed vector control with insecticides is used to contain the spread of many viruses, other substances are also effective. Certain types of mineral oil (Simons 1981) are used commercially and vegetable oils show promise. Nevertheless, the integration of cultural practices with those which limit vector efficiency can achieve considerable benefits (Zitter and Simons 1980).

18.4 The application of chemicals for plant disease control

Decisions concerning chemical control of plant diseases often have to be made in advance of the growing season and in many ways the use of chemicals for disease prophylaxis can be regarded as a type of insurance which will give protection against the possibility of disease. Routine application to protect the crop against disease is equivalent to paying a full premium but in some situations may not be necessary; the grower may decide in the light of available knowledge to evaluate the risks more precisely and to plan his control programme accordingly. Previous experience of disease occurrence in particular situations is a major factor in determining these risks, and advances in disease epidemiology and forecasting techniques have enabled considerable refinements in the timing of application according to the likelihood of infection. The increasing battery of fungicides with various modes of action have also enabled increased flexibility in timing and placement, to the extent that control of some diseases can now be based on a threshold approach analogous to the economic damage threshold widely used in control of insect pests (see Chapter 7). This relies on stopping the disease epidemic when it has reached a critical level beyond which significant damage to the crop would result, and depends upon the use of chemicals with some therapeutic as well as protectant effects. Cook *et al.* (1981) show how risk assessment, disease prediction, and other criteria are used to determine the economic feasibility for fungicidal control of cereal diseases.

The application of chemicals to seed, soil, or growing crops depends upon the epidemiology of diseases requiring control and the characteristics of the pesticide to be used. It is therefore convenient to consider these basic methods separately.

18.4.1   Application to soil

The application of chemicals to soil for plant-protection purposes dates back to about 1820 when carbon disulphide was shown to control *Phylloxera* on vines in France. The subsequent development of soil fumigation techniques leading to the use of more selective compounds against specific diseases has been reviewed by Wilhelm (1966); a comprehensive account of modern soil disinfestation procedures is given by Mulder (1979).

Chemicals applied to soil are usually aimed at the source of root-infecting fungi which often consists of thick-walled resting spores, sclerotia, or dormant mycelium embedded within pieces of old plant debris. Because these are scattered throughout the soil, and are resistant to chemical treatment, methods which 'saturate' the soil with fairly powerful fungicidal materials are often required. Easily volatilized liquids having a fumigant action and general biocidal properties, such as formaldehyde, chloropicrin, methyl bromide, and other chlorinated hydrocarbons are very suitable materials for this purpose. Chloropicrin and methyl bromide are particularly efficient soil fumigants for disease-control purposes and are often formulated together as a mixture. Methyl isothiocyanate and compounds which generate it when mixed with soil (e.g. dazomet) are also useful fumigants. As these substances are phytotoxic they have to be applied well before crops are planted.

The efficiency of soil fumigants depends upon their vapour pressure and the physical and chemical properties of the soil which determine its ability to adsorb the fumigant and the amount of free air space between soil particles allowing diffusion. Goring (1967) has reviewed the action of soil fungicides relative to physical aspects of the soil. Generally, wet compacted soils are more difficult to fumigate efficiently, whereas diffusion in light dry soils is more rapid. The toxicity of soil fumigants depends on their concentration and the time that pathogens are exposed to them. Therefore, if diffusion is very rapid, the fumigant may disperse too quickly, and this can be a problem in light soils under tropical conditions. For control of pathogens which can be present in the deeper layers of field soils such as those occurring on the roots of trees, the depth to which the fumigant penetrates can be a critical factor.

Because plant roots in soil are not directly accessible, conventional fungicides are often applied to the soil to protect crops from root diseases, and some such as thiram, pentachloronitrobenzene (PCNB), ethazol, chloroneb, and others can be specifically formulated for this purpose. Soil application of systemic compounds, such as benomyl, prothiocarb, and pyroxychlor, as drenches to the rooting zone of plants, can permit uptake by the plant roots and translocation to other parts of the plant.

The major advantage of using specific fungicides rather than fumigants for controlling soil-borne pathogens is that they are not phytotoxic and can be applied to soil beneath growing crops or when seed is sown. Against this has to be offset the disadvantage that they do not penetrate soil very effectively and large doses have to be used to protect roots against disease, much of which may be wasted.

All chemicals applied to soil and many applied to aerial plant parts affect the soil microflora and this subject has received much attention (Munnecke 1972: Wainwright 1978; Rodriguez-Kabana and Curl 1980). Frequently this has effects on the growth of the plant, some beneficial and some harmful, which are quite unrelated to the control of soil-borne pathogens and this aspect will be discussed later. Other differential effects of pesticides on the soil microflora influence disease control in a variety of ways, as reviewed by Munnecke (1972). Volatile fumigants generally seem to be more toxic to plant pathogenic fungi so that the relative proportion of the saprophytic fungal population of the soil is increased, but bacterial and actinomycete populations are often reduced. Partial soil fumigation may select a soil microflora antagonistic to pathogens. The control of *Armillaria mellea* in citrus soils is achieved by fumigation with sublethal doses of carbon disulphide (Bliss 1951) and this has been shown to be caused by the increased susceptibility of the fungus to an enhanced *Trichoderma* flora (Garrett 1957).

Plant pathogens vary in their sensitivity to different chemicals. Although generally biocidal fumigants will eradicate all pathogens when soil treatment is complete, those such as *Verticillium dahliae* which produce sclerotia often survive if exposure is limited in any way. Many of the volatile chlorinated hydrocarbons developed primarily as nematicides (see Chapter 17) have some fungicidal activity, especially against oomycetes. General soil fungicides such as thiram and captan are effective against a wide range of soilborne pathogens but have little effect on soil bacteria or actinomycetes. PCNB is particularly effective against *Rhizoctonia* but not against *Fusarium*, while fenaminosulf and ethazole are effective against oomycetes such as *Phytophthora* and *Pythium*. Several systemic compounds may also be applied as drenches to the rooting zone of plants to control these diseases (Hoitink and Schmitthenner 1975; Zentmyer 1978).

Because of the cost of application of soil fumigation techniques they are more frequently used in intensive horticultural production, such as glasshouse crops, where volumes of soil requiring treatment are limited and the produce valuable. Soil fumigation is also an important control method for preventing replant diseases of perennial crops. These are usually caused by the enhanced populations of parasitic nematodes and pathogenic fungi in old rooting zones to levels against which young plants are not resistant and are often specific to each crop (Savory 1966). The apple and cherry replant diseases, caused primarily by fungi, can be prevented by fumigation of orchard sites with chloropicrin before replanting (Jackman 1979).

Application of fungicidal drenches to soil is a method sometimes used to check root diseases of vegetables and nursery stock. For seed sown crops, infurrow application of substances such as PCNB, often mixed with captan or a dithiocarbamate, have been used to control snow mould of cereals in the USA (Sprague 1961) or *Rhizoctonia* root rot of cotton (Maier and Staffeldt 1963). In modern nursery practice, the incorporation of fungicides into peat blocks in which young plants are raised can be undertaken (Ryan and Doyle 1981). Fungicides can also be applied in solution to plants being grown by the nutrient film technique (Price and Dickinson 1980).

Direct application of fungicides to roots can be carried out when plants are transplanted and is a method used for club root (*Plasmodiophora brassicae*) control on brassica crops. Root dips with mercuric chloride at transplanting are very effective but the use of more efficient, less toxic chemicals, particularly of a systemic nature (Buczacki 1973) is preferable and allows greater flexibility of application. Control of Dutch Elm Disease has been claimed by drenching the roots with 8-quinolinol benzoate and by drenching the soil with a suspension of benomyl (Smalley 1971). However, this has been superseded by the more accurate placement of the chemical by direct injection into the tree. The application and placement of soil pesticides in relation to their biological activity has been discussed by Wheatley (1977).

18.4.2 Application to seed

This was the first method to be used in applying chemicals for disease control, largely because of its simplicity, and this advantage still holds today. Treating seed is economical both in monetary terms and in the amount of pesticide used (100–200 g/ha); seeds provide an ideal carrier for the chemical which is accurately placed so that environmental disturbances are minimal.

Recent reviews of various aspects of chemical application to seed are by Neergaard (1977) and Jeffs (1978); application methods are also mentioned in Chapter 4 of this book. Recent advances in techniques have aimed mostly at increased adherence to the seed coat and at achieving an even coverage throughout the seed lot. A major disadvantage of dusts is that they tend to rub off the seed whereas chemicals applied as liquids or slurries adhere more firmly. Phytotoxicity problems can occur when large concentrations of the pesticide come into contact with the embryo, another reason why even coverage is essential. Fungicides and other chemicals may also be applied to seeds when they are 'pelleted'. One of the limitations of seed application is the restricted duration of effectiveness against soil-borne pathogens. Chemicals are degraded in the soil and as the plant grows, roots and shoots extend beyond the sphere of influence of the chemical immediately around the seed. Nevertheless, they are effective in protecting the vulnerable early stages of plant growth and their use specifically for this purpose has largely replaced the more cumbersome method of soil application. Systemic fungicides which are taken up by the seedling can protect it against pathogens which may attack at or above soil level but translocation of most of these substances towards root apices is very limited.

There are several ways in which pathogens can be associated with seeds (Baker and Smith 1966), and Richardson (1979) lists pathogens which have been recorded as seed-borne. Firstly, spores may contaminate the seed coat, and this is a frequent way whereby saprophytic fungi are carried, but important pathogens such as bunts (*Tilletia* spp.) and some smuts (*Sphacelotheca* spp.) of cereals and *Fusarium* spp. on pulses are also carried in this way. Pathogenic fungi may also accompany seed in debris or as descrete spores in dust, etc. Fungicidal treatment of seed can control these fungi relatively easily. More frequently, pathogens invade the seed and are carried either as resistant mycelium within the seed coat, as with *Septoria* of cereals and celery or *Ascochyta* on legumes, or as dormant systemic infections of the embryo, e.g. loose smut (*Ustilago* spp.) of cereals. Chemical treatments to eradicate these sources of infection are more difficult. Organomercurial seed applications have some eradicative action as they can diffuse through the seed coat and often have vapour phase activity; despite their toxic qualities they are still used for control of many seed-borne cereal pathogens. Ulfvarson (1969) has reviewed the uses and characteristics of these substances.

Maude *et al.* (1969) showed that a wide range of deep-seated seed-borne diseases of many vegetable crops could be eradicated by soaking for 24 hours in aqueous suspension of 0.2% a.i. thiram at 30°C. The use of systemic compounds as seed fungicides has enabled internal seed infections of certain fungi to be efficiently controlled, e.g. the carboxithiin fungicides applied to seed infections of smuts, e.g. *Ustilago nuda* on wheat (Moseman 1968). Most systemic compounds are rather selective in action so that their uses in this respect are often limited, particularly against oomycetes and many dermatiaceous hyphomycetes. Some important bacterial diseases (e.g. *Xanthomonas campestris* pv. *malvacearum* on cotton, *Pseudomonas syringae* pv. *phaseolicola* on beans) are seed-borne on or in the seed coat and can be controlled with copper, mercuric, or antibiotic compounds. Bronopol is a specific bactericide used on cotton seed.

The selective nature of the systemic fungicides restricts their use for control of soil-borne diseases, as most offer little protection against many of the important pathogens which attack young seedlings. Non-systemic broad-spectrum fungicides such as thiram have to be added if protection from a wide range of soil-borne organisms is required. Some recently developed systemic compounds show useful activity against many oomycetes including seed- or soil-borne downy mildews, e.g. *Sclerospora graminicola* on pearl millet (Williams and Singh 1981). Paulus *et al.* (1977) have compared the use of systemic compounds active against oomycetes when applied to seed, soil, or as a spray to foliage.

A new dimension to the chemical control of plant diseases was opened up by the use of seed-applied systemic fungicides for the control of pathogens subsequently attacking shoots or young plants, and this has been widely developed for the control of cereal diseases in Europe such as powdery mildew of barley which is controlled by seed application of ethirimol (Jenkyn and Bainbridge 1978). This chemical is taken up by the plant and persists long enough in the plant tissue to prevent early infection, reduce the infection rate, and allow the yielding potential of the crop to be realized.

Fungicides are also applied to vegetable planting material such as bulbs, sugarcane setts, etc. mainly as protection against soil-borne diseases. These are usually applied by dipping the material into a suspension of the fungicide. Organomercurials previously used for this purpose are now being replaced by thiram preparations, often combined with one of the methyl benzimidazole carbamate (MBC)-generating systemic fungicides.

The potential use of seed fungicides as a method of chemical control is very great. The attractions of the method far outweigh any disadvantages, and systemic fungicides giving prolonged protection during early growth offer many possibilities on a wide range of crops. Because of the relative simplicity of application and the economy with which seed fungicides can be used, application to seed could be particularly useful in developing countries, especially as it is in these areas where climatic and soil conditions so readily predispose young plants to attack by soil-borne pathogens.

18.4.3 Application to standing crops

Pathogens which infect the aerial parts of plants frequently spread rapidly throughout crops causing polycyclic diseases, and may produce devastating epidemics. Relatively simple control measures which only reduce initial inoculum sources are often ineffective against these diseases, and continuing protection of the plant by repeated applications during the growing period is the basic method of chemical control for these diseases. There has, however, been considerable development in the methodology of control, particularly in relation to the epidemiology of these diseases so that more efficient and economical techniques are continually being developed.

Apart from choosing the correct pesticide with appropriate activity against particular pathogens, problems relating to placement and timing are particularly critical in the chemical control of diseases on growing crops. These are influenced not only by the type of pesticide used and its method of application but also by factors relating to the epidemiology of the disease and growth of the crop. Efficient disease control depends upon an adequate appreciation of these variables.

The target at which the pesticides are aimed is usually the susceptible part of the plant which requires protection from infection, but systemic fungicides can be applied so that they are translocated to the site of the pathogen within the plant tissue. Efficient coverage of the target is essential for adequate control of most diseases particularly where protectants are used. The leaves of herbaceous crops may be fairly easily covered by spray deposit, but dense canopies of perennial tree crops and objects such as blossom, fruit, or shoot tips present more difficult targets. Special application measures may be required to reach vascular pathogens.

The target usually changes with growth of the crop; leaf areas expand, new leaves are produced, and foliage canopies grow; it may also be transitional in that only certain stages of the crop may be susceptible, e.g. blossoms, fruit. Particular growth stages of the crop which are critically related to yield production may require extra protection. Activity of the pathogen is often related to prevailing weather conditions and this also influences spray timing. Techniques of disease prediction may permit the application of chemicals to be pin-pointed to critical periods favouring infection, thereby reducing the number of applications required for successful control.

The duration of effective fungicidal cover determines the interval between applications and is related to crop growth, weather conditions, and fungicide persistence. New areas of growth will need protection and the removal of chemical deposits by rain will require renewed applications. Thus many vegetable and fruit crops require routine applications of protective fungicides against foliage diseases when grown in areas of intense disease pressure, e.g. tomatoes require continuous fungicidal protection against a range of leaf pathogens in tropical areas. Control of tea blister blight (*Exobasidium vexans*) in S.E. Asia and coffee rust (*Hemileia vastatrix*) in Brazil also require routine fungicide spraying at frequent intervals during the rainy seasons.

The value of the crop must be able to bear the costs of fungicide application which on this scale are considerable. Therefore improvements in pesticide performance, application techniques and timing can lead to considerable savings. These three basic aspects fundamentally influence the efficiency and economy of disease control on growing crops and are worthy of separate consideration.

18.4.4 Pesticide performance

The biological performance of a pesticide results from a combination of its intrinsic activity with the various factors which take it to or remove it from the target (Shephard 1981). With protectant fungicides, the balance between persistence and redistribution by rainfall or as vapour is critical, and the more successful foliar protectants such as captafol, chlorothalonil and some copper based fungicides are both persistent yet able to be redistributed in effective quantities from their original site of application. Bordeaux mixture is probably still the most tenacious of fungicides under tropical conditions, but this hampers its redistribution properties.

The formulation of pesticides can profoundly influence their biological activity and their suitability for new application techniques such as vascular injection, ULV application, etc. This aspect has recently been reviewed by Backman (1978). Although oil-based sprays can allow smaller droplets and lower doses, as used in ULV application, their redistribution potential in rain-water can be adversely affected.

Advances in fungicide chemistry have led to the production of new and more potent materials often having different modes of action from earlier materials and which may be applied in different ways. Siegel and Sisler (1977) have reviewed the wide variety of anti-fungal compounds whch have been developed and the chemical control of many diseases has been revolutionized by commercial availability of new compounds. Many of these have been aimed at pathogens which present difficult control problems such as *Botrytis*, powdery mildews (Bent 1978a), *Phytophthora*, and other oomycetes (phycomycetes) (Schwinn 1981).

The increasing selectivity of fungicides is a predominant feature of modern developments and this has had both advantages and disadvantages; for example it is especially critical when used for the control of mushroom diseases (Gandy and Spencer 1978), but where a range of crop diseases require control, mixtures may have to be used to get a sufficiently broad spectrum of activity. When used correctly, fungicides selectively active against particular pathogens may be expected to do less damage to non-target organisms than broad-spectrum compounds. Byrde and Richmond (1976) have reviewed the various aspects of fungicidal selectivity.

The systemic nature of many recent fungicides has probably had the most important impact on chemical control techniques. Marsh (1977) has produced a comprehensive review covering the activities and chemistry of systemic fungicides and the results of their use to control diseases of temperate crops. As these compounds have appreciable eradicant properties and allow greater flexibility of placement and timing, they have enabled completely new application methods to be used. Chemicals which can penetrate and move systemically in the vascular tissue are redistributed inside the plant so that persistence and redistribution in rainfall, critical for the surface protectants, are replaced by uptake and translocatability as the major criteria influencing performance. The vast majority of systemic fungicides currently used only travel significant distances in the apoplast (intercellular spaces and xylem tissue) of the plant (Dekker 1977). Therefore movement is mainly acropetal from root to shoot or by translaminar diffusion across leaves, as it follows the general water movement in the plant influenced primarily by transpiration. Consequently, the systemic redistribution of most fungicides to roots and organs with low transpiration rates such as fruits and flowers is insufficient for this to have a major effect on controlling diseases of these organs. Apoplastic translocation is particularly suited to root or seed application where the fungicide is moved to the vascular tissue and eventually to stems and leaves following uptake. Thus a notable achievement of systemic fungicides has been to enable shoot diseases of young plants to be controlled by seed application. Translocation in woody plants is affected by absorption in tissues; this is correlated with their lignin content which tends to impede the translocation of lyophillic substances to the greatest extent (Barak *et al.* 1981).

Recently certain compounds have been shown to be transported basipetally in the phloem, and foliar application of these has been shown to control *Phytophthora* root rots of woody plants (Zentmyer 1978). Table 18.1 compares the basic characteristics of systemic and protectant fungicides.

18.4.5 Application and placement

Refinements in the techniques of spray application have been dealt with elsewhere in this book. As far as disease control is concerned, low-volume techniques have been widely adopted, but a major limitation to using much reduced volumes, especially on tree crops, has been the need to ensure adequate fungicide cover of susceptible crop surfaces. Morgan (1972) has

TABLE 18.1. *Basic features of protectant and systemic fungicides*

	Protectant	Systemic
Activity	Protectant, prophylactic	Protectant and therapeutic (eradicant)
Action	Multisite toxins affecting several metabolic pathways	Usually only single-site toxins
Application timing	Preinfection and critical with maximum coverage	Pre- or post-infection, so less critical
Movement and persistence	Redistributed over plant surface and removed by weathering	Distributed within plant tissues but mainly acropetally; absorbed and rendered ineffective by plant tissues
Resistance (insensitivity)	Not a problem	Often a major problem
Environmental effects	Broadly toxic to microflora, seldom hazardous to higher plants or animals	Selectively toxic, less hazardous to microflora, retained in plant and degraded

outlined the basic requirements of tree crop spraying and the techniques used to assess its efficiency for various purposes.

Many diseases are adequately controlled by techniques which do not initially achieve complete coverage of the target. Martin (1940) showed that Bordeaux mixture applied by watering can could give adequate control of potato blight, and Brandes (1971) has reviewed the successful use of low-volume sprays, including those applied by air, against crop diseases. Redistribution of fungicides over the outside of the plant by rain-water has been shown to be largely responsible for this. Control of coffee rust which only penetrates through the underside of leaves was shown to be due to the redistribution of fungicides in rain or dew (Hislop 1969). Similarly, coffee berries can be protected by infection from *Colletotrichum coffeanum* by the application of fungicides only to the tops of trees (Waller 1972) or by aerial spraying (Pereira 1970). This is due to the way in which fungicides are redistributed by rain throughout the canopy of the coffee trees (Pereira *et al.* 1973). Control of *Scirrhia pini,* the cause of a needle blight of pines, is also achieved in this way (Gibson 1974). Sigatoka disease of bananas is ideally suited to aerial spraying because the target is the young leaves situated in the crown of the plant and these are exposed in such a way that they are more easily covered by aerial spraying than by ground-based equipment.

Ground-based ULV methods have been slow to be adopted for disease control, although they have been shown to be effective against diseases of herbaceous crops (Quinn *et al.* 1975); control of coffee leaf rust is also possible using ULV techniques and is under investigation in S. America (Waller 1982). The major advantages of ULV application are that lower doses are possible as the controlled-sized droplets impact more efficiently and selectively on the target (see Chapter 2) and it obviates the need to carry large volumes of water. Against this must be set some practical disadvantages. The wind drift used to carry ULV droplets to their target is often erratic in both intensity and direction in hilly tropical areas, and both the oil-based nature of the spray droplets and the small doses used tend to restrict the redistributive potential of protectant fungicides.

Fogging machines or smoke generators have been little used in disease control except in glasshouses where they can be confined over the crop, but their greater penetrating power is a useful property in the control of foliage diseases of large trees such as rubber (Lim 1978). Dusting is also used for the control of rubber leaf diseases, but except when used to apply sulphur, has generally proved less efficient than spraying due to inferior retention of dusts on crop surfaces.

Direct injection of systemic fungicides into the xylem has been used for therapeutic control of vascular diseases of trees, such as Dutch Elm Disease (Gibbs and Clifford 1974), and vascular injection of antibiotics has also been used to treat diseases such as Coconut Lethal Yellowing, caused by mycoplasma-like organisms (MLOs) and similar pathogens (Markham 1977). Root and trunk diseases of tree crops can be treated by direct application of fungicide preparations to wounds or infected tissue, but is restricted to situations where these relatively laborious methods are justified. Perennial cash crops such as citrus (*Phytophthora* gummosis) and coffee (*Fusarium* bark diseases) are examples. Wastie (1975) summarizes the use of these techniques for the control of root and panel diseases of rubber.

18.4.6 Timing

Knowledge of the basic epidemiology of diseases enables critical periods in the epidemic cycle to be identified so that application can be pin-pointed more accurately. Few diseases increase uniformly (as implied by the van der Plank models); most increase in steps depending on the occurrence of certain climatic conditions favouring infection or susceptible stages in crop development.

The application of chemical control measures against diseases which only affect seedlings, flowers, or developing fruits can be limited to times when these susceptible stages are present, and present few problems of timing as this can usually be planned in advance. Some foliage diseases of perennial crops, such as coffee, tea, and rubber, are only important on very young plants so that chemical control is limited to the nursery. Detailed analysis of the epidemiology of cereal powdery mildews in relation to the development of yield potential has enabled single sprays of systemic compounds timed critically according to the amount of disease on the crop (Jenkins and Storey 1975) or the activity of the pathogen (Jenkyn 1978) to give adequate control.

Predicting when infection is likely to occur is the basis of several methods which aim to deploy fungicides more efficiently and economically. Krause and Massie (1975) have reviewed modern approaches to this form of control. Accurate prediction and spray timing are now less critical with systemic eradicant sprays which are effective when applied after infection. Disease prediction based on weather patterns which permit infection have led to substantial savings in the cost of spraying. The occurrence of 'Mills periods' favouring apple scab infection enables control with eradicant post-infection sprays to be concentrated at times when these conditions have occurred (Lewis and Hickey 1972). Similarly, the knowledge that specific climatic conditions were essential for spore release and infection of *Phytophthora infestans* (Beaumont periods) enabled routine spraying against potato blight to be replaced by more economic schedules. In Japan, the recording of conditions favourable to rice blast development has enabled chemical control of this disease to be made economically viable by application at the few times when conditions are critical (Ono 1965). The spread of virus diseases depends upon conditions favourable to vector activity, so that timing of insecticidal sprays aimed at the vector is critical (Hull and Heathcote 1967). Rainfall is one of the most important factors which limits fungicidal persistence, and a flexible approach relating the interval between applications to the amount of rainfall has been advocated for some crops (Steiner 1973).

The development of powerful eradicant chemicals has enabled some polycyclic diseases to be controlled by destroying sources of overwintering inoculum. This goes against the general principle that sanitation measures have little effect on such diseases, but it is effective on deciduous fruit trees. Dormant-season eradicant sprays of surfactant chemicals can control apple mildew (*Podosphaeria leucotricha*) (Hislop and Clifford 1976), and the control of apple scab (*Venturia inaequalis*) is greatly facilitated by the chemical eradication of the overwintering perfect state on old apple leaves (Burchill 1975).

18.4.7   The integration of control techniques

The history of Sigatoka leaf spot control on bananas illustrates very well how advances in fungicide performance, application techniques, and disease prediction contributed to improved disease control.

Chemical control of Sigatoka has advanced through several distinct phases. Until the late 1950s the standard control was to apply high-volume sprays of Bordeaux mixture which was the most successful of the fungicides then available, largely because of its superior tenacity under tropical conditions. Attempts were made to develop low-volume application techniques (copper in oil) and this led to the use of banana spray oil when it was discovered that certain oil fractions (346–354°C distillation range) controlled Sigatoka when applied by air at about 11 litres/ha (1 gallon/acre). The oil acted by arresting the development of the disease in its early stages. Oil sprays were replaced by oil/water emulsions with dithiocarbamates such as maneb during the 1960s, as these caused less phytotoxicity problems than oil alone. Subsequently, the dithiocarbamates were replaced by systemic fungicides in the 1970s and control was improved further. Nowadays, systemic fungicides such as benomyl or thiophanate-methyl are applied by air using fixed-wing aircraft equipped with micronair rotary atomizers at about 500 g.a.i./ha in 15 litres of oil/water emulsion.

A better understanding of the epidemiology of the disease led to the development of disease-prediction techniques which enabled further refining of disease control. Sporulation, infection, and subsequent disease development are all favoured by prolonged wet conditions; sunshine also has an effect. Meteorological data coupled with continuous disease assessment and the use of systemic fungicides in oil/water emulsions enabled Ganry and Meyer (1972, 1973) to time the interval between sprays according to the rate of disease progress. The spraying cycle was reduced from a 2–4 week interval when oil alone was used to 5–7 weeks so that total applications per year have now been reduced by almost half in many countries (Cronshaw 1982).

Most crops are threatened by several diseases, and in practice the different control measures effective against them are integrated together. Plants usually possess resistance to most pathogens, others may be controlled by cultural conditions, but where resistance 'breaks down', fungicides may be needed to resolve the situation. This happened with the epidemic of Southern Corn Leaf Blight in the USA (Ulstrupp 1972) and of *Puccinia recondita* on the wheat variety Jupateco in Mexico during 1976/77 (Fuentes 1978).

Sometimes, where simultaneous control of several diseases is required, it may be necessary to use different fungicides, either mixed or on an integrated schedule. Control of coffee diseases in Kenya illustrates this aspect very well. Rust control is achieved by copper sprays applied before the rainy season, whereas maximum control of coffee berry disease requires application of captafol or similarly effective fungicides throughout the rainy season. Furthermore, the recommended schedules vary according to altitude and location which affect the relative severity of these two diseases. In higher areas the situation is further complicated, as bacterial blight (*Pseudomonas syringae*) may also require chemical control (Anon. 1978).

The modern approach to cereal disease control in Europe utilizes different chemicals applied in different ways. Seed dressings of ethirimol used against barley mildew are sometimes backed up by a summer foliar spray which may need to contain several fungicides if other leaf diseases such as rust or *Septoria* are important (Jenkins 1977). Where eyespot (*Pseudocercosporella herpotrichoides*) is prevalent, autumn or spring applications of systemic MBC-generating fungicides and growth-regulating chemicals, such as chlormequat, which stiffen the stems can be beneficial (Lescar 1977).

18.5   Secondary effects of chemical control of plant diseases

18.5.1   Non-target organisms

Fungicides may affect other 'non-target' organisms including the host plant itself. Phytotoxicity effects are probably the most widely known of these. Many of the simple inorganic fungicides such as sulphur and mercury compounds can cause phytotoxicity on certain crops, and copper fungicides are known to have stunting effects when used in large doses or applied over long periods on perennial crops. Sometimes phytotoxicity is a result of formulations, oils being particularly well known for this. Weather conditions may also influence phytotoxicity.

The ecological effects of pesticides on organisms not directly involved in the host/parasite relation are now receiving wide attention. Their effects on soil microflora have been reviewed by Wainwright (1978) and Rodriguez-Kabana and Curl (1980); some of these may provide indirect control of soil-borne pathogens, as was mentioned earlier. Some beneficial effects of soil fumigation on plant growth can be attributed to an increase in total soil nitrogen (and other elements) caused by the release of these substances from dead organisms but often there is a shift to greater amounts of ammoniacal nitrogen as nitrifying bacteria are killed. Harmful effects are also evident; rhizobial bacteria are killed by soil fumigation so that nodulation of legumes may be inhibited. Vesicular-arbuscular mycorrhizal fungi are also affected by a wide range of fungicides and this can explain the negative effects on plant growth sometimes seen after soil treatment (Gerdemann 1968). Soil drenches of systemic fungicides have been shown to prevent vesicular-arbuscular mycorrhiza infections with a corresponding decrease in phosphate uptake by test plants (Boatman *et al.* 1978). MBC-generating compounds such as benomyl have been shown to reduce earthworm populations in orchard soils (Stringer and Lyons 1977).

Many foliar fungicides have been shown to affect pests such as mites (Cranham 1971) and insects. MBC fungicides give some control of aphids and may help to reduce the incidence of virus diseases transmitted by them (Russell 1977); they also appear to suppress symptom expression of some virus diseases (Tomlinson 1977). In other cases an increase in pest incidence has occurred after the use of fungicides. Coffee leaf miner (*Leucoptera* spp.) became more prevalent in Brazil after copper fungicides were used for rust control (Paulini *et al.* 1976).

Iatrogenic plant diseases (Horsfall 1979; Griffiths 1981), which are basically those induced or exacerbated by attempted control measures, are usually due to the indirect effect of fungicides. This may act through an effect on the host which reduces its resistance, on the pathogen itself perhaps by increasing its relative inoculum potential, sporulation, etc. or by an effect on the ecosystem which may block a naturally occurring biological control mechanism, for example. Incorrect fungicide usage may exacerbate some diseases. Coffee berry disease in Kenya was increased by the application of fungicides on a schedule which only protected the early stages of berry growth. Subsequent disease development was much greater than in unsprayed coffee and several possible reasons for this have been advanced (Griffiths 1971). Changes in the type of fungicide used has also affected disease incidence. The replacement of Sigatoka leaf spot (*Mycosphaerella musicola*) by black leaf streak (*M. fijiensis*) on bananas in Fiji has been attributed to the use of banana spray oil (Firman 1970). The change from broad-spectrum protectants to more specific systemic compounds has resulted in the upsurge of several diseases such as bacterial blight of coffee (*Pseudomonas syringae*) in Kenya (Ramos and Shavdia 1976) and *Phytophthora* fruit rots of apple (Upstone 1977).

Fungicides may have so called 'tonic' effects on crops in that they increase the vigour and yield of crops above that which can be explained by the control of visible diseases. This effect may be caused by the eradication of minor pathogens, especially those which hasten the death of senescent leaves. The pronounced tonic effect of fungicides on coffee is mostly due to increased leaf retention, an effect which is probably influenced by the balance of growth-regulating substances produced by the leaf and bark surface flora (Van der Vossen and Browning 1978).

18.5.2 *Resistance to fungicides*

Since the advent of systemic fungicides, the problem of pathogens becoming insensitive to them has become of increasing importance in disease control. Nowadays, most disease-control programmes using systemic, or non-systemic but selective, fungicides should be designed to avoid or cope with this problem, popularly known as fungicide 'resistance' or 'tolerance'. Previously, this had not been a major problem in chemical control of plant diseases, although resistance to some aromatic hydrocarbons, dodine, and mercurial compounds had been noted. Strains of *Pyrenophora avenae* resistant to organomercurial compounds are apparently able to fix and inactivate toxic ions before they interfere with the metabolism of the fungus (Greenaway 1971).

Resistance to many systemic fungicides is now common and has severely limited their effectiveness against several major pathogens; it has consequently been the subject of several recent publications (Georgopoulos 1977; Delp 1981; Dekker and Georgopoulos 1982).

Broad-spectrum, protectant fungicides are multisite toxins which act on a range of different enzyme systems and metabolic pathways. They owe their potency against fungi primarily because they are absorbed through fungal cell walls and not through plant cell walls. Systemic or other selective fungicides, however, are often single-site toxins acting on a discrete but critical enzyme system or metabolic pathway and differential absorption through cell walls is less critical, which is why they are systemic. Single mutations or other simple genetic changes which enable the affected pathway to be bypassed, can therefore readily occur. Insensitivity to one fungicide may be effective against others if these have the same toxic principle; such cross-resistance is common among the methyl benzimidazole carbamate (MBC)-generating fungicides such as benomyl, carbendazim, thiophanate, etc.

Wolfe (1975) has clarified the factors which influence the development of fungicide resistance. Firstly, the genetic change which enables the toxic action to be bypassed has to be possible. For those systemic fungicides which inhibit sterol synthesis, this is difficult, or is accompanied by adverse effects, so resistance to these fungicides is not a problem. Resistance to multisite toxins would require many simultaneous changes to bypass the many pathways which these fungicides block, and those rare cases where this has occurred seem to depend on immobilizing the chemical outside the fungus cell. The probability of resistance developing in a pathogen population depends upon the frequency of genetic changes able to mediate it and the degree of selection pressure applied. Where chemicals are used intensively in large doses over long periods, and wide areas, selection pressure is greater and the emergence of resistance is more likely. The speed with which resistant strains spread to become a serious problem in chemical control depends upon the reproductive potential of the pathogen and the epidemiology of the disease. On annual crops, where seasonal carryover of inoculum is very limited, or with monocyclic diseases, where spread of the pathogen is slow, the development and spread of resistant strains will be restricted.

Resistance in some pathogens seems to be associated with decreased fitness in other respects so that when the fungicide is withdrawn, the proportion of the population exhibiting resistance rapidly declines. Barley mildew isolates resistant to ethirimol were shown by Holloman (1978) to be less competitive than sensitive strains and despite the increase in the occurrence of pathogen isolates resistant to ethirimol in treated fields, this has not reduced the efficiency of control, possibly because of the limited exposure of the pathogen to the fungicide and the low frequency of survival of mildew populations between successive spring barley crops (Bent 1978b). On the other hand, resistance among populations of *Cercospora beticola* on sugarbeet in S. Europe to benomyl appears to be more stable. This has not regressed despite the withdrawal of MBC fungicides (Dovas *et al.* 1976). The stability of resistance may well be greater where there is prolonged exposure thus enabling strains to develop with both resistance and greater survival fitness.

In practical plant disease control, the development of resistance to fungicides needs to be prevented. This depends largely on reducing selection pressure and includes such general principles as limited use of single-site toxins in space and time, maximum sanitary precautions to prevent carryover of resistant strains between crops and seasons, and the inclusion of multisite toxins in the spraying schedule either as mixtures with the fungicide at risk or as alternative applications. Monitoring the pathogen populations for the development of resistance will also allow problems of loss of field control of diseases to be foreseen and appropriate action such as a change of fungicide to be taken in good time.

18.6 Future prospects

The International Society of Plant Pathology (ISPP) Committee on Chemical Control (1980) have assessed the need for developing new chemical control techniques and improving existing ones in the light of the problems facing contemporary fungicide usage. There are several groups of pathogens against which chemical control is still not very effective.

Control of virus diseases with chemicals is still in its infancy. Although there are several compounds which act in different ways to inhibit the development of virus diseases (Hirai 1977), they have not developed to the stage of commercial application. White and Antoniw (1981) have categorized potential methods into those which affect the infection, multiplication, spread, and symptom expression of virus diseases; they suggest that materials which prevent the spread of virus in the plant and which will localize infection seem to be the most promising.

There are inadequate chemical control techniques for bacterial pathogens; apart from copper compounds, few fungicides have anti-bacterial activity; problems with antibiotics prevent their use except in very limited circumstances, and there is a major requirement for an effective systemic bactericide. The increasing use of non-bactericidal fungicides as a replacement of copper compounds has resulted in an upsurge of some bacterial diseases. Diseases caused by mycoplasmas and similar organisms also require better chemical control.

The development of compounds which can move in the symplast of plants and can be translocated from leaves to roots, fruit, and juvenile tissues would dramatically improve the control of many diseases and a few compounds with these characteristics are now appearing. Substances with better activity against dermatiaceous hyphomycetes, basidiomycetes (especially Corticiaceae and similar fungi) and *Phytophthora* fruit diseases, are needed. At present strategies for dealing with fungicide resistance are based on restricting selection pressure by limiting the exposure of fungicides to which resistance may be developed. Other strategies that are being studied include those based on integration with host resistance (Wolfe 1981), thus increasing the diversity which the pathogen has to overcome and improving the durability of both fungicide and genetic resistance.

Novel methods of chemical control which do not rely directly on fungitoxic substances are also feasible. Growth-regulating substances have been known for some time to influence plant resistance (Dimond and Rich 1977) and some are used in disease-control practices, e.g. cycocel for eyespot of wheat. Other substances have now been found which stimulate the production of phytoalexins, e.g. cyclopropene compounds can affect phytoalexin production in rice plants when challenged with *Pyricularia oryzae,* thus inducing a resistant response in cultivars susceptible to virulent isolates (Cartwright *et al.* 1977). Langcake (1981) has recently reviewed the potentiality of these methods which have many advantages in being capable of rational design by biochemists, being non-toxic to non-target organisms, and being less likely to favour the selection of resistance.

18.7   References

Anon. (1978). *Kenya Coffee* **43**, 38.
Backman, P. A. (1978). *Annu. Rev. Phytopathol.* **16**, 211.
Baker, K. F. and Smith, S. H. (1966). *Annu. Rev. Phytopathol.* **4**, 311.
Barak, E., Dinoor, A., and Jacoby, B. (1981). *Proc. 1981 Br. Crop Prot. Counc. Conf. – Pests Dis.,* 129.
Bent, K. J. (1978a). In *The powdery mildews* (ed. D. M. Spencer), p. 259. Academic Press, London.
—— (1978b). In *Plant disease epidemiology* (eds. P. R. Scott, and A. Bainbridge), p. 177. Blackwell, Oxford.
Berger, R. D. (1977). *Annu. Rev. Phytopathol.* **15**, 165.
Birt, P. E., Broadbent, L., and Heathcote, G. D. (1960). *Ann. Appl. Biol.* **48**, 580.
Bliss, D. E. (1951). *Phytopathology* **41**, 665.
Boatman, N., Paget, D., Hayman, D. S., and Mosse, B. (1978). *Trans. Br. Mycol. Soc.* **70**, 443.
Brandes, G. A. (1971). *Annu. Rev. Phytopathol.* **9**, 363.
Buczacki, S. T. (1973). *Ann. Appl. Biol.* **74**, 85.
Burchill, R. T. (1975). *Proc. 5th Symp. Integrated Control Orchards,* 249.
Byrde, R. J. W. and Richmond, D. V. (1976). *Pestic. Sci.* **7**, 372.
Cartwright, D., Langcake, P., Pryce, R. J., and Leworthy, D. P. (1977). *Nature (London)* **267**, 511.
Cook, R. J., Jenkins, J. E. E., and King, J. E. (1981). in *Strategies for the control of cereal diseases* (eds. J. F. Jenkin and R. T. Plumb), p. 91. Blackwell, Oxford.
Cranham, J. E. (1971). *East Malling Res. Stn. Rep. 1971,* 133.
Cronshaw, D. K. (1982). *Trop. Pest Manag.* **28**, 136.
Dekker, J. (1977). In *Plant disease: an advanced treatise* (eds. J. G. Horsfall and E. B. Cowling), Vol. 1, p. 307. Academic Press, New York.

—— Georgopoulos, S. G. (eds) (1982). *Fungicide resistance in crop protection. 265 pp.* Pudoc, Wageningen.

Delp, C. J. (1981). *Proc. 1981 Br. Crop Prot. Counc. Conf. – Pests Dis.,* 865.

Dimond, A. E. and Rich, S. (1977). In *Systemic fungicides* (ed. R. W. Marsh), p. 115. Longmans, London and New York.

Dovas, C., Skylakakis, G., and Georgopoulos, S. G. (1976). *Phytopathology* **66,** 1452.

Ebbels, D. L. (1976). *Rev. Plant Pathol.* **55,** 747.

Evans, E. (1968). *Plant diseases and their chemical control.* Blackwells, Oxford.

Firman, I. D. (1970). *Nature (London)* **225,** 1161.

Fuentes, S. (1978). *CIMMYT Review 1978,* 126.

Gandy, D. E. and Spencer, D. M. (1978). *Ann. Appl. Biol.* **90,** 355.

Ganry, J. and Meyer, J. P. (1972). *Fruits d'Outre Mer* **27,** 665.

—— —— (1973). *Fruits d'Outre Mer* **28,** 671.

Garrett, S. D. (1957). *Can. J. Microbiol.* **3,** 135.

Georgopoulos, S. G. (1977). In *Plant disease: an advanced treatise* (eds. J. G. Horsfall and E. B. Cowling), Vol. 1, p. 327. Academic Press, New York.

Gerdemann, J. W. (1968). *Annu. Rev. Phytopathol.* **6,** 397.

Gibbs, J. N. and Clifford, D. R. (1974). *Ann. Appl. Biol.* **78,** 309.

Gibson, I. A. S. (1974). *Eur. J. For. Pathol.* **4,** 89.

Goring, C. A. I. (1967). *Annu. Rev. Phytopathol.* **5,** 285.

Greenaway, W. (1971). *Trans. Br. Mycol. Soc.* **54,** 127.

Griffiths, E. (1971). *Proc. 6th Br. Insectic, Fungic. Conf. 1971,* 817.

—— (1981). *Annu. Rev. Phytopathol.* **19,** 69.

—— Gibbs, J. N., and Waller, J. M. (1971). *Ann. Appl. Biol.* **67,** 45.

Harrison, B. D. (1977). *Annu. Rev. Phytopathol.* **15,** 331.

Hirai, T. (1977). In *Plant disease: an advanced treatise* (eds J. G. Horsfall and E. B. Cowling), Vol. 1, p. 285. Academic Press, New York.

Hislop, E. C. (1969). *Ann. Appl. Biol.* **63,** 71.

—— Clifford, D. R. (1976). *Ann. Appl. Biol.* **82,** 557.

Holloman, D. W. (1978). *Ann. Appl. Biol.* **90,** 195.

Hoitink, H. A. J. and Schmitthenner, A. F. (1975). *Phytopathology* **65,** 69.

Horsfall, J. G. (1979). In *Plant disease: an advanced treatise* (eds J. G. Horsfall and E. B. Cowling), Vol. 4, p. 343. Academic Press, New York.

Hull, R. and Heathcote, G. D. (1967). *Ann. Appl. Biol.* **60,** 469.

ISPP Committee on Chemical Control (1980). *FAO Plant Prot. Bull.* **28,** 92.

Jackman, J. E. (1979). In *Soil disinfestation* (ed. D. Mulder), p. 184. Elsevier, Amsterdam.

Jeffs, K. A. (ed.) (1978). *Seed treatment – CIPAC monograph no. 2,* i–vi + 101 pp.

Jenkins, J. E. E. (1977). *Proc. 1977 Br. Crop Prot. Counc. Conf. – Pests Dis.* **3,** 785.

—— Storey, I. F. (1975). *Plant Pathol.* **24,** 125.

Jenkyn, J. F. (1978). *Ann. Appl. Biol.* **88,** 369.

—— Bainbridge, A. (1978). In *The powdery mildews* (ed. D. M. Spencer), p. 284. Academic Press, London.

Kaars Sijpesteijn, A. (1977). In *Systemic fungicides* (ed. R. W. Marsh), p. 131. Longmans, London and New York.

Krause, R. A. and Massie, L. B. (1975). *Annu. Rev. Phytopathol.* **13,** 31.

Lamberti, F. (1981). *Plant Dis.* **65,** 113.

Langcake, P. (1981). *Philos. Trans. R. Soc. London Ser. B.* **195,** 83.

Large, E. C. (1952). *Plant Pathol.* **1,** 109.

Lescar, L. (1977). *Proc. 1977 Br. Crop Prot. Counc. Conf. – Pests Dis.,* 763.

Lewis, F. H. and Hickey, K. D. (1972). *Annu. Rev. Phytopathol.* **10,** 399.

Lim, T. M. (1978). *3rd Int. Plant Pathol. Congr., Munich, Abstracts,* 362.

Lukens, R. J. (1971). *Chemistry of fungicidal action.* Springer-Verlag, Berlin and New York.

Lyr, H. (1977). In *Plant disease: an advanced treatise* (eds J. G. Horsfall and E. B. Cowling), Vol. 1, p. 239. Academic Press, New York.

Maier, C. R. and Staffeldt, E. E. (1963). *New Mexico Agric. Exp. Stn. Bull.* **474,** 1.

Markham, P. G. (1977). *Proc. 1977 Br. Crop Prot. Counc. Conf. – Pests Dis.,* **3,** 815.

Marsh, R. W. (ed.) (1977). *Systemic fungicides,* 401 pp. Longmans, London and New York.

Martin, H. (1940). *Ann. Appl. Biol.* **27,** 433.

Maude, R. B., Vizor, A. S., and Shuring, C. G. (1969). *Ann. Appl. Biol.* **64,** 245.

Morgan, N. G. (1972). *PANS* **18(3),** 316.

Moseman. J. G. (1968). *US Dept. Agric. CR42–68.*

Mulder, D. (ed.) (1979). *Soil disinfestation. Developments in agriculture and managed forest ecology,* Vol. 6, Elsevier, Amsterdam.

Munnecke, D. E. (1072). *Annu. Rev. Phytopathol.* **10,** 375.

Neergaard, P. (1977). *Seed pathology,* Vols 1 and 2, MacMillan, London.

Nutman, F. J. and Roberts, F. M. 1961). *Trans. Br. Mycol. Soc.* **44,** 511.

Ono, K. (1965). The rice blast disease. *Proc. Symp. IRRI, 1963,* 173.

Paulini, E. A., Matiello, J. B., and Paulino, A. J. (1976). *Quarto Congresso Brasileiro de Pesquisas Cafeeiras, Resumos,* 48.

Paulus, A. O., Nelson, J., Gafney, J., and Snyder, M. (1977). *Proc. 1977 Br. Crop Prot. Counc. Conf. – Pests Dis.,* 929.

Pereira, J. L. (1970). *Agric. Av.* **12,** 17.

—— Mapother, R., Cooke, B. K., and Griffiths, E. (1973). *Exp. Agric.* **9,** 209.

Price, D. and Dickinson, A. (1980). *Acta Hort.* **98,** 277.

Quinn, J. G., Johnstone, D. R., and Huntingdon, K. A. (1975). *PANS,* **21,** 388.

Ramos, A. H. and Shavdia, L. D. (1976). *Plant Dis. Rep.* **60,** 831.

Richardson, M. J. (1979). *Phytopathol. Pap.,* no. 23.

Rodriguez-Kabana, R. and Curl, E. A. (1980). *Annu. Rev. Phytopathol.* **18,** 311.

Russell, G. E. (1977). *Proc. 1977 Br. Crop Prot. Counc. Conf. – Pests Dis.,* 831.

Ryan, E. W. and Doyle, A. A. (1981). *Proc. 1981 Crop Prot. Counc. Conf. – Pests Dis,* 489.

Savory, B. M. (1966). *Research review no. 1.* Commonwealth Bureau of Horticulture and Plantation Crops, Maidstone.

Schwinn, F. J. (1981). In *The downy mildews* (ed. D. M. Spencer), p. 305. Academic Press, New York.

Shephard, M. C. (1981). *Proc. 1981 Br. Crop Prot. Coun. Conf. – Pests Dis.,* 711.

Siegel, M. R. and Sisler, H. D. (eds.) (1977). *Antifungal compounds.* Dekker, New York.

Simons, J. N. (1981). *Proc. 1981 Br. Crop Prot. Counc. Conf. – Pests and Dis.,* 413.

Smalley, E. B. (1971). *Phytopathology* **61,** 1351.

Sprague, R. (1961). *Recent advances in botany.* University of Toronto Press, Toronto.

Steiner, K. G. (1973). *Z. Pflanzen.* **80,** 671.

Stringer, A. and Loyns, C. H. (1977). *Pestic. Sci.* **8,** 647.

Taylor, J. D. and Dudley, L. C. (1977). *Ann. Appl. Biol.* **85,** 223.

Tomlinson, J. A. (1977). *Proc. 1977 Br. Crop. Prot Counc. Conf. – Pests Dis,* 807.

Torgeson, D. C. (ed.) (1967). *Fungicides: an advanced treatise,* Vol 1. Academic Press, New York.

—— (1969). *Fungicides: an advanced treatise,* Vol 2. Academic Press, New York.

Ulfvarson, U. (1969). In *Fungicides: an advanced treatise* (ed. D. C. Torgeson), Vol. 2, p. 303. Academic Press, New York.

Ullstrup, A. J. (1972). *Annu. Rev. Phytopathol.* **10**, 37.

Upstone, M. E. (1977). *Proc. 1977 Br. Crop. Prot. Counc. Conf. – Pests Dis.*, p. 197.

Van der Plank, J. E. (1963). *Plant diseases: epidemics and control.* Academic Press, New York.

—— (1967). In *Fungicides: an advanced treatise* (ed. D. C. Torgeson), Vol. 1, p. 63.

Van der Vossen, H. A. M. and Browning, G. (1978). *J. Hort. Sci.* **53**, 225.

Wainwright, M. (1978). *Soil Sci.* **29**, 287.

Waller, J. M. (1972). *Ann. Appl. Biol.* **71**, 1.

—— (1982). *Crop Prot.* **1**, 385.

Wastie, R. L. (1975). *PANS* **21**, 268.

Wheatley, G. A. (1977). *Proc. 1977 Br. Crop Prot. Counc. Conf. – Pests Dis.*, 973.

White, R. F. and Antoniw, J. F. (1981). *Proc. 1981 Br. Crop Prot. Counc. Conf. – Pests dis.*, 759.

Wilhelm, S. (1966). *Annu. Rev. Phytopathol.* **4**, 53.

Williams, R. J. and Singh, S. D. (1981). *Ann. Appl. Biol.* **97**, 263.

Wolfe, M. S. (1975). *Proc. 8th Br. Insectic, Fungic. conf. 1975*, 813.

—— (1981). *Philos. Trans. R. Soc. London Ser. B.* **295**, 175.

Zentmyer, G. A. (1978). *3rd Int. Plant Pathol. Congr., Munich. Abstracts,* 362.

Zitter, T. A. and Simons, J. N. (1980). *Annu. Rev. Phytopathol.* **18**, 289.

19
Weed control

A. J. LACEY

19.1 Introduction

Some people define weeds as plants growing in the wrong place, in the wrong quantity, at the wrong time; while others define them as plants competing with valued crops. This chapter aims to clarify the meaning of competition between weeds and crops, to outline the methods of measurement of the effect of competition, and to describe the techniques which are available to protect crops from weeds. Examples of both weeds and crops are drawn from the World scene. Weed problems and their control are listed for cereal crops, other arable root and vegetable crops, perennial and rangeland crops, and those in aquatic situations. Control by chemicals is surveyed, and the need for the integration of cultural practices, and particularly the need for control of weed seed dispersal as a complement to chemical control, is emphasized. The control of weeds to minimize their effects on the crop in which they are growing is the short-term need; the prevention of the spread of weeds from field to field, area to area, country to country, hemisphere to hemisphere, or from one agricultural system to another is the long-term requirement.

19.2 Competition

The meaning and significance of competition between crops and weeds may be illustrated by an experiment by Hewson (1971) on the growth of lettuce in populations of the weed *Chenopodium album*. Using 11 plants of lettuce per square metre, he showed that over a range of 0–38 plants/m² of *Chenopodium* weeds, although the combined biomass of lettuce and weeds produced per unit area remained about the same, the proportion of aerial fresh weight realized in the form of lettuce was reduced until, in the presence of 38 weed plants/m², the lettuce yield was only 10% of what it was in the absence of weeds, the balance being made up by the increasing weight of the weeds. He further found that the presence of only two *Chenopodium* plants/m² greatly reduced the number of marketable lettuces (by 58%) while reducing the lettuce weight by 22%. Thus the potential production is shared between crop and weeds but the yield is also determined by market factors.

The experimental measurement or evaluation of competition has often shown that intraspecific competition is greater than interspecific, although a notable exception is the comparative aggressiveness of wild oats, *Avena* sp. to other plants (Haizel and Harper 1973; Holroyd 1976; Wellbank 1963). The literature gives some details of yield losses for a variety of arable crops from a variety of climates and a selection is given in Table 19.1. Few data are available on the significance of weeds to pasture productivity, but King (1966) claimed that 8% of cattle were lost as a result of eating poisonous weeds of pastures, and Noda (1977) estimated that between 8 and 18% yield loss resulted from the presence of weeds in pastures. The position in grasslands is complicated by the comparisons that can be made between a permanent unimproved grassland with a range 0.5–3.0 g/m²/day dry matter production to that obtained by ploughing up and replacing with a single species grassland sward of *Lolium perenne, Anthoxanthemum odoratum,* or *Cynosurus cristatus* which will give a maximum productivity of 10 g/m²/day or 500–700 g/m²/year in temperate climates (Duffey *et al.* 1974). Available data suggest that a sward composed of a number of species is more ecologically stable and may give a yield higher than a rye grass/white clover

TABLE 19.1. *Yield losses (%) due to weeds in agricultural crops*

Crop	Yield loss (%)	Country, region, or climate	Weeds	Reference
Beets (sugar)	78–93*	Texas	Pigweed	Winter and Wiese (1982)
Cassava	92	Venezuela	Mixed	Moody and Ezumah (1974)
Cereals (general)	24	Temperate	Mixed	Kormso quoted in Hewson (1971)
Rice	30–73 (54)	Colombia	Mixed	Lange (1970) quoted by Holm (1976)
Sorghum	50–70	Tanzania/ Nigeria	*Striga*	Doggett (1970)
Cotton	90	Sudan	*Cyperus rotundus*	Hamdoun and El Tigani (1977)
Groundnuts	60–90	Sudan		Hamdaum and El Tigani (1977)
Onions	99	UK	Mixed	Hewson (1971)
Sweet Potatoes	78	W. Indies	Mixed	Moody and Ezumah (1974)
Yams	72	Nigeria	Mixed	Moody and Ezumah (1974)

* Loss of sucrose yield.

association even though some of the components may only be inferior grasses (Mathews 1977).

The measurement of the effect of weeds on yield is a complex one largely because the basis for comparison of yield is so difficult to determine. If the yield per unit area of a weed-free crop is to be determined, then a value may be obtained by extrapolation from yields of weedy crops of different densities of weed and constant density of crop plants as in the lettuce experiment referred to earlier. A totally weed-free crop achieved either by hand-weeding [see Peters (1972)] or by chemical means may not however give the best possible yield because of the density of the crop itself. The base line can be the 'totally' weedy crop, and then the improved yield obtained in the absence of weeds can be regarded as the percentage gain from weeding; the converse can be described as the loss of crop yield due to weeds.

Attempts have been made to find a common basis for the estimation of crop losses (FAO, 1971), and some authors have even considered the use of models to predict the effect of weed densities on crop yields (Chisaka 1977). Although considerable emphasis has been placed on the yield aspects of crop responses to the presence of weeds, the yield crop will be the final summation of a number of physiological factors which are themselves altered by the presence of weeds. Evans (1972) has shown that the morphology of the individual plant varies with its environment and that the plant's proportions, such as the distribution of dry weight into the stems, roots, leaves, and the ratio of leaf area to such parameters as plant weight or leaf weight, may be altered. Harper (1960) has shown that some weedy plants are very plastic in their response either to the presence of their own species or to the other species, varying in such parameters as numbers of tillers or capsules per plant and the number of seeds per capsule. There appears, however, so far to be little attempt to measure the proportionality of growth in crop plants in spite of the likelihood that responses to the presence of weeds would show up much more quickly than if the final yields are used as the measure of that response. Again the yield of a crop is the quantity produced per unit of area of land but the physiological responses to competition are likely to be in terms of weights, areas, numbers, etc. per plant or per unit dry weight of plant. The crops which are sold as individuals, for example lettuces, will be larger if the leaf weight is spread in as large a leaf area as possible, while in cabbages the weight, by which they are sold, will be the result of the proportion of plant weight in the form of leaves.

The impact of weeds on the leafiness of these two crops may be quite different. The investigation of the competitive ability of plants has been reviewed by Harper (1977) and he summarizes the experimental design used by various authors to separate root and shoot 'competition'. Burrill *et al.* (1977), Truelove (1977), Mercado and Manuel (1977), and FAO (1971) offer methods of evaluating competition between crops and weeds. The concept of 'critical period' in competition studies of Nieto *et al.* (1968) does not appear to have been developed very far, possibly for reasons not related to the effect on the immediate yield of crops but more for those concerning the long-term build up of a weed flora or even the avoidance of a social stigma in having a weedy field. The mechanism of interaction and competition ranges from direct parasitism, by such weeds as *Cuscuta* and *Striga*, through competitive uptake of nitrogen or the shading out of incoming light, to the exudation of growth inhibitors (allelopathy) by such weeds as *Pennisetum clandestinum* (Baker 1974; Harper 1977; Putnam and Duke 1978). The new short-strawed varieties of wheat and rice are less competitive than the earlier forms and may lead to increased problems (Parker 1977).

The ability of crop plants to control weeds is, however, well understood particularly in the case of plantation crops where growers confine weeding to the period before canopy closure. The environment in which the crop–weed interaction takes place is known to affect the relative competitiveness of the plants and the status of fertilizers, water balance, and grazing or other predatory pressures, is manipulated to the crop's advantage wherever possible. Soil disturbance is a feature of agricultural cropping and alien species aided by disturbance have often been seen to be aggressive to the disadvantage of native flora where agriculture has been newly introduced into an area. If the disturbance is reduced or removed the competitive ability of the native flora may reassert itself and suppress the alien one, provided sufficient seed sources of the native species remain.

19.3 World weeds

Weed floras are published for a number of areas of the world and are listed in a supplement to *PANS* Volume 17 (1971). Several countries or states declare lists of noxious weeds which are the subject of legislation such as the Vermin and Noxious Weeds Act 1958, Australia, and the Weeds Act 1959, UK. The word noxious appears, in the final analysis, to be applied to a weed which is the subject of control under the regulations. Any weed may be declared noxious and the lists vary in length—five in the UK to 93 in Victoria, Australia.

Cyperus rotundus has been declared the World's worst weed (Holm and Herberger 1970). Holm has also suggested that 206 species distributed in 59 plant families may have this title (Holm 1976) with about 80 of them, in three families, being of primary importance in that they devastate fields and waterways. The water weed *Eichhornia crassipes* has been blamed for losing 10% of the water in the Nile, equivalent to 7×10^9 cubic metres annually, through increased evapo-transpiration [Dissogi (1974) quoted by Hamdoun and El Tigani 1977], and this loss is in addition to its deleterious effects on irrigation systems, fishing activity, navigation of the river, and health by harbouring vectors of human disease organisms, all of which, together with its remarkable pan tropical spread, have earned it the name 'the million dollar' weed. The basis for grading the importance of weeds varies from assessment of the relative damage they do to food crops, waterways or grazing, through the extent of World distribution, to the impact they have on areas of unique vegetation. Hall and Boucher (1977) have stated that introduced trees and shrubs in South and S.W. Cape Province have in some cases reached exponential rates of growth, infesting and displacing on a vast scale the wild vegetation of that part of South Africa—a vegetation which is one of the six major subdivisions of the World's flora.

Some tropical weeds may be spread throughout the warm areas of the World, e.g. *Echinochloa crus-galli*, but fail to colonize significantly the temperate regions. Some temperate weeds, however, such as *Avena fatua* and *Polygonum convolvulus*, have become firmly established in such non-temperate places as East Africa. *Oxalis pes-caprae*, a native of South Africa, has become a significant World weed in temperate and subtropical areas of Europe, Eastern Asia, Africa, Australia, South America and western North America (Baker 1965). It is perhaps appropriate to note in the light of Hall and Boucher's (1977) comment above that another South African species *Chrysanthemoides monifera* (South African bone seed) has been declared a noxious weed in Victoria, Australia, and is regarded as a serious threat to the native vegetation of the Dartmouth area of Australia (Amor and Stevens 1976).

19.4   Chemical control of weeds

The numbers, quantities, and modes of use of chemicals for weed control rise annually, a trend which can be seen by whatever basis is used for comparison. In 1976 there were about one hundred chemically different herbicides commercially available around the World (Blair *et al.* 1976), while 450 officially approved herbicide formulations were applied annually to between 10.5 and 13.5 million acres (4.2–5.5 million hectares) of agricultural land in the UK (Fryer 1977). In 1982 in the UK alone there were 80 chemicals used in 437 formulations (Roberts 1982). The value of the World's market in herbicides was 1131 million dollars in 1971, rising to 2190 in 1974, and to a predicted level of 3422 in 1980, based on 1974 prices and user levels (Fryer 1977). There is still an exponential growth pattern in the number of patents filed for herbicide uses growing, for example from 50 in 1950 to 4700 in 1980 (Thomas 1982). The increases in herbicide usage is due to the application of established chemical controls to an increased area of new crops being treated together with an increase of total herbicides used, for example, in multispray programmes. There is, however, a decline in the rate of increase due perhaps to the increasing cost of herbicides as a result of, among other factors, the greatly increased costs of development (Hill 1982). Another factor which may play a part in the future is the greater precision in application by such developments as the controlled-droplet applicators (CDA) and the electro-statically charged droplets enabling about one-tenth to one-fortieth of the volume to be used to produce the same level of control (Parham 1982). In addition to the use of herbicides in new arable, glasshouse, and plantation crops, there is an increasing belief in the value of herbicide applications to grasslands; both for direct seed crops and also as crops for grazing, particularly in developed countries. Fryer (1977) estimated that only 10% of the area of the UK grassland was treated at that time. Mathews (1977) pointed out that 25% of all herbicides used in New Zealand were for brush control in pastoral lands. The potential herbicide usage for grasslands is therefore very large.

19.4.1. Weed-control techniques

Weed control encompasses:

    (1) the reduction of the competitive ability of an existing population of weeds in a crop,

    (2) the establishment of a barrier to the development of further significant weeds within that crop,

    (3) the prevention of weed problems in future crops either from the existing weed reservoir or from additions to that weed flora.

Chemical techniques are used primarily for the first of these objectives but also for the second, whilst cultural techniques have developed during the history of agriculture to keep weed growth to a minimum. Legal regulations concerning such aspects as crop seed purity, and importation of weed-seed-contaminated soils, have attempted to achieve the third. Cultural practices have been or are being modified or replaced to prevent the development of weeds but they are also being seen to interact with chemical procedures and the need for integration of control measures is increasingly being recognized (Fryer and Matsunaka 1977).

Modern herbicides began to appear in the 1930s and dinitrophenol (DNOC) was patented in 1933 in France for the control of weeds in cereals. At that time in most arable agricultural systems cereals were regarded as dirty crops, that is they led to the development of weeds, while other crops were rotated with them to facilitate weed control. The early development of herbicides was therefore to destroy selectively the broad-leaved weeds in cereal crops, the selection being based on the differential uptake of the chemicals when applied to the foliage. In the 1950s and 1960s a large proportion of the cereal crops in Britain, USA, and elsewhere was being sprayed with an increasing range of herbicidal chemicals to achieve control of a widening variety of weeds. In the 1960s developments of herbicides for use in crops other than cereals occurred, for example beans, potatoes, sugar beet, carrots, fruit and brassica crops. This development has continued and in the late 1970s it was probably true to say that there was a chemical which would give some weed control in almost any crop in the World. Comprehensive lists of herbicides and information about the chemistry, manufacture, principal use, toxicity (LD_{50}) and methods for analysis are to be found in the British Crop Protection Council (1977) *Pesticide Manual* or *Herbicidal Chemicals* published by the Weed Science Society of America. Summaries of chemical data are published by Green *et al.* (1977) and Fryer and Makepeace (1977). Statements on the uses of herbicides classified by both crop and weed with recommended dose, timing, and method of application are brought together in Fryer and Makepeace (1978). Additional specific weed control by the use of specific chemicals is given by commercial literature, by research journals and by such journals as *Tropical Pest Management* (formerly *PANS*). The major sources of information on the biochemistry and mode of action are those by Ashton and Crofts (1973), Audus (1976), and Corbett (1974) in which there are also accounts of the methods for the screening and development of herbicides and discussions on the basis of selectivity.

Herbicides may be applied to the soil for the control of weeds even before they emerge as recognizable plants, or to their foliage after they have appeared, either as germination inhibitors, or toxic to the underground stages of the weed or as contact or translocated destructive agents. Where there is no crop present and provided that there is no residual effect left in or on the soil, there need be no selectivity. If, however, the crop seed is also in the soil, or the crop plant has emerged, then there must be a selective effect damaging to the weed but not significantly damaging to the crop. The variation of applications for weed control in crops is summarized in Fig. 19.1. The terms are classified by reference to the crop; thus pre-sowing, pre-emergent or post-emergent refer to the crop. Clearly the basis of selectivity can take account of different relative stages of the crop and weed; for example, pre-emergent contact herbicide treatment can make use of the fact that the crop is not exposed to the chemical being protected by the soil, or a germination inhibitor may be used on the soil after the crop has emerged. Selectivity on the basis of differential uptake, or differential biochemical mode of action, will be needed if the herbicide is applied post-emergence to the foliage. The use of directed application relies on the selectivity achieved by the operator and may involve the use of masking guards for the crop plants. The herbicide may be very short lived, being either rendered inactive after contact with the soil or lost by leaching, volatilization or biodegradation, or it may be a long-lasting residual, clearly again requiring differential effects on the crop and weed. Total vegetation control, for example, in railway tracks or industrial yards would require long-lasting but non-selective herbicides. The persistence of the herbicide can be altered by its

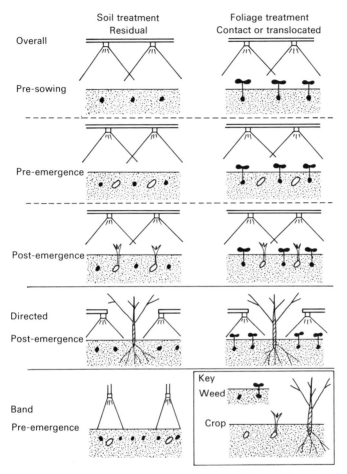

Fig. 19.1. Situations in which herbicides may be used for weed control in crops (after Fryer and Makepeace 1977).

formulation and whether or not it is incorporated into the soil. Climatic factors such as temperature, degree of soil moisture and the proximity and intensity of rainfall to the application may greatly alter the effectiveness of a herbicide applied either as a soil or foliage treatment. The tolerance of crops is important at the varietal level [see for example Table 9 in MAFF (1978)] for treatment of that particular crop, but residues of herbicides from those crops must also not damage any succeeding crops.

19.4.2   Weed control in cereals

Lists of herbicides for specific crops and even for specific weeds are available from a number of sources including, of course, the manufacturers, the agricultural advisory services, and, in a more general form, in such publications as the *British Crop Protection Council's Weed Control Handbook* (Fryer and Makepeace 1977; Roberts 1982). Fryer and Makepeace (1977) list for example, one pre-drilling, seven pre-emergence, and 27 post-emergence herbicides for the control of weeds in British cereals—principally barley and wheat—while Noda (1977) gives 17 herbicides for use in rice on a World basis, and Romanowski (1977) cites 10 pre-plant, 13 pre-emergence, and 9 post-emergence herbicides for use in the World's maize crops.

The whole history of herbicide development since 1950, illustrated particularly in small-grain cereals in the UK, has been one of adjustment to changing weed flora. After 2,4-D and MCPA had been in use for some years, mecoprop and 2,3,6-TBA were developed to control *Stellaria media* and *Galium aparine*, and then dichlorprop and dicamba for control of *Polygonum* spp. Later still ioxynil was introduced to help control may-weeds (*Matricaria* and *Tripleurospermum* spp.). Meanwhile there was a continuous increase in the importance of grass weeds, particularly *Avena fatua*, necessitating the development initially of barban and triallate and later benzoylprop-ethyl, chlorfenprop, flamprop-methyl, and difenzoquat. However, not all these compounds control other grass weeds such as *Alopeurus myosuroides*, and substituted triazine and urea herbicides have been introduced to combat these (Parker 1977).

Comparable changes have taken place elsewhere and in other crops; the annual grass weeds *Digitaria sanguinalis*, Setaria *sp.*, and *Panicum dichotomiflorum* have become common in maize in the USA under regimes of triazine and now are controlled by alachlor. A further problem in both maize and sugarcane treated with both atrazine and alachlor is the build-up of the resistant weed, *Rottboelia exaltata*.

There have been changes in the annual weed flora of horticultural as in other crops in Britain over about 20 years, and Davison and Roberts (1976) have indicated that such changes, as seen in Table 19.2, have been brought about by factors such as the merging of vegetable cultivation into rotation with cereals in arable farming. *Matricaria matricarioides* and *Galinsoga parviflora* have tended to increase in intensive vegetable holdings due to weed control based primarily on chlorpropham formulations to which they are resistant. Paraquat, introduced in 1958, was a most significant development in herbicides since it allowed new methods of cultivation such as direct-drilling and minimum cultivation to be established with their resultant effects on weed seed dynamics which are discussed in a later section. Paraquat is a contact herbicide, and, although effective against nearly all green aerial parts of plants, is not effective against their underground portions. Glyphosate, introduced in 1971, although acting against the chlorophyll system of plants, is more effective in the control of perennating weeds such as *Agropyron repens* because it is translocated to the underground parts. Its full potential as

TABLE 19.2. *Annual weeds of vegetable crops in the UK in descending order of frequency*

1953*	1959–63†	1970–75‡
Stellaria media	Stellaria media	Poa annua
Urtica urens	Poa annua	Matricaria spp.
Chenopodium album	Urtica urens	Stellaria media
Poa annua	Senecio vulgaris	Polygonum aviculare
Senecio vulgaris	Capsella bursa-pastoris	Chenopodium album
Polygonum persicaria	Chenopodium album	Urtica urens
Capsella bursa-pastoris	Matricaria spp.	Capsella bursa-pastoris
Veronica spp.	Veronica persica	Veronica persica
Matricaria spp.	Polygonum aviculare	Viola arvensis

* Questionnaire survey (Roberts 1954).
†Viable seeds in soils, 58 fields (Roberts and Stokes 1966).
‡Viable seeds in soils, 83 fields (H. A. Roberts and P. M. Lockett unpublished).
From Davison and Roierts (1976).

a herbicide has yet to be achieved, and its mode of action yet to be fully understood (see for example Kitchen *et al.* 1981).

When cereals undersown with clovers or other legumes, or pastures containing these as an integral part of the sward, are to be treated, the number of herbicides available is drastically reduced, for example, to only five in the list given earlier by Fryer and Makepeace (1977). Parker (1977) suggested that the changes in weeds in rice had been slower than in some crops because propanil being used to control an annual weed flora had been controlling the annual grasses as well. The red and wild rice *Oryza punctata* has begun to emerge as a weed problem in rice cultivation, however, and such developments require an even greater degree of selectivity in the herbicide as the weed is physiologically closely allied to the crop.

The degree of selectivity can be improved by the use of crop protectants or 'safeners' and the chemistry and mode of action of these substances is reviewed by Pallos and Casida (1978), while Ko Wakabayashi and Matsunaka (1982) review their use in Japan. Romanowski (1977) includes in his list of herbicides for use in maize the compound EPTC (*S*-ethyldipropylthiol carbamate) which is a germination inhibitor. It was originally used as a preplant herbicide for a wide range of crops but now can be used selectively if the crop is protected. The maize can be protected by R 25788 (*NN*-diallyl 2,2-dichloroacetamide) against EPTC, and the sown crop even sprayed with a tank-mix of the EPTC with antidote. A seed dressing of 1,8-naphthalic anhydride (NA) can give a similar protection to rice to enable molinate and alachlor to control selectively the rice weed *Oryza rufipogon* (Fryer 1977) and to protect sorghum, maize, and oats from other herbicides used for the control of annual grass weeds (Blair *et al.* 1976).

Immunity from herbicide damage can be put into a crop at a much earlier stage by the use of Seed-Parent Treatment (SPT). Terbutryne is used in Israel as a pre- and post-emergent herbicide to control *Phalaris paradoxa* and *P. brachystachys* in wheat, but causes some crop damage. Resistance can be conferred to the crop by treatment of the parents of the wheat seed with chlormequat at the soft-dough stage at a rate of 16 kg/ha (Pinthus 1972).

The tolerance of cereal crop plants to growth-regulator herbicides is related to the stage of development reached by the apical meristems. For example, some herbicides, such as 2,4-D and MCPA, can cause ear deformities if applied before the spikelets of the young ears have been determined. Sprayed at a later stage, when the cells are dividing to form the pollen and ovules, they sometimes lead to sterility. Other herbicides, the benzoic acid derivatives in particular, can cause drastic depression in yield when sprayed during jointing while the anther and carpel primordia are developing (Tottman 1976).

The change of crop sensitivity to herbicides has been recognized for a considerable time and has led to the development of recommendations for the application of herbicides to increasingly precise stages of the crop. The decimal code description of cereal stages known as the ZCK scale is used in commercial literature and has been reviewed by Tottman and Makepeace (1977). MCPA should be applied on winter cereals at the stage from pseudostem erection (ZCK 30) to that of the detection of the first node (ZCK 31) and on spring cereals after the fifth leaf is unfolded (ZCK 15) but before the first node is detected. The 1978 edition of the *Weed Control Handbook* recommends the use of terbutryne (at 1.5 kg/ha) on winter-sown wheat and barley as a pre-emergent herbicide when the crop is at the stage between the radicle emerging from the caryopsis and the coleoptile doing so, i.e. ZCK 06. The time intervals between stages of any crop are of course dependent on weather and other factors and can be very short in some cases and therefore are particularly prone to problems of suitable weather for spraying. Some success in overcoming the latter is likely however with the introduction of controlled drop applications (CDA) in which the much reduced volumes require much less heavy equipment, thereby allowing access to wet land (Wilson and Taylor 1978) (see also Chapter 4).

Improved control, with less damage to the cereal crop, of volunteer potatoes, or ground keepers, has been achieved by applying glyphosate or aminotriazole particularly with ammonium sulphate in the previous autumn. The potatoes otherwise do not appear until May or early June by which time the crop is at a stage sensitive to these herbicides (Lutman and Richardson 1978).

The stimulation of weeds to grow so that they may be destroyed, discussed more fully in the section on weed seeds, has found some application in the control of *Cyperus esculentus* a weed of rice, soy-beans, and many other crops. Naptalam affects the formation and germination of tubers of this plant, the physiological changes being also related to the photoperiod. Tuber formation may be inhibited by application of naptalam at the end of the summer, and a second application of naptalam in the following spring, after new shoots have emerged, may prevent new rhizome formation and induce further shoot development which would be destroyed by glyphosate applied in mid-summer (Appleby and Paller 1978).

The early success of herbicides for weed control in cereals led to the abandonment of rotation of crops and the development of cereal monocultural systems of agriculture. The experience, however, described by Burnside (1978), for sorghum and for soy-beans illustrates the need for integrated thinking to achieve satisfactory weed control and the usefulness of reintroducing rotations to effect that control of weeds. Atrazine was first recommended in 1965 to Nebraska, USA farmers as a herbicide applied pre-plant, or pre-emergence, to sorghum for control of most annual broad-leaved weeds and some annual grass weeds. Trifluralin was also first recommended in 1965 to those farmers as a pre-plant and soil-incorporated control of most annual grass and some annual broad-leaf weeds in soy-bean production. With expanded use of atrazine in sorghum fields and trifluralin on soy-bean fields, shifts in annual weed population began to occur. Some annual grass weeds were increasing in the sorghum and some, particularly large-seeded broad-leaf weeds, were increasing in the soy-bean. A rotation of these crops was suggested in order to achieve a high level of weed control in both crops, and experiments to test this control were carried through from 1968 to 1974 successfully.

The use of rotation for arable crops in Britain has been reviewed by Cussans (1976), and he points out that the use of oil-seed rape in which carbetamide, propyzamide, and dalapon can be used to control wild-oats and black-grass, is a herbicidal rotation crop with cereals, although the rape itself is a valuable crop.

If the crops are selected only on value, however, very short rotations may result. The problem of volunteer crops then arises and is particularly well known in the case of potatoes in rotation with sugar beet and cereals. A promising rotation of winter wheat and spring barley for the control of wild-oats and black-grass weeds as well as diseases was quickly rejected because of the volunteer barley becoming a serious weed of that wheat (Cussans 1976). The development of herbicides to control volunteer crops has of course, all the problems of selectivity but additionally the use of them in short rotations has the added risk of carry over into the next crop.

19.4.3 *Herbicides for other annual crops*

To summarize the use of herbicides in annual crops other than cereals presents a formidable task in a short article. Romanowski (1977) listed the selective herbicides used in root and vegetable crops in Great Britain, USA and the tropics. Kasasian (1971), in reviewing the control of weeds in tropical areas, has included extensive comments on both the principal weeds and the chemicals used for their control.

Treatments of crops with a sequence of herbicides is sometimes needed because of the narrow range of activity of some chemicals, and sequential herbicide treatments are commonly used in the culture of carrots and also on onion and sugar beet. In carrot crops, linuron is used for pre-emergence control and when carrots have reached the pencil stage a mixture of chlorbromuron–metoxuron is applied to control emerged knot-grass and mayweed. Problems of synergistic effects or accumulative effects of chemicals from different sources are largely unknown (Scarr 1976).

Chemical weed control in potatoes (*Solanum tuberosum*) came late because in rotational agricultural systems this crop has been regarded as a cleaning crop whereby weeds could be cleared by inter-row cultivations. In recent years, however, chemical control has developed in which glyphosate, TCA, and EPTC are used to control *Agropyron repens* and other weeds in pre-planting treatments, while barban, diclofop-methyl, MCPA salt, and metribuzin are used post-emergence. A large number of herbicides are now available for use on different soil types, weeds, and varieties of crop, pre-emergence, including paraquat for contact and barban, dalapon sodium, diclofop-methyl for translocated, chlorbromuron, dinoseb in oil, dinoseb acetate plus monolinuron, linuron on its own or in mixture either with cyanazine monolinuron or trietazine, and metribuzin, prometryne, terbutryne with terbuythylazine for contact and residual foliar and soil activity. Additional herbicidal chemicals are for haulm desiccation, using diquat dinoseb in oil or sulphuric acid, and for post-senescence weed control dalapon sodium as a translocatable herbicide. It has already been mentioned that potatoes may themselves be weeds in such following crops as cereals, beans, or carrots, and in such cases glyphosate and aminotriazole can be used to control them.

Cotton (*Gossypium hirsutum* and *G. barbardense*), although a perennial plant, is mostly grown as an annual, thus requiring removal after harvest. It carries the doubtful distinction of having more pesticides applied to it than any other crop, principally in the form of insecticides but also herbicides for weed control. Kasasian (1971), reviewing the control of weeds in cotton, indicates that the crop appears to have a critical period between 20 and 40 days after sowing in which the presence of weeds may reduce the first yield by as much as 75%. The value of the crop is also reduced by the presence of weeds at harvesting, debris from which interferes with the preparation of the fibre.

The use of defoliants of the crop plants themselves, as well as for the control of weeds in those crops, is now widespread. The presence of leaves on cotton, vine, and potato can be a nuisance in harvesting, and chemicals to bring about pre-harvest leaf fall have been described by Thomas (1982) for cotton and Costa and Intrieri (1982) for vine, while those for potato have already been mentioned. The desiccation of potato haulms pre-harvest can be used to restrict the size of the tubers produced but is more frequently done to prevent the contamination of the tubers by the late blight fungus disease organism.

The weeds of horticultural crops are often annuals but of indeterminate growth, beginning to flower when quite small, flowering and setting seed simultaneously and continuing to do so until some extrinsic factor, such as frost or drought, causes them to die, *Veronica*, *Poa annua*, and *Senecio vulgaris* are three examples. They appear to be particularly well adapted to the wide range of cultivation rotations that are a feature of such crops. Harper (1977) has implied that the population dynamics of such species would make an interesting comparison with those of determinate annuals. It is not altogether clear as to whether there is a rapid turnover of generations in such plants or whether there are ecotype variations within the populations.

19.4.4 Weed control in plantation and perennial crops

The weed problem in a plantation crop may be divided into those associated with the preparation of the ground prior to planting, those in the establishment stages of the crop before the canopy has closed and those in the long-standing crop itself, particularly if planted in widely spaced rows. Conditions in localities used for plantation crops are often those in which soil erosion by wind or water is a serious problem, and the areas may be large or with difficult access. In tropical plantation crops, the use of cover plants between rows is widely practised either to prevent erosion or loss of soil condition. 'Soft' grass component of the natural vegetation, or legumes such as *Pueraria phaseoloides*, *Centrosema pubescens*, and *Calapogonium mucunoides* are sown in order to enhance the nitrogen component of the soil, while in Malaysia and Papua/New Guinea, the planting of cocoa in the inter-row areas of coconut enables further financial returns to be made on the crop area. Chemical weed control in the legumes has been developed by the use of a paraquat/diuron mixture known as Para-Col, and by directed spraying between grouped drills of legumes, considerable improvement in weed control has been affected (Seth 1977). Stalder *et al.* (1977) suggested that the annual weeds *Stellaria media*, *Veronic apersica*, and *Lamium purpureum* are helpful in supplementing the MCPB, glyphosate, and paraquat control of *Calystegia* spp. in Swiss vineyards. Clearly there are economic advantages if the groundcover can be a useful crop.

Residual herbicide treatment of main crops has involved the use of simazine and atrazine in tea, coffee, and oil palm, while diuron has been used in rubber, oil palm, and sugarcane. However, some phytotoxicity to young crops has been reported, for example, with diuron on cocoa, tea, and bananas grown under certain soil conditions; these effects may not be visible for some considerable time.

The use of paraquat as a post-emergent herbicide has increased since the mid-1960s in a wide range of crops particularly because of its wide spectrum of activity and rapid action, its inactivation on contact with the soil, the tolerance of trees and shrubs protected by their bark, its rainfastness, and safety in use (Seth 1977). Glyphosate is likely to be an even more valuable herbicide particularly for the control of perennial weeds in these conditions. The development of mixtures of herbicides based on paraquat with photo-synthesis inhibitors such as diuron has shown some promise in the control of *Paspalum conjugatum* and other weeds in Malaya.

Urea and triazine are the principal herbicides used in the control of weeds in pineapple, particularly in Hawaii. Soil applications to prevent weed seedling growth based on long-residual 'broad-spectrum' (diuron, atrazine, and ametryn) and foliar treatments to remove emerged weeds are often sprayed from aircraft over irrigated areas and from tractor-mounted sprayers on non-irrigated areas of sugar cane in that country (Romanowski 1977).

In permanent plantations the concept of crop rotation is not a practical one and the continued use of a few herbicides may well bring about the development of new weeds. *Hypericum humifusem* has become a dominant weed in some apple plantations in Ireland because of its tolerance of simazine and paraquat. A rotation of herbicides is perhaps more appropriate in place of a rotation of crops (Robinson 1978). Weed control in fruit crops has been reviewed by Davison (1978) for British situations, and he has suggested that the use of herbicides has permitted some new growing techniques such as early strawberry production in low tunnels and blackcurrants in beds. The development of mechanical harvesters in strawberries, apples, and raspberries will necessitate the control of unwanted vegetation at that stage of the crop. Simazine is the most important component of most programmes but changes from spring applications to autumn ones have occurred to control late-germinating annual weeds while remaining at sufficient concentration to control the following spring-germinating weeds. An added advantage resulting from this timing is that foliage-applied herbicides for perennial weeds can be sprayed at a later more effective time. One of the most difficult to control is *Agropyron repens,* particularly in strawberries, but alloxydin-sodium is a possible herbicide for this situation. *Convolvulus arvensis, Cirsium arvense,* and *Equisetum arvense* are long-standing troublesome weeds, but perhaps new weed problems involving *Sedum acre* and *Endymion nonscriptus* may develop in cane fruits (Davison 1978).

Weed control in forestry systems has many aspects similar to that of other plantation crops, particularly those associated with establishment, although the value of the final crop timber may be reduced if there is a heavy weed ground flora presenting difficulties of access and harvesting. The economics of weed control in forestry has the additional problem of a long interval before the realization of the cost-benefit can be attained. The use of natural regeneration systems in forestry is attractive on ecological grounds although less versatile and flexible in choice of crop than systems based on either chemical or manual weeding control. Elimination of previous crop trees with their competitive coppice growth from the newly planted crop involves the spot application of herbicides. Again large areas, difficult terrain, and shortage of water are factors influencing choice of species, feasibility, and value of weed control. The review by Wittering (1974) for British situations takes a work study approach, while Fryer and Makepeace (1978) include recommendation of herbicides for weed control in forest nurseries. The thinning of a plantation is, of course, a requirement to reduce intraspecific competition and to realize the site potential in the highest-value crop possible.

19.4.5   Weed control in grasslands

The formulation and solution of weed problems in grasslands can involve a number of ecological principles including those of assessment of trends and underlying causes of changes in plant populations. Examples are chosen here to illustrate some of those principles, and for further information the reader should see Auld (1978), Duffey *et al.* (1974), and Harper (1977). The presence of a weed in a pasture is often considered to indicate a factor of the ground or of the type of grazing regime. For example, *Juncus* spp. indicates poor drainage, *Senecio jacobaea* and *Rumex* sp. indicate overgrazing, *Pteridium aquilinum* in mountain regions indicates higher soil fertility and absence of frost or wind damage, while *Urtica* is said to be linked with loose soil. The development of coarse grasses and scrubland in an otherwise herb-rich chalk grassland indicates either undergrazing or wrongly timed grazing. The response of species mixtures in grasslands to pH adjustment by liming and to fertilizer treatments has been the subject of the long-term experiments done at Park Grass in Rothamsted. In some plots, the original sixty species of grasses and herbs have been reduced to a single-species stand. This response to fertilizer and other management regimes of mixtures of species can be considered in two ways; if the species-richness is an undesirable feature, say in a short-term *Lolium*/clover ley, then the new species coming in, in response to reducing fertility, will be regarded as weeds. If, however, the species-richness is a desirable feature, for example in a nature reserve, conserved either for the plants themselves or the fauna which they support, the fall in numbers of species will be seen as unfortunate, although the actual dry weight productivity of the remaining few species may have risen.

The benefits of weed control in agricultural grassland have been discussed by Drummond and Soper (1978) and Williams (1980) by reference to *Pteridium aquilinum* in upland areas of Britain where this species is, in Scotland, a serious nuisance over 160 000 hectares. Aerial application of the sodium salt of asulam at 4 kg/ha applied to six farms showed some very good control with recognizable improvement in profitability in the rearing of both sheep and cattle, particularly where follow-up treatment with lime and fertilizers occurred. In New Zealand barley grass (*Hordeum* spp.) is known to damage lambs physically by affecting eyes, nostrils, lip, gum, and skin, which will in turn lead to depression of the lamb growth rate or death, but despite these findings, selective control of barley grass with herbicides has not led to obvious increases in lamb weight gains or in the dry matter production [Hartley and Atkinson (1972) reported in Mathews 1977].

Although docks (*Rumex* sp.) have always been regarded as serious weeds and often seen as indicators of poor management of pastures, increasing in numbers due to poor competitive ability of the rest of the sward or due to avoidance by animals, they are linked in intensively used grass with high nitrogen levels in soil (Davies 1976) particularly in dairy farm situations. There is some evidence that *Rumex* is not entirely without value as the dairy animal might take it in, perhaps to redress some imbalance in the diet in high-grade ryegrass pastures. The animal may, however, be merely responding to the morphology of the *Rumex* as it becomes stemmed and rises above the ryegrass (Courtney and Johnston 1978). MCPA salt or 2,4-D amine or the two together have been recommended for the control of *Rumex* spp. in grassland; but mecoprop, dicamba, and asulam, often in mixtures, are also used. Where clovers are present in grass/legume swards, *Rumex* may be controlled with benazolin salt 0.42 kg/ha with MCPA 0.63 kg/ha and 2,4-DB 3.5 kg/ha, but care in controlling the relative heights of the crop/weeds by mowing is required for success in this programme (Fryer and Makepeace 1978).

The chemical control of the poisonous weed *Senecio jacobaea* by picloram, sodium chlorate, 2,4-D, dicamba, asulam, dichlobenil, chlorthiamid, 2,4,5-T, and diquat has been mentioned for New Zealand conditions (Mathews 1977), and where selective control is required in grassland in the UK, the MCPA salt and 2,4-D amine at 2.2 kg/ha or 2,4-D ester at 1.6 kg/ha have been considered to be satisfactory up to the early bud stage of growth of the ragwort. This plant is widely distributed and is a serious weed in rangelands of Australia, South Africa, South America, and North Western USA, as well as in New Zealand and the UK. Populations of the weed may fluctuate enormously from year to year and the factors causing these changes have been the subject of experiments in biological control. Partially successful control of ragwort (*Senecio jacobaeae*) has been achieved by the introduction of the insects *Tyria jacobaeae* (cinnabar moth), the ragwort fly *Hylemya seneciella* (Meade), and the ragwort flea beetle *Longitarsus jacobaeae* (Waterhouse) [see Andres (1977), for a review], while Dempster and Lakhani (1979) have shown, by the use of a model, that the relative populations of this weed species and its insect predators are modified in the UK by land drainage, climate, and grazing pressures on the pastures. *Hypericum perforatum* is a serious weed of rangelands, due partly to its ability to push out desirable grasses but also to its ability, reviewed by Giese (1980), to photosensitize the skin of grazing animals; it has been successfully controlled by insects in some areas of the World but not in others. Between 1900 and 1944, this weed infested an area of about 800 000 ha in California; several species of controlling insects were released between 1944 and 1950 and the weed was reduced to a road-side plant in about 10 years, principally by the attacks of the beetle *Chrysolina quadrigemina*. The population dynamics of this insect, together with *C. hyperici, Agrilus hyperici,* and Zeuxidiplosis giardi on *Hypericum* have been described by Harper (1977) as 'perhaps providing the most exciting experiment in the whole of the science of plant–animal relationships'. Raghavan *et al.* (1982) have reviewed the use of plant pathogens for biological control of weeds, and Bruzzese (1982) has suggested that the fungus *Phragmidium violaceum* shows promise for the control of *Rubus* sp. in Australia.

In grassland or rangeland, shrubby or woody plants represent some of the most difficult problems. Very large areas of rangelands may be affected, and the weeds themselves are often extensive, frequently forming a mosaic of thicket with grassland, sometimes of considerable height presenting difficulties of coverage with herbicide. Combellack estimated in 1978 that about 31% of the work time of the weed-control programme of the Lands Department Victoria, was spent in spraying 20.1 million litres of herbicide annually as a spot treatment of the weed *Rubus fruticosus* in that area of Australia. The sagebrush *Artemisia tridentata* is an undesirable plant in several millions of hectares in the sagebrush/grasslands of Nevada, USA, presenting difficulties for the upgrading of these pasturelands (Young and Evans 1978).

Rubus fruticosus is a problem in Britain in forestry but also occurs wherever grasslands are left undergrazed or unoccupied. In Britain ammonium sulphamate is used to control *Rubus* spp. either as a spray or as crystals applied to cut stumps in uncultivated land. In forestry, aerial applications including the use of 'invert' emulsions (water drops in an oil phase) to reduce drop break-up, have been tried. Rogers (1975) has reviewed the use of CDA for the application of 2,4,5-T as a control for this and other woody species in forestry, while Combellack (1978) has similarly studied its use in control of *Rubus, Ulex europaeus,* and *Rosa rubigenosa* in Victoria rangelands, with 2,4,5-T iso-octyl ester and 2,4,5-T butyl ester. Considerable savings of herbicide result from such application which achieve control comparable with other application techniques.

Scifres et al. (1978) describe the aerial application of karbutilate for the control of honey mesquite (*Procopis juliflora*) in Texas, dropped in the form of spheres 1.34 cm in diameter; although these fall at random they form a grid of herbicide concentration to which the deeper rooted mesquite will be susceptible, while damage to the shallow-rooted desirable grasses will be confined to the areas immediately alongside the spheres.

Perhaps one of the best-known rangeland weed problems has been that in Queensland, Australia, where prickly pear *Opuntia* sp. occupied a strip of land approximately 320 km wide extending from 20 to 33° South Latitude, i.e. about 1450 km, at an infestation rate of up to 1200–2000 tonnes/hectare fresh weight. Its area of infestation was by 1925 about 24 million hectares, often forming a thicket between 1 and 2 m high, impenetrable to Man or animal (Dodd 1940). We have already mentioned the use of biological control for *Senecio jacobaea* and *Hypericum* in grasslands: in the case of scrub, the control of *Opuntia* sp. by the larva of the moth *Cactoblastis cactorum*, a native of Argentina and South America in general, has been perhaps the greatest success story, achieving virtually complete suppression by 1940.

Because of the huge areas involved and because the weeds in rangelands are in a dynamic vegetation system often influenced by fire, grazing, drought, and other ecological factors, the use of biological methods of control is always likely to be preferable on grounds of cost, but very great care must be shown in selecting the agent of control, as once released it should confine its effect entirely to the weeds.

The weed floras of grasslands show changes; the control of one weed seems to expose the need for the control of another. Mathews (1977) has indicated that the annual *Carduus tenuiflorus* is being replaced in New Zealand by *Carduus pycnocephalus*, which is more difficult to control as a biennial or perennial than the one it has replaced. The new spread of weeds into rangelands is recognized as a possible problem arising from tourist use of roads and camping grounds. Amor and Stevens (1976) describe the spread of *Rubus* in Australia by this means; and Ivens (1974) suggested that the spread of *Eupatorium odoratum* has been similarly facilitated by new roads in Nigeria. Changes in agricultural systems may bring about new weed problems, Mathews (1977) suggesting that higher fertility levels in grassland and the change from sheep rearing to cattle or dairy grazing has bought about changes in rangeland weeds of New Zealand, while Harper (1977) has suggested that the spread of *Rubus fruticosus* in the New Forest area of Britain is a result of changed grazing patterns associated with the genetic improvement of the ponies roaming there by the introduction of Arab stock.

The upgrading of grassland by changing the species composition can be achieved by a one-pass technique, developed by the Weed Research Organization for the UK, but perhaps having considerable potential in other areas of the World. A slot 2.5 cm wide is cut in the centre of a 10 cm band and laid to one side, and seed of the species to be introduced, together with fertilizer and slug pellets, are placed in the slot. The band, having been sprayed by a herbicide, allows the seed to establish without immediate competition. The technique is cheaper than total reseeding and allows grazing to recommence almost at once (Anon. 1977).

Where grassland is part of an embankment for railways or roadsides there is a requirement, for visibility reasons, not only to suppress the growth of tall dicotyledonous plants but also to suppress the growth of the grass itself. Maleic hydrazide has been tried in combination with 2,4-D in order to suppress the growth and to remove the dicotyledonous species. An additional consideration of importance in the management of extensive grasslands may be the need for the conservation of wildlife. In this situation a suppression of growth consistent with safe transport is required but must be achieved with a maximum diversity of species and habitat. The herbicides used and the spraying and mowing may be adjusted to facilitate conservation aspects (Way 1973). The control of scrub in a grassland nature reserve is an extension of this concept of the maintenance of diversity, and is discussed by Duffey *et al.* (1974).

19.4.6   *Herbicides in aquatic situations*

The greatest care is required in the use of herbicides for the control of vegetation in or near water as the water itself may be utilized for drinking by Man and animals, as well as being itself a habitat for wildlife. The use of herbicides in water catchment areas may affect the environment some distance downstream, a danger, of course, not unique to herbicides but applies to all pesticides. A further problem of the use of herbicides and mechanical cutting of weeds in water is the depletion of oxygen which can be caused by the dead vegetation.

Soerjani (1977) listed 18 herbicides as potentially useful in aquatic weed control in the Rawa Pening area of Indonesia, while British recommendations for control of emergent weeds and waterside plants listed by Fryer and Makepeace (1978) include dalapon-sodium, paraquat, 2,4-D amine, maleic hydrazide, and, for floating weeds and algae, diquat, dichlobenil, chlorthiamid, 2,4-D amine, terbutryne, and glyphosate, but warnings are given that these herbicides should be used only in formulations which bear labels specifically permitting their use in or near water.

The increase of plant growth in water enriched by nutrients (eutrophic waters), often derived from Man's agricultural land fertilizer programmes but also from increasing quantities of domestic and other sewage effluent flowing into rivers, has led to problems of filtering in water works where phytoplankton may block filters or cause taste problems. Other areas of the World have massive problems associated with water weeds, such as the water hyacinth *Eichhornia crassipes* and *Salvinia molesta*, which reduce water-storage capacity and cause water loss by increased evapotranspiration. Interference in fishing, transport, and drainage are three other ways which these and other aquatic weeds can cause trouble.

Chemical control of aquatic weeds is often ineffective because of the rapid dispersion of herbicides. Considerable interest has been shown in slow-release systems in which the herbicide is lodged in floating or sinking pellet media. Floating strips of natural rubber matrix containing dissolved 2,4-D butoxyethanol eliminated *Eichhornia crassipes* in 7 months in one experiment (Thompson 1975).

The change of species of plankton or macrophytes has been the subject of much concern to nature conservationists in many areas of the World (for example Lund 1971; National Academy of Sciences 1969) with, at present, little hope of direct control.

19.4.7. Herbicides as growth regulators

Herbicides acting by hormonal control, such as 2,4-D, 2,4,5-T and maleic hydrazide, are members of a group of chemicals becoming increasingly important in manipulating the growth of plants in which the death of the plant is the extreme effect. A number have been used as growth suppressants for some time. One example mentioned earlier is the use of 2,4-D and maleic hydrazide for the control of roadside vegetation. Maleic hydrazide has also been used to suppress lateral shoots in tobacco after the flowering heads have been nipped out (Green *et al.* 1977). Compounds which partially destroy plants by, for example, inhibiting apical growth are known as morphactins. Fryer and Makepeace (1977) have listed some of the growth regulators, and their potential and practical uses are reviewed by Thomas (1982). Such compounds often require to be minimally translocated so that the effect in the plant is essentially local.

19.5   Weed seeds

In a few cases weed problems are due to the seeds or fruits themselves as, for example, in the case of the puncture vine *Tribulus terrestris* which has hard sharp fruits, while *Hordeum* spp. (Barley grass) causes damage to the lambs through its spikelets (fruits), and Holm (1976) has reported that direct human poisoning is suspected from the eating of the seeds of *Heliotropium eichwaldii* consumed as a contaminant of wheat.

An important contribution to the understanding of long-term weed control, however, is the understanding of the nature, number, and behaviour of weed seeds both in and on the soil. The seed population levels or 'seed banks' in the soil have been extensively described by Harper (1977) (see, for example, Table 19.3) and Roberts (1981), who, with others, have used flow diagrams and mathematical models to depict the changes and factors

TABLE 19.3. *Some examples of buried seed populations [data from Harper (1977) with permission]*

Area studied	Vegetation	Seeds/m^2	Reference*
Prairies of Great Plains, USA	Native prairie	300–800	Lippert and Hopkins (1950)
	Disturbed areas (mainly *Sporobolus*)	20 000	
Meadow grassland, USSR	*Bromus inermis* dominant	280–2450	Rabotnov (1946)
	Geranium pratense grass meadow	16980	
Bunch-grassland, California, USA	Ungrazed area (top 5 cm only)	8230	Major and Pyott (1966)
	Grazed area (top 5 cm only)	12 200	
Arid grassland, Saskatchewan, Canada		4000–15 000	Budd *et al.* (1954)
Sown grassland (UK)	Newly sown leys	4940–18 800	Champness (1949) Champness and Morris (1948)
Lowland grassland (UK)	Managed for hay	38 000	Chippindale and Milton (1934)
Arable land (UK)	Continuous wheat	34 100	Brenchley and Warrington (1933)
	Mainly cereals	56 500	Roberts (1958)
	Vegetable crops	1600–86 000	Roberts and Stokes (1966)
	Continuous wheat (27 800 *Papaver* spp.)	40 000– 75 000	Brenchley and Warrington (1930)

*References can be found in Harper (1977).

affecting rates of change in these populations. Fig. 19.2 is such a flow diagram and the letters used symbolize quantities in the mathematical model, devised by Cohen and quoted by Harper (1977), given below:

$$ST + 1 = St - StG - D(St - StG) + GYtSt$$

where Yt is the number of seeds produced in time, t, per germinated seedling, D is the fraction lost by death, G is the fraction lost by germination, and St is the number of seeds present in the soil at time t.

We have already mentioned the enormous range of values for Yt as the ability of the plant to respond to competition.

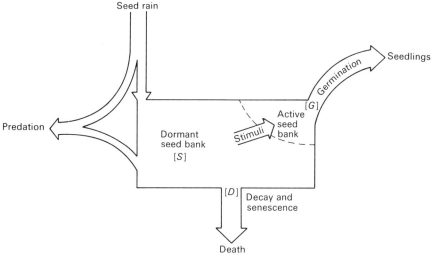

Fig. 19.2. Diagrammatic flow chart for the dynamics of the population of seeds in the soil. The symbols, *G*, *S*, and *D* refer to the model of Cohen (after Harper 1977).

Values for that fraction lost by death (*D*) have been estimated as 25% per annum for *Avena fatua* and *A. ludoviciana* under a long ley, i.e. without soil disturbance (Thurston 1966), while with 3 years continuous maize cropping they have been stated to decline by 67% from an initial population of 1225 million per hectare (Schweiger and Zimdahl 1979). This equation takes no account of the 'seed rain' which comes in from outside the field. The value of *Yt* would be sufficient in itself to produce an infestation of weeds in a few years from a very few initial seeds, yet important contributions, particularly of new species, can be made from outside the field. There are a number of agricultural practices which are used to cut these contributions from the outside to a minimum, including, particularly, the use of uncontaminated crop seed and the use of scrupulously clean farm machinery. There is an ever-improving seed-cleaning technology in some countries which enables strict regulations to be achieved and observed in the commercial sales of crop seed. The tyres of machinery and even the clothing of operatives can carry weed seed from one area to another. Contributions to the seed bank made by wild animals such as foxes and emus have been described by Brunner *et al.* (1976), those by domestic animals by Piggin (1978), and those by irrigation water by Kelley and Bruns (1975).

The loss of seed by predation has been described by Harper (1977) and this predation by animals can occur in the parent plant before the seed is shed and is an important component of the biological control of weeds.

The biggest factors in the model are however likely to be the value of *G* and the converse, that is the number of active seeds which are made dormant. Both are the result of stimuli arising from soil disturbance, the burial of seed resulting from soil turnover inducing dormancy and prolonging the life of many seeds, while the uncovering of seeds by raising them to the surface stimulates them to germinate. It has been claimed that the avoidance of soil disturbance, so much a feature of traditional agricultural practice, by the introduction of such techniques as direct-drilling and minimum cultivation has been the most significant change in the cultural pattern of arable agriculture in Great Britain for 200 years (Roberts 1982). A traditional saying that 'one year's seeding is seven years' weeding' aptly describes the part played by weed seeds in causing problems for succeeding crops. The new techniques based on the destruction of existing weeds by total herbicides, such as paraquat or glyphosate, followed by the direct-drilling of the new crop seed into an undisturbed or minimally disturbed soil, prevents the germination of any seed the soil might contain. Considerable additional advantages are achieved by this method including greater flexibility, reduction of cultivation inputs and, perhaps most important of all at least in some countries, the reduction of the bare soil phase with its inherent risks of soil erosion. The burial of weed seed is also to be avoided as it can lead to the development of a reserve of weed genetic material protected from surface predation or destruction through germination and subsequent kill by herbicides. Thus the deliberate stimulation of germination can in partnership with herbicides be used for weed control. The stimulation of seeds of the cereal parasites *Striga* and *Orobanche* by the synthetic analogue of the chemical Strigol I [British Patent 16990 (1974)] can occur at concentrations as low as 10^{-9} M (Johnson *et al.* 1976). The weeds can be destroyed subsequently by a contact herbicide. *Avena* spp. seeds can be stimulated in the field to germinate by treatment with sodium azide (Fay and Gorecki 1978). The stimulation of rhizomes of perennials such as *Agropyron,* to be subsequently destroyed by herbicides, has been achieved by cultivation techniques, and naptalam stimulates the production of new shoots from rhizomes of the yellow nutsedge *Cyperus esculentus* L as mentioned earlier.

The germination of both seeds and other plant propagules, such as rhizomes, corms, or bulbs, varies in time and percentage not only for the population of seed within the soil but also within the population produced by a single plant. In the case of seeds this variability may be under genetic or physiological control and represents a variation of considerable evolutionary potential as well as having experimental significance (Andersen 1968).

The periodic episodes of germination and other events in the life of the plant are the subject of phenology, a study which has been somewhat neglected in recent years, but which is now recognized as making a contribution to such phenomena as the weed associations in certain crops, occurrence of weeds with certain soils, or with the timing of agricultural methods. The British Crop Protection Council has published diagrammatic representations of the seed germination cycles of 37 annual species and of the flowering and growth periods of 19 perennial species of British weeds (see Fryer and Makepeace 1977). A comparison of the phenological data of populations of single species drawn from weed floras of different areas of the World could well provide valuable information on physiological variation within those global populations. The comparison of such data with that for the species in its original native area will perhaps contribute further to our understanding of why some species become weedy when introduced into new areas (Baker and Stebbins 1965).

19.6   The future

Parker (1976) reviewed weed research requirements in developing countries, indicating that the development of new herbicides for minor tropical crops, simpler application techniques, herbicides suitable for application after dry-planting, cultivation systems for perennial weeds, crop varieties resistant to parasitic weeds, and improved weed control in minimum cultivation situations and in groundcover legumes were particular research needs. He noted also that attention should be given to the possibility of changing weed floras occurring in these countries, as has happened elsewhere in the developed World. The setting up in 1972 of the International Crops Research Institute for the Semi-Arid Tropics (ICRISTAT) in Hyderabad, India, should provide on-going help also for Africa as well as India. Doll (1976) has indicated that an education programme for farmers in Latin America must include the concepts of sound agronomic practices to increase crop competitiveness, the significance of preventive weed-control measures and the importance of timely weeding. However, any 'expert' suggesting a new technique of weed control to a small farmer must remember that that farmer is there by virtue of his ability to survive as far as he has done, and his resistance to change may well be because he is not convinced that the new method is better!

The proportional use of agrochemicals in different cultivated arable crops was described by Cramer (1975) and reported in Aebi (1976), and those figures show that up to that time, cotton used 21%, maize 20%, deciduous and citrus fruits 16%, rice and potatoes each 7.5%, wheat 7%, soya-beans 4.3%, and others between them 16.7%. This ranking is particularly significant when it is recalled that rice feeds 60% of the World's population. Aebi (1976) has pointed out that on a worldwide basis, herbicide consumption has outgrown insecticide consumption due in part to the replacement of expensive human labour for weeding by chemical control and in part by the move towards mechanical harvesting. Fletcher (1974) suggested that the use of chemicals had by 1974 reduced the cost of weeding strawberries by a factor of nine, and for aquatic situations the cost is ten times less per mile than for other methods. If, however, this chemical weeding is to retain its economic advantage, not only will care have to be taken to preserve the effectiveness of established herbicides by, for example, wise rotational use, but also new uses will have to be found for existing ones as the cost of developing new herbicidal chemicals is very high indeed (Ivens 1980). The direction of development in weed control in the 1980s will be towards greater efficiency as with all aspects of farm technology with the accent on economic return from investment (Hill 1982). Improved efficiency will result from the use of weed control by herbicides over a wider area, of better application methods of those herbicides, better choice of weed control systems interlocking with the use of fertilizers or maintaining nitrogen-fixation processes, and prevention of soil erosion. If fertilizers are to be used then it is sensible to use herbicides as well because the weeds may well take up the fertilizers faster than the crop and in so doing may become increasingly luxuriant to further shade the crop or compete with it for water. In highly developed countries such as Germany herbicides account for 50% of the total crop-protection chemicals while insecticides are 5–6%. In India only 5% of the treatable area is treated with herbicides. Thus, although in tropical countries insecticides may well remain the dominant pesticide, the potential growth of herbicide usage is very great indeed.

Finally Parker (1977) has outlined the factors which might give rise to new weed problems, particularly in the developing world; he noted particularly that genetic changes do occur in plants either to bring about weediness in an otherwise non-weedy species, or, and increasingly recognized, to give rise to herbicide resistance in a selected population and thus lead to loss of effective chemical control. It is likely that, in spite of phytosanitary practices, weeds will continue to spread into new areas as they have done in the past, perhaps particularly through the movements of contaminated soils but also perhaps by the deliberate introduction of plants for good agricultural reasons. The climber *Mikania micrantha,* distributed within Malaysia as a groundcover, has been recognized now as a serious crop competitor, and the kudzu vine (*Phaseolus lobatus*), introduced as an anti-erosion species into the Southern United States, may be too successful in its growth and become a weed problem.

Whatever criteria are adopted it is clear that weeds do affect the performance of crops, or, put in another way, crops need protection from weeds. Traditionally that protection has been achieved by hand, removing the weeds or attempting to reduce their growth or reduce their effect by mechanically hindering them. These techniques are still employed over a very large area of the globe both in less-developed countries and in developed ones. The scale of operations in arable crops has required the development of chemical weed control, but the scale of operation in rangelands and the risks involved in adding chemicals to water has made biological control a desirable method. The understanding of vegetation as a dynamic response to environmental conditions has encouraged the integrated approach to weed control in the context of agricultural systems, soil factors, and ecological principles.

The protection of crops from existing or from new weeds involves, as we have seen, a combination of knowledge of the source of weeds, and of the relative sensitivity of weed and crop plants to a range of chemicals presented at a range of doses and over a range of morphological and physiological stages of their life history, firstly to control the weeds and secondly to avoid damage to the crop. Weed control and crop protection must therefore be looked at in both a short- and a long-term context.

19.7   References

Aebi, H. (1976). *Proc. 1976 Br. Crop Prot. Counc. Conf.—Weeds* **3**, 817.

Amor, R. L. and Stevens, P. L. (1976). *Weed Res.* **16**, 111.

Andersen, R. N. (1968). *Germination and establishment of weeds for experimental purposes.* Weed Science Society of America, Urbana, Illinois.

Andres, L. A. (1977). In *Integrated control of weeds* (eds J. D. Fryer and S. Matsunaka), p. 153.

Anon. (1977). *PANS* **23**, 433.

Appleby, A. P. and Paller, E. C. (1978). *Weed Res.* **18**, 247.

Ashton, F. M. and Crofts, A. S. (1973). *Mode of action of herbicides.* Wiley – Interscience, London.

Audus, L. J. (1976). *Herbicides: physiology, biochemistry, ecology* (2nd edn). Academic Press, London.

Auld, B. A. (1978). *PANS* **24**, 67.

Baker, H. G. (1965). In *The genetics of colonizing species.* (H. G. Baker and G. L. Stebbins), p. 147. Academic Press, New York and London.

—— (1974). *Annu. Rev. Ecol. System.* **5**, 1.

—— and Stebbins, G. L. (eds) (1965). *The genetics of colonizing species.* Academic Press, New York and London.

Blair, A. M., Parker, C., and Kasasian, L. (1976). *PANS* **22**, 65.

Brunner, H., Harris, R. V., and Amor, R. L. (1976). *Weed Res.* **16**, 171.

Bruzzese, E. (1982). *Proc. 1982 Br. Crop Prot. Counc. Conf.—Weeds* **2**, 787.

Burnside, O. C. (1978). *Weed Sci.* **26**, 362.

Burrill, L. C., Cardenas, J., and Locatelli, E. (1977). *Field manual for weed control research.* International Plant Protection Centre, Oregon.

Chisaka, H. (1977). In *Integrated control of weeds* (eds J. D. Fryer and S. Matsunaka). University of Tokyo Press, Tokyo.

Combellack, J. H. (1978). *Proc. 1st Conf. Counc. Austr. Weed Sci. Soc., Melbourne.* p. 15.

Corbett, J. R. (1974). *The biochemical mode of action of pesticides,* Academic Press, London.

Costa, G. and Intrieri, G. (1982). *Proc. 1982 Br. Crop Prot. Counc. Conf.—Weeds* **2**, 585.

Courtney, A. D. and Johnston, R. (1978). *Proc. 1978 Br. Crop Prot. Counc. Conf.—Weeds* **1**, 325.

Cussans, G. W. (1976). *Proc. 1976 Br. Crop Prot. Counc. Conf.—Weeds* **3**, 1001.

Davies, T. H. (1976). *Proc. 1976 Br. Crop Prot. Counc. Conf.—Weeds* **3**, 955.

Davison, J. G. (1978). *Proc. 1978 Br. Crop Prot. Counc. Conf.—Weeds* **3**, 897.

—— Roberts, H. A. (1976). *Proc. 1976 Br. Crop Prot. Counc. Conf.—Weeds* **3**, 1009.

Dempster, J. P. and Lakhani, K. H. (1979). *J. Anim. Ecol.* **48**, 143.

Dodd, A. P. (1940). *Biological campaign against prickly pear.* Commonwealth Australian Prickly Pear Board, Brisbane.

Doggett, H. (1970). *Sorghum.* Longmans Green, London.

Doll, J. D. (1976). *Proc. 1976 Br. Crop Prot. Counc. Conf.—Weeds* **3**, 809.

Drummond, J. M. and Soper, D. (1978). *Proc. 1978 Br. Crop Counc. Conf.—Weeds* **1**, 317.

Duffey, E., Morris, M., Sheail, J., Ward. L. K., Wells, D. A., and Wells, T. C. E. (1974). *Grassland ecology and wildlife management.* Chapman and Hall, London.

Evans, G. C. (1972). *The quantitative analysis of plant growth.* Blackwells, Oxford.

FAO (1971). *Evaluation and prevention of losses due to pests, diseases and weeds.* FAO/CAB Manual.

Fay, P. K. and Gorecki, R. S. (1978). *Weed Sci.* **26**, 323.

Fletcher, W. W. (1974). *The pest war.* Blackwell, Oxford.

Fryer, J. D. (1977). In *Ecological effects of pesticides (Linn. Soc. Symp. 5)* (eds F. H. Perring and K. Mellanby). Academic Press, London.

—— Makepeace, R. J. (eds) (1977). *Weed control handbook, Vol. I. Principles* (6th edn). Blackwells Scientific Publications, Oxford.

—— —— (eds) (1978). *Weed control handbook, Vol. II. Recommendations* (8th edn). Blackwells Scientific Publications, Oxford.

—— Matsunaka, S. (eds) (1977). *Integrated control of weeds.* University of Tokyo Press, Tokyo.

Giese, A. C. (1980). *Photochem. Photobiol. Rev.* **5**, 229.

Green, M. B., Hartley, G. S., and West, T. F. (1977). *Chemicals for crop protection and pest control.* Pergamon Press, Oxford.

Haizel, K. H. and Harper, J. L. (1973). *J. Appl. Ecol.* **10**, 23.

Hall, A. V. and Boucher, C. (1977). *Proc. 2nd Natl. Weed Conf. S. Afr.* (Weed Abstracts 1977 vol. 26, No. 3885).

Hamdoun, A. M. and El Tigani, K. B. (1977). *PANS* **23**, 190.

Harper, J. L. (1960). In *Biology of weeds (1st Symp. Br. Ecol. Soc.).* Blackwells Scientific Publications, Oxford.

—— (1977). *Population biology of plants.* Academic Press, London.

Hewson, R. T. (1971). *Studies on weed competition in some vegetable crops.* Ph.D. Thesis, Brunel University.

Holm, L. (1976). *Proc. 13th Br. Weed Cont. Conf.* **3**, 754.

—— Herberger, J. (1970). *Proc. 10th Br. Weed Cont. Conf. 1970* **3**, 1132.

Hill, G. D. (1982). *Weed Sci.* **30**, 426.

Holroyd, J. (1976). In *Wild oats in World agriculture* (ed. D. Price Jones). Agricultural Research Council, London.

Ivens, G. W. (1974). *PANS* **20**, 76.

—— (1980). In *Perspectives in World agriculture*, p. 181. Commonwealth Agricultural Bureau, Slough.

Johnson, A. W., Roseberry, G., and Parker, C. (1976). *Weed Res.* **16**, 223.

Kasasian, L. (1971). *Weed control in the tropics.* Leonard Hill, London.

Kelley, A. D. and Bruns, V. F. (1975). *Weed Sci.* **23**, 486.

King, L. J. (1966). *Weeds of the World.* Interscience, New York.

Kitchen, L. M., Witt, W. W., and Rieck, C. E. (1981). *Weed Sci.* **29**, 271.

Ko Wakabayashi and Matsunaka, S. (1982). *Proc. 1982 Br. Crop Prot. Counc. Conf.— Weeds* **2**, 439.

Lund, J. W. G. (1971). In *The scientific management of animal and plant communities (11th Symp. Br. Ecol. Soc.)* (eds E. Duffey and A. S. Watt), p. 225. Blackwells Scientific Publications, Oxford.

Lutman, P. J. W. and Richardson, W. G. (1978). *Weed Res.* **18,** 65.

MAFF (1978) *Weed control in cereals 1979.* ADAS (UK) CG 21.

Mathews, L. J. (1977). In *Integrated control of weeds* (eds J. D. Fryer and S. Matsunaka), p. 89. University of Tokyo Press, Tokyo.

Mercado, B. L. and Manual, J. A. (1977). *Laboratory manual on tropical weed biology.* College of Agriculture, University of Philippines.

Moody, K. and Ezumah, H. C. (1974). *PANS* **20,** 292.

National Academy of Sciences (1969). *Eutrophication: causes, consequences and correctives.* Publication 1700. National Research Council, Washington.

Nieto, J., Brondo, M. A., and Gonzalez, J. T. (1968). *PANS* **14,** 159.

Noda, K. (1977). In *Integrated control of weeds* (eds J. D. Fryer and S. Matsunaka), p. 17. University of Tokyo Press, Tokyo.

Pallos, F. M. and Casida, J. E. (1978). *Chemistry and action of herbicide antidotes.* Academic Press, London.

Parham, M. R. (1982). *Proc. 1982 Br. Crop Prot. Counc. Conf.—Weeds* **3,** 1017.

Parker, C. (1976). *Proc. 1976 Br. Crop Prot. Counc. Conf.—Weeds* **3,** 801.

—— (1977). In *Origins of pest, parasite, disease and weed problems (18th Symp. Br. Ecol. Soc.)* (eds J. M. Cherrett and G. R. Sagar). Blackwells Scientific Publications, Oxford.

Peters, N. C. B. (1972). *Proc. 11th Br. Weed Cont. Conf.* **1,** 116.

Piggin, C. M. (1978). *Weed Res.* **18,** 155.

Pinthus, M. J. (1972). *Weed Res.* **12,** 241.

Putnam, A. R. and Duke, W. B. (1978). *Annu. Rev. Phytopathol.* **16,** 431.

Raghavan, R. R., Charudatton, R., and Lynn Walker, H. (1982). *The biological control of weeds with plant pathogens.* Wiley, London.

Roberts, H. A. (1954). *MAAF Q. Rev.* **24,** 139.

—— (1981). *Adv. Appl. Biol.* **6,** 1.

—— (1982). *Weed control handbook: principles* (7th edn). Blackwells Scientific Publications, Oxford.

—— and Stokes, F. G. (1966). *J. Appl. Ecol.* **3,** 181

Robinson, D. W. (1978). *Proc. 1978 Br. Crop Prot. Counc. Conf.—Weeds* **3,** 800.

Rogers, E. V. (1975). *Ultra low volume herbicide spraying.* Forestry Commission Leaflet 62. HMSO, London.

Romanowski, R. R. (1977). In *Integrated Control of Weeds* (eds J. D. Fryer and S. Matsunaka), p. 47. University of Tokyo Press, Tokyo.

Scarr, P. W. (1976). *Proc. 13th Br. Weed Cont. Conf.* **3,** 1029.

Schweiger, E. E. and Zimdahl, R. L. (1979). *Proc. West Soc. Weed Sci.* **32,** 74. Scifres, C. J., Mutz, J. L., and Meadows, C. H. (1978). *Weed Sci.* **26,** 139.

Seth, A. K. (1977). In *Integrated control of weeds* (eds J. D. Fryer and S. Matsunaka), p. 69. University of Tokyo Press, Tokyo.

Soerjani, M. (1977). In *Integrated control of weeds* (eds J. D. Fryer and S. Matsunaka), p. 121. University of Tokyo Press, Tokyo.

Stalder, L., Potter, C. A., and Barben, E. (1977). *Weed Abstr.* **27,** No. 711.

Thomas, T. H. (1982). *Plant growth regulator potential and practice.* British Crop Protection Council, London.

Thompson, W. F. (1975). *Weed Abstr.* **27/2,** no. 732.

Thurston, J. M. (1966). *Weed Res.* **6,** 67.

Tottman, D. R. (1976). *Proc. 1976 Br. Crop Prot. Counc. Conf.—Weeds* **3,** 791.

—— Makepeace, R. J. (1977). In *Weed control handbook. Vol. I. Principles* (eds J. D. Fryer and R. J. Makepeace), Appendix I. Blackwells Scientific Publications, Oxford.

Truelove, B. (1977). *Research methods in weed science* (2nd edn), Southern Weed Society.

Way, J. M. (1973). *Road verges on rural roads; management and other factors.* Natural Environment Research Council, Occasional Report No. 1.

Wellbank, P. J. (1963). *Ann. Appl. Biol.* **51,** 107.

Williams, G. H. (1980). *Bracken control: a review of progress 1974–1979.* Res. and Dev. Publ. No. 12, West of Scotland Agricultural College.

Wilson, B. J. and Taylor, W. A. (1978). *Weed Res.* **18,** 215.

Winter, S. R. and Wiese, A. F. (1982). *Weed Sci.* **30,** 620.

Wittering, W. O. (1974). *Weeding in the Forest. A Work Study Approach.* Forestry Commission Bulletin No. 48, HMSO, London.

Young, J. and Evans, R. (1978). *Weed Sci.* **26,** 27.

Index